D1601047

Fixation for Electron Microscopy

Fixation for Electron Microscopy

M. A. HAYAT

Department of Biology
Kean College of New Jersey
Union, New Jersey

1981

ACADEMIC PRESS

A Subsidiary of Harcourt Brace Jovanovich, Publishers

New York   London   Toronto   Sydney   San Francisco

COPYRIGHT © 1981, BY ACADEMIC PRESS, INC.
ALL RIGHTS RESERVED.
NO PART OF THIS PUBLICATION MAY BE REPRODUCED OR
TRANSMITTED IN ANY FORM OR BY ANY MEANS, ELECTRONIC
OR MECHANICAL, INCLUDING PHOTOCOPY, RECORDING, OR ANY
INFORMATION STORAGE AND RETRIEVAL SYSTEM, WITHOUT
PERMISSION IN WRITING FROM THE PUBLISHER.

ACADEMIC PRESS, INC.
111 Fifth Avenue, New York, New York 10003

United Kingdom Edition published by
ACADEMIC PRESS, INC. (LONDON) LTD.
24/28 Oval Road, London NW1 7DX

Library of Congress Cataloging in Publication Data

Hayat, M. A., Date.
    Fixation for electron microscopy.

    Bibliography: p.
    Includes index.
    1. Fixation (Histology) 2. Electron microscopy--
Technique. I. Title.
QH235.H39   578'.62   81-12745
ISBN 0-12-333920-0   AACR2

PRINTED IN THE UNITED STATES OF AMERICA

    83 84   9 8 7 6 5 4 3 2

"I cannot bring a world quite round
Although I patch it as I can.
I sing a hero's head, large eye
And bearded bronze, but not a man,
Although I patch him as I can
And reach through him almost to man.
If to serenade almost to man
Is to miss, by that, things as they are,
Say that it is the serenade
Of a man that plays a blue guitar."

Wallace Stevens

Contents

Preface   xvii

1 Introduction

Text   1

2 Factors Affecting the Quality of Fixation

Tissue Specimen Size   11

   Preparation of Tissue Blocks   12

Automatic Specimen Processing   15

Osmolarity   16

   Ionic Composition of Fixative Solution   20
   Vehicle Osmolality   23
   Effects of Added Substances   26
     Specific Effects of Nonelectrolytes   27
     Specific Effects of Electrolytes   28
     Undesirable Effects   32
   Methods for Adjusting the Osmolarity   33
   Recommended Osmolality   34
   Measurement of Osmolality   35

Buffers   39

   Buffer Types   44
     Phosphate Buffers   45
     Collidine Buffer   46
     Veronal Acetate Buffer   46

Tris Buffer 47
Cacodylate Buffer 47
Na-Bicarbonate Buffer 47
PM Buffer 48
Zwitterionic Buffers 48
Preparation of Buffers 50
pH 54

Fixative Penetration 58

Temperature of Fixation 60

Duration of Fixation 62

Fixative Concentration 63

3 Aldehydes

Glutaraldehyde (Glutaric Acid Dialdehyde) 66

Uses of Glutaraldehyde 68
Nature of Commercial Glutaraldehyde 72
Effect of pH and Temperature on Glutaraldehyde 76
Reaction with Proteins 77
Formation of Schiff's Bases 82
Mechanism of Cross-Linking 83
Reaction with Lipids 87
Reaction with Nucleic Acids 88
Reaction with Carbohydrates 90
Effects of Aldehydes on Tissue Physiology 90
Osmolality 92
Osmolarity of Glutaraldehyde 92
Temperature 94
Concentration 94
pH 96
Fixation Method with Glutaraldehyde at Higher pHs 96
Rate of Penetration 97
Shrinkage Caused by Aldehydes 98
Heated Glutaraldehyde 99
Fixation Method Employing Glutaraldehyde at Higher
Temperatures 99

Limitations of Glutaraldehyde 99
Purification of Glutaraldehyde 102
The Charcoal Method 103
The Distillation Method 103
Determination of Glutaraldehyde Concentration 104
Storage of Glutaraldehyde 105
Glutaraldehyde-Containing Fixatives 106
Glutaraldehyde—Alcian Blue 106
Glutaraldehyde—Digitonin 107
Glutaraldehyde—Formaldehyde 110
Glutaraldehyde—Hydrogen Peroxide 111
Glutaraldehyde—Lead Acetate 113
Glutaraldehyde—Malachite Green 113
Glutaraldehyde—Phosphotungstic Acid 114
Glutaraldehyde—Trinitro Compounds 114
Glutaraldehyde—Potassium Dichromate 115
Glutaraldehyde—Potassium Ferricyanide—Osmium
Tetroxide 118
Glutaraldehyde—Potassium Ferrocyanide—Osmium
Tetroxide 120
Glutaraldehyde—Ruthenium Red 120
Glutaraldehyde—Spermidine Phosphate 124
Glutaraldehyde—Tannic Acid 124
Glutaraldehyde—Uranyl Acetate 129

Formaldehyde 129

Reaction with Proteins 130
Reaction with Lipids 133
Reaction with Nucleic Acids 134
Reaction with Carbohydrates 136
Preparation of Paraformaldehyde 136
Formaldehyde Fixatives 136

Acrolein 137

Reaction with Proteins 140
Reaction with Lipids 140
Precaution in the Handling of Acrolein 141
Acrolein Fixatives 142

Polyaldehyde   145

Role of Aldehydes in Quantitative Electron Microscopy   145

Tissue Storage   146

4   Osmium Tetroxide

Introduction   148

Reaction with Unsaturated Lipids   150
   Trilaminar Appearance of Membranes   154
Reaction with Saturated Lipids   156
Reaction with Proteins   157
Reaction with Lipoproteins   159
Reaction with Nucleic Acids   160
Reaction with Carbohydrates   163
Reaction with Phenolic Compounds   164
   Mechanism of Reaction   164

Loss of Lipids   166

Loss of Proteins   168

Swelling Caused by Osmium Tetroxide   170

Parameters of Fixation   171

Concentration of Osmium Tetroxide   171
Temperature of Fixation   171
Rate of Penetration   172
Duration of Fixation   174

Osmium Tetroxide as a Primary Fixative   175

Osmium Tetroxide as a Vapor Fixative   175

Removal of Bound Osmium from Sections   176

Osmium Blacks   176

Osmeth   177

**Preparations for and Precautions in the Handling of Osmium
Tetroxide**   178

Regeneration of Used Osmium Tetroxide   180

5 Permanganates

Potassium Permanganate Reactions   184

    Reaction with Membranes   184
    Reaction with Monoamines   185
    Reaction with Other Cellular Substances   186

Contrast and Loss of Cellular Substances   188

Comparison between Potassium Permanganate and Osmium Tetroxide   189

Fixation Procedures   190

    Permanganate Fixatives   192
        Potassium Permanganate   192
        Lanthanum Permanganate   192
        Sodium Permanganate   192
        Lithium Permanganate   193

6 Miscellaneous Fixatives

Ruthenium Tetroxide   194

Dimethylsuberimidate   195

Carbodiimides   198

Trioxsalen   199

7 Methods of Fixation

Double Fixation   200

Glutaraldehyde and Osmium Tetroxide Mixture   201

Simultaneous Fixation for Light and Electron Microscopy   206

Anhydrous Fixation   207

8 Modes of Fixation

Vascular Perfusion   209

  Considerations in Achieving Satisfactory Vascular Perfusion   211
    Perfusion Pressure   213
  Apparatus for Vascular Perfusion   215
  Methods of Vascular Perfusion   223
    General Method of Vascular Perfusion   223
    Aorta   225
    Arteries   226
    Central Nervous System, Method I   227
    Central Nervous System, Method II   228
    Central Nervous System, Method III   230
    Embryo   232
    Fish (Whole)   233
    Heart   235
    Kidney, Method I   237
    Kidney, Method II   237
    Kidney (Large Animals), Method III   238
    Liver, Method I   240
    Liver, Method II   242
    Liver, Method III   242
    Liver (Embryos and Very Small Animals), Method IV   244
    Liver (Biopsy), Method V   244
    Lung, Method I   245
    Lung, Method II   246
    Lung (Small Animals), Method III   247
    Lung (Postmortem Fixation of Human Lung), Method IV 248
    Lung (Human) Biopsy Specimens, Method V   249
    Muscle   249
    Ovary   251
    Spinal Cord   251
    Spleen   252
    Testes, Method I   253
    Testes, Method II   254
    Testes, Method III   255

Immersion Fixation   256

  Dripping Method   258
  Injection Method   259

Anesthesia   260

9 Effects of Fixation

Effect of Fixation on Actin   262

Effect of Fixation on Brain Tissue   265

Effect of Fixation on Kidney Tissue   269
  Renal Medulla   270
  Biopsy Specimens   271

Effect of Fixation on Membrane Fracture   271

Effect of Fixation on Mesosomes   274

Effect of Fixation on Microtubules and Microfilaments   275

Effect of Fixation on Mitochondria   277

Effect of Fixation on Myelin   282

Effect of Fixation on Plant Virus   283

Effect of Fixation on Plasma Membrane   284

Effect of Fixation on Staining   287

Effect of Fixation on Synaptic Vesicles   291

Effect of Fixation on Tight Junctions   294

Effect of Fixation on Specimens for X-Ray Microanalysis   294

Effect of Fixation on Vascular Endothelium   296
  Artifacts   297

10 Changes in Specimen Volume

Text   299

11 Postmortem Changes

Text   307

12 Plant Specimens

Text   314

13 Fixation for Scanning Electron Microscopy

Cleaning of Specimens   321

    Preservation of Microbial Association to Intestinal Epithelium   331
    Relaxation Procedures   332

Fixation Process   333

Parameters of Fixation   334

    Mode of Fixation   334
    Specimen Size   335
    Temperature   336
    Duration of Fixation   336

Specimen Shrinkage   337

Fixatives   338

14 Fixation for Enzyme Cytochemistry

Comparison of Different Aldehydes   346

Effect of Aldehydes on Enzymes   348

Factors Affecting the Preservation of Enzyme Activity   350

    Specimen Size   350
    Concentration of Aldehyde   351
    Duration of Fixation   351
    Temperature   352
    pH   352
    Buffers   353
    Mode of Fixation and Incubation   353

Single-Cell Specimens   355

Subcellular Fractions   357

Cryoultramicrotomy   359

15 Fixation for Immunoelectron Microscopy

Fixatives   370

16 Use of Dimethylsulfoxide in Fixation

Effect on Enzymes 374

Effect on Ultrastructure 375

17 Criteria for Satisfactory Specimen Preservation

Text 378

18 Artifacts

Transmission Electron Microscopy 382

Scanning Electron Microscopy 385

19 Specific Fixation Methods

Text 389

Appendix

Balanced Salt Solutions 411
Commonly Used Salts and Their Physicochemical Properties 412
Commonly Used Chemicals 413
References 415
Index 473

Preface

In spite of its inherent limitations (loss, displacement, and alteration of cell components), chemical fixation is the most widely used method for preserving biological specimens for both transmission and scanning electron microscopy. No other method of specimen preparation can match chemical fixation in regard to the basic ease and simplicity with which the latter can be carried out. With the continued improvements in instrumentation, including the resolving power, it is essential to have a better understanding of the effects of fixatives upon the macromolecular structure of the cell. In this volume, I have attempted to consider almost every aspect of fixation including chemical interactions between fixatives and individual cellular substances. The data on these interactions have been obtained primarily by studying model systems. All the commonly used fixatives are discussed.

As stated above, the chemistry of fixative interactions discussed here is based primarily on the reactions of a fixative with isolated proteins, lipids, nucleic acids, and carbohydrates. The conditions in such studies are simpler than those existing in cellular components *in situ*. Although the application of results obtained with *in vitro* studies to the fixation of cells and tissues has limitations, such studies are of profound importance in molecular biology and provide an accurate and sound foundation for the eventual understanding of mechanisms of tissue fixation and development and choice of appropriate fixatives. It is hoped that the reader will become aware that correct interpretation of the information retrieved from electron micrographs is dependent, in part, on an understanding of the principles underlying the fixation procedure. It is imperative that an electron micrograph be interpreted with respect to the treatments that the specimen has undergone.

The fixation of both eukaryotic and prokaryotic specimens is presented. The fixation of plant specimens frequently presents special problems, and the optimal protocols may differ from those used for animal specimens. With a few exceptions, the treatment of the former specimens has received only superficial coverage in texts on electron microscope techniques. Therefore, special fixation conditions required for plant specimens are pointed out, not only throughout the text but also discussed in detail in a separate chapter.

In order to understand the modifications of cell structures necessarily introduced during their processing, the connection between morphology and biochemical aspects of preparatory treatments is emphasized. Special effort has been made to explain the chemical basis of formation of artifacts. A useful guide for recognizing and minimizing major artifacts and fixation faults commonly encountered is presented.

Applications of fixation procedures to both biology and clinical medicine are described. Most of the methods presented have been tested for their reliability, and are the best of those currently available. However, in order to achieve the best results, it is desirable that an attempt be made to optimize even a basic method. This book departs from the tradition that books on methodology present only the contemporary consensus of knowledge. Relatively new methods have also been presented provided that they show potential usefulness. Some of these methods are in the developmental stage. This book contains new viewpoints with particular regard to current problems. Changing views on methodology and areas for future experimentation are suggested.

A concerted effort has been made to present a logical, stepwise description of the fixation process. The reader is provided with both certainties and gaps in our knowledge. Areas of disagreement are pointed out. Each chapter contains numerous cross-references to relevant sections dealt with elsewhere in the volume. Each chapter also offers a guide to a wide range of useful literature. I have attempted to incorporate the latest available information.

Although the advent of fixation procedures for electron microscopy goes back to almost three decades, their standardization has been unnecessarily delayed. One of the main reasons is that scientific journals do not require detailed description of the methodology used. It will be helpful to have the author of an article specify the exact pH, osmolarity, concentration, and composition of all the solutions as well as the duration and temperature of all the treatments employed.

I appreciate that present-day methodology moves faster than any printing press and that the useful half-life of any tome dealing with preparatory methods may be depressingly brief. Nonetheless, I believe that communication is well served by periodic compilations of existing knowledge, especially when such knowledge involves multiple disciplines and groups of researchers that otherwise might not interact. With many important and fascinating problems still at hand, I trust that this volume will do more than transiently satisfy the bibliophile. It is my hope that it will serve and inspire my friends, students, colleagues, and co-workers—to whom it is offered and dedicated.

I hope that this volume will be a valuable record and resource for further progress in this very active interdisciplinary field. It may provide the necessary impetus to encourage research workers to explore and develop new and better methodology. It is my genuine and abiding desire to continue to compile information which might help to improve the quality of specimen preservation as well as interpretation of the fine structure.

I am grateful to Professors Jacob Hanker, Nick Sperelakis, and Michael Forbes for their invaluable suggestions.

M. A. Hayat

1

Introduction

In spite of its inherent limitations, chemical fixation is the most extensively used method for preserving biological specimens for both scanning and transmission electron microscopy. Some of the reasons for its universal usage are the adequate preservation of many cellular components including enzymes, the clarity of structural details shown by electron micrographs, and the ease of application to both prokaryotes and eukaryotes. No other fixation method presently available can claim these advantages. For example, physical methods of fixation such as freeze substitution and freeze-drying (Rebhun, 1972), freeze-etching (Koehler, 1972; Bullivant, 1973), and inert dehydration (Pease, 1966, 1973a) are relatively tedious, are time-consuming, and may be accompanied by serious problems such as ice crystallization and recrystallization. Moreover, the effects of cryoprotectants on the ultrastructure are not clear. Considerable skill and experience are needed to obtain satisfactory results by using these methods. However, these methods are a useful adjunct to chemical fixation. It is encouraging that an understanding of the chemistry of chemical fixation is beginning to emerge. Such an understanding is imperative in order to interpret cellular details revealed by electron micrographs correctly.

The main objectives of fixation are to preserve the structure of cells with minimum alteration from the living state with regard to volume, morphology, and spatial relationships of organelles and macromolecules, minimum loss of tissue constituents, and protection of specimens against subsequent treatments including dehydration, embedding, staining, vacuum, and exposure to the electron beam. Fixation should also minimize the alteration in the chemical reactivity

1

of cellular substances such as enzymes and preserve and protect them. The role of fixation in enzyme cytochemistry is discussed later.

Ideally, the aim of a desirable fixation method is satisfactory preservation of the cell as a whole and not merely the best preservation of only a small part of it. In practice, however, fixation is usually of selective rather than of general effectiveness in the sense that the objective of the study determines the type of fixation method used. For example, in studying lipids, one would select OsO_4 or glutaraldehyde containing digitonin, whereas for the study of distribution and translocation of water-soluble substances, one would select a freeze-drying method over OsO_4 solution. Various methods employed for freezing cells and tissues for transmission electron microscopy (Rebhun, 1972; Nermut, 1977) and scanning electron microscopy (Boyde and Echlin, 1973; Hayat, 1978) have been presented. Freezing methods for X-ray microanalysis have been presented by several authors in the volumes edited by Erasmus (1978) and Hayat (1980).

Another example of the dependence of the choice of fixative on the objective of the study is illustrated by the study of helical proteins. Since glutaraldehyde or OsO_4 produce extensive transitions from helical to random coil conformations, these fixatives are unsuitable for the study of helical proteins. In contrast, ethylene glycol is the ideal fixative for these proteins because it preserves helical conformations. The importance of the preservation of α-helix in the study of myofilamentous system is apparent. The effect of various fixatives on the conformation of helical proteins is reviewed by Puchtler *et al.* (1970).

For some studies approaches other than conventional fixation are desirable. One such approach is the critical point drying method (Anderson, 1951). This method used in combination with inert dehydration is useful for the electron microscopy of whole-mounted cell organelles. The advantage of this method is that specimens are preserved in three dimensions and can then be observed by their natural contrast without chemical fixation, staining, or sectioning. This method, however, has limited application. Detailed methodology of critical point drying for scanning (Cohen, 1974, 1977, 1979; Hayat, 1978) and transmission electron microscopy (Hayat and Zirkin, 1973) has been presented.

The separation of the liquid phase from the solid phase of protoplasm is an essential effect of fixation. The degree of smoothness and rapidity of this separation depend on the type of fixation method used. As a general rule, freeze-drying achieves a more rapid and smooth separation than that obtained by chemical fixation. The net physical dislocation of the solids after fixation may or may not be visible. Nevertheless, even in tissues fixed with buffered glutaraldehyde or OsO_4, some movement of solids does take place. Most other changes in the properties of protoplasm are due to the chemical reaction of the fixative with various organic substances, especially active groups of protoplasmic proteins (e.g., amine) and lipids.

Prior to embedding in water-insoluble resins, free water present in the tissue must be replaced by a solvent during dehydration. Since important chemical bonds in living tissue are dependent for their stability on the presence of water, a fixative should provide stable bonds that will hold the molecules together during dehydration and subsequent treatments so that they will not be translocated or extracted. This is accomplished primarily by the formation of inter- and intra-molecular protein cross-links. Generally, fixatives form cross-links not only between their reactive groups and the reactive groups in the tissue, but also between different reactive groups in the tissue. It is recognized that chemical fixation unmasks or frees certain reactive groups in the tissue for cross-linkage that otherwise may not be involved in intermolecular bonding.

As stated earlier, an important feature of fixation is a change in the appearance and "nature" of cellular proteins. This change in the physical configuration of conformation of the protein molecule is called denaturation (weakening or dis-ruption of secondary or tertiary structure, or both). Denaturation is followed by the loss of or change in many properties (e.g., solubility, specific gravity, and crystallization) of the protein molecule. Some of these changes are responsible, in part, for the stabilization of proteins. For example, the denatured protein is less soluble in aqueous solutions, since during denaturation the unfolding of proteins exposes hydrophobic groups that repel water. Denatured proteins are also less susceptible to precipitation than those in the native state.

Although in many cases the denaturation is essentially irreversible, it may also be reversible. The permanency of denaturation depends on the type and concen-tration of the fixative used as well as on the duration of fixation. In the case of temporary or partial denaturation, protein may subsequently either revert to its original state or remain unstable until completely fixed. The unstable state of the protein is an obstacle in achieving satisfactory preservation. On the other hand, fully denatured proteins become coagulated and the changes involved are irrever-sible. Permanently denatured proteins are generally well stabilized and are less susceptible to extraction during dehydration than those denatured temporarily. For example, the protein of egg white (albumin) is globular and water soluble in the native state, but becomes fibrous and water insoluble in the denatured state.

Fixatives differ in their effects on the optically homogeneous ground substance of the living protoplasm. The reagents that coagulate proteins into an opaque mixture of granular or reticular solids suspended in fluid are called coagulant fixatives (e.g., ethanol). It must not be thought that coagulant fixatives flocculate all proteins; for instance, ethanol does not flocculate nucleoproteins. Essentially, coagulant fixatives permanently flocculate molecules of most of the protoplasmic proteins, and thus cause a considerable change in the protein structure. This results in distortion of the fine structure. The coagulant fixatives are, therefore, unfit for use in electron microscopy. Most of the fixatives of this type are also called nonadditive, for they fix proteins without becoming a part of them.

In contrast, noncoagulant fixatives transform proteins into a transparent gel.

This results in the stabilization of proteins without much structural distortion of the original state. Noncoagulant fixatives cause very little dissociation of protein from water, and proteins retain at least some of their reactive groups. Another desirable effect of these fixatives is that they can render some proteins noncoagulable by subsequently used coagulant reagents including fixatives. The importance of this change becomes apparent when the fixed tissue is subsequently dehydrated with ethanol. Glutaraldehyde, a noncoagulant, has become the most widely used prefixative because of its capacity to stabilize most proteins without coagulation. Other commonly used noncoagulants are OsO_4, acrolein, and formaldehyde. On the basis of their chemical affinity for cellular proteins, these fixatives are also called additive fixatives, for they chemically become a part of the proteins they fix. These fixatives may also add themselves onto cell constituents other than proteins; for example, OsO_4 stabilizes certain lipids by additive fixation. Some of these fixatives (OsO_4 and ruthenium tetroxide) are heavy metals and as a result impart density to the tissues they fix.

Fixation tends to cause dimensional changes in the intercellular spaces. Various tissues respond differently in this respect to the fixative used. The dimensions of the intercellular spaces in the central nervous tissue are especially sensitive to the method of fixation. Torack (1965, 1966) demonstrated that immersion fixation with glutaraldehyde or OsO_4 produced intercellular spaces of ~15 nm. On the other hand, fixation by perfusion with glutaraldehyde followed by postfixation with OsO_4 produced much larger intercellular spaces. Larger intercellular spaces in the tissue were also produced by rapid freeze substitution (Van Harreveld *et al.,* 1965). Since cell structure preserved by freeze techniques is considered to be a more faithful representation of the *in vivo* structure, fixation by controlled vascular perfusion has been suggested to preserve the intercellular space distribution more accurately (also see p. 265). The magnitude of the space is also affected by the solvent used for dehydration. The use of ethanol may result in a somewhat larger space than that obtained after dehydration with acetone.

Although the exact nature and dimensions of intercellular spaces present in living tissue remain uncertain, the artifactual effect of fixation on the dimensions of the intercellular spaces is obvious. Drastic changes in the tissue fluid distribution during fixation preclude any direct relationship between the intercellular spaces in the living and fixed tissues.

Since the extraction phenomenon greatly influences the contrast and general appearance of an electron micrograph, increasing importance is being given to the problem of extractability or solubility of tissue constituents (Table 1.1) during and after fixation. Experimental evidence indicates that various amounts of carbohydrates, proteins, lipoproteins, nucleic acids, and lipids are lost during fixation. For example, a loss as high as 25% of phospholipids from amoeba has been reported using various fixatives and dehydration solutions (Korn and

TABLE 1.1

Solubilities of Compounds in Cells[a]

Classes of compounds	Subclass	Examples	Solubilities of examples
Protein	Albumin	Egg albumin	Dissolves in distilled water; coagulates on heating
	Histone	Thymus histone Globin	Water soluble; relatively low molecular weight; basic; forms scum on heating
	Protamin	Sperm protamin	Water soluble; relatively low molecular weight; basic
	Globulin	Serum globulin	Soluble in salt solution
	Prolamine	Gliadin (wheat)	Soluble in 80% alcohol
	Glutelin	Glutenin (wheat)	Soluble in acid or basic solution
	Scleroprotein (albuminoid)	Horns, nails, hair	Insoluble in most reagents
Lipids	Triglycerides	Tripalmitin, tristearin	Soluble in fat solvents
	Sterols	Cholesterol	Soluble in fat solvents
	Waxes	Beeswax	Soluble in fat solvents
	Phospholipids	Cephalin, lecithin	Soluble in both fat solvents and water to some degree
Carbohydrates	Monosaccharides		
	Hexoses	Glucose Fructose	Water soluble
	Pentoses	Ribose Deoxyribose	Water soluble
	Disaccharides	Sucrose Maltose Lactose	Water soluble
	Polysaccharides		
	Pentosans	Xylans in wood	Insoluble in fat or water solvents
	Hexosans	Starch, glycogen	May be prepared in colloidal state in water
		Cellulose	Insoluble in water or fat solvents
	Chitin	Insect chitin	Insoluble (glucosamine condensate)

(Continued)

TABLE 1.1

(*Continued*)

Classes of compounds	Subclass	Examples	Solubilities of examples
	Mucopolysaccharides		
	Hyaluronic acid	Intercellular cement	Insoluble (attacked by enzyme hyaluronidase)
	Chondroitin		
	Sulfuric acid	In matrix of connective tissue	Insoluble in water or fat solvents
	Nitrogenous neutral hetero-polysaccharides	Blood group substances	
Conjugated compounds	Nucleoproteins	Nucleohistone	Soluble in strong salt solutions
	Chromoproteins	Hemoglobin	Soluble in water
	Lipoproteins	Rhodepsin (visual purple)	
	Phosphoproteins	Casein of milk	Water soluble
	Flavoproteins	Yellow enzyme	Water soluble
	Glycoproteins	Protein of connective tissue matrix	Water soluble
	Adipocellulose	On plant cell walls	Insoluble

[a] From Giese (1968).

Weisman, 1966). The lipid loss appears to be related to the degree of saturation of fatty acids present in the lipid fraction of the tissue.

Among various fixatives currently in use, double fixation with glutaraldehyde followed by OsO_4 is the most effective in reducing the loss of cell constituents. These two fixatives are employed in order to stabilize the maximum number of different types of molecules. It should be noted, however, that information concerning the interaction between the heavy metal and the organic fixative is insufficient, and the resultant changes in the reactivity and structure of cell components are not fully known. It is apparent that, presently, the information on the chemical reactions between the fixative solution and specific cellular constituents is inadequate. The correct interpretation of the electron micrographs is dependent on a better understanding of the chemistry of fixation. Such an understanding, in turn, depends on the study of the reaction mechanism and the kinetics involved. In this regard, the most important questions are (1) what the reaction products are, (2) what kinds of organic substances are fixed or precipitated and at what rates, and (3) what the nature of the changes is in the arrange-

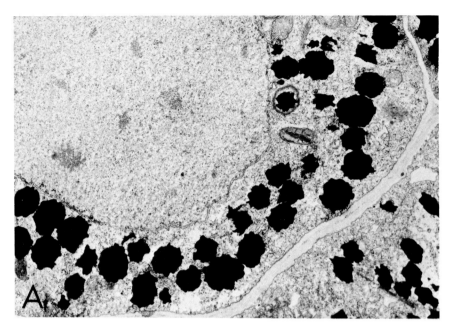

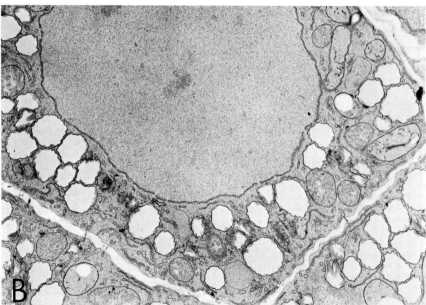

Fig. 1.1. Cortical cells of the root tip of leafy spurge showing the effect of postfixation with OsO₄. A, fixed with glutaraldehyde followed by OsO₄. The black structures are reaction products of osmium primarily with unsaturated lipids. B, fixed with KMnO₄; lipids are lost and the empty-looking structures are comparable to the black ones in the upper micrograph.

ment of molecules of the reaction products. The physiochemical properties of almost all the fixatives used in electron microscopy are discussed.

In the past not enough attention has been paid to the use of more than one fixation method in studies of a particular tissue. In some cases, this limited approach has led to erroneous interpretations. Rosenbluth (1963) aptly pointed out that "even in tissues which look well fixed and in which there are no obvious architectural disruptions, there may nevertheless be systematic artifacts which can be identified as such only through comparison with specimens fixed by alternative methods." A case in point is the presence of slime, a lipoprotein, in sieve tubes. Slime appears fibrillar when the tissue has been fixed with OsO_4, glutaraldehyde, or formaldehyde, or with some combination of these, whereas it appears granular or amorphous when it has been fixed with $KMnO_4$. The granular material is mostly an artifactual precipitate produced by the reaction of $KMnO_4$ with the sieve tube contents, especially with sucrose, and the fixative vehicle and dehydration solvents. In embryonic chick heart, microtubules measure 15 nm in diameter after fixation with OsO_4, whereas after glutaraldehyde fixation the microtubules measure 24 nm in diameter (Rash *et al.,* 1970).

The preceding data and other evidence indicate the desirability of applying not only different fixation methods but also various durations of fixation. It is advisable, therefore, to fix the tissue, when possible, in at least two different fixing agents. In this way the validity of the results obtained with one fixative can be easily and quickly checked by those gained by the use of the other fixative (Fig. 1.1). This method does require more time, but its value cannot be minimized. An example of this approach is the use of glutaraldehyde, OsO_4, $KMnO_4$, and lithium permanganate for the classification of different types of nerve endings in frog median eminence according to the size and internal structure of the granules and vesicles (Nakai, 1971). Examination of similar specimens with the transmission electron microscope (TEM) and the scanning electron microscope (SEM) may be desirable in some cases. The effects of differences in composition of the fixation vehicle on the quality of tissue preservation are discussed later.

Factors Affecting
the Quality of Fixation

It is recognized that the pH, total ionic strength, specific ionic composition, dielectric constant, osmolarity, temperature, length of fixation, and method of application of the fixative may be critical factors in determining the quality of tissue fixation. In addition, a fixative may have a damaging effect on the structure of an organelle, although it may not be the primary cause of the damage. For example, plastids from corn plants, kept in the dark prior to fixation, tend to swell during $KMnO_4$ fixation, and this swelling leads to disruption of the fretwork (Paolilli *et al.*, 1967). In this respect, the exact relationship between photoperiod and fixation is not known.

In certain tissues, even under similar fixation techniques, the ultrastructural morphology of an organelle may show variations. These variations are related to the functional state of the cells at the time of fixation. Circadian rhythm, for example, is known to affect the development of certain organelles. The Golgi complex is an organelle that shows circadian variations in its morphology. The morphology of the Golgi complex of STH cells in the pars distalis of male mice, for instance, differed, depending on the time of the day at which the mice were killed (Gomez Dumm and Echave Llanos, 1970). In mice killed at midnight the Golgi complex was small and well defined, whereas the mice killed at noon showed a marked hypertrophy of the Golgi complex.

Diurnal variation in the fine structure of endoplasmic reticulum membranes has been established. There are also regional differences in the distribution of smooth and rough endoplasmic reticulum within the hepatic lobule (Chédid and Nair, 1972). The amount of smooth endoplasmic reticulum varies with respect to the time of the day. These variations reflect changes in numerous processes such as DNA synthesis, mitotic activity, and protein (enzyme) and hormone synthesis. Enzyme rhythm can be correlated with the diurnal rhythm in the endoplasmic reticulum.

Another example is mitochondria, the ultrastructure of which is markedly affected by different physiological states prevalent just prior to fixation. There is evidence indicating that isolated mitochondria (Hackenbrock, 1966) as well as mitochondria *in situ* (Meszler and Gennaro, 1970) show ultrastructural changes with different physiological states. It has been demonstrated, for instance, that in the radiant heat receptors of certain snakes, free nerve endings show mitochondria having a swollen configuration if the organ is unstimulated immediately before fixation, whereas they appear condensed if stimulated by infrared radiation prior to fixation (Meszler and Gennaro, 1970).

Certain axonal membranes of crayfish abdominal nerve cords display ultrastructural changes if the axons are fixed during electrical stimulation with glutaraldehyde followed by OsO_4. Peracchia and Robertson (1971) demonstrated an increase in osmiophilia and thickness of membranes' dense strata on the axon surface, endoplasmic reticulum, and outer mitochondrial membranes when the axons were fixed during electrical stimulation. The unmasking of SH groups in membrane proteins as a result of electrical stimulation is thought to be responsible for the increased osmiophilia. It is known that SH groups are very reactive with OsO_4.

The appearance of cross-bridges on microtubules has been suggested to reflect a change in physical state. It has been demonstrated that cytoplasmic microtubules even of interphase cells show cross-bridges when HeLa cells are cooled and fixed at 4°C (Bhisey and Freed, 1971). The microtubules in cooled cells differ from those in the control material.

It is well known that changes in temperature for growth of unicellular organisms are accompanied by modifications in their fine structure. It has been shown, for example, that the lowering of temperature from 37° to 15°C induces mesosome deterioration and double cell wall formation in gram-negative bacteria, *Bacillus subtilis* (Neale and Chapman, 1970). The changes in the morphology of mesosomes in this organism affected by chemical and physical fixation and by the stage of development have been discussed by Nanninga (1971). (See p. 274 for further discussion on this topic.) Another example is that of *Escherichia coli,* which becomes filamentous when transferred from just above to just below the minimum temperature for growth (Ryter, 1968). The temperature at which cells are grown has a pronounced effect on the amount of unsaturated fatty acids in their membranes. For instance, bacteria grown at a low temperature have membranes with a greater proportion of unsaturated fatty acids than those grown at a higher temperature (also see p. 336).

TISSUE SPECIMEN SIZE

A simple and reasonable requirement for satisfactory fixation for transmission electron microscopy is the uniformity of fixation throughout the tissue specimen, since most tissue specimens are fixed through successive layers. Glutaraldehyde penetrates slowly into the tissue, proceeding more slowly the farther it advances. It is apparent that small size of the specimen is of the utmost importance in achieving uniform fixation; yet many workers inadvertently fail to take advantage of the small size of the specimen. The origin of most of the failures in achieving satisfactory fixation lies in the large size of the tissue block. An example of artifacts resulting from using specimens of large size is the presence of apical extrusions in the epithelial cells of the midgut of insects. These cells show extrusions presumably because the fixation of apical part of the cell is too slow to prevent its swelling because of inflow of excess water. Such extrusions are absent when very small tissue pieces are used (Brunings and Priester, 1971).

Rapid and uniform fixation with OsO_4 can occur only to a depth of ~ 0.25 mm in most tissues. Consequently, the ideal size of the tissue block should not exceed 0.5 mm if it is to be uniformly fixed. Although specimens of larger size can be adequately fixed with aldehydes, especially with formaldehyde, it is not encouraged. Moreover, larger blocks have to be cut into smaller pieces for postosmication. According to Rømert and Matthiessen (1975), 2.5% glutaraldehyde penetrates ~ 130 μm into fetal pig liver, whereas a mixture of glutaraldehyde and formaldehyde penetrates ~ 500 μm. Generally, the smaller the size of the specimen, the better and more uniform will be the quality of fixation irrespective of the type of tissue and fixative used. If for some reason the tissue cannot be cut into small cubes, it may be cut into thin strips of ~ 0.5 mm or less thickness. When using a hand-held razor blade, the cutting should be accomplished by one quick slashing motion; several back-and-forth movements of the razor blade will result in extensive physical damage, especially to soft tissues.

In practice, a tissue block fixed by immersion usually shows nonuniform fixation: Figure 2.1 shows three regions of varying fixation quality. The inward diffusion of the fixative from the surface of the tissue block in direct contact with the fixative produces a gradient of fixative concentration. The surface layers of the block are well fixed and contain the highest concentration of the fixative, but may show cellular changes resulting from mechanical injury and extraction of tissue constituents. The extraction of tissue constituents is greatest in the surface layers of the block. Distortion of cell components in the surface layers can also

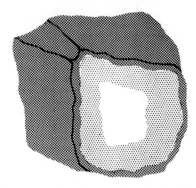

Fig. 2.1. Diagrammatic representation of a tissue block (1–2 mm³) partially penetrated by OsO₄ and showing a peripheral stratification with visibly different layers at varying depths from the surface. Note the unfixed core.

occur in pathological tissues. Great caution should be exercised, therefore, in the interpretation and evaluation of diseased tissues after fixation by immersion, and it is advisable to trim off a few surface layers of cells. Obviously, no such problem is encountered in the study of whole organisms such as bacteria, algae, or fungi.

Immediately beneath the surface layer lies the region of well-fixed cells that are ideal for study. The cells constituting the core of the tissue block are incompletely fixed and may show autolysis and displacement of cellular materials though devoid of mechanical injury and excessive extraction of cellular materials. Very nearly uniform fixation, however, can be obtained when the perfusion technique is possible. The nonuniform fixation discussed earlier is encountered more commonly in tissue blocks that are larger than 0.5–1.0 mm³ in size (Fig. 2.2). Because the importance of obtaining extremely small tissue blocks with minimum mechanical damage is understood, mechanical sectioners are in use, which are discussed later.

Preparation of Tissue Blocks

To facilitate rapid and uniform fixation, tissue blocks should be as small as possible (less than 1 mm³) for transmission electron microscopy. It is especially important to obtain the thinnest possible tissue blocks. The easiest way is to dice the tissue into small pieces with a fresh, clean razor blade or with a pair of fine dissecting scissors. For certain studies it is desirable to slice the tissue in such a way that the precise orientation after embedding can be recognized. However, dicing with a razor blade or scissors has limitations in that slices smaller than 1 mm³ are difficult to cut and excessive mechanical damage is likely. Therefore, various types of instruments are used to cut tissue slices of small size with

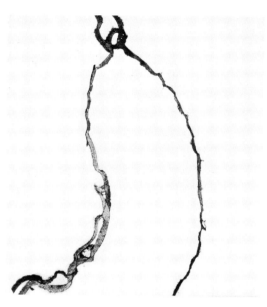

Fig. 2.2. A thick section of liver tissue fixed by immersion with 2% glutaraldehyde for 1 hr. Limited penetration of the tissue by the aldehyde is apparent. Only a thin layer of cells to a depth of 0.4 mm from the surface is preserved and shows positive acid phosphatase activity. The cells in the core are lost during the buffer wash and/or incubation possibly as a consequence of the continuous shaking. ×8. (Saito and Keino, 1976; © 1976, The Histochemical Society, Inc.)

minimum mechanical damage. These instruments include tissue chopper, Vibratome, hand microtome, and a modified rotary microtome.

A simple and inexpensive hand microtome consists of a glass microscope slide (1 mm thick) onto which two cover slips are cemented with a resin with space between them for the tissue (Lewis and Knight, 1977). The tissue either in the buffer or in the fixative is held firmly against the slide and a razor blade is drawn across the tissue, with its ends placed flat against the cover slips. The size of the tissue slice is determined by the thickness of the cover slips. Slices of 150–250 μm thickness can be cut by using this device.

A modified rotary microtome can be used to cut slices from the tissue supported in agar (Snodgrass and Peterson, 1969). The agar is held on a cutting platform clamped in place of the microtome knife. A modified chuck holds a razor blade that cuts vertical slices from the tissue supported by agar.

At least two tissue choppers are available commercially. The TC-2 tissue sectioner was designed by Smith and Farquhar (1965) and is manufactured by the Du Pont Co.; Sorvall Operations; Newton, Connecticut. The Vibratome (Smith, 1970) is manufactured by Oxford Instruments, Inc.; San Mateo, California. The TC-2 has a cutting arm that moves in a fixed vertical plane. A single-edge razor blade is held by the cutting arm, which is adjusted so that it barely fails to hit the platform at the bottom of the cutting stroke. A horizontal tissue platform moves automatically sideways by a controllable distance after each cutting stroke. This distance determines the thickness of the slice. With practice, slices ranging in thickness from 25 to 250 μm can be obtained. Slices can be cut from the tissue either supported by agar or glued directly to the platform with methyl cyanocrylate or a similar adhesive. A large number of slices can be obtained rather rapidly, but it is difficult to maintain serial order of slices.

A modified procedure to improve the established method for cutting slices with TC-2 was described by Shannon (1974). The advantages of this procedure are (1) the time previously required to set up for sectioning and cleaning is almost eliminated, (2) fresh and briefly fixed tissues are more easily sliced, and (3) the possibility of mixing the tissue slices and filter paper slices is eliminated. According to this procedure, the polyethylene disk is lightly roughened with a piece of very coarse silicon carbide abrasive sandpaper. The tissue is placed on the disk which has been precooled to 4°C, and the specimen and disk are covered with 5% agar solution at 43°C. The disk is immediately cooled for 5 min at 4°C to gel the agar. The preparation is removed to room temperature and the gelled agar is trimmed with a razor blade to provide a broad base of adherence and support for the tissue. A minimum amount of agar must be left on the sides perpendicular to the blade. Fresh or briefly fixed tissues are sectioned more easily if additional slight pressure is exerted by the middle finger on the cutting arm toward the end of the cutting stroke. After the sectioning has been completed, the agar can be left to dry and then peeled off, leaving the disk ready for reuse.

The specimen stage of the Vibratome is located in the middle of the bath that is filled with a buffer solution. A razor blade is held horizontally by the cutting arm, which advances the razor blade across the face of the tissue block. The cutting arm vibrates across the line of advance, resulting in a sawing movement. The desired rate of advance as well as the desired amplitude of vibration of the razor blade can be obtained. The specimen stage can be raised after each return stroke of the razor blade.

The Vibratome is considered to be superior to any other tissue chopper; uniformly thin slices in the range of 5–20 μm can be obtained. Both fixed and unfixed and frozen or unfrozen tissue blocks can be sliced readily over a wide range of thickness, and serial slices can be obtained. Another advantage is that tissue blocks as large as 2 cm^2 can be used. Minor limitations are that slices have to be cut individually, which is time-consuming, and it will not cut certain fixed tissues, such as intestine, pancreas, and muscle (also see p. 350).

If it is a problem to keep tissue elements together while cutting them into small slices, gelatin can be used as a glue. For example, spinal roots become separated from the spinal cord when slices of 1 mm are cut for processing. As a result, the normal association of the roots with the cord cannot be studied. Such an association can be maintained by surrounding the tissue with molten gelatin (20%) prior to preparing 1 mm tissue slices (Stirling, 1978). The gelatin is pipetted over the tissue, which is gently moved so that it is completely surrounded by molten gelatin. This step is performed rapidly before the gelatin sets. The gelatin is allowed to gel for 5–8 min, and then fresh fixative is applied to make the gelatin firmer. After 4–8 min, any excess gelatin may be trimmed off, and the tissue surrounded by gelatin can be cut into very small slices for standard processing for electron microscopy. This method is compatible with viewing thick sections with the light microscope or thin sections with the TEM.

AUTOMATIC SPECIMEN PROCESSING

Irrespective of the method used to obtain the desired type and size of specimens, they must be processed immediately and uniformly. The actual method used for changing various solutions and transferring the specimens may affect their final quality of preservation. A considerable amount of time is spent in processing specimens. This problem becomes serious when large numbers of specimens need to be prepared in a busy laboratory. Manual processing of numerous batches of small numbers of specimens is expensive and wasteful. Furthermore, variables introduced during manual processing are a hindrance in achieving reproducible and comparable results. Manual processing also carries with it the risk of a certain degree of harm to the worker. An automatic specimen processor can eliminate most of the problems mentioned earlier. Large numbers of specimens can be prepared simultaneously under controlled and uniform conditions of fixation, washing, dehydration, and infiltration. Other conditions, such as pH, temperature, humidity, pressure, and agitation, can also be controlled precisely. If desired, specimens can be processed at night or at any other time when the technician is busy performing other duties.

To ensure trouble-free operation, an ideal automatic specimen processor should be relatively simple in design, with a minimum number of moving parts. It should provide an accurate process repeatability and be corrosion-proof and economical. The rate of flow should be independent of any changes in the viscosity or surface tension of the effluent and specimens must not be allowed to dry.

Bernhard (1955) described a device that accomplishes specimen dehydration by continuous drip method. Goldfarb *et al.* (1977) described a device for dehydration that was based on the generation of simple concentration gradients in a continuous, rather than stepwise, fashion. The rate of dehydration can be regulated by geometry and/or outflow rate. It is known that by altering the geometry of the reservoirs, a variety of linear, nonlinear, and logarithmic concentration gradients can be produced (Hegenauer *et al.*, 1965). Devices that require the generation of simple concentration gradients in a continuous fashion are widely employed in chromatographic separation and density gradient centrifugation (Peterson and Sober, 1962). A prototype model has been described for electron microscopy, which compactly incorporates the solvent reservoirs and a tissue tray that can accommodate 10 tissue specimens in stainless steel mesh baskets (Goldfarb *et al.*, 1977).

An automatic tissue processor for electron microscopy was developed by Aihara *et al.* (1967, 1972, 1978): commercial models are available from Miles, Lab-Tek Div.; Naperville, Illinois. This processor can carry out fixation, washing, dehydration, and resin infiltration in a single programmed schedule without resort to manual manipulation. The temperature is regulated by circulating water; the humidity, by the circulation of air. In the newest model (Type IV), duration of each step is controlled by photosensoring or computerized unit. Future improvements of this processor are likely to facilitate quantitative electron microscopy and immunocytochemical studies.

Another automatic processor was developed by Norris *et al.* (1967) and Banfield (1970). This processor can hold 50 specimens in 10 chambers. Moisture is removed from the air entering the reservoirs by caps filled with a drying compound. Kölbel (1978) developed a processor that incorporates up to eight specimen chambers. The time is controlled by the electronic schedule timer, which controls 16 steps with time intervals ranging from 1 sec to 99 min each. The processor can control up to 19 different functions, which are indicated by light signals when operating. This processor is not yet on the market.

C. Reichert Optische Werke AG, Wien, Austria, has introduced an EM tissue processor that operates through electronic programming. Up to 46 specimen capsules may be accommodated simultaneously. A motorized agitator produces a continuously revolving and vertical motion of the specimen rack. Any channel not required for processing of a specimen can be omitted, or a channel may be repeated up to three times. The processor is distributed by the American Optical Company. All the previously mentioned processors are quite complicated and await further testing.

OSMOLARITY

It is instructive to define some terms used commonly in discussing the subject of osmolarity; abbreviations used are according to the SI system.

Molarity (mol/l). Amount of substance in moles divided by the volume of the mixture (solution).

Molality (mol/kg). Amount of substance in moles divided by the mass of the solvent.

Osmolarity (Osmol/l). The molarity that an ideal solution of a nondissociating substance must possess in order to exert the same osmotic pressure as the test solution.

Osmolality (Osmol/kg). The molality that an ideal solution of a nondissociating substance must possess in order to exert the same osmotic pressure as the test solution.

The term osmolarity is used here strictly with regard to the response of cells immersed in a solution. Solutions should be defined in osmolarity measurements because they are independent of temperature. Since nonelectrolyte solutions do not ionize, the molar and osmolar concentrations are equivalent. On the other hand, electrolyte solutions dissociate into ions and thus exert a greater osmotic pressure than their molar concentration indicates. However, because of ion interaction, the osmotic pressure is lower than would be expected on the basis of number of particles. As a result, two solutions of identical molarities can have different osmolarities.

An isotonic solution exerts an osmotic pressure equal to that exerted by cell cytoplasm. A fixative solution is isotonic with a cell (or tissue specimen) if the cell neither shrinks nor swells when immersed in it. Two solutions are isosmotic with each other if both exert the same osmotic pressure; in other words, both solutions possess the same solute particle concentration. An isosmotic fixative solution may or may not be isotonic.

The actual molecular weight of salts differs, depending on the degree of their hydration. The presence or absence of several molecules of water results in a difference in the molecular weight. For example, $Na_2HPO_4 \cdot 2H_2O$ and $Na_2HPO_4 \cdot 12H_2O$ have molecular weights of 177.99 and 358.17, respectively. Since salts used in the preparation of solutions are available as different hydrates, it is important to know the exact formulas of the chemicals in order to obtain solutions of known concentrations.

The osmolarity (tonicity) of a fixative has a direct effect on the appearance of the fixed specimen (Fig. 2.3). Cell size and shape may be affected by a change in buffer osmolarity (Fig. 2.4). It appears, though, that various tissue types differ in the degree of their response to the difference in osmolarity between the fixative solution and the cell's normal external environment. The effect produced is probably partly correlated with the degree of "density" (compactness) of the tissue. The effect of osmolarity generally appears to be minimal on relatively compact or hard tissues, whereas osmolarity exerts a marked effect on "less compact" tissues.

A classic example of the difference in response to the osmolarity of the fixative solution is exhibited by brain tissue. The brain of the newborn rat shows a profound sensitivity to changes in the osmolarity of glutaraldehyde solutions, whereas the mature brain is affected less when the osmolarity is varied. The

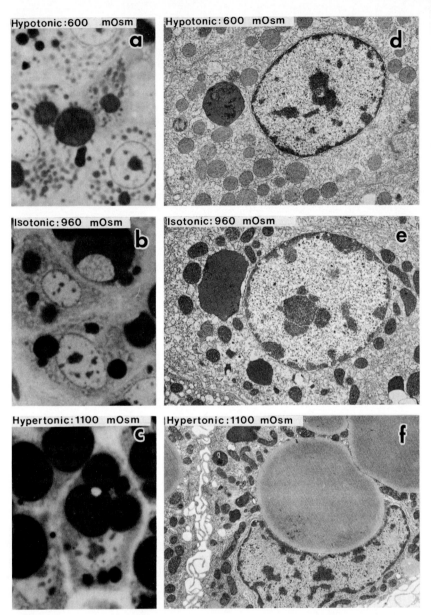

Fig. 2.3. Effects of different osmolarities of fixative solutions on the liver of a fish (*Triakis scyllia*). Photomicrographs a, b, and c correspond to electron micrographs d, e, and f, respectively. The fixatives used were a and d, 2% glutaraldehyde in 0.2 *M* cacodylate buffer caused swelling; c and f, 1% glutaraldehyde in marine water caused shrinkage; b and e, 2% glutaraldehyde in 0.1 *M* buffer with 1.68% NaCl caused minimum changes. Mitochondria clearly show the effects of osmolarity. (Saito and Tanaka, 1980.)

18

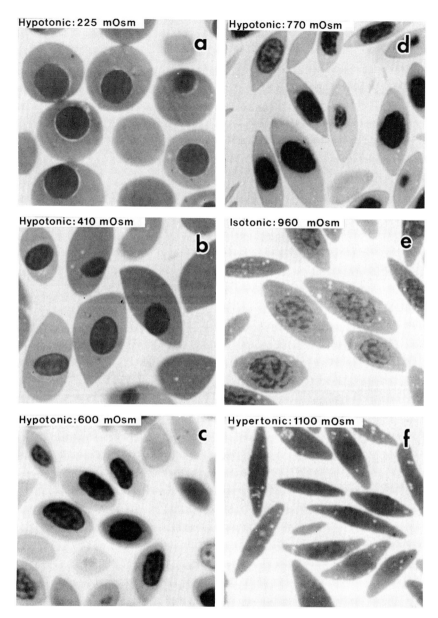

Fig. 2.4. Photomicrographs of thick sections of Epon-embedded peripheral blood cells of marine fish, *Triakis scyllia*, treated with fixatives of different osmolarities. The fixatives used were a, 1% glutaraldehyde in 0.05 M cacodylate buffer (pH 7.4); b, 2% glutaraldehyde in 0.1 M buffer; c, 2% glutaraldehyde in 0.2 M buffer; d, 1% glutaraldehyde in 0.2 M buffer with 0.92% NaCl; e, 2% glutaraldehyde in 0.1 M buffer with 1.68% NaCl; f, 1% glutaraldehyde in marine water. Note swelling of nucleated erythrocytes fixed with hypotonic solutions (a to d), shrinkage with hypertonic solutions (f), and satisfactory preservation with isotonic solutions (e). (Saito and Tanaka, 1980.)

sensitivity of the former is due, in part, to its less compact and more hydrated nature. The immaturity of the capillaries of the blood–brain barrier in the developing brain may also be related to the sensitivity.

Maser *et al.* (1967) indicated that the morphology of newborn rat epidermis is affected by a difference as little as 50 mosmols. Some of the other cell systems that have shown the effects of changes in the fixative osmolarity are grasshopper spermatocytes and spermatids (Tahmisian, 1964), rat muscle (Fahimi and Drochmans, 1965b), neuromuscular junctions (A. W. Clark, 1976), and membranes of *Limnaea* eggs (Elbers, 1966).

Various components of a cell seem to differ in their sensitivity to the changes in fixative osmolarity. The ultrastructure of mitochondria in chick embryo heart is sensitive to a change in buffer osmolality as small as 20 mosmols, whereas the endocardial lining of the organ is less sensitive to a change in the osmolarity within reasonable limits (Pexieder, 1977). (It should be noted that chick embryo heart is isotonic to 250 mosmols.) The relatively lower sensitivity shown by the endocardium may be due to the instantaneous diffusion of the fixative into this lining by vascular perfusion.

Detailed studies correlating the changes in mitochondrial optical density, volume, and ultrastructure with osmotically induced swelling have been undertaken using osmolalities ranging from 3 to 250 mosmols (Stoner and Sirak, 1969). The fixation of swollen, spherical mitoplasts requires adjustment of the fixative vehicle to lower osmolalities ($\sim$45 mosmols) than those used routinely (Hackenbrock, 1972). Apparently, mitoplasts are more sensitive to extrinsic variations in osmotic pressure than are intact mitochondria, and the former respond to the osmolarity faster than the penetration and fixation of their membranes.

Ionic Composition of Fixative Solution

It is known that the maintenance of a delicate balance among various ions (e.g., Na^+, K^+, Ca^{2+}, and Mg^{2+}) in the immediate environment of a living cell is necessary for the normal function and structure of the cell. An upset in this balance is likely to bring about a change in metabolism, even if the osmolarity were kept constant. This change in turn could cause changes in the structure of the cells. Sufficient data are available indicating that many of the fixation vehicles presently in use do not meet the physiological requirements of the tissues. Although the degree to which a fixative solution should be "physiological" is still in question, it is reasonable that an approximation of the fixation conditions with the normal extracellular environment would be helpful in obtaining a more successful fixation. It must be pointed out that although the effect of osmolarity has been recognized in many studies, less attention has been paid to the ionic composition of the fixative solutions. The type of nonelectrolyte or ion (especially the valence of cations) and the dielectric constant of the fixative vehicle play an important role in the preservation of cell fine structure.

The ionic composition of the body fluids of some representative animals is given in Tables 2.1 and 2.2. It should be noted that although the great majority of marine invertebrates (e.g., *Echinus*) are isotonic to seawater, some of the crabs and prawns show an osmotic pressure lower than that of their environment. Marine vertebrates show greater osmotic differences with their environment. For example, *Muraena* is greatly hypotonic to seawater (Table 2.1). Most of the freshwater animals possess higher concentrations of Na^+ and K^+ than those in the surrounding environment. In freshwater animals, Na^+ and Cl^- comprise the major part of the total ion concentration in extracellular fluids. From the preced-

TABLE 2.1

The Composition of the Body Fluids of Some Marine Animals[a,b]

Animal	Freezing-point depression	Na	K	Ca	Mg	Cl	SO$_4$
Echinoderms							
Echinus esculentus	100	100	102	101	100	100	101
Holothuria tubulosa	100	101	103	102	104	100	100
Annelid							
Arenicola marina	100	100	103	100	100	98	92
Mollusks							
Mytilus edulis	100	100	135	100	100	101	98
Eledone cirrosa	100	97	152	107	103	102	77
Arthropods							
Homarus vulgaris	100	110	85	131	14	101	32
Carcinus maenas	98–103	110	118	108	34	104	61
Pachygrapsus marmoratus	91	94	95	92	24	87	46
Palaemon serratus	74	79	70	115	22	74	9
Vertebrates							
Cyclostome							
Myxine glutinosa	102	110	90	56	33	97	22
Elasmobranchs	102–110	—	—	—	—	—	—
Mustelus canis	109	61	47	37	6	45	7
Teleosts	34–44	—	—	—	—	—	—
Muraena helena	—	38	16	31	4	29	17

[a] After Shaw (1960).

[b] Values are expressed as a percentage of the corresponding values in seawater or in the body fluid dialyzed against seawater.

TABLE 2.2

The Composition of the Body Fluids of Some Freshwater Animals[a,b]

Animal	Total concentration	Na	K	Ca	Mg	Cl	SO₄
Crustacea							
Telphusa fluviatilis	695	—	—	—	—	301	—
Eriocheir sinensis	645	—	5	10	3.5	282	—
Astacus fluviatilis	480	—	5.2	10.4	2.7	194	—
(river water)	12	—	0.05	0.7	0.4	0.3	—
Teleost							
Cotegonus clupeiodes	315	141	3.8	5.3	1.7	117	213
Amphibian							
Rana esculenta	261	104	2.5	2.0	1.2	74	1.9
Insect							
Aedes aegypti	287	110	3	—	—	43	—
Annelid							
Lumbricus terrestris							
(in tapwater)	184	—	—	—	—	47	—
Mollusks							
Limnea peregra	130	—	—	—	—	—	—
Anodonta cygnea	46	15.5	0.5	8.4	0.2	11.7	0.8
Coelenterates							
Hydra viridis	40–50	—	—	—	—	—	—
Pelmatohydra oligactis	—	2.3	14	—	—	0.15	—
Poriferia							
Spongilla sp.	50–60	—	—	—	—	—	—
Protozoa							
Spirostomum ambiguum	50	1	7	—	—	0.6	—

[a] After Shaw (1960).
[b] Concentrations of ions in millimoles per liter. Total concentrations in milliosmols per liter equivalent sodium chloride solution.

ing discussion it should be evident that a fixative having a specific osmolarity will vary in its rate of penetration into different tissues.

As mentioned earlier, changes in cell structure occur since the ionic composition of fixative solutions differs from that of the natural extracellular fluids. These changes may result in the extraction of cellular materials or in the deposition of fixative components, especially during the initial period of fixation. The problem of extraction is more serious when OsO_4 or paraformaldehyde is used as a primary fixative rather than glutaraldehyde. The possible reason is that OsO_4 penetrates slowly and paraformaldehyde reacts slower than glutaraldehyde. Thus, when OsO_4 or paraformaldehyde is used as a primary fixative, the extraction caused by the vehicle may precede the actual fixation by the fixative. This is the reason that ionic composition of the vehicle is of greater importance when

OsO_4 or paraformaldehyde is used as primary fixative than when glutaraldehyde is used, which reacts with and stabilizes proteins very efficiently.

The extraction of cellular materials can be minimized by stabilizing them by maintaining an approximate Ringer-type of balance among the various ions surrounding the cells until the fixing agent has penetrated in sufficient concentration to bring about permanent fixation. In practice, this is accomplished by making the fixative solution somewhat similar to the extracellular fluids. This is done by adding nonelectrolytes or electrolytes to the vehicle. The advantages of the addition of these substances outweigh the possible disadvantages (which are presented later).

Vehicle Osmolality

The vehicle of a fixative usually consists of a buffer to which may be added electrolytes or nonelectrolytes. Although the buffer is partly responsible for the vehicle osmolarity, salts or other substances when added increase the osmolarity. Since most buffers, when used at the physiological range of pH, are hypotonic, the desirability of adding other substances to increase the osmolarity is apparent.

During fixation, specimens are exposed to the osmotic effect of the vehicle of the fixative, the osmotic effect of the fixative, and the fixation effect. The last effect is presented later, along with each of the fixatives. The osmotic effect of the vehicle plays a dominant role in determining the quality of specimen preservation during primary fixation with glutaraldehyde (Fig. 2.5). A plausible explanation for this phenomenon is in order. The total number of molecules and ions other than water in a fixative solution, rather than the nature of these particles, determines the total osmotic pressure. The effective osmotic pressure, on the other hand, refers to the pressure caused by those particles that do not freely pass through the cell membrane. Thus, the effective osmotic pressure is different from the total osmotic pressure as measured by the freezing point depression method. The effective osmotic pressure is primarily a function of the fixative vehicle because its components do not pass freely through the membrane. Since the total osmotic pressure of a fixative solution is less important than its effective osmotic pressure, the osmolarity of the vehicle becomes more important than that of the glutaraldehyde in the fixative solution.

Glutaraldehyde also exerts some osmotic effect on cells. That an isotonic vehicle will cause shrinkage of cells in the presence of glutaraldehyde indicates an osmotic contribution by the dialdehyde. This is confirmed by the finding that a hypertonic solution of glutaraldehyde either in an isotonic or a hypotonic buffer causes shrinkage damage on lung tissue, as evidenced by changes in the shape of erythrocytes in capillaries and small blood vessels examined with the TEM or the SEM (Mathieu *et al.*, 1978). Thus, the importance of glutaraldehyde concentra-

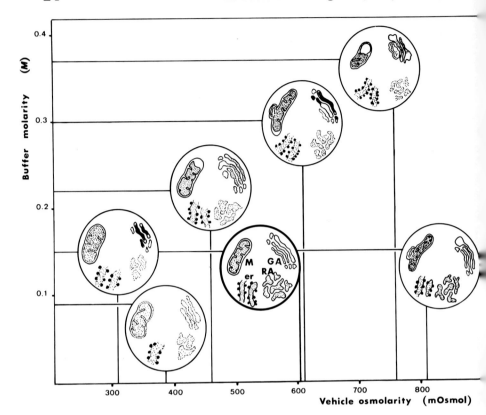

Fig. 2.5.   Diagram showing changes in ultrastructure of the organelles in prothoracic gland cells of an insect (*Rhodnius prolixus*) (during the mid-molting cycle) fixed under different vehicle osmolarities. Changes in the appearance of these organelles are apparent. The optimum preservation was obtained with 2–5% glutaraldehyde or a mixture of glutaraldehyde and formaldehyde (1:1) in 0.15 mol/l phosphate buffer (pH 7.2). A vehicle osmolarity of 510–560 mosmols (by adding sucrose) was judged satisfactory for aldehyde concentrations ranging from 2 to 5%. During the fasting cycle, the optimal vehicle osmolarity was 295–310 mosmols. Mitochondria (M), golgi (GA), and rough (er) and smooth (RA) endoplasmic reticulum. (Beaulaton and Gras, 1980.)

tion in determining the optimal osmolarity of the fixative solution for tissues in general and cells in culture, or in suspension in particular, cannot be disregarded. The extent of the contribution of an osmotic effect by glutaraldehyde is dependent on its concentration as well as upon the type of the specimen; changes in glutaraldehyde concentration ranging from 2 to 5% may not show any significant effect on the ultrastructure of intact tissues, especially in the presence of an optimal vehicle osmolarity.

As stated earlier, glutaraldehyde itself contributes less than the vehicle to the effective osmotic pressure of the glutaraldehyde fixative solution. An over-simplified relationship between the osmotic contributions by the vehicle and glutaraldehyde in the fixation of isolated cells can be determined (Barnard, 1976):

$$EOP = \frac{Mg}{2} + Mv,$$

where EOP = effective osmotic pressure, Mg = milliosmols of glutaraldehyde, and Mv = milliosmols of vehicle. Isotonicity, for example, for isolated hepatocytes and multiocular adipocytes was obtained at an effective osmotic pressure of 280 mosmols.

However, for intact tissues, the osmotic contribution by glutaraldehyde is much less than half of that by the vehicle. Moreover, the osmolarity of the vehicle and the concentration of glutaraldehyde are not independent of each other in their effect on cell ultrastructure.

The osmolarity of the buffer used for rinsing and for preparing OsO_4 solution is important because fixation with aldehydes does not destroy the osmotic activity of cellular membranes. Presently, most workers use the same buffer for rinsing and preparing OsO_4 solution as in the glutaraldehyde fixative solution. This practice is undesirable because the osmolarity of the buffer is lower than that of the glutaraldehyde fixative solution. This lower osmolarity may cause hypoosmotic damage. Therefore, it is desirable that the osmolarity of the rinsing buffer be increased to the value of the total osmolarity of the primary fixative solution. Similarly, OsO_4 solution should be prepared in the buffer having an increased osmolarity so that the total osmolarity of the OsO_4 solution is equal to that of the glutaraldehyde solution. In other words, the total osmolarity of the glutaraldehyde solution, rinsing buffer, and OsO_4 solution should be equal. This practice, however, will result in differences in the effective osmotic pressure of the three solutions; consequently, it needs further study. It is convenient to remember that 1% pure glutaraldehyde is 0.1 mol/l and contributes ~ 100 mosmols, whereas 1% OsO_4 is 0.04 mol/l and contributes 40 mosmols to the solution.

Studies on the fine structure of tissues undergoing morphogenesis should take into account physiologically controlled osmotic variations. These variations require corresponding changes in the vehicle osmolarity. An example is the alterations in the osmolarity of the hemolymph during the last two larval stages of the insect *Rhodnius prolixus* (Beaulaton and Gras, 1980). They found that optimum preservation of the fine structure of prothoracic gland cells during the fasting stage required the vehicle osmolality of 300 mosmols, whereas during the midmolting stage, the vehicle osmolality had to be raised to 510–560 mosmols.

Unfortunately, either the osmolality of the fixative solution is not given or only total osmolality measured by the freezing point depression method is specified in the published articles. Although fixation of intact tissues always involves changes in the osmolality of the fixative solution due to limitations in the penetration by the aldehyde and the dilution of the fixative by the tissue fluid (which cannot be completely controlled even with the best perfusion technique), specification of the total as well as vehicle osmolality should be helpful to the readers to compare and standardize the methodology. Editors of scientific journals should request that both osmolalities be specified in the articles.

Effects of Added Substances

The addition of an electrolyte or a nonelectrolyte or both to the fixative solution seems to be necessary in order to minimize the distortion by fixative particles that differ from those of the tissue fluids in mobility and electrostatic charge. The addition of these substances is useful because fixatives used for electron microscopy penetrate relatively slowly.

The effects of the addition of substances to the fixative vehicle can be explained not only by osmolarity but also by the nature of the substance used. There may be interactions between the added substances and the cell constituents, on the one hand, and the fixative agent, on the other hand. It is known, for example, that the denaturation of proteins and nucleic acids is affected by the salts added to the fixative solution. Von Hippel and Wong (1964) found that neutral salts affect the stability of the native form of macromolecules as diverse as ribonuclease, DNA, collagen, and myosin. They indicated that $CaCl_2$ and potassium thiocyanate act as strong general structural denaturants and that ammonium sulfate and potassium dihydrogen phosphate stabilize the native conformation of the proteins. Biochemical studies indicate that calcium and lithium ions act as denaturants of proteins through interacting with peptide groups (Bello *et al.*, 1966). It has been suggested that in addition to direct ion–peptide interactions, salts may enhance the denaturation ability of water.

Robinson and Grant (1966) have shown that Na salts of trichloracetate, thiocyanate, perchlorate, and iodide had a potent effect on DNA. They found that anions had a strong denaturing effect, whereas cations had only a small effect.

Sucrose and polyvinylpyrrolidone (PVP) are known to affect enzymatic activity. Even very low concentrations of sucrose and PVP tend to inhibit the activity of certain cerebral enzymes (e.g., mitochondrial succinate dehydrogenase) (Lisý *et al.*, 1971). According to Lisý *et al.*, the inhibition does not seem to be related to the increased osmolarity caused by the addition of nonelectrolytes, because even isotonic solutions have an inhibitory effect. Hinton *et al.* (1969) have discussed possible reasons for the inhibitory effect of sucrose on enzyme activity. It was pointed out that nonelectrolytes are more than just osmotic particles to which the cell is impermeable; their presence greatly increases the effect of salts subsequently added. The possibility of sucrose reaction by hydrogen bonding with groups on the membrane has been proposed (Bernheim, 1971).

However, an agreement on the inhibitory effect of sucrose and the nature of this effect on enzyme activity is lacking. Evidence is available from both eukaryotic (Hayashi and Freiman, 1966) and prokaryotic (Burnham and Hageage, 1973) organisms that indicates that the addition of sucrose to the aldehyde fixative solutions reduces their inhibitory effect on enzyme activity. Moreover, according to Burnham and Hageage (1973), the increase in enzyme activity is possibly due to the increased osmolarity imparted to the fixative solution by the addition of sucrose. From the foregoing discussion, it is apparent that a number of varied and complex factors are involved in the total effects of added particles on the preservation of cell components, including enzyme activity.

Specific Effects of Nonelectrolytes

The addition of nonelectrolytes generally minimizes the extraction of certain cellular constituents. A significant reduction in the extraction of chlorophyll in isolated chloroplasts by adding sucrose to the fixative is well known. Klein and Shochat (1971) demonstrated that the presence of sucrose in OsO_4 prevented the extraction of almost 50% of the chlorophyll present in isolated chloroplasts, whereas use of sorbitol resulted in 100% extraction of the chlorophyll. The latter loss of chlorophyll seems to be related to the interference with fixation of chloroplast membranes.

The addition of dextran to glutaraldehyde improves the preservation of myelin and node of Ranvier and prevents swelling of brain slices during incubation *in vitro* and subsequent fixation. Dextran or PVP even offsets the swelling of chloroplasts caused by detergent treatment (Deamer and Crofts, 1967). In perfusion fixation the addition of dextran to glutaraldehyde prevents the artifactual enlargement of extravascular space, especially in pancreas (Bohman and Maunsbach, 1970). One explanation of this enlargement is that the extraction of plasma proteins results in a disturbed colloid osmotic balance over the vascular wall; this in turn causes an increased flow of capillary fluid into extravascular spaces.

The effectiveness of nonionizing, osmotically active substances in preventing swelling is related, in part, to the extremely slow-penetrating or nonpenetrating property of these substances. The distribution of water in cells is normally a function of the relative concentrations of internal and external osmotically active species. During fixation the concentration of these species is constant inside the cell, and the external concentration of these species can be controlled. It is apparent, therefore, that the amount of water in the cell, and thus cell volume, is controlled primarily by the external concentration of nonpenetrating, neutral molecules of substances such as sucrose, dextran, etc.

Immediately after coming in contact with OsO_4, cell membranes lose their relative impermeability to the smaller molecules and ions in the fixing solution;

thus, an equilibrium is soon reached and thereafter the direction of water movement is dependent on the difference in the osmolarity of the nonpenetrating, osmotically active species present inside and outside the cell. If this osmolarity is greater outside, then the water will move outward. Since larger molecules of nonelectrolytes are unable to pass through the fixed membranes, they exert an osmotic pressure that tends to prevent inward flow of water. Although the diffusion gradient for water is the main force that determines the swelling, if the osmotic pressure is too high, the swelling rate is very slow despite the high diffusion gradient.

In order for the cell not to shrink or swell, the external concentration of nonelectrolytes has to be equal to the internal concentration of the fixed, osmotically active species. As an example, the external concentrations of dextran and PVP that allow neither shrinkage nor swelling of isolated chloroplasts in 0.1 mol/l salt solution are 0.05 mosmols and 0.15 mosmols, respectively (Deamer and Crofts, 1967). They estimated that 0.1 mosmols is the internal concentration of the fixed species in the isolated chloroplasts. It is emphasized that although the influx or outflux of a nonelectrolyte is related to its concentration at the opposite side of the membrane, the fluxes of ions are generally not such simple functions of the concentrations, but depend also on the potential difference across the membrane and on the ionic pumps in the membrane.

Because of their extremely slow rate of penetration through membranes, nonelectrolytes may also act as osmotic stabilizers where the permeability properties of membranes persist, as, for example, after fixation with glutaraldehyde. Furthermore, the addition of nonelectrolytes may help to prevent the collapse of certain cell structures by keeping them expanded when their active transport and permeability properties are destroyed by the fixative. This is related to the hydration property of nonelectrolytes. It has been demonstrated, for instance, that the addition of sucrose to unbuffered OsO_4 leads to the preservation of well-defined vesicles bound by clear unit membrane in the bacterial mesosomes (Burdett and Rogers, 1970). The protective action of PVP or dextran against damage associated with freezing and thawing in biological systems in well known (Ashwood-Smith and Warby, 1971).

Specific Effects of Electrolytes

The addition of electrolytes (especially $CaCl_2$) to the fixative solutions has many beneficial effects including (1) decrease in the swelling of cell components, (2) maintenance of cell shape, (3) reduction in the extraction of cellular materials, and (4) membrane stabilization. These and other beneficial effects and the mechanisms involved in these effects are discussed here.

Electrolytes seem to be superior over nonelectrolytes in the control of swelling and maintenance of cell shape. The presence of an electrolyte at isotonic concentration is an essential condition for shape preservation during the fixation of red blood cells (Morel *et al.*, 1971). It was reported that the swelling of human erythrocytes was prevented by the addition of 0.5% $CaCl_2$ to the fixative, whereas the presence of sucrose did not inhibit swelling (Millonig and Marinozzi, 1968). Elbers (1966) also reported the prevention of swelling of *Limnaea* egg cells by the addition of $CaCl_2$. It is also known that in the absence of Ca^{2+}, OsO_4 solutions cause chromosomal and nuclear swelling. Although the exact role played by Ca^{2+} in the prevention of swelling is not completely clear, the following explanation seems plausible (Tooze, 1964).

Since the pH of fixatives is usually above the isoelectric point of most proteins, the fixed proteins carry a net negative charge, which gives rise to a double-layer repulsive force. The divalent ions are capable of reducing this repulsive force by cross-linking the separate, negatively charged protein chains. This would prevent any increase in the osmotic pressure because of Donnan equilibrium, and thus the result is cellular shrinkage. It is known that after fixation with OsO_4, cell membranes become permeable to Ca^{2+} but remain impermeable to protein macromolecules. The divalent cations are firmly bound to the protein gels and are not removed during dehydration. This repulsive force can also be reduced by lowering the dielectric constant of the fixative vehicle or by lowering its pH to the isoelectric point of the fixed proteins. At a low pH, the net negative charge on the fixed proteins is less than at a high pH. The rate of cellular swelling is dependent on the net negative charge of the cytoplasmic colloids.

Electrolytes are added to the fixation solution not only to prevent swelling but also to lessen the amount of extraction of cellular materials. The effect of electrolytes on the solubility of proteins has long been known, and the capacity of these salts to precipitate or to dissolve various proteins is a familiar phenomenon in protein chemistry. The salts that combine chemically with amino acids are the ones that either increase or decrease the aqueous solubility of polypeptides. Different amino acids differ in their response to the addition of a specific salt to the fixation vehicle. It is known, for example, that the aqueous solubilities of leucine, tryptophan, and phenylalanine are decreased in the presence of NaCl, whereas those of glycine and cysteine are unaffected. It is also known that a 0.01 mol/l solution of $CaCl_2$ increases the solubility of glycine as much as 3.95%, whereas 0.01 mol/l solution of KCl increases its solubility by only 1.0%. The solubility of leucine, on the other hand, is decreased in the presence of $CaCl_2$. Apparently, the type and concentration of electrolyte ions added to the fixative vehicle profoundly affect the solubility of amino acids.

Numerous attempts have been made to lessen the amount of extraction of cellular materials by the addition of electrolytes, especially divalent cations. It was demonstrated by Tooze (1964), for example, that the addition of Ca^{2+} ions to a final concentration of 0.01 mol/l in the fixative completely suppressed extraction of hemoglobin and stabilized erythrocytes, whereas the addition of monovalent sodium ions had no stabilizing effect. That the ultrastructure of chromosomes is affected by Ca^{2+} has been shown (Barnicot, 1967). The addition of Ca^{2+} solution prevented the precipitation artifact in the nuclei of frog erythrocytes (Davies and Spencer, 1962), and spindle filaments were stabilized by either using a pH of 6.1 or adding Ca^{2+} and Mg^{2+} (Harris, 1962).

Studies by Robbins (1961) indicated that OsO_4 in isotonic NaCl, to which $MgCl_2$ was added to a final concentration of 0.02 mol/l, prevented chromosomal dispersion in cells in metaphase. Addition of $MgSO_4$ to the fixative solution seems to stabilize the microtubules (Shigenaka *et al.*, 1974). Trump and Ericsson (1965) indicated that the addition of NaCl to the fixative produced a slightly greater electron density in the cytosomal matrix. The available information on the exact role played by electrolytes in reducing the extraction of cellular materials is insufficient. Stabilization of membranes especially by divalent ions possibly plays some role in decreased loss of cellular materials, which is discussed later.

The ionic environment of cells during and after fixation with an aldehyde may affect membrane structure. Calcium added to the primary fixative generally binds to membranes, resulting in the stabilization of membranes in various cell and tissue types. Although the exact mechanism responsible for this stabilization is not known, the following explanation seems feasible. The most likely possiblity is that $CaCl_2$ reduces the extraction of phospholipids by tightly complexing with fatty acids and/or phosphate groups, presumably as an insoluble calcium salt. This bonding may be responsible for retention of phospholipids in the membranes of cells fixed in the presence of $CaCl_2$. Further explanation of this point is in order.

Phospholipids are thought to have greater affinity for Ca^{2+} than for most other divalent ions. Calcium ions may bind to the negative groups of phospholipids and induce phase separations, resulting in the modification of various properties of the lipid bilayer. The lipid bilayer may become more condensed in the presence of Ca^{2+}, which may result in a decreased fluidity of hydrocarbon chains of the membrane. These modifications are likely to affect the rate of movement of molecules and ions across the lipid bilayer.

Another possibility is that the retention of proteins in the presence of Ca^{2+} may be responsible for the stabilization of membranes. This decreased leakage of proteins allows more cross-linking to occur. It has been shown that the addition of $CaCl_2$ to the fixative solution suppressed extraction of proteins in erythrocytes (Tooze, 1964). As stated earlier, divalent ions are assumed to be effective in reducing the protein repulsive force in the membrane system and provide some cross-linking between the separate, negatively charged protein chains. This cross-linking probably causes membrane stabilization, for proteins are anchored more firmly. It has been suggested that when Ca^{2+} ions react with protein, these ions combine with free carboxyl groups of dicarboxylic acid and hydroxyl groups of amino acids, resulting in a cross-linking between Ca^{2+} and adjacent polypeptide chains (Greenberg, 1944; Roth et al., 1963; Trump and Ericsson, 1965). It has also been suggested that Ca^{2+} ions denature proteins to various degrees by interacting with amides on the peptide chain. (Some workers, however, are of the opinion that Ca^{2+} does not seem to denature membrane proteins.) Thus, the retention of proteins may encourage their cross-linking with lipids that otherwise may also gradually leach out.

The presence of divalent ions in the fixative solution may cause alterations in the surface charge of membranes. Not only intra- but also intermembranous linkage by divalent ions may occur. If the previously mentioned series of events is operative, the expected result is stabilization of membranes. A brief comment on the possible role of aldehydes in the binding of Ca^{2+} to reactive groups in membranes is pertinent. Glutaraldehyde, for example, may alter the conformation of membrane proteins by exposing negatively charged sites or by reacting with basic residues, which may render proteins more acidic. Thus, Ca^{2+} could bind to sites that are not normally available under physiological conditions (Oschman et al., 1974). However, Ca^{2+} may alter the solubility of nonpolar groups on the membrane constituents and thereby affect the membrane structure without actually physically binding to it. The specific changes and beneficial effects caused by the additon of electrolytes to the fixative solutions follow.

Addition of Ca^{2+} to the OsO_4 vehicle facilitates the preservation of cellular structures such as spindle fibers, myelin figures, membranes (including those of red blood cells), and phospholipids. Improved preservation of these structures is due to decreased leaching of lipids in the presence of Ca^{2+}. This point has already been explained earlier. Similarly, the presence of Ca^{2+} in glutaraldehyde solution and in the dehydration solutions results in the stabilization of cytomembranes of red algae (Lin and Sommerfeld, 1978). These membranes often project negative images following conventional double fixation. Studies of the effects of divalent ions on the structure of plasma membranes and the inner and outer acrosomal membranes of human spermatozoa indicated that Ca^{2+} produced an increase in thickness of the outer acrosomal membrane by $\sim20\%$ but did not affect the plasma membrane or inner acrosomal membrane (Roomans, 1975). Magnesium did not change the thickness of any of the membranes studied.

Sodium or K^+ produces structures on both the inside and outside the plasma membrane in the rat central nervous tissue (Karlsson et al., 1975). The presence of Na^+ induces osmiophilia at the extracellular side of the membrane, whereas K^+ induces a widening of the intracellular lamina of the membrane. These changes are seen only when OsO_4 is used. Aldehyde fixation alone does not reveal these deposits caused by Na^+ or K^+. It appears that certain membranous or cytoplasmic constituents are stabilized by complexing with these cations and that these complexes in turn reduce OsO_4. Whether these structures are artifacts or real is not clear.

Calcium-containing fixative solutions facilitate the submicroscopic viewing of Ca-binding sites, known as electron-dense spots (EDS), in insect intestinal membranes (Oschman and Wall, 1972). EDS are visible only when Ca^{2+} ions are added to the fixative whether OsO_4 is used or not. Similarly, Ca-containing fixatives facilitate the demonstration of a subpopulation of synaptic vesicles fused to the nerve terminals (Boyne *et al.*, 1975).

Probably, the most dramatic effect of the presence or absence of divalent ions (e.g., Ca^{2+}) during fixation is exhibited by bacteria. The morphological pattern of bacterial membranes is influenced by Ca^{2+} deficiency during fixation. Phospholipid losses during dehydration increase in bacteria when Ca^{2+} is omitted from the fixative. Mesosome configuration is profoundly affected by the ionic composition of the OsO_4 fixative. It has been demonstrated that the fixation of bacteria under conditions of Ca^{2+} deficiency results in symmetrical membranes and lamellar mesosomes, whereas in the presence of Ca^{2+}, the membranes appear asymmetrical and mesosomes vesicular (Silva, 1971) (also see p. 274).

The addition of Ca^{2+} to the fixative solution appears to minimize the diffusion of enzymes from membranes during incubation for cytochemical studies, which may be explained on the basis of interactions between Ca^{2+} and phosphate and/or fatty acids in the membrane. For additional information on the effects of divalent ions on membrane structure, the reader is referred to Radda (1975), Wallach (1975), Williams (1975), and Hardonk *et al.* (1977).

Undesirable Effects

Possible deleterious side effects of substances added to the fixative solutions cannot be overlooked. Available data indicate that the rate of fixative penetration into the tissue may be retarded by the addition of substances to the fixation vehicle. The reasons for this decreased rate are not known, although the effect could be due to a greater conformational stability of diffusion barriers when other molecules or ions are present. The addition of electrolytes, especially divalent ions of Ca and Mg, however, results in only a slight decrease in the rate of penetration of the fixative, in comparison to the decrease caused by the addition of nonelectrolytes.

That nonelectrolytes decrease the rate of penetration of OsO_4 has been demonstrated. Hagström and Bahr (1960) reported as much as a 50% decrease in the penetration rate when 0.25 M sucrose was added to OsO_4. The reasons for this decrease in the rate of penetration are not known. Increased viscosity and competition with the fixative molecules could be a partial explanation. According to Sitte (1960), however, the rate of OsO_4 penetration into plant tissues (spruce needles and storage parenchyma of potatoes) was not affected significantly by the addition of buffer, glucose, or urea.

In some cases an undesirable effect of the addition of nonelectrolytes to the OsO_4 may be an increase in the extraction of cellular materials. Wood and Luft

(1963) reported an increase in the extraction of proteins in the presence of sucrose or dextran in the OsO_4 fixative, and Millonig (1966) demonstrated increased extraction of albumin in the presence of sucrose.

The addition of divalent cations may cause protein precipitation and excessive granularity. For these reasons it is generally preferable to use only low concentrations of these ions to adjust osmolarity. Monovalent ions impart only negligible granularity, although fixative solutions containing only monovalent cations are generally less effective in suppressing the extraction of cellular materials. Divalent cations are 10 to 100 times more effective than are monovalent ions in this respect. A disadvantage of adding Ca^{2+} is considerable movement of chromosomes relative to each other (Skaer and Whytock, 1976). The addition of Ca^{2+} may bring about physiological effects such as the stimulation of myofibrils or triggering of secretion (Glauert, 1974).

Calcium ions seem to disrupt the regular pattern of microtubules (Schliwa, 1977). The extraction of microtubules in the presence of bicarbonate ions has been shown. Phosphate buffers may cause precipitation. This is exemplified by the appearance of spherical dense granules on the alveolar surfaces of the lung fixed primarily with phosphate-buffered OsO_4. It should be remembered that artifactual precipitation is not easy to detect. Although both Ca^{2+} and Mg^{2+} tend to prevent excessive swelling, they may cause shrinkage. Therefore, the role of Li^+ and Be^{2+} in reducing specimen shrinkage during dehydration deserves exploration.

Methods for Adjusting the Osmolarity

Several methods are available for adjusting the osmolarity of the fixative solution. In addition to the buffer and fixative, osmolarity can be adjusted by adding appropriate nonelectrolytes or electrolytes. The most commonly used nonelectrolytes are sucrose, glucose, dextran, and PVP. These substances can be used in aldehydes and secondary OsO_4 fixatives, but not in primary OsO_4 fixatives because they may cause leaching of soluble cellular materials during fixation. The addition of these substances to aldehyde solutions is safe, for there seems to be no appreciable binding between the two. However, since these substances may not be available in a pure form, caution is warranted in their use for cytochemical studies. Commercially available PVP, for instance, usually contains small amounts of oxidizing substances (e.g., H_2O_2) that may interfere in enzyme reactions.

Alternatively, osmolarity can be adjusted by adding electrolytes such as NaCl and $CaCl_2$. These substances are added to the buffer before adjusting the pH, since they may change the final pH. Care is required in selecting the type and concentration of the substance to be added. If there is any indication that ionizing substances will have an adverse effect, nonelectrolytes should be used. In some cases, monovalent ions are preferred over divalent ions, for the latter are more likely to produce ion effects. In some cases, Mg^{2+} is preferred over Ca^{2+}, since the former does not precipitate proteins (Millonig, quoted by Glauert, 1974). It is convenient to remember that 0.03 mol/l glucose contributes 30 mosmols to the solution.

Although the addition of sucrose to primary OsO_4 fixative is generally not recommended, in certain cases its addition might minimize extraction and swelling of some cellular components. It has been reported that sucrose can be used in a secondary OsO_4 fixative, after primary aldehyde fixation (Glauert, 1974). However, such addition can cause distortion of certain cell types. This problem can be solved by adding an electrolyte in place of sucrose.

Recommended Osmolality

Different cell types require different osmolalities of fixative solution for optimal preservation. This has to be determined by trial and error except in the case of a few cell types for which the optimal osmolality is known. It should be noted that any tissue organ having more than one cell type may require different osmolalities of the fixative for each cell type. This is exemplified by kidney. Indeed, it is difficult to preserve kidney tissues uniformly because a fourfold range of osmolarities may be present between cells. The tubular elements and the collecting system of the medulla contain fluid markedly hypertonic to plasma, and the interstitial fluid is likewise hypertonic. Renal tubules exposed to fixatives hypertonic to plasma show widening of intercellular spaces, whereas glomeruli are well preserved with such fixatives. The satisfactory fixation of the latter is probably due to the rapid speed of fixation occurring at this site. Satisfactory preservation of the renal cortex can be obtained by using fixatives isotonic to the plasma. Interstitial cells in the cortex are well preserved in such fixatives, but interstitial cells in the medulla show signs of swelling and loss of intracellular organelles.

Renal medulla is especially difficult to fix uniformly well. Fixatives that yield satisfactory preservation of other cell types in the kidney cannot be used for the renal medulla because of the specific osmotic conditions in this tissue. Renal cortical cells of rats and cats are well preserved with fixative solutions containing 1.0–1.5% glutaraldehyde with a total osmolality of 300–320 mosmols. However, the cells of the inner stripe of the center zone of the medulla show swelling. The cells of the inner medulla exhibit extreme swelling and empty-looking cytoplasm. Each level of the medulla requires a fixative solution of a specific osmolality.

The problem of determining optimal osmolality of the fixative for different cell types is also encountered, to somewhat lesser extent, in the preservation of not only other complex animal tissues but also plant tissues. Almost every plant part consists of many cell types; for example, a leaf typically has cell types such as epidermis, spongy, palisade, xylem, and phloem. Each of these cell types possesses a specific osmolarity. Even similar cells at different developmental stages have different osmotic pressures. In fact, the osmolarity of a cell may change moment to moment, and various organelles or compartments within a cell may possess different osmolarities. For instance, mitochondria, myofibrils, and sarcoplasmic reticulum system in the striated muscle have different osmotic pressures. Average osmolalities for plant and animal specimens follow.

For transmission electron microscopy, a total (fixative and vehicle) osmolality of 800 mosmols is recommended for mature, vacuolated plant cells (e.g., mesophyll), whereas 400 mosmols is suggested for meristermatic cells (e.g., root tip). As an example, 3% glutaraldehyde in 0.025 mol/l PIPES buffer (pH 7.2) has the same osmolality as that of the sap from a 120 mm internode of the flowering stem of a grass; this osmolality is higher than that in a 350 mm internode (Lawton and Harris, 1978).

For most mammalian tissues, the recommended total osmolality ranges from 500 to 700 mosmols, whereas the vehicle osmolality is 300 mosmols. Lower total osmolalities (300–350 mosmols) are desirable for delicate specimens. Certain sea organisms and viruses are preserved at higher osmolalities; for example, the core of fowlpox is best fixed with unbuffered glutaraldehyde (0.5%) having a total osmolality of 700 mosmols (Hyde and Peters, 1970). Sea organisms such as *Amphioxus* need a total osmolality of 1000 mosmols; the osmolality of seawater may be 1025 mosmols. Generally, higher osmolalities are needed when fixation is accomplished by vascular perfusion. The classic fixative solution of Karnovsky (1965) has a total osmolality of 2010 mosmols.

For scanning electron microscopy the vehicle osmolality should be isotonic to the specimen under study. Such a vehicle should be used to prepare 1–3% solutions of glutaraldehyde or a mixture of glutaraldehyde and formaldehyde. For embryonic tissues the total osmolality should be 300–400 mosmols. Relatively higher vehicle osmolalities are used for fixing tissue blocks for transmission electron microscopy because total osmolality changes because of limitations in the penetration of the aldehyde and because of the dilution of the fixative solution during fixation.

Although a fairly wide range of total osmolality can, in certain cases, be employed without apparent damage to the specimen, an optimal osmolality yields consistently reliable results. The desired osmolality of a vehicle can be selected from Table 2.3, and the total osmolality of various fixative solutions can be chosen from Tables 2.4 to 2.10.

Measurement of Osmolality

Osmotic pressure is one of the colligative properties of a solution. Depression of the freezing point is another colligative property of solutions, which can be used to determine their osmolarity. Freezing point depression is related to the

TABLE 2.3

Approximate Osmolalities of Vehicles (mosmols)

0.1 mol/l cacodylate + 0.05% CaCl$_2$ + 0.07 mol/l sucrose	250
0.1 mol/l cacodylate + 0.05% CaCl$_2$ + 0.1 mol/l sucrose	300
0.1 mol/l cacodylate + 0.05% CaCl$_2$ + 0.18 mol/l sucrose	350
0.1 mol/l cacodylate + 0.05% CaCl$_2$ + 0.23 mol/l sucrose	400
0.2 mol/l cacodylate	350
0.15 mol/l Sörensen	300
0.18 mol/l Sörensen	350
0.2 mol/l Sörensen	400
0.2 mol/l collidine + 0.16 mol/l sucrose	400

TABLE 2.4

**Average Osmolalities of Buffered Glutaraldehyde Formulations
of Different Concentrations**

	Milliosmols	pH
Phosphate buffer (0.1 mol/l)	230	7.4
1.2%	370	7.2–7.3
2.3% glutaraldehyde	490	7.2–7.3
4% glutaraldehyde	685	7.1–7.3

TABLE 2.5

**Approximate Osmolalities (mosmols) of Buffers and Glutaraldehyde Fixatives
(pH 6.85)[a]**

Buffer	Buffer only	Buffer + 1% GA	Buffer + 2% GA	Buffer + 4% GA
0.1 mol/l phosphate	212	345	472	710
0.1 mol/l cacodylate	220	340	470	720
Isotonic phosphate	310	441	575	800
Caulfield's	280	390	504	730
Dalton's dichromate	345	503	630	895

[a] 0.1 mol/l phosphate or cacodylate buffer is hypotonic.

TABLE 2.6

Approximate Total Osmolalities of Glutaraldehyde Fixative in Various Vehicles

Glutaraldehyde concentration	Vehicle	Osmolality (mosmols)
1.5%	0.07 mol/l Sörensen + 0.015% $CaCl_2$	300
1.8%	0.05 mol/l Sörensen	250
1.8%	0.05 mol/l Sörensen + 0.05 mol/l sucrose	300
1.8%	0.05 mol/l Sörensen + 0.1 mol/l sucrose	350
1.7%	0.1 mol/l Sörensen	350
1.7%	0.1 mol/l Sörensen + 0.05 mol/l sucrose	400
2.5%	0.05 mol/l Sörensen + 0.08 mol/l sucrose	400
2.5%	0.1 mol/l Sörensen + 0.13 mol/l sucrose	450
2.0%	0.05 mol/l Sörensen	300
1.8%	0.1 mol/l Sörensen	350
2.0%	0.1 mol/l Sörensen	400
1.8%	0.08 mol/l Sörensen + 0.015% $CaCl_2$	350
2.2%	0.08 mol/l Sörensen	400
1.8%	0.1 mol/l cacodylate + 0.05% $CaCl_2$	350
2.0%	0.1 mol/l cacodylate	400
1.5%	0.1 mol/l collidine + 0.05% $CaCl_2$	350
3.0%	0.1 mol/l collidine + 0.05% $CaCl_2$	400
2.5%	0.1 mol/l collidine	350

number of particles, osmolarity is determined by referring to the effects of a perfect nondissociating solute in an ideal solution.

Several methods have been employed for measuring the osmolarity of fixative solutions. The freezing point depression (Δ) method seems to be the most accurate one, considering that different salts dissociate to different degrees. As mentioned earlier, a convenient method to determine fixative osmolarity is by

TABLE 2.7

Osmolality of 1% Aldehyde Solutions in 0.1 mol/l Cacodylate Buffer with or without 8% Sucrose

Aldehydes	Without sucrose (mosmols)	With sucrose (mosmols)
Acetaldehyde	415	739
Crotonaldehyde	413	702
Formaldehyde	693	835
Formalin	1081	1300
Glutaraldehyde	362	573

TABLE 2.8

Approximate Total Osmolalities of 2% OsO₄ Fixative in Various Vehicles

Vehicle	Osmolality (mosmols)
Distilled water	75
0.1 mol/l cacodylate + 0.05% 0.05% CaCl₂	250
0.1 mol/l cacodylate + 0.05% CaCl₂ + 0.05 mol/l sucrose	300
0.1 mol/l cacodylate + 0.05% CaCl₂ + 0.09 mol/l sucrose	350
0.1 mol/l cacodylate + 0.05% CaCl₂ + 0.14 mol/l sucrose	400

determining its freezing-point depression with the aid of a commercial osmometer. The fixative solution is then adjusted by adding saline or water. A depression equivalent to that of 0.8% NaCl is considered to be isotonic to most mammalian cells. The freezing point depression of mammalian blood plasma is $\sim -0.6°C$. The contribution of 1% OsO_4 to the freezing-point depression is only $-0.05°C$. Glutaraldehyde by itself, on the other hand, is capable of exerting considerable osmotic effect because of relatively high concentrations generally used and other reasons. It is apparent, therefore, that in the case of OsO_4 formulations, the vehicle is primarily responsible for freezing-point depression, whereas the concentration of glutaraldehyde contributes significantly to the freezing point depression of the fixative solution.

TABLE 2.9

Average Osmolalities of Different Concentrations of Glutaraldehyde and Osmium Tetroxide in Various Buffers[a]

	Glutaraldehyde (%)	Osmolalities (mosmols)
Krebs buffer	—	280
Millonig's buffer	—	275
Krebs buffer[b]	0.1	320
Krebs buffer[b]	2.0	550
Millonig's buffer	0.1	305
Millonig's buffer	2.0	95
Collidine buffer + OsO₄	—	170
Millonig's buffer + 1% OsO₄	—	305
Millonig's buffer + 2% OsO₄	—	360

[a] Modified from Arnold et al. (1971).
[b] These fixatives were proved best for preserving mouse red blood cells.

TABLE 2.10

Osmolalities of Some Buffers and Fixative Solutions[a]

Buffer		Osmolality (mosmols) pH 6.8	pH 8.0	Buffer + fixative	Osmolality (mosmols) pH 6.8	pH 8.0
Phosphate	0.4 M	730	800	Phosphate 0.1 M + GA 5%	770	820
	0.3 M	510	560	Phosphate 0.1 M + GA 3%	620	670
	0.2 M	360	420	Phosphate 0.1 M + GA 2.5%	530	580
	0.1 M	190	210	Phosphate 0.2 M + OsO_4 2%	500	530
				Phosphate 0.4 M + OsO_4 2%	783	820
Bicarbonate	0.5 M	670	830	Bicarbonate 0.1 M + GA 5%	690	800
	0.2 M	256	320	Bicarbonate 0.1 M + GA 3%	430	540
	0.1 M	130	160	Bicarbonate 0.45 M + OsO_4 2%	710	850
PIPES	0.3 M	760	813	PIPES 0.08 M + GA 5%	787	805
	0.15 M	380	420	PIPES 0.03 M + GA 3%	390	425
	0.12 M	330	350	PIPES 0.12 M + OsO_4 2%	400	420
	0.2 M	660	680	PIPES 0.2 M + OsO_4 2%	730	750

[a] Salema and Brandão (1973).

BUFFERS

The capacity of buffers occurring naturally in cells for maintaining a constant pH in the presence of weakly buffered or unbuffered fixatives is rather limited. Also, it is recognized that tissues fixed in unbuffered fixative solutions form artifacts, since acidification of tissues precedes their fixation. Tissues in unbuffered OsO_4 solution, for instance, undergo a pH drop from 6.2 to 4.4 in a 48 hr period. It is known that a shift to lower pH is usually associated with cell death. The acidification may be explained at least partly on the basis of an irreversible dissociation of protein macromolecules to low-molecular-weight proteins and the consequent increase of ionizable carboxyl groups. Osmium tetroxide is an anhydride of a very weak acid and, to that extent, can also cause acidification of the tissue. By buffering the fixing solution the probable acidification can be neutralized, and thus damage to the tissue is minimized.

Buffer solutions contain a weak acid and its salt or a weak base and its salt, and they resist changes in hydrogen ion concentration when small amounts of a strong acid or base are added to it. Thus, during fixation the acidification changes very slowly the pH of the buffer, since the salt suppresses the dissociation of the acid by the common ion effect. The necessity of maintaining a proper pH during fixation is, therefore, apparent, especially while using slowly penetrating fixatives such as glutaraldehyde or OsO_4.

The relative tolerance of some specimens to variations in pH does not mean that there is no optimal pH for these specimens. In general, not only do various tissue types respond differently to pH variations, but different cells and organelles are influenced differently by changes in the pH. Available data are quite sufficient to indicate the importance of pH in specimen preservation. It is indeed possible that inadequate preservation of deep layers of tissue specimens is a result not only of autolysis and slow penetration of the fixative but also of inadequate buffering. It should be noted that although dilution of a buffer changes the pH slightly, it proportionately decreases its buffering capacity, since the concentration of the buffer determines its capacity to absorb acid or base at any pH.

Temperature has a significant effect on the pK_a values of many buffers; for example, Tris buffer having a pH of 8.4 prepared in a cold room shows a pH of 7.8 at room temperature and a pH of 7.4 at 37°C (Good et al., 1966). This change in the pH is important since the efficiency of a buffer system varies at different pH levels. For instance, phosphate buffer has poor buffering capacity above pH 7.5, whereas Tris(hydroxymethyl)aminomethane has poor buffering capacity below pH 7.5. It is, therefore, recommended that the pH be measured immediately prior to fixation.

Comparatively little attention has been given to the effect of the specific ionic composition of buffers on the preservation of fine structure. The quality of fixation is influenced not only by the pH but also by the type of ions present in the buffer, since buffers are active reagents in the fixation process. In other words, the colligative nature of fixing fluids has a significant effect on the fine structure. Presumably, ions in the buffer vehicle interact with certain chemical groups within the tissue and thus affect the fixation quality. Variations in the specific constitution of the buffer produce significant differences in the stainability and morphological appearance of cells and organelles. The chemical composition of the buffer, for instance, affects the density of the cytosomal matrix in renal proximal tubular cells of the rat kidney (Ericsson et al., 1965). It has been shown that the amount of proteins extracted during fixation in the presence of Ca^{2+} is different from that in the presence of Na^+, even though in both cases the osmolalities were nearly equal (Wood and Luft, 1965). Table 2.11 shows the difference in sectioning properties, stainability by heavy metals, and general appearance of cell organelles resulting from the effects of specific ions of the vehicles used with OsO_4.

The efficiency of buffer solution is also dependent on the concentration of the solute, that is, concentrated solutions resist changes in the pH more than do the less concentrated ones. This is the major advantage of collidine buffer. Solutions of various concentrations of this buffer can be easily prepared, and thus isotonic or hypertonic fixing solutions can be prepared without adding other solutes. Veronal acetate, on the other hand, is relatively insoluble in water, and solutions of high concentrations of this buffer are difficult to prepare.

Buffering efficiency of a buffer during fixation is also influenced by the nature of the interacting cellular materials. Any interaction that can affect the hydrogen ion concentration of a buffer solution will tend to affect the buffering capacity of the solution. Buffering efficiency is also adversely affected when the buffer reacts with the fixative; for example, barbiturate buffers tend to react with aldehydes (Holt and Hicks, 1961).

The use of two different buffers, one during prefixation with glutaraldehyde and the other during postfixation with OsO_4, may result in the formation of artifactual electron-dense granules scattered throughout the specimen, but especially along the membranes. This artifact may occur when cacodylate and Millonig buffers are used during prefixation and postfixation, respectively. It has been suggested that the artifact is a complex compound composed of the decomposition products of glutaraldehyde, substances dissolved from the tissue being fixed, cacodylate, some component of the Millonig's buffer, and OsO_4 (Kuthy and Csapó, 1976). Such artifacts, however, can be prevented by using an extremely pure glutaraldehyde, by using the same buffer throughout the fixation procedure, or by using veronal as a second buffer during postfixation. The addition of divalent cations to the buffer may also result in the formation of fine electron-dense granules scattered throughout the specimen. Calcium, for instance, can form several insoluble phosphate salts that are not removed during rinsing, postfixation, and dehydration.

Different tissue types differ in their reaction to the same buffer, and a single cellular component responds differently to different buffers. This is exemplified by nuclear components, the fine structure of which is affected differently by different buffers. Various buffers and their calcium content affect differently the size of the threads in the artifactual network formed in the nuclear sap during fixation with glutaraldehyde (Skaer and Whytock, 1977). In cacodylate the threads range from 5 to 20 nm in diameter, whereas in nonchelating HEPES containing 1.25 mM calcium, the size is mostly 10 nm. In HEPES, nuclear sap generally forms a finer network. Studies of the influence of various buffers on rat salivary gland secretory granules indicated that with phosphate buffer, filaments appeared aggregated into large, fibrillar structures; with cacodylate buffer the filaments tended to be dispersed; and with collidine buffer the filaments as well as vesicles were present within the granules (Simson, 1977).

Certain cell organelles are more sensitive to a certain type of buffer than are others. For instance, neuronal microtubules are lost in the presence of bicarbonate buffer even when the tissue is fixed by vascular perfusion, whereas these microtubules are preserved in the presence of phosphate, collidine, or cacodylate buffer (Schultz and Case, 1968). All other morphological features are preserved well in the presence of any one of these four buffers. These data also indicate that the effect of bicarbonate ions is very specific. It is apparent that variations in the specific constitution of the buffer produce significant differences in the

TABLE 2.11

Comparative Effect of Buffers[a,b]

Buffer	Sectioning properties of Epon embedded blocks	Stainability with lead and uranyl salts	Appearance
Collidine	Excellent and uniform. Consistently better than any other buffer tested	Stainability with lead tends to be erratic. Not particularly intense. Uranyl acetate stains well	Mitochondria have smooth profiles and dense matrix. Ribosomes apparent but not too numerous. Glycogen dispersed. Cytoplasmic matrix light. Bile canaliculi fairly large. Nuclear pores prominent
Phosphate	Noticeably more difficult than s-collidine	Intense and uniform with lead. Clumped chromatin stains intensely with uranyl acetate	Mitochondria irregular in profile with light matrix. ER membranes straight, undilated. Ribosomes closely packed. Glycogen granules clumped. Organelles closely packed in cytoplasm. Bile canaliculi less patent. Nuclear pores not usually obvious
Arsenate	Comparable to phosphate	Comparable to phosphate	Comparable to phosphate. Nuclear membrane less distinct with little suppres-

carbonate	than phosphate	with lead but less intense than phosphate. Chromatin stains intensely with uranyl acetate	ular in profile, matrix of intermediate density. ER not dilated. Ribosomes closely packed. Glycogen granules less clumped than phosphate. Nuclear membrane smooth and pores obvious
Veronal acetate	Noticeably more difficult than s-collidine. Better with higher concentration of OsO_4	Stains well with lead and uranyl acetate	Mitochondria irregular in profile with relatively dense matrix. ER partially dilated. Ribosomes closely packed. Glycogen granules clumped, but not as tightly as with phosphate. Nuclear membrane smooth with easily detectable pores
Chromatin–dichromate	Similar to $NaHCO_3$	Inherent contrast better than with other buffers. Stains well with both lead and uranyl acetate	Mitochondria irregular in profile and ER not markedly dilated. Ribosomes closely packed. Glycogen granules clumped as with veronal–acetate

[a] From Wood and Luft (1965).
[b] Based on studies on rat liver and pancreas tissues fixed in osmium tetroxide in cold at pH 7.4–7.45 and embedded in Epon 812. No attempt was made to make the fixative solutions isotonic.

morphological appearance of cells and organelles. The modifying effects of different buffer systems on fine structure are, however, less apparent if the tissue is prefixed *in situ* with glutaraldehyde. This stability is due to the strong cross-linking effect of glutaraldehyde.

Some of the published data on the effects of different buffers on ultrastructure should be interpreted with much caution, since a change in buffer system also invariably involves a change in osmolarity. Thus, the osmolarity differences between the buffer system must be taken into consideration in interpreting the observed differences in the fixation of ultrastructure. More meaningful information, in this respect, can be obtained by keeping the osmolarity, pH, and other properties of the fixative vehicles nearly constant.

Although all the buffers currently used in electron microscopy have fairly desirable dissociation constants, solubilities, and reactivation, no single buffer can claim the universal superiority over the others. Each buffer system has certain advantages and disadvantages, which are given later with each of the buffers. The most efficient buffering action of various buffers cannot be predicted from a constant but must be determined for each buffer for a specific pH range and for a specific type of specimen. Most probably there is no one ideal buffer for a given pH range. It cannot be assumed that various particulate systems and soluble enzyme systems in cells respond the same way to different buffer systems. It is apparent that the objective of the study best determines the type of buffer to be used. It should be remembered that many experiments may have failed only because of the imperfections of the buffer used.

The two most important aspects of the evaluation of a buffer are (1) the buffering capacity in the desired pH range and (2) the side effects. The side effects probably vary in different tissue types. Some of the criteria of a satisfactory buffer system given by Good *et al.* (1966) are (1) pK_a (i.e., the pH of the midpoint of the buffering range) between 6.0 and 8.0, since most biological reactions occur within this range; (2) maximum solubility in water and minimum solubility in all other solvents; (3) reduced penetration of biological membranes; (4) reduced ion effects (if desired, suitable ions can be added); (5) lack of any primary amines that are expected to react with glutaraldehyde and interfere with fixation; (6) dissociation of the buffer least influenced by the buffer concentration, temperature, and ionic composition; (7) resistance to oxidation (stability); and (8) inexpensive and easy preparation and storage.

Buffer Types

A number of buffers are in use in electron microscopy, and several of these buffers are called *physiological buffer systems* because they are quite effective in the pH range of 7.2–7.4. The advantages and limitations of these buffers follow.

Phosphate Buffers

Phosphate buffers are "more physiological" than any other buffer because they are found in living systems in the form of inorganic phosphates and phosphate esters. The extracellular fluid of animal tissues contains mainly Na^+. These buffers are nontoxic to cells grown in culture. The pH of the buffers seems to change little at different temperatures and can be stored for several weeks in the cold, provided glucose or sucrose has not been added. However, these buffers gradually become contaminated and precipitates may appear during storage.

In spite of the fact that sodium mono- and diphosphate are present as effective buffer systems in animal tissues, the use of phosphate buffers is not always desirable. These buffers, for instance, produce artifacts in the form of electron dense, spherical particles of different sizes on the luminal surfaces of the types I and II pulmonary epithelial cells in the lung. Postincubation of specimens fixed in phosphate-buffered glutaraldehyde in the same buffer may cause a decrease in the nuclear mass (Kuran and Olszewska, 1974). Phosphate buffers decrease nuclear mass more than that caused by veronal buffer. Phosphate buffers extract nonchromosomal proteins in the nucleus, whereas veronal buffer removes part of the chromatin proteins. There is some evidence indicating that phosphate buffers may cause swelling of intracellular organelles, whereas cacodylate does not seem to exert this effect. Phosphate buffers may precipitate polyvalent cations and lead and uranium salts, which may result in general artifactual precipitation.

Phosphate buffers are undesirable when the addition of certain concentrations of calcium to the fixation solution is necessary. The activity of glucose-6-phosphate dehydrogenase is completely inhibited in the presence of phosphate buffers (Löhr and Walker, 1963). These buffers must be removed from the tissue prior to applying many of the hydrolytic enzyme techniques. Phosphate buffers have been reported to be unsuitable for fixing plant tissues at pH 6.0, but are satisfactory at pH 8.4 (Salema and Brandão, 1973).

Increased contrast of certain cell structures (actin, myosin, and membranes) may be obtained by rapid dehydration after specimens have been rinsed with phosphate buffer. This is accomplished by starting dehydration with 50% ethanol instead of lower concentrations. It is known that phosphate ions are precipitated in concentrations of ethanol equal to or higher than 50%. These precipitates adhered to cellular structures are thought to attract uranyl and lead staining cations (Colquhoun and Rieder, 1980). However, if the first step of dehydration begins in 50% ethanol, it may invite insoluble artifactual precipitates of phosphate. In conventional dehydration, rinsing in lower concentrations of ethanol results in the removal of much of the phosphate.

According to Carson et al. (1973), formalin in a modified Millonig's buffer (monobasic sodium phosphate and sodium hydroxide) preserves ultrastructural details in relatively large pathological specimens (fixed for light and electron microscopy) better than that obtained by using neutral buffered formalin utilizing both monobasic and dibasic sodium phosphate salts. It was suggested that the superiority of the former buffer is due to its higher osmolality ($\sim$ 290 mosmols), which is close to that of plasma. For rat red blood cells, 0.05 mol/l phosphate buffer is strongly hypotonic, 0.2 mol/l is strongly hypertonic, and 0.1 mol/l is slightly hypertonic. Recommended chemicals for the preparation of phosphate buffers are sodium dihydrogen phosphate dihydrate, $NaH_2PO_4 \cdot 2H_2O$ (mol. wt. 156.0), disodium hydrogen phosphate dihydrate, $Na_2HPO_4 \cdot 2H_2O$ (mol. wt. 178.0), and sodium hydroxide pellets.

Collidine Buffer

Collidine buffer, when half neutralized by HCl, is very efficient in the neighborhood of pH 7.4. Its buffering capacity is in the biological range of pH (6.0–8.0). It neither reacts nor complexes with OsO_4. When OsO_4 is added to this buffer, the pH does not change any more than when an equivalent amount of distilled water is added. It is biologically stable at room temperature almost indefinitely.

Collidine buffer proved best for lung tissue among six different buffers tested (Gil and Weibel, 1968). Collidine is a pyridine derivative, and since pyridine is a classical phospholipid extracting agent, an excessive extraction of cellular substances in the presence of this buffer is likely. According to Luft and Wood (1963), OsO_4 buffered with collidine extracted more proteins from rat liver during and after fixation than it did with seven other buffers. Because collidine causes excessive extraction, certain tissue specimens exposed to this buffer are thought to section easier than those exposed to other standard buffers (Wood and Luft, 1965). Because of its extraction capability, collidine facilitates penetration by fixatives and embedding plastics into very dense specimens such as seeds. Collidine is considered superior over phosphate buffers for tissue storage in 10% paraformaldehyde solution. The former is effective in the fixation of large tissue blocks because it extracts cellular materials and thus facilitates the penetration of the fixative. Membranes may be damaged when paraformaldehyde is used with collidine (Carson *et al.*, 1972). This buffer is toxic and has a strong smell. Only pure collidine should be used. Collidine cannot be recommended for routine electron microscopy.

Veronal Acetate Buffer

Veronal acetate buffer is most effective between 4.2 and 5.2, and thus inoperative at pH 7.2–7.5. It should not be used with aldehydes, since it reacts with these fixatives and the reaction product has no buffering value in the physiologically important range of pH. Veronal acetate cannot be stored in the absence of the fixative because it easily becomes contaminated by bacteria and mold. It has the advantage of not precepitating uranyl acetate. Membranes appear to be preserved better when OsO_4 is buffered with veronal than with other buffers.

Tris Buffer

Tris buffer has a poor buffering capacity below pH 7.5 and is a biological inhibitor. Tris buffer, like other primary amines, appears to react with glutaraldehyde and, therefore, to be avoided except when no alternative is available.

Cacodylate Buffer

Cacodylate buffer is quite effective in the pH range of 6.4–7.4. Cacodylate avoids the presence of extraneous phosphates that may interfere with cytochemical studies. This buffer does not seem to decrease the nuclear dry mass, and there is little removal of acid-soluble proteins from nuclei fixed with glutaraldehyde in cacodylate (Olins and Wright, 1973). Cacodylate buffer is superior at least over either Tris–HCl or Tris–maleate for preserving the activity of enzymes such as glycosyltransferase and sialyltransferase, especially in tissue homogenates and cells *in vitro*. Cacodylate-buffered glutaraldehyde is thought to be a desirable fixative for autoradiography. It is resistant to bacterial contamination during specimen storage. Calcium can be added to the fixative solution in the presence of cacodylate without the formation of precipitates. Recommended chemicals for the preparation of cacodylate–HCl buffer are sodium cacodylate trihydrate, $Na(CH_3)_2AsO_2 \cdot 3H_2O$ (mol. wt. 214), and concentrated HCl (36–38%).

Sodium cacodylate contains arsenic and thus is a health hazard. If inhaled or absorbed through skin, it can cause dermatitis, liver and kidney inflammation, etc. Hands should be protected by gloves and fume hoods should be used for weighing out the reagent and preparing the buffer solution. The reagent should not come in contact with acids in order to avoid the production of arsenic gas (also see Weakley, 1977). Since cacodylate reacts with H_2S, the two should not be used in the same solution. The buffer is incompatible with uranyl acetate. Since this arsenic buffer has toxic effect on the cell prior to fixation, cellular response to this buffer might alter membrane permeability, leading to redistribution of material along osmotic gradients of changed chemical reactivity (Schiff and Gennaro, 1979). Thus, the preceding impairment by toxic conditions will affect subsequent cellular preservation by the fixative.

Na-Bicarbonate Buffer

Although bicarbonate buffer is not used commonly for routine electron microscopy, it may be preferred for some studies. It has been reported to be superior to phosphate buffers in the comparable pH range for plant tissues (Salema and Brandão, 1973). Studies of mouse pancreas embedded in water-miscible methacrylates at low temparatures indicate that Tris and phosphate buffers are less efficient in preserving cellular integrity than is the bicarbonate buffer (Cope, 1968). The buffer is compatible with uranyl salts.

PM Buffer

The PM buffer is a nonionic buffer used initially for *in vitro* polymerization of microtubules (Weisenberg, 1972) and stabilizes microtubules when used during fixation with glutaraldehyde. This buffer provided a twofold increase in the number of microtubules and their contour lengths in HeLa cells when compared with the results obtained by using phosphate or cacodylate buffer (Luftig *et al.*, 1977). It is suspected that prefixation with cacodylate- or phosphate-buffered glutaraldehyde may result in the destabilization of microtubules. The reason for the destabilization is presumed to be the slow penetration by glutaraldehyde. When the PM buffer is used, on the other hand, it presumably stabilizes the organelle until penetration and fixation by glutaraldehyde occur.

PM buffer also permits enhanced visualization of 10 nm filaments in HeLa cells (Luftig *et al.*, 1977). This buffer has proved effective as a vehicle for the fixation of fungus cells (Handley and Ghosh, 1980). Preliminary studies indicate that guanosine triphosphate component of the buffer is essential for the beneficial effects listed previously. The buffer contains the following reagents:

Guanosine triphosphate	1 mM
MgSO$_4$	1 mM
Ethylene glycol-bis (β-aminoethyl ether)-N,N-tetraacetic acid	2 mM
Piperazine-N,N'-bis(2-ethanesulfonic acid)	100 mM

Zwitterionic Buffers

Twelve relatively little-used hydrogen ion buffers covering the pK_a = 6.15–8.35 were reported by Good *et al.* (1966). These compounds are amines or N-substitued amino acids. These buffers are compatible with most media of biomedical and physiological interest and are of particular use in the preparation of buffer solutions for the control of acidity in the pH range of physiological interest. The compatibility of these buffers (PIPES) with life functions has been demonstrated in tissue culture media. Massie *et al.* (1972) indicate that HEPES and TES are nontoxic substitutes for bicarbonate buffers in cell culture media. Zwitterionic buffers have been combined in the pH range of 6.4–8.3 for stabilizing the pH of mammalian cell cultures (Eagle, 1971). Although these buffers have been employed in cell cultures and virus assays, their application in electron microscopy has not become popular.

There is evidence indicating that both plant and animal specimens fixed in the presence of these buffers show relatively high electron density because of increased retention of cellular materials, especially of proteins and phospholipids. The extraction of chlorophyl a and b, phospholipids, and proteins is significantly reduced when PIPES is used as a vehicle with both glutaraldehyde and OsO_4 (Salema and Brandão, 1973). The plasma membrane of cells fixed with glutaraldehyde in PIPES seems to resist the deleterious changes produced by solvents used for dehydration (Schiff and Gennaro, 1979). This beneficial effect may be due to higher retention of lipids in the cells fixed in the presence of PIPES buffer. This buffer probably facilitates rapid and extensive cross-linking of cellular materials.

These buffers (PIPES) may be preferred when long durations of fixation are necessary. Biopsy specimens, sometimes by necessity, are kept in the fixative for hours or days before being processed. Such specimens show less extraction and do not show artifactual precipitous materials when fixed in the presence of PIPES. Another advantage is that, unlike some other buffers (phosphate), PIPES does not adversely affect staining with uranyl and lead salts. In the author's laboratory, PIPES is used routinely and yields excellent results with the TEM and the SEM.

Since these buffers do not seem to contribute extraneous ions to specimens, relatively accurate determination of elemental composition may be performed by means of X-ray microanalysis. The only cationic contribution that may occur during the use of PIPES, for instance, would be Na^+, a component of this buffer. However, the use of these buffers in X-ray microanalysis requires further study and caution because it is not clear why specimens fixed in the presence of PIPES show relatively low elemental concentration (Baur and Stacey, 1977). Perhaps this buffer causes a greater loss of ions and electrolytes than that caused by the more commonly used buffers. Some of these buffers have definite advantages over the more conventionally used buffers, and should be further tested for use in electron microscopy; N-(2-hydroxyethyl)piperazinesulfonic acid (EPPS), for instance, exhibits a lower tendency to bind heavy metal ions and has a pK_a that is less temperature dependent than Tricine. Some of these buffers with their pK_a values follow and are available commercially:

Chemical name	Acronym	pK_a at 20°C
N,N-bis(2-hydroxyethyl)-2-amino- ethanesulfonic acid	BES	7.15
N,N-bis(2-hydroxyethyl)glycine	Bicine	8.35
3-(Cyclohexylamino)propanesulfonic acid	CAPS	10.40
N-(2-hydroxyethyl)piperazinesulfonic acid	EPPS	8.00

(Continued)

Chemical name	Acronym	pK_a at 20°C
N-2-hydroxyethylpiperazine-N'-2-ethanesulfonic acid	HEPES	7.55
2-(N-morpholino)ethanesulfonic acid	MES	6.15
3-(N-morpholino)propanesulfonic acid	MOPS	7.20
Piperazine-N,N'-bis(2-ethanesulfonic acid)	PIPES	6.80
N-tris(hydroxymethyl)methyl-2-aminoethanesulfonic acid	TES	7.50
N-tris(hydroxymethyl)methylglycine	Tricine	8.15

It should be noted that one of the intermediates for the synthesis of the aminosulfonic acid buffers is potentially carcinogenic and may become commercially unavailable. PIPES and other buffers of this series (Good *et al.*, 1966) may soon be difficult to obtain. Ferguson *et al.* (1980) have prepared a new series of zwitterionic buffers having similar properties. These new buffers with pK_a's between 6.9 and 7.9 are worthy of testing for electron microscopy.

Preparation of Buffers

1. *Cacodylate* (0.05 *M* or mol/l)
    Solution A:
    Sodium cacodylate [Na(CH$_3$)$_2$AsO$_2$·3H$_2$O]   42.8 g
    Distilled water to make   1000 ml
    Solution B: 0.2 mol/l HCl
    Conc. HCl (36 to 38%)   10 ml
    Distilled water   603 ml
    The desired pH can be obtained by adding solution B to 50 ml of solution A and diluting to a total volume of 200 ml with distilled water:

Solution B (ml)	pH of Buffer
18.3	6.4
13.3	6.6
9.3	6.8
6.3	7.0
4.2	7.2
2.7	7.4

The buffer should be prepared under a fume hood.

2. *Collidine* (0.2 mol/l)
   Stock solution
s-Collidine (pure)	2.67 ml
Distilled water to make	50.0 ml

   Buffer
Stock solution	50.0 ml
1.0 mol/l HCl	~9.0 ml
Distilled water to make	100.0 ml

   The pH is 7.4, which can be adjusted with HCl.

3. *Phosphate* (Karlsson and Schultz, 1965)
$NaH_2PO_4 \cdot H_2O$	3.31 g
$Na_2HPO_4 \cdot 7H_2O$	33.77 g
Distilled water to make	1000 ml

   The pH is 7.4 and the osmolality is 320 mosmols, which is equal to that of the cerebrospinal fluid of rats.

4. *Phosphate* (0.135 mol/l) (Maunsbach, 1966)
$NaH_2PO_4 \cdot H_2O$	2.98 g
$Na_2HPO_4 \cdot 7H_2O$	30.40 g
Distilled water to make	1000 ml

5. *Phosphate* (Millonig, 1961)
   Solution A: 2.26% $NaH_2PO_4 \cdot H_2O$ in water
   Solution B: 2.52% NaOH in water
   Buffer (0.13 mol/l)
Solution A	41.5 ml
Solution B	8.5 ml

   The pH is 7.3. The desired pH can be obtained with solution B without changing the molarity. The buffer is stable for several weeks at 4°C.

6. *Phosphate* (Millonig, 1964)
$NaH_2PO_4 \cdot H_2O$	1.8 g
$Na_2HPO_4 \cdot 7H_2O$	23.25 g
NaCl	5.0 g
Distilled water to make	1000 ml

7. *Phosphate* (Sörensen) (0.1 mol/l)
   Solution A: 0.2 mol/l of dibasic sodium phosphate
$Na_2HPO_4 \cdot 2H_2O$	35.61 g
or $Na_2HPO_4 \cdot 7H_2O$	53.65 g
or $Na_2HPO_4 \cdot 12H_2O$	71.64 g
Distilled water to make	1000 ml

Solution B: 0.2 mol/l of monobasic sodium phosphate

$NaH_2PO_4 \cdot H_2O$	27.6 g
or $NAH_2PO_4 \cdot 2H:_2O$	31.21 g
Distilled water to make	1000 ml

Prepare the buffer by mixing the two solutions as given here and diluting to 100 ml with distilled water:

pH at 25°C	Solution A	Solution B
5.8	4.0	46.0
6.0	6.15	43.85
6.2	9.25	40.75
6.4	13.25	36.75
6.6	18.75	31.25
6.8	24.5	25.5
7.0	30.5	19.5
7.2	36.0	14.0
7.4	40.5	9.5
7.6	43.5	6.5
7.8	45.75	4.25
8.0	47.35	2.65

The osmolality of 0.1 mol/l buffer (pH 7.2) is 226 mosmols; addition of 0.18 mol/l sucrose raises it to 425 mosmols.

8. *Piperazine* (PIPES) (0.3 mol/l)

Distilled water	50 ml
Piperazine N,N-bis-2-ethanol sulfonic acid	9 g

Add enough 0.1 mol/l NaOH (0.4%) to adjust the pH; at pH 5.5–6.0 the powder is completely dissolved. After the required pH has been reached, more distilled water is added to make up 100 ml. The stock solution is stable for several weeks at 4°C.

9. *Tris(hydroxymethyl)aminomethane maleate* (0.2 mol/l)

Solution A

Tris(hydroxymethyl)aminomethane	30.3 g
Maleic acid	29.0 g
Distilled water to make	500 ml
Charcoal	2 g

The preceding solution is shaken, allowed to stand for about 10 min, and filtered.

Solution B: 4% NaOH

NaOH	4 g
Distilled water	96 ml

The desired pH can be obtained by adding solution B to 40 ml of solution A and diluting to a total volume of 100 ml with distilled water

Solution B (ml)	pH of Buffer
15.0	6.4
18.0	6.8
19.0	7.0
20.0	7.2
22.5	7.6
24.2	7.8
26.0	8.0

10. *Veronal acetate* (Zetterqvist, 1956)

   Stock solution

Sodium veronal (barbitone sodium	2.94 g
Sodium acetate	1.94 g
Distilled water to make	100 ml

The stock solution is stable for several months at 4°C.

   Ringer's solution

	8.05 g
Potassium chloride	0.42 g
Calcium chloride	0.18 g
Distilled water to make	100 ml

   Buffer

Stock solution	10.0 ml
Ringer's solution	3.4 ml
Distilled water	25.0 ml
0.1 mol/l HCl	∼ 11.0 ml

The HCl is added dropwise until the required pH is obtained. The buffer cannot be stored.

11. *Veronal acetate* (Ryter and Kellenberger, 1958)

   Stock solution

Sodium veronal (barbitone sodium)	2.94 g
Sodium acetate (hydrated)	1.94 g
Sodium chloride	3.40 g
Distilled water to make	100 ml

   Buffer

Stock solution	5.0 ml
Distilled water	13.0 ml
1.0 mol/l $CaCl_2$	0.25 ml
0.1 mol/l HCl	∼ 7.0 ml

The HCl is added dropwise until the required pH is obtained. The buffer should be prepared fresh.

12. *Veronal acetate* (Palade, 1952)

    Stock solution

Sodium veronal (barbitone sodium)	2.89 g
Sodium acetate (anhydrous)	1.15 g
Distilled water to make	100 ml

The stock solution is stable for several months at 4°C.

    Buffer

Stock solution	5.0 ml
Distilled water	15.0 ml
0.1 mol/l HCl	∼ 5.0 ml

The HCl is added dropwise until the required pH is obtained.

pH

It is known that proteins are charged both positively and negatively, and their isoelectric pH is dependent on the relative number of acidic and basic groups present in the molecules. At their isoelectric pH, proteins are electrically neutral, that is, they show minimal solubility and viscosity, the lowest osmotic pressure and the least swelling. In other words, various properties exhibited by proteins are minimal at the isoelectric pH. Any shift from the isoelectric point (which is defined as the hydrogen ion concentration where the mean charge on the protein is zero) affects the physicochemical properties of a protein. In general, proteins are capable of combining with cations on the alkaline side of the isoelectric point and with anions on the acid side.

Under normal conditions the pH of cells is usually on the alkaline side of the isoelectric pH of proteins. For instance, the pH of cytoplasm in various types of animal cells has been found to be in the close neighborhood of 7.0, whereas for nuclei it appears to be in the range of 7.6–7.8. These values are higher than the isoelectric pH (5.0–6.0) of the majority of the natural tissue proteins. This means that majority of the cytoplasmic proteins would carry a net negative charge. Since a pH shift in either direction from the isoelectric point results in increased ionization, the osmotic pressure of the protein solution increases. This increase in osmotic pressure is partly responsible for the osmotic attraction of fixing fluids into the cells. The isoelectric points of a number of proteins have been compiled by Young (1963).

Since the isoelectric point of a protein depends on the dissociation constants of acid and basic groups, it is readily affected by the nature of the fixing fluid and is altered by fixation. It is known that OsO_4 lowers the isoelectric point of proteins, probably because of the destruction of basic amino acids. In other words, the net negative charge of the proteins is significantly increased by the reaction with

OsO_4. Such a lowering of the isoelectric point has been reported for amphibian erythrocytes (Tooze, 1964) and other proteins. Tooze (1964) indicated that isoelectric pH was lowered from a range of 6.9–7.2 to 5.0–5.2. Other fixatives, such as glutaraldehyde, formaldehyde, and potassium dichromate, also lower the isoelectric point of proteins (Hopwood et al., 1970). An increase in the negative surface charge of human red blood cells treated with glutaraldehyde has been demonstrated (Vassar et al., 1972). This increase is thought to result from the appearance of additional ionogenic groups in the peripheral zone of these cells. The pH of the fixative must remain close to the pH of the tissue because a change in the tissue pH brought about by the fixative is bound to alter radically the structure and behavior of tissue proteins in solution.

Buffer solutions are universally present in living cells. Although the inherent buffer system of protoplasm neutralizes to some degree the penetrating ions of a fixative, electrolytes in a fixative exert a profound effect not only on the structural relations of protein molecules but also on the activity of proteins as enzymes. For example, β-lactoglobulin has a molecular weight of ~35,000 at a pH above its isoelectric point; the molecule aggregates to a weight of ~70,000 at a pH close to its isoelectric point but then breaks down at a lower pH to units of 17,000 weight. Disaggregation of polyribosomes in L cells grown in cultures at pH 6.0 has been shown (Perlin and Hallum, 1971). Even during the normal growth of cells in a culture, the pH rises first followed by acidification as a result of cell metabolism. Not only the growth of cells but other parameters of cellular metabolism may also be affected by pH variations. Another example is the occurrence of albuminoid (scleroprotein) degradation after linkages between its amino acids are reduced under alkaline conditions.

Suskind (1967) showed that the nuclear protein labeled with uridine -^{3}H is diminished in acetic ethanol fixative or in formalin at acid pH but is fixed in phosphate-buffered formalin or glutaraldehyde at pH 7.3. On the other hand, erythrocytes are stabilized and hemoglobin extraction is reduced when the pH of fixative is 6.2 instead of 7.4 (Tooze, 1964). Wrigglesworth and Packer (1969) demonstrated that pH-dependent conformational modifications of protein (the term "conformational change" is defined as a change in the secondary, tertiary, or quaternary structure of the protein molecule) occur in mitochondrial membranes, which lead to reversible alterations in membrane ultrastructure. Small changes in the pH of a fixative bring about large changes inside the cell, including the consistency of protoplasm, selectivity of the cell membrane, and activity of enzymes. It is known that the combination of proteins with anions or cations to form insoluble compounds is closely related to the pH of the fixing fluid. Since proteins are responsible for the characteristic structure of a cell and the molecular weight of proteins is strongly pH dependent, the importance of pH of the fixative is obvious.

As stated previously, a knowledge of not only the pH of proteins but also the actual hydrogen ion concentration of the cell interior is important in order to control the quality of fixation. It is not uncommon to find differences between the pHs of the body fluids and the interior of the cells. This is exemplified by red blood cells; the pH of these cells is slightly higher than the extracellular pH. This difference increases at 4°C compared with that at room temperature. The intracellular pH shifts to the alkaline side with a decrease in temperature. There are differences in the pH level even among various parts of a cell. A case in point is the difference in the pH between the vacuoles and the cytoplasm of a cell. The difference is apparent especially in plant cells, since they possess more prominent vacuoles. The sap is relatively weakly buffered so that the penetrating fixative readily shifts its pH. Since the major part of mature cells of plants consists of a large vacuole or vacuoles surrounded by a thin layer of cytoplasm, these cells are more susceptible to shifts in their pH during fixation than most animal cells.

Since the average pH value of most animal tissues is 7.4, the best preservation of the fine structure is obtained by keeping the pH of the fixative within narrow limits (i.e., 7.2–7.4) of this value. The pH of body fluids of 17 species of marine invertebrates has been presented by Mangum and Shick (1972). A more alkaline pH (i.e., 8.0–8.4) is preferred for certain highly hydrated tissues, such as protozoan, invertebrate, and embryonic. Helander (1962), for instance, indicated that gastric mucosa is preserved much better at pH 8.5. The improved preservation obtained under alkaline conditions was presumably due to the more efficient inhibition of the proteolytic action of the pepsin. A fixative of relatively high tonicity is also useful for fixing these tissues because they are characterized by a high plasma osmotic pressure. Plant cells with their relatively large sap vacuoles also possess a relatively high internal osmotic pressure.

The average pH of plant cells is lower than the average pH of animal cells. The pH of the vacuolar sap in plant cells ranges from 5.0 to 6.0, whereas the pH of the cytoplasm of the same cells is usually between 6.8 and 7.1. Although these pH values are relatively low, the recommended pH during prefixation with glutaraldehyde (in PIPES buffer) is 8.0, and a pH of 6.8 during postfixation with OsO_4 (in PIPES buffer) is desirable. The reason for using pH 8.0 is to compensate for the acidity contributed by the vacuolar sap that is released into the cytoplasm immediately after the tissue block comes in contact with glutaraldehyde. The release of the vacuolar sap not only lowers the pH but also dilutes the fixative solution. These damaging effects can be transmitted through plasmodesmata to other cells situated deep in the tissue block.

In certain cases acidic pHs during fixation seem to be desirable. Some of the examples are cited here. Gastrin cell granules show greater electron density when fixation is carried out at pH 5.0–6.0 (Mortensen and Morris, 1977). Neurphysin–hormone complex, dense core of granules in type II cells of the pars intermedia of the pituitary of the eel (Thornton and Howe, 1974), and *in vitro* insulin granules (Howell *et al.*, 1969) are better preserved at pH 5.0–7.0. Studies of the effect of fixative pH on the appearance of neurosecretory granules indicate that at pH 7.0 the granules are pleomorphic with respect to electron density, whereas at pH 5.0–6.0 all granules remain electron dense (Morris and Cannata, 1973; Nordmann, 1977). It has been suggested that improved preservation of the granules at an acidic pH is due to the precipitation of their protein; proteins are the least soluble at their isoelectric point. It is possible that at an acidic pH, the dense core of granules remains osmotically inactive, whereas at neutral pH the core dissolves and an influx of extragranular fluid, caused by osmotic pressure, may give rise to the observed loss of electron density in the neurosecretory granules (Nordmann, 1977).

Lower pHs (~6.0) have been recommended for stabilizing nuclear constituents, including delicate fibrils of mitotic spindle (Claude, 1962; Roth *et al.*, 1963). Coupland (1965) indicated that the most intense chromaffin reaction occurred when the tissue was fixed at pH 5.8. According to Tooze (1964), the extraction of hemoglobin was reduced when the erythrocytes were fixed with OsO_4 at pH 6.2 instead of at 7.4. Thiéry (1971) has recommended vascular perfusion of short duration by glutaraldehyde at pH 4.0–5.0 to facilitate the penetration of metals (e.g., manganese) into the tissue. Optimum fixation of fowlpox virus core was obtained with unbuffered glutaraldehyde at pH 3.0 (Hyde and Peters, 1970). It is, however, pointed out that the selection of the optimal pH for a certain type of tissue is determined only after experimentation. Quite obviously no definite statement as to the exact intracellular pH can be made, for pH varies from cell to cell and from moment to moment in the same cell. In addition, possible errors in measurements of intracellular pH cannot be disregarded. Butler *et al.* (1967) have questioned, for example, the validity of intracellular pH of skeletal muscle obtained through microelectrode measurements. Various methods used to determine the intracellular pH of a wide variety of cells and tissues have been reviewed by Waddell and Bates (1969). Table 2.12 gives the pH of body fluids of marine invertebrates.

TABLE 2.12

Approximate pH of Body Fluids of Marine Invertebrates

Species	pH
Cnidaria	7.3
Annelida	7.4
Sipunculida	7.4
Echinodermata	6.9
Mollusca	7.5
Arthropoda	7.4–7.8

FIXATIVE PENETRATION

An ideal fixative should kill specimens quickly, causing minimum shrinkage or swelling. Since the speed of killing is dependent primarily on the rate of pentration of a fixative into the tissue, chemicals of low molecular weight such as formaldehyde are usually most effective. However, a direct relationship between molecular weight and rate of penetration of a fixative does not always exist. A case in point is mercuric chloride, which penetrates rapidly, although it has a relatively high molecular weight. The rate of penetration is also affected by the number of reacting radicals in the molecule of a fixative independent of its molecular weight. Some of the intrinsic properties of a fixative that influence the rate of penetration are its solubility in lipids and the polarity of its molecule.

The penetration of a fixative into the tissue is generally indicated by its coefficient of diffusibility denoted by K. The following expression describes the phenomenon:

$$d = K\sqrt{t}$$

where d is the depth of penetration by the fixative in time t. According to Dempster (1960), the relationship between K, d, and t is derived more accurately by

$$t = Kd^e$$

where e is an exponent, the value of which is ~ 2, related to the rate of diffusion of the fixative. The rate of fixative penetration into the tissue can be measured easily and the value of K determined.

Among the commonly used fixatives, formaldehyde penetrates faster than either glutaraldehyde or OsO_4, and glutaraldehyde penetrates slightly faster than OsO_4. According to Johannessen (1978), OsO_4 penetrates into human liver somewhat faster than glutaraldehyde during prolonged fixation (Table 2.13). The K values for OsO_4, glutaraldehyde, and formaldehyde are 0.2, 0.34, and 2.0,

TABLE 2.13

Depth of Fixative Penetration in Human Liver[a]

Fixative (buffered)	Time (hr)	Penetration (mm)
Formaldehyde (5%)	4	2.5
Glutaraldehyde (3%)	4	0.5
OsO_4 (2%)	4	0.8
Formaldehyde (5%)	24	10.0
Glutaraldehyde (3%)	24	1.1
OsO_4 (2%)	24	1.4

[a] From Johannessen (1978).

respectively. In 24 hr OsO_4 alone penetrates deeper than does a mixture of OsO_4 either with formaldehyde or with glutaraldehyde; the latter mixture penetrates the least. Tissue turns pale yellow because of glutaraldehyde penetration, whereas OsO_4 penetration into the tissue is indicated by a brown to black color. However, penetration by OsO_4 may not always show up as a change in color.

Other factors that influence the rate and depth of penetration are tissue type, temperature, duration, and mode of fixation, concentration of the fixative, osmolality of the fixative vehicle, and the addition of electrolytes or nonelectrolytes to the vehicle. All fixatives penetrate faster at room temperature than in the cold. Up to certain limits, an increase in the fixative concentration and duration of fixation results in enhanced penetration of tissue specimens. Penetration is expedited by vascular perfusion.

Compact or dense tissue permits relatively slow penetration. The presence of air in the tissue (e.g., lung) and/or waterproof substances such as cutin and waxes on the tissue surface (e.g., leaf) hinder the penetration of a fixative. As a result, these tissues float instead of being submerged during immersion fixation. Floatation of specimens is highly undesirable. Rate of penetration into these tissues, however, can be enhanced by employing vascular perfusion, vacuum, or dilute solutions of common detergents prior to fixation. Hard exoskeletons of certain organisms (e.g., mite) prevent penetration of the fixatives. This difficulty can be overcome by immobilizing the organism in a stendor dish by carefully melting, with a hot needle, a small amount of paraffin around each leg. The organism is then chilled and flooded with a chilled fixative. Using microscalpels the exoskeleton is removed and the internal organs exposed to the fixative. Rate of penetration can also be enhanced by rupturing or damaging the cell membrane and/or cell wall. Somewhat rapid and uniform penetration results when specimens are stirred during fixation. The factors affecting the rate of penetration of individual fixatives into tissue specimens are discussed later.

Contrary to the situation in tissue specimens, penetration and fixation of single cells are rapid. The time required actually to immobilize the cytoplasm of a cell is dependent on the type of the cell under study and the fixative used. A few examples will suffice. In the presence of glutaraldehyde, all intracellular movement in cultured chick cells stopped in 30–45 sec, although some vesiculation continued for up to 2.5 min (Buckley, 1973a). Glutaraldehyde stopped all intracellular movement in *Haemanthus* endosperm cells in 10–20 sec (Bajer and Molé-Bajer, 1971); during this period chromosomes moved as much as 0.5 μm. The cytoplasm of highly vacuolated cells of petiolar hairs of tomato was immobilized with glutaraldehyde–acrolein or with formaldehyde in 6 min and 30 min, respectively (Mersey and McCully, 1978). It should be noted that the longer time required for the hair cell is due to the presence of cutinized cell wall, which acts as a barrier to fixative penetration. In fact, the progress of fixation through this cell is so slow that cytoplasm in the tip of the cell may still be streaming even when the cytoplasm at the base of the cell has been immobilized for 10 min (O'Brien *et al.*, 1973). The progress of a "fixation front" as it moves along the length of these cells can be watched under the phase microscope.

TEMPERATURE OF FIXATION

Generally, higher temperatures enhance the speed of chemical reactions between the fixative and the cell constituents including enzymes. Higher temperatures increase the rate of penetration (diffusion) of the fixative into the tissue and simultaneously the rate of autolytic changes. Lower temperatures depolarize the plasma membrane and increase its resistance to ion permeation. It is also most likely that a longer fixation time at higher temperatures is accompanied by excessive extraction of cellular materials. It appears logical to assume, therefore, that effects on the quality of preservation due to a change in temperature are modified by all these factors. It should be noted, however, that various fixatives respond differently to changes in temperature in terms of the net effect on the quality of preservation. Also, various tissues differ in their response to changes in temperature of fixation.

The use of low temperatures during fixation is justified by decreased extraction and a slower rate of autolysis. It is likely that low temperatures affect the rate of autolysis more than the rate of diffusion. In certain types of specimens where the penetration of the fixative in the cold is a problem, fixation at higher temperatures may be necessary. The fixation of bacterial spores, for example, is facilitated at 40°C. Cold labile structures such as cytoplasmic microtubules are best preserved at room or body temperature. It is well known that swelling is slightly more pronounced when the tissues are fixed in the cold in hypertonic solutions. On the other hand, there is some evidence that indicates that hypotonic fixative solutions cause more swelling at room temperature than at 4°C, whereas the effect of hypertonic solutions is less temperature dependent.

Generally, the effects of higher temperatures on the quality of preservation with rapidly acting fixatives differ from the effects with slower-acting fixatives. This is explained by the fact that with the former fixative, a reasonable increase in temperature results in an extremely rapid fixation that more than compensates for the increased extraction effect. On the other hand, similar increase in temperature with the latter fixatives would lead to excessive leaching because the speed of actual fixation is too slow to compensate for the increased extraction.

The fixatives that penetrate and fix rapidly (e.g., acrolein) commonly do not require higher temperatures or longer durations of fixation. The fixatives that penetrate and fix slowly (e.g., OsO_4) tend to require longer durations of fixation but lower temperatures to minimize extraction of cellular constituents. The fixatives that penetrate slowly but fix rapidly (e.g., glutaraldehyde) may be used at

4°C or at higher temperatures, provided the fixation time is reduced. The fixatives that penetrate rapidly but fix slowly (e.g., formaldehyde) are not preferable for the preservation of fine structure except when they are employed in combination with other fixatives or for special studies. In any case, longer fixation at higher temperatures should be avoided. Motion picture analysis of fixation of amoeba indicates that a short fixation time helps to minimize distortion of cell structures (Griffin, 1963).

It is probable that the structural appearances of various cell organelles differ in their response to changes in temperature. It is known, for instance, that microtubules are not always preserved if fixation in glutaraldehyde is carried out in the cold, although other cytoplasmic details are preserved satisfactorily. Lower temperatures also affect some metabolic processes of organelles necessary to maintain their normal structure; these temperatures affect, as far as is known, mainly the appearance of mitochondria. Baker and McCrae (1966) reported that rough endoplasmic reticulum showed disruption when exocrine pancreatic cells of the mouse were fixed with formaldehyde at high temparatures (37 or 45°C), whereas the attached ribosomes were undamaged and the Golgi apparatus was preserved exceptionally well. They reported that the Golgi apparatus was not very clearly seen when these cells were fixed at 1°C.

The effect of changes in temperature even during dehydration may differ on various cell organelles. Ito (1961) reported that changes in temperature during dehydration affect the preservation of delicate structures such as smooth endoplasmic reticulum. No satisfactory explanation is available to account for the difference in response of various cell organelles to changes in temperature of dehydration agents, but lipid extraction could be a factor.

It must be pointed out that the effect of rate of penetration on the quality of tissue preservation has been recognized in many of the studies, but too little attention has been given to the speed at which the actual process of fixation takes place. Similarly, although the role of temperature in the rate of diffusion has been explored, almost no information is available on the optimal temperature required to expedite actual fixation. The importance of the latter temperature becomes clear when one considers that although most fixatives kill the cell instantaneously at conventional temperatures used, they do not instantly fix it. Thus, an appreciable period of time may elapse between exposure of the intracellular to the extracellular environment and complete fixation of cell macromolecules by the fixative. This brief prefixation period of interaction between the intracellular and extracellular environments most likely has a profound influence on the final quality of tissue preservation.

It is erroneous to presume that fixation occurs at a macromolecular level because the structural organization of cell components is fixed. Park *et al.* (1966), for example, showed that chloroplasts isolated from spinach leaves infiltrated with 6% glutaraldehyde possessed substantial Hill activity. These chloroplasts were considered fixed on the basis of their structural stability. It is apparent, therefore, that a satisfactory fixative should not only penetrate and kill rapidly but also fix rapidly and that an optimal temperature should be provided to expedite actual fixation.

DURATION OF FIXATION

The optimal duration of fixation for most tissues is not known. For most purposes an arbitrary standardized duration of 1–4 hr at room temperature or 4°C is currently in use. The optimal duration of fixation for a specific tissue is controlled by the type of the fixative, specimen size, type of the specimen, temperature, buffer type, staining method to be used, and objective of the study. Very little is known of the effects of overfixation except that it results in the extraction of tissue constituents. Since the extraction caused by chemical fixation is progressive in time, the need to determine the optimal duration is obvious. A few examples of the effects of overfixation follow.

Synaptic vesicle flattening is affected by the preparatory procedure used, including duration of fixation, type of fixative, osmolality of the fixative and the vehicle, temperature, and degree of stimulation before fixation. Paula-Barbosa (1975) indicated that prolonged (up to 20 hr) immersion in glutaraldehyde during fixation (with or without vascular perfusion) may cause flattening of synaptic vesicles in the rat cerebellum. The elucidation of the mechanism of flattening is important, since the shape of the vesicle is a valuable criterion for the classification of synapses. The appearance of hormone granules in gastric cells is affected by the duration of fixation. The proportion of dense-cored granules decreases when the prefixation time exceeds 30 min (Mortensen and Morris, 1977). Although no change in the transverse fascicular area of rat peroneal nerve occurs when the duration of fixation is increased from 1 hr to 12 hr, shrinkage (~20%) of axis cylinders does take place with prolongation of fixation beyond 1 hr (Ohnishi et al., 1974).

Duration of fixation is more critical with OsO_4 than with glutaraldehyde, since the former does not cross-link many proteins that are probably being extracted by the vehicle while the tissue is being fixed. The implications of fixation time with respect to the quality of tissue preservation by each of the fixatives are discussed in detail later. It is pointed out, however, that the fixation times presently employed are safe but longer than necessary. It seems, therefore, that modifications to the present conventional fixation schedules will have to be evolved to retain and render visible relatively small amounts of cellular materials that are extracted because of longer than necessary exposure to fixatives. It is anticipated that in general if the duration of fixation was reduced from the present conventional fixation periods, the tissue would be capable of retaining an increased range of cellular materials in a demonstrable form.

FIXATIVE CONCENTRATION

In general, low concentrations of fixatives require longer durations of fixation. Longer durations of fixation tend to cause diffusion of enzymes, extraction of cellular materials, and shrinkage or swelling of the tissue. Higher concentrations of fixatives are also undesirable, since concentrated solutions destroy enzyme activity and damage the cellular fine structure. Oxidizing fixatives such as OsO_4 are effective cross-linking agents only when used in correct concentrations; higher concentrations can cause oxidative cleavage of protein molecules. This could result in the loss of peptide fragments (Hopwood, 1969a.)

Different cell components differ in their response to variations in the concentration of a fixative. Baker and McCrae (1966) have indicated that low concentrations (0.25%) of formaldehyde disrupted the endoplasmic reticulum, whereas mitochondria were insensitive to changes in concentration. However, mitochondria generally are more sensitive to changes in the osmolarity of the buffer system.

The effect of varying the concentration on the tissue differs, depending on the type of fixative employed. Glutaraldehyde has a wide range of effective concentration, provided optimal osmolarity of the buffer system is maintained. The effective concentration of OsO_4, on the other hand, is of a rather limited range. The optimal concentration of a fixative apparently varies within an appropriate range, depending on the specimen under study. In general, higher concentrations are needed to preserve satisfactorily hydrated tissues such as the embryonic tissue. The most suitable concentration of each of the fixatives for general purposes is indicated later.

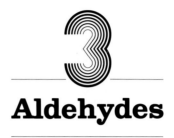

Aldehydes

The use of aldehydes other than formaldehyde as fixatives is a comparatively recent development. In 1959 Luft demonstrated that a monoaldehyde, acrolein, could be usefully applied to solve certain problems in the preservation of fine structure. However, it was the extensive studies by Sabatini *et al.* (1963) that brought the usefulness of aldehydes, especially glutaraldehyde, to the attention of the electron microscopists. These studies indicated that primary fixation with glutaraldehyde followed by secondary fixation with OsO$_4$ yields satisfactory preservation of fine structure and enzyme activity in a wide variety of specimens. This double fixation has become the standard procedure for preserving both plant and animal specimens including prokaryotes. A mixture of glutaraldehyde and formaldehyde (freshly prepared by depolymerizing paraformaldehyde) has the additional advantage of more rapid penetration into the tissue specimen and is preferred for many specimens. Acrolein in combination with glutaraldehyde has also been employed for certain studies.

Because OsO$_4$ is a heavy metal oxide with oxidizing activity that destroys enzymatic activity, it is not used as a primary fixative in enzyme cytochemistry; on the other hand, many cytochemical reactions can be performed on tissue specimens after aldehyde fixation. Also, relatively large tissue blocks can be fixed in glutaraldehyde and still larger blocks can be preserved with acrolein or formaldehyde. Prior to further treatment, aldehyde-fixed tissue blocks can be cut into small pieces with minimal mechanical damage, since cross-links introduced by aldehydes impart some degree of consistency to the tissue. Prefixation with aldehydes thus lessens the distortions introduced by mincing fresh tissues to be fixed subsequently with OsO$_4$. It is known that aldehydes form both intra- and intermolecular cross-links with protein molecules, which result in the formation of more rigid heteropolymers.

The reaction of aldehydes with proteins proceeds through aldehyde condensation reactions with amino groups to yield α-hydroxyamines which can condense with additonal amino groups to affect cross-linking. Formation of a methylol or substituted methylol derivative is the first step. These reactions are discussed in detail later. The major functional groups in proteins that may react with aldehydes are as follows:

Protein group	Formula
Amido (glutamine, asparagine)	$-\overset{\displaystyle }{\underset{\displaystyle \|}{C}}-NH_2$ O
Amino (lysine)	$-NH_2$
Aromatic (phenylalanine, tyrosine, tryptophan)	 OH
Carboxyl (aspartic acid, glutamic acid)	$-\underset{\underset{\displaystyle O}{\|}}{C}-OH$
Guanidino (arginine)	$-\underset{\underset{\displaystyle H}{\|}}{N}-C\overset{\displaystyle \nearrow NH}{\searrow NH_2}$
Hydroxyl (serine, threonine)	$-OH$
Imidazole (histidine)	
Peptide linkage	$-\underset{\underset{\displaystyle O}{\|}}{C}-\underset{\underset{\displaystyle H}{\|}}{N}-$
Sulfhydryl (cysteine)	$-SH$

Various aldehydes differ greatly in the rate and extent of their reaction with the reactive groups of proteins given above. The extent of protein reaction is influenced by many factors, including the duration of fixation, pH, temperature, conformation of the protein molecule, and type and concentration of aldehyde used. The variations in the interaction between proteins and aldehydes are considered to be related to the effects of the latter on enzyme activity.

GLUTARALDEHYDE (Glutaric Acid Dialdehyde)

$$\underset{H}{\overset{O}{\diagdown}}C-CH_2-CH_2-CH_2-C\underset{H}{\overset{O}{\diagup}}$$

Glutaraldehyde is a five-carbon dialdehyde of relatively simple structure. The molecular formula shows that a straight hydrocarbon chain links two aldehyde moieties. It is synthesized in a two-step process: (1) acrolein is interacted with vinyl ethyl ether in a typical Diels–Alder reaction to produce an ethoxydihydropyran, (2) ethoxydihydropyran is then hydrolyzed with water to form glutaraldehyde and ethanol. This reaction is as follows (Sanders et $al.$, 1958):

$$CH_2{=}CH{-}\overset{O}{\overset{\|}{C}}H \;+\; CH_2{=}CH{-}O{-}C_2H_5 \longrightarrow \;\; \substack{HC^{\diagup C\diagdown}CH_2 \\ \| \quad\; | \\ HC_{\diagdown O \diagup}CHOC_2H_5}$$

| Acrolein | Vinyl ethyl ether | 2-Ethoxy-3,4-dihydro-2H-pyran |

$$\Big\nearrow \quad H_2O$$

$$\overset{O}{\overset{\|}{H}}C{-}CH_2{-}CH_2{-}CH_2{-}\overset{O}{\overset{\|}{C}}H \;+\; C_2H_5OH$$

| Glutaraldehyde | Ethanol |

Glutaraldehyde has a molecular weight of 100.12 and is characterized by a relatively low viscosity. The vapor pressure is 17 mmHg at 20°C. It is convenient to remember that 1% glutaraldehyde is ~0.1 mol/l. After glutaraldehyde treatment, the gain in dry weight of plant tissues is ~1.1% of the fresh weight (Morré and Mollenhauer, 1969). It is regarded as slightly to moderately toxic and noncorrosive to stainless steel.

Comparative studies by Sabatini et $al.$ (1962, 1963, 1964) and Barrnett et $al.$ (1964) demonstrated conclusively that of all the aldehydes tested, glutaraldehyde proved to be the most effective for preserving fine structure. Glutaraldehyde is effective in preserving both prokaryotes and eukaryotes, including fragile specimens such as marine invertebrates, embryos, diseased cells, and fungi. The dialdehyde stabilizes intracellular systems of labile microtubules, rough and smooth endoplasmic reticulum, mitotic spindles, platelets, and pinocytotic vesicles better than does any other known fixative. Glutaraldehyde stabilizes blood plasma with little shrinkage of blood clots and attendant collapse of vessel walls (Chambers *et al.*, 1968).

Tissue specimens can be left in this fixative for many hours without apparent deterioration; this flexibility of the pretreatment is one of its major assets. Presently, glutaraldehyde is the most efficient and reliable fixative for the preservation of biological specimens for routine electron microscopy.

Among the commonly employed fixatives, glutaraldehyde causes the least protein conformational changes, although some significant structural modification of the protein molecule does occur, especially that of the α-helix. The molecular weight of proteins tends to increase somewhat after fixation with glutaraldehyde (Davies and Stark, 1970). Fixation of erythrocytes with 0.1% glutaraldehyde does not significantly alter the receptor groups on these cells for viruses of the ortho- and para-myxo groups and rubella virus (Dalen, 1976). Insolubilization of papain (Jansen and Olson, 1969) and subtilisin (Ogata *et al.*, 1968) with glutaraldehyde does not destroy their activity. Glutaraldehyde-fixed erythrocytes can be freeze-dried and reconstituted.

That Hill activity of chloroplasts isolated from glutaraldehyde-fixed leaves is retained indicates that this aldehyde has little effect on chloroplast proteins (Park *et al.*, 1966). Treffry (1969) has also indicated that the inhibition of photoconversion of protochlorophyll to chlorophyll in glutaraldehyde-fixed leaves is only partial. Fixation with glutaraldehyde prevents conformation and volume changes in isolated mitochondria and chloroplasts and allows electron transport to continue. That electron transport is quite resistant to glutaraldehyde treatment of mitochondria has been shown (Utsumi and Packer, 1967). It has been suggested that cytochrome oxidation is not affected when bacterial cells are fixed with glutaraldehyde. This is in contrast to studies in certain other tissues where mitochondrial cytochrome oxidase has been shown to be quite sensitive to glutaraldehyde (Hanker, 1975). Dielectric techniques have shown that even though membrane resistances of glutaraldehyde- or acrolein-fixed cells are generally much lower than those of normal cells, a major portion of the fixed membrane still acts as an insulating barrier (Carstensen *et al.*, 1971). It should be remembered that membrane resistance is considered to be an accurate criterion for determining membrane condition.

Studies on the shape of agranular synaptic vesicles indicate that even brief storage of aldehyde-perfused nervous tissue pieces in cacodylate buffer, prior to osmication, has a severe flattening effect on the vesicles of peripheral cholinergic axon endings (Bodian, 1970). By observing the wave bands of the light reflected by the reflecting cells on the scales of herring, Bone and Denton (1971) found that fixation with glutaraldehyde does not destroy the osmotic activity of these cells.

The previously mentioned and other available data indicate that proteins are not denatured to any marked extent by fixation with glutaraldehyde; protein tertiary structure seems to remain intact after glutaraldehyde fixation. Conformation and biological activity of proteins remain mostly unimpaired after their moderate cross-linking by glutaraldehyde. A possible explanation for this phenomenon is that amino groups, which are the primary target of glutaraldehyde, are usually abundant on the surface of proteins.

Uses of Glutaraldehyde

The major uses of glutaraldehyde are (1) in the fixation of specimens for electron microscopy, (2) in immunology by attaching enzymes as markers to proteins that retain their immunological activity, (3) in leather tanning, and (4) in sterilization and disinfection. In addition, unlike formaldehyde or OsO_4, glutaraldehyde-fixed specimens survive the rigors of paraffin-embedding for light microscopy rather well. Therefore, the use of coagulant fixatives for light microscopy should be discontinued whenever possible.

Glutaraldehyde is an excellent fixative for ultracentrifugal analysis of cell organelles. It it useful for the preservation of ribosomes for CsCl gradient centrifugation (Ceccarini *et al.*, 1970) or for electron microscopy (Nonomura, 1971). The use of 0.1–0.2% glutaraldehyde is quite effective in the fixation of ribosomes and polyribosomes *in vitro* (Voronina *et al.*, 1977). The use of higher concentrations (6%) may lead to the formation of a film of aggregated ribosomes. Fixation with this dialdehyde prevents dissociation of free ribosomes by the high hydrostatic pressure generated during ultracentrifugation or by low concentrations of Mg^{2+}. Glutaraldehyde is tolerant to limited variations in pH and Mg^{2+} concentration and is effective over a wide range of ribosome concentrations (Subramanian, 1972). Some loss of ribosomes by aggregation under certain conditions can be prevented by adding a protective agent such as bovine serum albumin during fixation. The integrity of glutaraldehyde-fixed ribosomes might be due to cross-linking between its subparticles by this reagent.

The dialdehyde is effective in preferential cross-linking of large ribosomal subunits of bound ribosomes to the microsomal membranes in attempts to recognize membrane proteins that may contribute to the binding sites (Kreibich *et al.*, 1978). Cross-linking with low concentrations of glutaraldehyde (0.02%) completely prevents the release of ribosomes from microsomal membranes caused by puromycin in media of high ionic strength. At even lower concentrations (0.005%), only large subunits remain associated with the membranes, while small subunits are released. The ribosome–membrane junctions are more readily cross-linked than are other membrane proteins by glutaraldehyde under the same conditions. Ribosomes treated with low concentrations (0.1%) of glutaraldehyde retain substantial functional activity and normal sedimentation properties (Kahan and Kaltschmidt, 1972). Since such treatments do not completely disrupt the tertiary or quaternary structure of ribosomes, the reactivity of proteins within ribosomes can be studied (Reboud et al., 1975).

The cross-linking property of glutaraldehyde has been used in X-ray crystallography to stabilize protein crystals without severe conformational changes in the protein structure (Quiocho et al., 1967). The catalytic activity of glutaraldehyde cross-linked, water-insoluble catalase was decreased 10 times when compared with the soluble native enzyme (Schejter and Bar-Eli, 1970). The insoluble enzyme remained stable for several months at 4°C. In contrast to this report, Herzog and Fahimi (1974) found that glutaraldehyde treatment enhanced the peroxidatic activity of catalase toward aromatic amines such as diaminobenzidine (DAB), the substrate ordinarily used for its light and electron microscopic demonstration. Freshly isolated peroxisomes from rat liver did not stain with DAB, whereas glutaraldehyde-fixed peroxisomes oxidized it readily. On the other hand, glutaraldehyde inhibited both the catalatic and peroxidatic activity of catalase toward ethanol and pyrogallol.

Ferrier et al. (1972) cross-linked microcrystalline catalase with glutaraldehyde and treated it with bisulfite to form an insoluble enzyme with activity nearly equal to the original crystals. Also, little change was observed in the helical content of myosin A fragments that had been cross-linked with glutaraldehyde.

Glutaraldehyde is useful in the study of single proteins (Quiocho and Richards, 1964), differential affinity of bacterial membrane-associated enzymes (Ellar et al., 1971), and preparation of insoluble enzymes (Quiocho and Richards, 1966; Schejter and Bar-Eli, 1970). Enzyme immobilization by cross-linking with glutaraldehyde is carried out routinely. Such enzymes are used as biocatalysts. Fairly high concentrations of glutaraldehyde are used for this purpose. For example, 25% and 10% glutaraldehyde was used for immobilizing glucoamylase on chitin (Stanley et al., 1978) and protease on anion exchange resin (Ohmiya et al., 1978), respectively.

Glutaraldehyde has been employed for preparing water-insoluble protein derivatives that are effective immunoadsorbents (Ternynck and Avrameas, 1972). These derivatives retain more than 70% of their biological activity, i.e., antigen or antibody binding capacity. The highly specific immunoadsorption of these derivatives allows the isolation of antigens and antibodies in high yields. This method involves essentially the coupling of antigens and antibodies to glutaraldehyde-activated beads of polyacrylamide gels.

Concanavalin A can be cross-linked with glutaraldehyde (Donnelly and Goldstein, 1970). The purpose of cross-linking is to obtain insoluble polymers. Such polymers or derivatives conserve many of their sugar binding sites. They are used for the isolation by affinity chromatography of polysaccharides or glycoproteins. Cross-linking of lectins with glutaraldehyde under mild conditions afford soluble high-molecular-weight polymers that possess many sugar binding sites. The effect of such a cross-linking of soybean agglutinin on its activities is assumed to be the result of an increase in valency of the divalent molecule by conversion into multivalent polymers (Lotan and Sharon, 1976). The latter are more efficient in forming multiple crossbridges between adjacent cells, which lead to cell agglutination.

Fixation with glutaraldehyde does not decrease the agglutinability of cells such as erythrocytes by lectins under conditions of low shear (Marquardt and Gordon, 1975). Similarly, glutaraldehyde-fixed erythrocytes remain sensitive to the hemagglutination and hemagglutination inhibition tests for arbovirus antigens and antibodies (Wolff *et al.*, 1977). Glutaraldehyde-fixed cells are as sensitive and specific as the fresh erythrocytes and are stable for 6 months at 4°C. Fixation with glutaraldehyde is superior to that with formaldehyde in that antibody titrations are more reproducible, the cells can be stored for longer periods of time, and cells coated with antigen react more specifically and yield higher antibody titers (World Health Organization, 1970).

The use of glutaraldehyde coupling followed by sodium borohydride reduction as a means of covalently labeling specific polypeptide binding sites has been explored. This approach has been proved successful for the affinity cross-linking of insulin and epidermal growth factor-urogastrone to specific binding sites in rat liver membranes (Sahyoun *et al.*, 1978). Probably, multimers of a basic receptor subunit are formed in this reaction. The method does not need the synthesis of specialized affinity-labeling reagents and should prove useful for the study of soluble receptors for other polypeptides.

Glutaraldehyde has been used for inducing tumor immunity by employing tumor cells treated with this aldehyde (Sanderson and Frost, 1974). Immunity to Marek's disease induced by glutaraldehyde-treated cells of Marek's disease lymphoblastoid cell line has been reported (Powell, 1975).

Covalently linked protein oligomers of varying molecular weight can be produced by using low concentrations of glutaraldehyde and high concentrations of protein (Payne, 1973). These polymers are useful as molecular-weight markers. Extensive polymerization of proteins can be achieved with minimal amino acid substitution. Consequently, polymerization can be achieved with little conformational perturbation. The inherent similarities of these polymers make them superior to commercial molecular-weight protein markers, which may have marked differences in composition and charge.

Glutaraldehyde at low concentrations (0.025%) is an effective antimitotic agent. The arrested mitosis is characterized by the disappearance of spindle and astral fibers, immobilization of chromosomes in the equatorial region, and blocking of centrospheres (Sentein, 1975). It is possible that at very low concentrations, only heterodimers of tubulin are cross-linked by glutaraldehyde. The dialdehyde seems to reduce greatly EDTA-lysosome lysis of gram negative bacteria (Russell and Haque, 1975). Possibly the dialdehyde strengthens the outer envelope by interacting with the lipoprotein layer. Similarly, glutaraldehyde reduces the action of lytic peptidase on the cell walls of *Staphylococcus aureus* (Russell and Vernon, 1975).

Differential staining of viable and nonviable cells with alcian blue is maintained after fixation with glutaraldehyde (Yip and Auersperg, 1972). The structural differences probably are responsible, in part, for this differential staining, which persists following fixation with this dialdehyde. The proportion of viable cells in a cell population can be estimated by the dye exclusion method (Hanks and Wallace, 1958). This method is based on the phenomenon that many stains are excluded by living cells but not by nonliving cells.

Since glutaraldehyde forms precipitates with primary but not with secondary biogenic amines, adrenalin- and noradrenalin-storing cells of the adrenal medulla can be differentiated by it. The dialdehyde tends to make cells resistant to osmotic stress of water immersion. It can prevent hemolysis of cells in suspension under hypotonic conditions. Spheroplasts pretreated with glutaraldehyde are better able to resist osmotic shock than are those exposed directly to a hypertonic solution. Cells fixed with this dialdehyde are quite stable in that they show little change in volume even when the ionic strength of their environment is decreased by a factor of 10.

Preservation and staining of neurotransmitter noradrenalin in the terminal vesicles of sympathetic nerves can be obtained by fixation with a mixture of glutaraldehyde and formaldehyde followed by treatment with potassium dichromate and OsO_4 (Tranzer and Snipes, 1968). Glutaraldehyde-fixed tissues can be used to demonstrate the presence of free cholesterol in cell membranes (Williamson, 1969). Glutaraldehyde is effective in preserving elementary particles attached to mitochondrial membranes. Clustering of intramembranous particles induced by glycerol or dimethylsulfoxide (DMSO) in mouse lymphocytes is prevented by a brief fixation with glutaraldehyde (McIntyre *et al.*, 1974). Glutaraldehyde causes reversal of polarity of membrane-associated particles in endothelial cells (Dempsey *et al.*, 1973) and in muscles (Bullivant *et al.*, 1972.

Treatment with low concentrations (0.4%) of glutaraldehyde facilitates penetration of relatively bulky tracer molecules into cells. Red blood cell ghosts fixed with 0.4% glutaraldehyde for 15 min at 4°C, followed by treatment with saponin (0.03 mg/ml), allowed penetration of ferritin into the ghost interior (Ohtsuki *et al.*, 1978). A similar treatment did not allow such penetration into intact red blood cells.

Glutaraldehyde is known to have a stabilizing effect on collagen. It has been proved that collagen fibrils fixed with glutaraldehyde become practically impervious to the solubilizing action of such dissociating solutions as 0.1 *M* acetic acid or 8 *M* urea. Fixed collagen also becomes unattackable by proteolytic enzymes including collagenase, even at low pHs (Bairati *et al.*, 1972). Collagen fixed with this dialdehyde shows an increased density in some intraperiodic bands and reduction in its intrinsic birefringence (Bairati *et al.*, 1972).

Cross-linking of histones (both free and a part of the nucleoprotein complex) by glutaraldhyde is irreversible. On the other hand, cross-linking of only free histones by formaldehyde is irreversible; cross-linking of histones that are a part of the nucleoprotein is reversible (Chalkley and Hunter, 1975). Glutaraldehyde is thought to generate bonds primarily between histones, whereas formaldehyde seems to form bonds between histone molecule and the bases of the DNA.

Certain macromolecules and even some low-molecular-weight constituents may be quantitated in specimens fixed with glutaraldehyde. This is possible, in part, because within a few minutes after the removal of the tissue from the body, energy reserves in the specimen are markedly depleted. The decline of high-energy phosphate is of special importance because biochemical reactions requiring energy would be slowed or completely stopped, resulting in the stabilization of certain low-molecular-weight cell constituents. For example, the amount of amino acids such as glutamate and aspartate does not seem to change because of fixation with an aldehyde, whereas alanine concentrations appear to increase with increasing time of fixation. Since a large increase in alanine occurs in the unfixed tissue, the alanine–glutamate ratio may reflect the time when biochemical fixation occurs (Conger et al., 1978). The limited available space does not permit the enumeration of many other uses of glutaraldehyde.

Nature of Commercial Glutaraldehyde

Preliminary to any study of the chemical nature of the cross-linking reaction, it is necessary to elucidate the nature of commercial glutaraldehyde. Furthermore, determination of at least the approximate extent of impurities in glutaraldehyde is necessary since these affect pH, osmolarity, and aldehyde concentration in the fixative solution, which in turn influence the rate of penetration and cross-linking of proteins. Another reason for knowing the nature of glutaraldehyde is that various impurities differ in the extent and nature of their reactions with OsO_4 during postfixation, which determines in part the final image of the specimen. For instance, the reaction rate for the formation of osmium *blacks* is greater with commercial glutaraldehyde than with pure glutaraldehyde.

Glutaraldehyde is generally supplied as a 25% solution having a pH of 3–6. In this state it exists as a hydrate and stays relatively stable for long periods of time if stored at subfreezing temperatures. With time, however, the solution turns yellow, which indicates an excessive polymerization of glutaraldehyde. As long as the color stays faintly yellow, there is not much polymerization or loss of aldehyde groups. Any glutaraldehyde solution with a pH less than 3 is suspect. Stock solutions of glutaraldehyde having a pH above 4 yield the best preservation of ultrastructure. However, pH is a rather poor indicator of the condition of a glutaraldehyde solution.

The rate of glutaraldehyde polymerization is dependent on many factors, including its concentration, pH, temperature, and age. Distilled glutaraldehyde tends to polymerize unless all traces of water are removed; even small amounts of water catalyze polymerization. In fact, at room temperature, concentrated solutions of commercial glutaraldehyde polymerize rapidly, whereas dilute solutions (4% or less) keep better. At high concentrations (e.g., 50%) it may gel spontaneously. It deteriorates rapidly at room temperature, in the presence of oxygen, acids, or bases, but remains unaffected by the presence of light. The polymerization is slow in diluted, slightly acidic aqueous solutions.

It is difficult to obtain the dialdehyde in a pure state as a monomer because most commercial lots contain an indeterminate amount of impurities. Commercially available "25% glutaraldehyde solution" may contain 79% water, 3% glutaraldehyde, and 18% derivatives (impurities) of high molecular weight, which can be broken down to glutaraldehyde. As stated earlier, the chemical nature of these impurities is of considerable importance in understanding the phenomenon of cross-linking of proteins by glutaraldehyde.

The impurities comprise mainly glutaric acid, acrolein glutaradozime, ethanol, methanol, and various polymers and products of oxidation and photochemical degradation. Recent evidence obtained from nuclear magnetic resonance spectroscopic and chromatographic studies indicates that commercial glutaraldehyde contains several types of species, which are shown below. Glutaraldehyde monomer (I) polymerizes into more stable hydrates of the types (II), (III), and (IV) in aqueous alkaline solutions. In the neutral and acid ranges, the glutaraldehyde molecule has a tendency to form different polymers with an acetal-like structure (V):

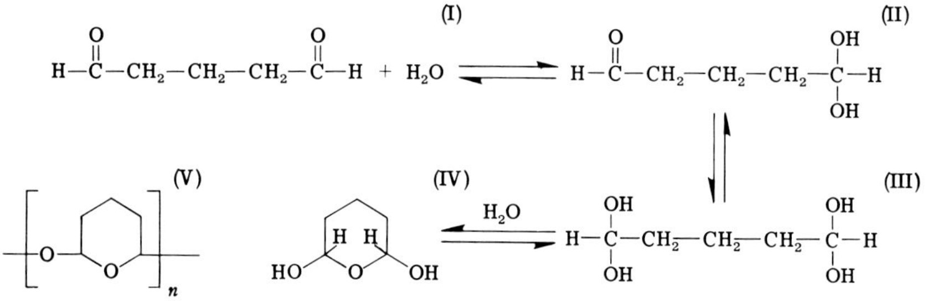

Heating of alkaline solutions of glutaraldehyde leads to accelerated irreversible polymerization. Because in the alkaline pH range, the hydrates such as (III) are irreversible and also because aldol-type condensations take place, there is little reversion to the monomers of glutaraldehyde. Heat or ultrasonics, on the other hand, facilitate the production of monomers in the acid range because the acetal-like polymers (V) are of the reversible type. Which of the preceding species plays a key role in protein cross-linking is an important question. Since agreement on this point is lacking, various points of view are presented.

Rubbo *et al.* (1967) suggested that the aldehyde existed as a monomer and that an equilibrium was established between the free aldehyde molecule (open chain) and the hydrated ring structure (cyclic hemiacetal of its hydrate) shown above. Whipple and Ruta (1974) indicated that in 25% glutaraldehyde solution, hemiacetal is present in equilibrium with free aldehyde, and that higher order oligomers contribute very little to the equilibrium mixture. Hardy *et al.* (1969, 1972) suggested that the monomer exists as a mixture of hydrated forms in aqueous solutions. They consider α,β-unsaturated aldehyde to be a minor component of the organic materials present in aqueous solutions of glutaraldehyde, provided the solutions are prepared from pure glutaraldehyde and analyzed at neutral pH. Glutaraldehyde solutions analyzed by Korn *et al.* (1972) had an absorption maximum at 280 nm, and they concluded that very small amounts ($\sim 1\%$) of unsaturated dimer are present in commercial glutaraldehyde.

Korn *et al.* (1972) suggested that aqueous solutions of glutaraldehyde consist of free glutaraldehyde, the cyclic hemiacetal of its hydrate, and oligomers in equilibrium. These authors further suggest that glutaraldehyde polymerizes in a different manner under alkaline conditions, and thus does not react efficiently with proteins. Boucher (1972, 1974) and Boucher *et al.* (1973) proposed that the monomeric glutaraldehyde is the active species in terms of cross-linking with proteins, and the ability of polymeric species to revert to the active monomer is dependent on pH. It is known that pH plays an important role in determining the type of polymers formed in glutaraldehyde solution. According to these authors, polymers formed at an alkaline pH cannot revert to the monomeric form because room temperature and time tend to encourage the formation of a more irreversible polymer. On the other hand, polymers formed under the neutral or acidic conditions are thought to revert easily to the monomeric form. On the basis of aforementioned studies, the protein cross-linking ability of glutaraldehyde can be attributed to the monomeric species or one of its hydrated forms and not to a condensation polymer (Gillett and Gull, 1972).

According to another point of view, polymeric species are the major component of commercial glutaraldehyde that plays a key role in protein cross-linking. There is no doubt that glutaraldehyde contains polymers. It may contain oligomers such as a dimer, cyclic dimer, trimer, bicyclic trimer, and high polymeric species. That glutaraldahyde polymerization increases with a rise in pH and that above pH 9.0 there is an extensive loss of aldehyde groups have been shown by Fein *et al.* (1959).

By polymerizing glutaraldehyde with cationic catalysts, Aso and Aito (1962) found that the final product contained few residual aldehyde groups. This suggests that ring formation occurs during the propagation process, apparently by an intramolecular mechanism. They also indicated that similar polymerization of glutaraldehyde may occur spontaneously in aqueous solutions at room temperature in the absence of a catalyst yielding a soluble pentamer or tetramer containing approximately one free aldehyde group per molecule. The polymer of glutaraldehyde proposed by these authors is as follows:

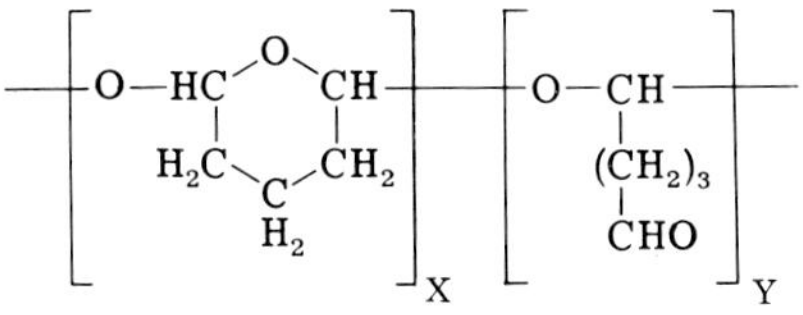

$$
\left[\mathrm{O{-}HC} \underset{\underset{\underset{H_2}{C}}{H_2C}}{\overset{O}{\diagup}} \underset{CH_2}{\overset{CH}{\diagdown}} \right]_X
\left[\mathrm{O{-}CH} \atop \underset{CHO}{(CH_2)_3} \right]_Y
$$

The ratio of X and Y in the preceding structure is 6:16.

Similarly, using an aluminum catalyst, the polymerization of glutaraldehyde resulted in the formation of a very-high-molecular-weight polyglutaraldehyde that was soluble in water but thermally unstable (Moyer and Grev, 1963). The structure of this polymer was similar to that proposed by Aso and Aito (1962). Yokota *et al.* (1965) indicated that glutaraldehyde existed as a trimer (see p. 76). Richards and Knowles (1968) also suggested that the formation of trimers in commercial glutaraldehyde as a result of reaction between a dimer and a molecule of glutaraldehyde is likely. It is unlikely, however, that these trimers are present in any large proportions in commercial glutaraldehyde.

The dimer thought to be the principal polymer in commercial glutaraldehyde is an α,β-unsaturated dimer. It forms slowly under ordinary storage conditions, but forms relatively rapidly when stored in a buffer at an alkaline pH. In other words, glutaraldehyde is somewhat in equilibrium with its cyclic hemiacetal and polymers of the cyclic hemiacetal at an acid pH, but at neutral or even slightly alkaline pH the dialdehyde undergoes an aldol condensation with itself. This is rather rapidly followed by dehydration, which results in the formation of the dimer. At a certain highly alkaline pH, the dimer will precipitate from solution.

As stated earlier, the dimer absorbs in the ultraviolet at 235 nm because of the α,β-unsaturated bond, and a linear relationship exists between glutaraldehyde concentration and absorption at 235 nm (Munton and Russell, 1970). The dimer can be distinguished from other impurities by differences in their absorption in the ultraviolet; glutaric acid, for instance, has been found to absorb maximally at 207 nm (Gillett and Gull, 1972). The polymerization of the dimer is enhanced by heat, acids, and bases.

The following scheme explains the formation of the dimer (Robertson and Schultz, 1970):

$$2CHO{-}(CH_2)_3{-}(CHO \xrightarrow{OH} CHO{-}(CH_2)_2{-}CH(CHO){-}CH(OH){-}(CH_2)_3{-}CHO \xrightarrow[H^-,\, \Delta H]{OH^-}$$

$$\text{(I)} \qquad\qquad\qquad\qquad\qquad \text{(II)}$$

$$CHO{-}(CH_2)_2{-}C(CHO){=}CH(CH_2)_3{-}CHO + H_2O$$

$$\text{(III)}$$

In the case of glutaradehye (I), the condensation product is a secondary alcohol (II), which then is able to eliminate H_2O forming an α,β-unsaturated aldehyde (III); the second reaction is endothermic (ΔH^θ). The initial oxidation of some aldehyde groups results in the formation of acids, and the presence of even small amounts of these acids, in turn, catalyzes an aldol condensation, forming a disubstituted α,β-unsaturated aldehyde. It is known that dilute alkaline solutions cause aldehydes containing at least one hydrogen in position α to undergo polymerization to hydroxyaldehydes as a result of aldol condensation (Fieser and Fieser, 1956).

Commercial glutaraldehyde solutions usually have two peaks (at 280 nm and 235 nm) in their ultraviolet absorption spectra. The 280 nm carbonyl peak is due to monomeric glutaraldehyde, whereas the 235 nm peak is attributed to the presence of a dimer. However, the size of these two peaks does not indicate the absolute content of monomeric and polymeric glutaraldehyde. The optimal ratio between the monomeric form and the dimer in glutaraldehyde solution to accomplish satisfactory fixation seems to be 1:1 to 2:1; lower or higher than this range may cause poor fixation. Since for different uses of glutaraldehyde, different grades of purity are needed, standardized supplies should be available; such supplies are also needed to achieve reproducible results. It is desirable that the suppliers give some indication of the impurities present in each batch of glutaraldehyde.

Effect of pH and Temperature on Glutaraldehyde

The structure of the glutaraldehyde molecule is profoundly affected by changes in both pH and temperature. Glutaraldehyde exists in cyclic forms probably due to intra- or intermolecular bonding, which are thought to be affected by changes in the pH. Polymerization under alkaline conditions accompanied by an extensive loss of aldehyde groups has been reported (Fein *et al.*, 1959; Fieser and Fieser, 1961). The formation of precipitates at higher pHs probably is an indication of polymerization. The higher the pH, the greater will be the degree of polymerization at the same temperature.

As stated previously, glutaraldehyde tends to polymerize when standing at room temperature or when heated. It has been shown that the structure of polymers obtained by polymerization of glutaraldehyde is temperature dependent between -60 and $+40°C$ (Yokota *et al.*, 1965). From -60 to $0°C$, 80% of glutaraldehyde is in form I, above $0°C$ form II predominates, and the amount of form III is rather small within the polymerization temperature range.

(I)

(II) —̵OCH₂(CH₂)₃CO—̵—̵OCH₂(CH₂)₃O—̵—̵CO(CH₂)₃CO—̵—

$$\text{(II)} \quad -\!\!\!-\!\!\text{OCH}_2(\text{CH}_2)_3\text{CO}-\!\!\!-\!\!\text{OCH}_2(\text{CH}_2)_3\text{O}-\!\!\!-\!\!\text{CO}(\text{CH}_2)_3\text{CO}-\!\!\!-$$

$$\text{(III)} \quad -\!\!\!-\!\!\text{CH}-\!\!\!-\!\!(\text{CH}_2)_3\text{CHO}-\!\!\!-\!\!\text{O}-\!\!-$$

The effect of changes in temperature (25–90°C) on the composition of glutaraldehyde has also been determined by nuclear magnetic resonance absorption (Korn *et al.*, 1972). These studies indicate the presence of monomeric glutaraldehyde as well as other species as oligomers at all temperatures.

Reaction with Proteins

No other fixative has surpassed glutaraldehyde in its ability to cross-link and preserve tissue proteins for routine electron microscopy. Although glutaraldehyde is used extensively as a cross-linking agent not only in electron microscopy but also in enzymology, X-ray crystallography, and tanning, the nature of its reaction with proteins remains uncertain. Since the reaction products of glutaraldehyde with tissue proteins have been neither isolated nor identified, the chemistry of its reaction with proteins has been the subject of much debate in the published literature. The presence of the numerous polymerized forms in glutaraldehyde discussed earlier is partly responsible for the difficulty in chemical identification of the reaction products. Whereas the reaction of proteins with formaldehyde is at present well understood, the mechanism of glutaraldehyde reaction remains uncertain. Nevertheless, an understanding of the reaction mechanism of this dialdehyde with amino acids is beginning to emerge. It must be pointed out at the outset that although simple chemical explanations are available for reactions between glutaraldehyde and amino acids or even peptides, relatively little is known about its effect on the secondary and tertiary conformation of peptides and proteins.

Glutaraldehyde is unable to cross-link low concentrations of protein. In other words, the presence of adequate concentrations of protein is a prerequisite for the aldehyde to fix cells. One established fact is that the reaction of glutaraldehyde with proteins involves α-amino groups of lysine side chains and results in the appearance of an absorption maximum at ~265 nm. A decrease of the lysine content by ~10% has been observed in the glutaraldehyde-treated wool (Ziegler and Liesenfeld, 1976). Protein modification obtained through cross-linking by glutaraldehyde is irreversible and withstands treatment with acids, urea, semicarbazide, and heat. At least two major controversies intimately related to protein cross-linking with glutaraldehyde exist: (1) the proportion of monomeric versus polymeric species in commercial glutaraldehyde solutions, discussed earlier, and (2) the extent of cross-linking carried out by the two types of species. Both sides of the latter controversy are now presented.

According to one point of view, the monomeric form of commercial glutaraldehyde plays only a minor role in cross-linking of proteins at neutral or slightly alkaline pH, but various polymeric species, especially α,β-unsaturated aldehydes, readily cross-link reactive groups located at different distances. When unpurified glutaraldehyde is used, these unsaturated compounds seem to play a key role in the cross-linking of proteins.

It should be noted that absorption at 235 nm (typical of polymeric species) increases at room temperature at an alkaline pH and with time, and that glutaraldehyde solutions used for fixation have an alkaline pH. One study indicates that during fixation of tissue specimens with glutaraldehyde, material absorbing at 235 nm increases in amount (Goff and Oster, 1974). Thus, it may be assumed that because fixation conditions promote the presence of longer oligomers of glutaraldehyde, they are more important in introducing cross-links than is the short molecule of monomeric glutaraldehyde. It has been shown that 8% polymerized glutaraldehyde yielded superior fixation of lung tissue than that obtained with purified glutaraldehyde (Wrench, 1970). It has also been shown that the condensed form of glutaraldehyde is more effective in maintaining the insulating properties of the erythrocyte membrane than those obtained after fixation with highly purified monomer (Carstensen et al., 1971).

The fact that purified glutaraldehyde (containing monomeric species only) preserves enzyme activity better than does unpurified glutaraldehyde indicates that polymeric species are more efficient in the introduction of cross-links than are monomeric forms. It is known that enzyme activity is preserved much better by mild cross-linking. However, the inhibitory effect of other impurities present in unpurified glutaraldehyde on enzyme activity cannot be disregarded. Because of its better protein cross-linking potential, polymerized glutaraldehyde is preferred for the preservation of ultrastructure. Nevertheless, the results of fixation by polymerized glutaraldehyde are unpredictable because even if the nature and effect of the glutaraldehyde polymers present are known, the nature, extent, and effect of other impurities (present in unpurified glutaraldehyde) on the quality of fixation are not known.

According to the second point of view, it is chiefly the monomeric glutaraldehyde present in solutions of glutaraldehyde that is responsible for the irreversible cross-linking of proteins (Blass et al., 1976). According to Korn et al. (1972), aqueous solutions of commercial glutaraldehyde consist of free glutaraldehyde (I), the cyclic hemiacetal of its hydrate (II), and oligomers of this (III) in

(I)   (II)   (III)

equilibrium with each other. On dilution, (III) is converted rapidly to (I) and (II). These workers suggest that the reactive species in these solutions is free glutaraldehyde and not a condensation polymer as has been suggested by Richards and Knowles (1968). Moreover, free glutaraldehyde is capable of reacting at acidic or neutral pH. Since aqueous solutions of purified glutaraldehyde do not contain unsaturated polymeric species and yet are quite effective in the cross-linking of proteins, it is apparent that at least the initial presence of unsaturated compounds is not necessary in the cross-linking phenomenon. However, the possibility remains that aldol condensation may occur under the influence of proteins immediately prior to cross-linking.

According to Hardy *et al.* (1969), on the other hand, very little free glutaraldehyde is left in its aqueous solution and the monomeric glutaraldehyde exists mostly as a cyclic monohydrate (II) in equilibrium with an open-chain monohydrate (III) and dihydrate (IV). Thus, these hydrated forms are responsible

for the irreversible cross-linking of proteins by monomeric glutaraldehyde. As early as 1964, Milch showed that only those saturated aldehydes, including glutaraldehyde, that exist in aqueous solution in a hydrated configuration are capable of cross-linking collagen. It may be pertinent to note that when the amount of polymers in solution increases, the biocidal activity of glutaraldehyde decreases (Boucher, 1972, 1974). The potency of glutaraldehyde in the 5–8 pH range can be increased by the addition of cationic (Stonehill, 1966) or anionic surfactants (Sidwell *et al.*, 1970; Wilkoff *et al.*, 1971).

The disagreement discussed here indicates that additional studies are needed to determine the species in unpurified solutions and in solutions prepared from pure glutaraldehyde that are most reactive with proteins. Also, it is important to know whether fixation pH (slightly alkaline) renders the reactive groups in proteins more accessible or less accessible to glutaraldehyde. Very little is known regarding the mode of action of this dialdehyde at alkaline pHs in the tissue, and even less is known about the nature of reactions between various impurities present in unpurified glutaraldehyde and OsO_4 during postfixation.

On the basis of available information it is concluded that the proportions of free monomer, hydrated monomer, and polymeric forms in an aqueous solution of glutaraldehyde are not known. Even if the predominance of a certain species in the solution could be demonstrated, the same species may not necessarily be the most active one. Sufficient evidence is lacking to implicate any of the species as initiating the cross-linking reaction.

The use of glutaraldehyde as a cross-linking agent in the study of collagen has provided a great deal of information (although some of it is contradictory) on the fixation of proteins by this dialdehyde (Bowes and Kenten, 1949; Cater, 1963, 1965; Bowes and Raistrick, 1964; Bowes et al., 1965; Bowes and Cater, 1966; Trnawska et al., 1966). Bowes and Cater (1966) indicated that although glutaraldehyde was very reactive toward the amino groups of lysine, it showed little reaction with tyrosine or guanidino groups of arginine. Studies by Korn et al. (1972) indicated that the major reaction of glutaraldehyde was with lysine residues in proteins and not with tyrosyl residues, as was suggested by Habeeb and Hiramoto (1968). Studies of wool also indicated that the lysine residue was the primary amino acid that reacted with glutaraldehyde (Happich et al., 1970). Limited amounts of other amino acids (e.g., histidine and tyrosine) present in wool also reacted with the dialdehyde. Incubation of glycogen phosphorylase b with glutaraldehyde also indicated that sulfhydryl groups did not react in the modified enzyme (Wang and Tu, 1969).

According to Hopwood (1968b), however, glutaraldehyde reacted with the amino acids tyrosine, tryptophan, and phenylalanine. Habeeb and Hiramoto (1968) also indicated that some reaction occurred with tyrosine, histidine, and sulfhydryl residues. Using model compounds such as amino acids and peptides, and protein solutions, Habeeb and Hiramoto (1968) showed that glutaraldehyde was very reactive toward the α-amino groups of amino acids, the N-terminal amino groups of some peptides, and the sulfhydryl group of cysteine. The imidazole group of histidine was partially reactive with glutaraldehyde, and the presence of a free amino group seems to be necessary for this dialdehyde to react with the former. The phenolic ring of tyrosine was also partially reactive in N-acetyltyrosine ethyl ester and glycyltyrosine. Nevertheless, overwhelming evidence indicates that lysine is the most important component of protein involved in the reaction with glutaraldehyde.

The amount of cross-linking of proteins by glutaraldehyde is dependent on several factors besides those already discussed (e.g., pH, temperature, and time). The differences in the shape of the protein molecule as well as the accessibility of the reacting groups to glutaraldehyde are important factors. Intramolecular cross-links predominate over intermolecular cross-links when the reacting amino acids are present within the molecule. Studies with proteins in solution indicate that the rate of formation of glutaraldehyde–protein links per protein molecule glutarated is ~ 1 sec^{-1} mol^{-1} (Hopwood et al., 1970).

The size of the glutaraldehyde molecule seems to be particularly suitable for binding the gap between the amino groups of the polypeptide chains. It is known that the most reactive groups in proteins are free amino groups. The cross-linking is facilitated when the reactive groups are evenly distributed over the surface of the molecule. On the other hand, when the residues are concentrated in one small portion of the protein molecule, the cross-linkage of more than a few molecules would be difficult. It appears that in cross-linking the distribution of the reacting amino acids is more important than the number of these reacting groups, except that the presence of a minimal critical number of free amino groups in proteins is a prerequisite for their cross-linking by glutaraldehyde. Since glutaraldehyde reacts principally with ϵ-amino groups of lysine, the proximity, accessibility, and location of specific lysine residues in the protein molecule are important factors in the rate and quality of cross-linking.

It has been suggested that on the average two or three molecules of glutaraldehyde are linked together in each cross-link and at each center of unipoint fixation. If two amino groups are involved in each cross-link, then the molecules of glutaraldehyde bound exceed a 1:1 ratio with the amino groups. That glutaraldehyde forms bipointal likages with polypeptide chain collagen structures has been indicated (Milch et al., 1965). Studies of collagen also indicate that three molecules of glutaraldehyde are bound per lysine residue. In wool the ratio of the moles of glutaraldehyde consumed to the moles of amino acid (lysine) appears to be constant and close to 4:1 (Happich et al., 1970). The same relationship has been found for protein solutions (Korn et al., 1972). Studies on the effect of glutaraldehyde on human red blood cells indicate that, depending on glutaraldehyde polymerization during fixation, between 100 and 300 sites in each hemoglobin molecule react with the dialdehyde (Morel et al., 1971). Approximately 200 molecules of glutaraldehyde may be involved in cross-linking each molecule of hemoglobin.

Glutaraldehyde is expected to react with amino-, imino-, and guanidinium groups of proteins, lipids, lipoproteins, and glycoproteins, and to mask their positive surface charges thereby increasing the negative charge on the cell surface. However, cells fixed with this dialdehyde bear positive fixed charges since all surface cations are not masked. Such a condition has been indicated in glutaraldehyde-fixed sarcoma 180 ascites cells (Skehan, 1975). Biochemical studies also indicate that the positive charge on the amino groups, that is normally present at neutral pH, is maintained in the reaction product. This may in part be responsible for the excellent preservation of proteins and cell structure by glutaraldehyde. The rapid cross-linking by glutaraldehyde also plays a part in *in situ* insolubilization of many proteins.

Formation of Schiff's Bases

A brief discussion of the formation of Schiff bases and their role in protein cross-linking is in order. A simple linear formula of five carbons with two carbonyl groups is attributed to monomeric glutaraldehyde in solution. The interaction of this molecule with free amino groups of proteins would lead to the formation of a Schiff base. Molin *et al.* (1978) proposed a two-step reaction explaining the reaction of glutaraldehyde with amino groups through the formation of Schiff's bases; this scheme is as follows:

Step 1:

$$\text{Protein—NH}_3^+ + \text{OHC —(CH}_2)_3\text{—CHO} \rightarrow \text{Protein—N}{=}\text{CH —(CH}_2)_3\text{—CHO}$$

Step 2:

$$\text{Protein—N}{=}\text{CH —(CH}_2)_3\text{—CHO} + {}^+\text{H}_3\text{N —(CH}_2)$$

Step 1 shows the reaction between a protein and the glutaraldehyde molecule. The resulting derivative is formation of a Schiff's base. In the second step the reacted protein is conjugated to the amino group of aminohexyl Sepharose 4b, through formation of a Schiff's base.

However, since the reaction resulting in the formation of a Schiff's base is reversible and the product does not withstand the treatments with acids, and since the protein modification obtained by glutaraldehyde is stable to acid treatment, a simple Schiff's base formation does not explain the mechanism of cross-linking. Schiff's bases may, therefore, only be intermediates in the formation of more stable structures. Fixation or cross-linking of proteins proceeds through more than one step or through several routes.

According to Peters and Richards (1977), the predominant reactive species of glutaraldehyde, when cross-linking is carried out at its usual pH, is the α,β-unsaturated aldehyde. Therefore, the Schiff's bases formed with lysine residues are in conjugation with double bonds:

$$\underset{\text{CH}_2-(\text{CH}_2)_2-\text{CH}}{\overset{\text{CHO}}{|}}{=}\left[\underset{\text{C}-(\text{CH}_2)_2-\text{CH}}{\overset{\text{CHO}}{|}}\right]_n{=}\underset{\text{C}-(\text{CH}_2)_2-\text{CHO}}{\overset{\text{CHO}}{|}} + \text{Protein—NH}_2$$

$$={=}\underset{\text{CH}-(\text{CH}_2)_2-\text{CH}}{\overset{\text{CHO}}{|}}\quad\underset{\text{C}-(\text{CH}_2)_2-\text{CH}}{\overset{\text{CH}{=}\text{N—Protein}}{|}}{=}{=}$$

The resonance interaction of the Schiff base with the double bond leads to the stability of cross-linked proteins (Monsan *et al.*, 1975). Since terminal aldehyde groups are not in conjugation with a double bond, they do not seem to be involved in the acid-stable cross-links. Although the predominant reaction product is the conjugated Schiff base, a small amount of the Michael addition product may be formed if the amine concentration is high (Peters and Richards, 1977).

Mechanism of Cross-Linking

An attempt to explain the general mechanism of the cross-linking of proteins by glutaraldehyde was made by Richards and Knowles (1968). They postulated a pathway involving a Michael-type addition of the side-chain amino groups to α, β -unsaturated aldehydes formed by aldol condensation of the glutaraldehyde. The probable mechanism and expected products from reaction between this bifunctional reagent and protein crystals are as follows:

(A) Aldol condensation:

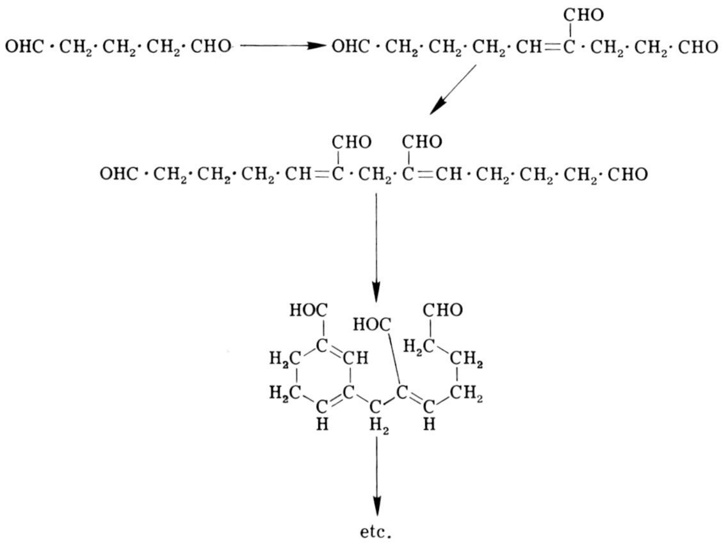

$$OHC \cdot CH_2 \cdot CH_2 \cdot CH_2 \cdot CHO \longrightarrow OHC \cdot CH_2 \cdot CH_2 \cdot CH_2 \cdot CH{=}\overset{\displaystyle CHO}{C} \cdot CH_2 \cdot CH_2 \cdot CHO$$

$$OHC \cdot CH_2 \cdot CH_2 \cdot CH_2 \cdot CH{=}\overset{\displaystyle CHO}{C} \cdot CH_2 \cdot \overset{\displaystyle CHO}{C}{=}CH \cdot CH_2 \cdot CH_2 \cdot CH_2 \cdot CHO$$

etc.

(I)

(B) Cross-linking reactions:

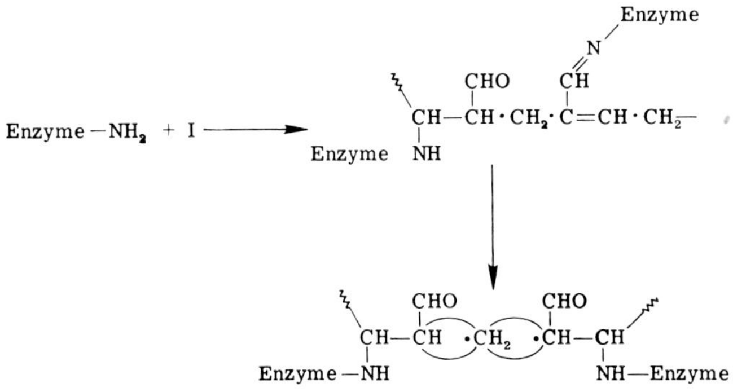

$$Enzyme{-}NH_2 + I \longrightarrow$$

(II)

It was suggested that the smallest and most common cross-link would be with five carbon atoms in a chain between the two nitrogen atoms of amino groups (as shown in II).

Lubig (1972) studied the reaction of glutaraldehyde with simple alkyl amines as model compounds. In the reaction mixture a polyether consisting of 2,6-dihydroxytetrahydropyran and *N*-alkyl-2,6-dihydroxypiperidine was isolated. It was suggested that the formation of such an ether is due to the reaction of the cyclic form of glutaraldehyde with the amine, resulting in an unstable inter-mediate, *N*-alkyl-2,6-dihydroxypiperidine which, upon losing water, would yield *N*-alkyldihydropyridine. Condensation of this with 2,6-dihydroxytetrahydropyran would give the following product:

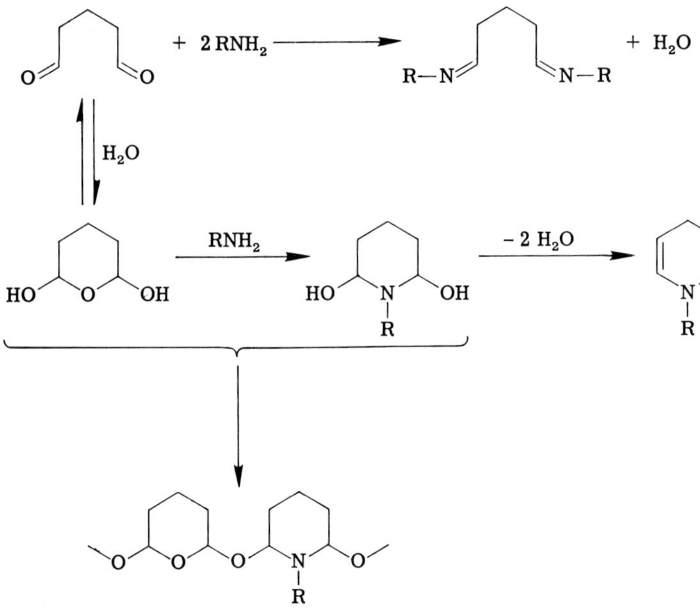

Hardy *et al.* (1976a) utilized 6-amino hexanoic acid as a model for lysinyl residues and proposed a structure for the reaction product. The initial product is

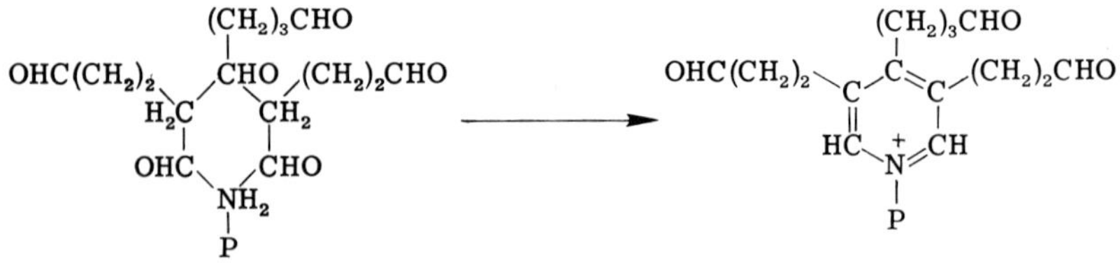

formed in the protein from a lysinyl residue, and three molecules of glutaraldehyde condense to form the heterocyclic ring compound as follows: The first
step involves the amine-to-aldehyde addition reaction, which is followed by
aldehyde condensation reactions. At a certain stage in the process, an oxidation
step is included to arrive at the pyridinium ring. Additional reactions of the same
type yield cross-linked products of various types, two of which are shown here
(A and B). Somewhat similar products of the interaction between glutaraldehyde
and proteins have been reported by Korn *et al.* (1972).

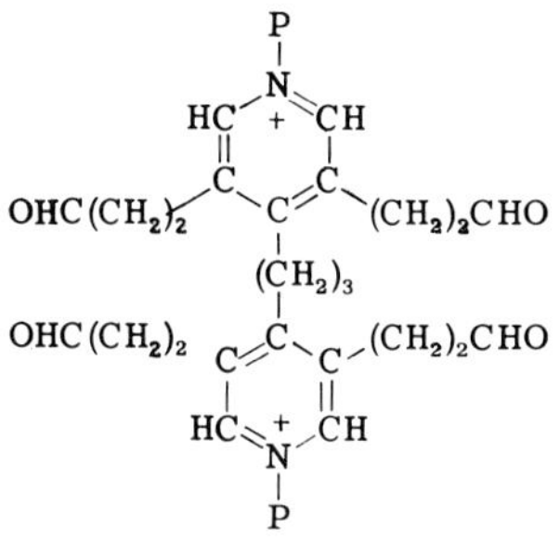

(A) 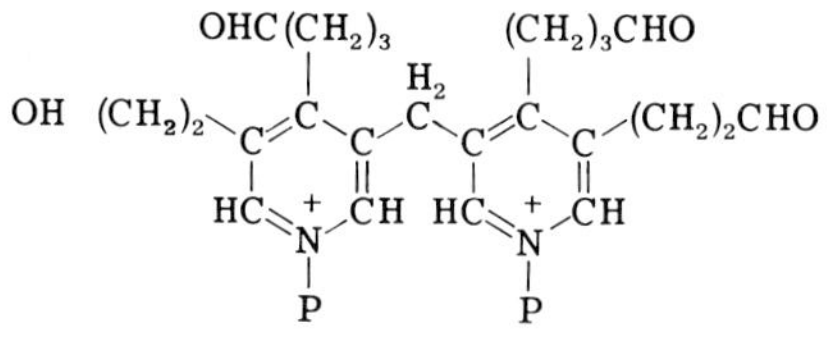   (B)

The first cross-linked entity, "anabilysine," 3-(2-piperidyl) pyridinium has
been isolated from glutaraldehyde-treated proteins (Hardy *et al.*, 1977):

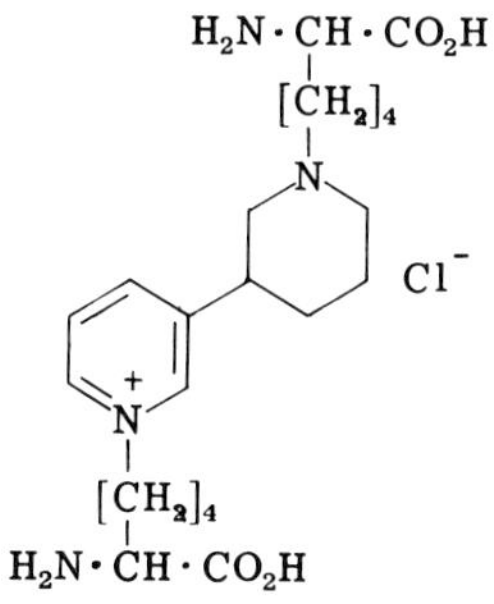

(F)

The isolation of this entity is thought to be evidence for the presence of cross-linkages of type (G), which could arise by internal oxidation-reduction of the isomeric
cross-linkages (H) derived from two lysine side chains and two molecules of
glutaraldehyde. However, the cross-linkages of type (G) are not the only ones
present in glutaraldehyde-treated proteins.

On the basis of model experiments it was suggested by Hardy *et al.* (1976b)

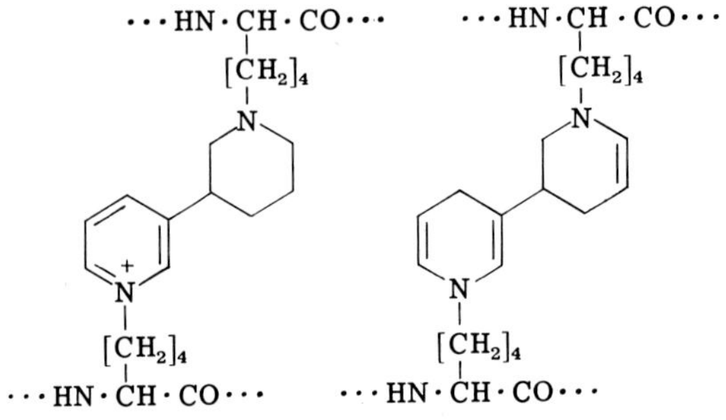

$$\cdots HN \cdot CH \cdot CO \cdots$$
$$[CH_2]_4$$

$$\cdots HN \cdot CH \cdot CO \cdots$$
$$[CH_2]_4$$

$$[CH_2]_4$$
$$\cdots HN \cdot CH \cdot CO \cdots$$

$$[CH_2]_4$$
$$\cdots HN \cdot CH \cdot CO \cdots$$

(G)   (H)

that the cross-linking action of glutaraldehyde on proteins is due to the formation of structures such as:

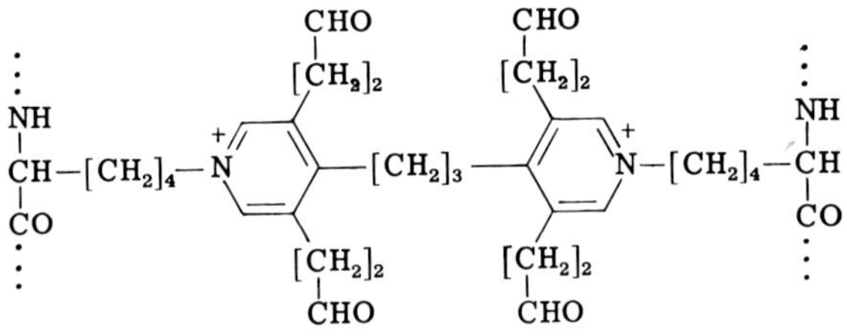

$$CHO$$
$$[CH_2]_2$$

$$CHO$$
$$[CH_2]_2$$

$$NH$$
$$CH-[CH_2]_4-N$$
$$CO$$

$$-[CH_2]_3-$$

$$N-[CH_2]_4-CH$$
$$NH$$
$$CO$$

$$[CH_2]_2$$
$$CHO$$

$$[CH_2]_2$$
$$CHO$$

(C)

The chromophore at ~265 nm in the reaction products of glutaraldehyde with proteins is ascribed to the quaternary pyridinium compound present in system (C). The pyridinium salt has been isolated:

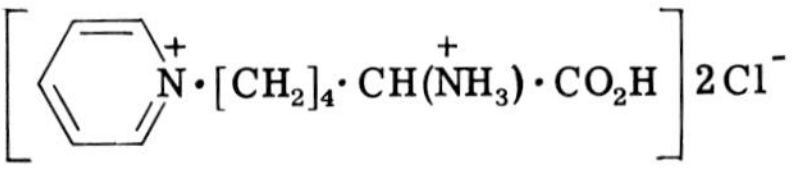

$$\left[\overset{+}{N} \cdot [CH_2]_4 \cdot CH(\overset{+}{N}H_3) \cdot CO_2H \right] 2Cl^-$$

(D)

This simple compound is formed by the oxidative cyclization (in the unhydrolyzed protein) of the Schiff's base:

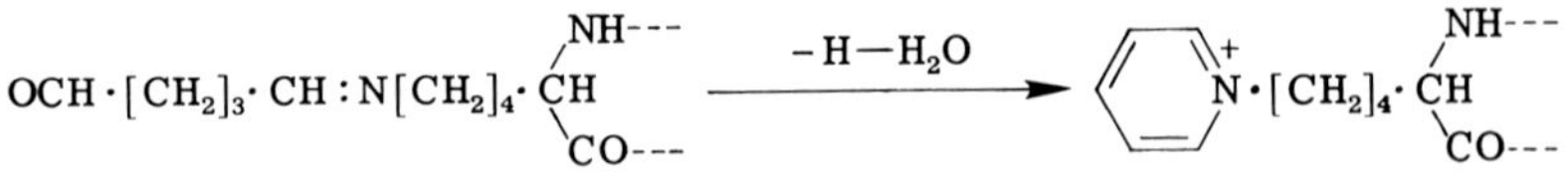

$$OCH \cdot [CH_2]_3 \cdot CH : N[CH_2]_4 \cdot CH \overset{NH---}{\underset{CO---}{}}$$

$$\xrightarrow{-H-H_2O}$$

$$\overset{+}{N} \cdot [CH_2]_4 \cdot CH \overset{NH---}{\underset{CO---}{}}$$

(E)   (C)

The Schiff's base is the first step on the way to more complex structures such as (C), by further reaction with lysine side chains and glutaraldehyde molecules.

Although the mechanism of cross-linking and the structure of reaction products resulting from glutaraldehyde interaction with proteins have been proposed by Hardy *et al.* (1976a,b, 1977) and Richards and Knowles (1968), additional studies are needed to understand the process of protein cross-linking, which is of vital importance in fixation for electron microscopy. The need for further investigation is apparent, since the results of the studies discussed here differ in important aspects.

Reaction with Lipids

The reactions between glutaraldehyde and lipids have not been extensively explored. Using the spin labeling technique, the fluidity of phospholipid bilayers and neuron membranes exposed to glutaraldehyde was studied by Jost *et al.* (1973). They found that glutaraldehyde reduced motion of maleimide spin labels covalently attached to protein, but had no effect on orientation and mobility in fluid bilayer regions. These observations indicate that glutaraldehyde does not prevent the potential loss in phospholipid bilayer. According to Levy *et al.* (1965), glutaraldehyde does not react with lipids in the brain and does not change their solubility characteristics during chloroform-methanol extraction. According to Korn and Weisman (1966), fixation of amoeba with glutaraldehyde causes no apparent change in the lipids extracted from this organism by organic solvents.

However, many studies indicate that glutaraldehyde probably reacts with those phospholipids containing free amino groups. A few examples will suffice. That glutaraldehyde reacts with phosphatidylserine and phosphatidylethanolamine has been indicated (Roozemond, 1969; Wood, 1973). Studies of the effect of glutaraldehyde fixation on the extraction of lipids from mouse brain, kidney, and liver indicated that phosphatidylethanolamine is cross-linked with protein and retained in the tissue; in the case of brain, the cross-linking is only partial (Gigg and Payne, 1969). Fixation of rod and cone membranes with glutaraldehyde rendered 38% of the phospholipids unextractable with chloroform–methanol (Nir and Hall, 1974). L'Hermite and Israel (1969) indicated the reactivity of glutaraldehyde with cerebroside, phosphatide, and proteolipid components of rat myelin. Preliminary studies by the author indicate that glutaraldehyde renders some of the phospholipids unextractable during dehydration, and very little phospholipids are extracted by this fixative. Glutaraldehyde could conjugate a phospholipid molecule to a protein.

A probable explanation for the retention of some lipids may be that certain phospholipids are fixed to proteins by the cross-linking action of glutaraldehyde involving free amino groups of proteins and phospholipids. It is possible that one end of a glutaraldehyde molecule reacts with the amino groups in a phospholipid and the other end finds one of the many amino groups available in proteins (Wood, 1973). Changes occur in the extraction of lipids on alteration of proteins by glutaraldehyde. Glutaraldehyde-fixed mitochondrial fractions dehydrated for electron microscopy retain more lipid phosphorus and cholesterol than do myelin fractions (Wood, 1973). This difference could be due to relatively low concentrations of protein in myelin, and thus less cross-linking by glutaraldehyde.

Glutaraldehyde does not seem to retain phosphatidylcholine (Roozemond, 1969), which may be the reason for the diffusion of catalase from peroxisomes fixed with glutaraldehyde. It is known that peroxisome membranes contain ~55% phosphatidylcholine (Donaldson et al., 1972), and thus the lability of this membrane in glutaraldehyde-fixed tissues. It should be noted that the leakage of enzymes from organelles, such as peroxisomes and lysosomes, may or may not be accompanied by any visible damage to their membranes. While interpreting electron micrographs, however, the selective retention of certain lipids should be kept in mind. Ljubešić (1970) noted osmiophilia of some plant cells and their plastids after fixation with aldehydes.

With the exception of certain phospholipids, the reaction of most lipids in the tissue with glutaraldehyde is slight. Even if lipids were retained completely during fixation in glutaraldehyde, most of them would be completely extracted during dehydration and embedding procedures. When glutaraldehyde-fixed cells are not osmicated prior to dehydration, lipoprotein membranes have negative images; osmication after dehydration does not help. Glutaraldehyde may cause some phospholipids to pass into solution, causing myelin-like figures to form when OsO_4 reacts with these lipoproteins during postfixation. These artifactual myelin figures are more abundant in large tissue blocks. Formation of these figures can be almost eliminated by adding $CaCl_2$ (1–3 mM) to the glutaraldehyde fixative. Addition of $CaCl_2$ also minimizes the loss of lipids during dehydration (Mitchell, 1969) and improves the preservation of mitochondria (Busson-Mabillot, 1971).

Reaction with Nucleic Acids

Only scanty information is available on the possible interactions between glutaraldehyde and nucleic acids, although in recent years some information on the use of this dialdehyde as a probe of chemical structure of nucleic acids has become available. In studies of the reactions of aldehydes with nucleic acids, however, formaldehyde is used more frequently (see p. 134).

Electron micrographs of eukaryotic nuclei fixed with glutaraldehyde show that some nucleic acids are retained in the specimen. Because dialdehyde-treated DNA shows considerable resistance to deoxyribonuclease, it is assumed that dialdehydes react with the amino groups of cytidine and guanine (Brooks and Klamerth, 1968). Since after treatment with glutaraldehyde, purified DNA can reduce the silver nitrate–methenamine solution, it has been assumed that this dialdehyde may react with DNA (Thiéry, see Millonig and Marinozzi, 1968).

However, the association of proteins with nucleic acids of eukaryotic nuclei cannot be disregarded in explaining the "retention" of nucleic acids. It is known that various polyamines and histones are associated with nucleic acids. [Elgin *et al.* (1971) have reviewed the biology and chemistry of chromosomal proteins.] The primary reason for "retention" of DNA seems to be cross-linking of the associated proteins in eukaryotic nuclei. Fixation of nucleoproteins occurs through the protein moiety, and nucleic acids are trapped in surrounding cross-linked protein.

Glutaraldehyde does not fix DNA of noneukaryotic nuclei (in blue-green algae, bacteria, bacteriophages, mitochondria, chloroplasts, and dinoflagellates) because either it does not contain an adequate amount of protein or protein is rapidly lost during the initial stage of fixation. In the absence of cross-linking, DNA in these structures aggregates randomly during dehydration (E. Kellenberger, personal communication). Thus, the DNA of noneukaryotic nuclei differs widely in its appearance in the electron micrographs. On the other hand, eukaryotic nuclei show only very minor differences in their appearance. That fixation and dehydration lead to a fundamentally different behavior of chromatin when compared to all noneukaryotic DNA-containing structures suggests basic differences in the amount and/or binding capacity of proteins of eukaryotic nuclei and those of noneukaryotic DNA-containing structures. It has been found that buffered or unbuffered glutaraldehyde does not gel DNA solutions even in the presence of Ca and amino acids (E. Kellenberger, personal communication). On the other hand, solutions of DNA are easily gelled by buffered $KMnO_4$, whereas they are gelled by buffered OsO_4 only in the presence of Ca and amino acids.

In vitro studies by Hopwood (1975) indicate that at temperatures up to 64°C no reaction occurs between native DNA and glutaraldehyde. At temperatures above 75°C, the reaction follows pseudo-first-order kinetics, proceeding more rapidly at higher temperatures. Similar reactions occur between the dialdehyde and RNA except that they start above 45°C. The probable reason for the lack of reaction of glutaraldehyde with nucleic acids at low temperatures and the reaction at higher temperatures is that the nucleic acid tertiary structure is maintained by hydrogen bonds that are weakened or broken only at higher temperatures. Doty *et al.* (1959) indicate that when nucleic acids are heated, hydrogen bonds are broken in direct proportion to the increase in temperature. Presumably, the disruption of hydrogen bonds facilitates the availability of purine and pyrimidine bases for reaction with glutaraldehyde (Hopwood, 1975). The ability of glutaraldehyde to react with nucleic acids at higher temperatures may be responsible for a limited interaction between the glutaraldehyde and nucleic acids in the specimen during polymerization of the embedding medium, which is carried out usually at ~60°C.

Studies by Langenberg (1980) indicate that glutaraldehyde does not cross-link and precipitate RNA even at RNA concentrations as high as 20%. However, this aldehyde does react rapidly with available bases in single-stranded RNA of brome mosaic virus and other plant viruses; the rapid reaction is followed by a slower reaction that may take several hours. This reaction is revealed by loss of infectivity by the virus.

To what extent the information derived from *in vitro* studies can be applied to *in situ* nucleic acids is uncertain, for the situation is infinitely more complex in tissue specimens compared with that in models. Although cross-linking of proteins with glutaraldehyde is well established, what role these glutarated compounds play in the stabilization of nucleic acids is not known. Thus, a definitive statement on the reaction of glutaraldehyde with RNA and DNA in tissue specimen during fixation cannot be made.

Reaction with Carbohydrates

Some carbohydrates, especially glycogen, are retained in the aldehyde-fixed tissue. Approximately 40–65% of the total glycogen is retained in tissues (e.g., liver and heart) after fixation with glutaraldehyde. Glycogen in relatively thick sections of glutaraldehyde-fixed tissue can be stained with ammoniacal silver solution, without pretreatment with periodic acid (Millonig and Marinozzi, 1968). In this reaction the silver solution brings out glycogen, presumably by reacting with the free aldehyde groups of glycogen-bound glutaraldehyde. Glutaraldehyde most likely reacts with polyhydroxyl compounds (e.g., pentaerythritol) to form polymers. Similar reactions may be responsible for cross-linking mucopolysaccharides:

$$HOCH_2-\underset{\underset{CH_2OH}{|}}{\overset{\overset{CH_2OH}{|}}{C}}-CH_2OH \longrightarrow \left[HC\underset{O-C}{\overset{O-C}{<}}\underset{\underset{H_2}{C}}{\overset{\overset{H_2}{C}}{<}}\underset{C-O}{\overset{C-O}{>}}CH \cdot CH_2 \cdot CH_2 \cdot CH_2 \right]_n$$

Additional information is needed to explain the reaction of glutaraldehyde with carbohydrates.

Effects of Aldehydes on Tissue Physiology

In order to understand the function, it is essential to relate it to the exact structure of the tissue. Moreover, the proper interpretation of structure is dependent on knowledge of physiological alterations in the tissue caused by the fixative. In other words, examination of the structure of fixed tissues is preceded by modifications of function of the tissue. It is apparent therefore that fixation brings about changes in the function, which, in turn, may alter the structure of the tissue. Although very little information on this subject is available, a few comments are in order.

On exposure to an aldehyde, not all chemical and physical properties of the living material are destroyed, although the function of the living material is modified in many ways. Certainly many enzymatic functions in the cytoplasm depend on the tertiary structure of various proteins, and protein structure is invariably altered by the action of aldehydes. This type of enzymatic inactivation would be due not to membrane effects, but to direct effect on the enzyme protein. Another major change involves interference with ion movements and membrane permeability. The following discussion deals primarily with changes in cell membrane permeabilities caused by aldehydes.

Since aldehydes are capable of binding to positive amino groups of proteins, fixation with these reagents results in increased fixed negative charge (by release of protons from the proteins) in the cell membrane as well as in the cytoplasm. This increase in the negative charge in the pores is expected to suppress anion movement and promote cation movement. Concomitantly, aldehydes probably cause the pores within the cell membrane to increase their dimensions, which might also be expected to facilitate the movement of small ions at least. However, there is considerable evidence to indicate that, in general, after fixation with aldehydes the movement of certain ions through the cell membrane is suppressed.

Reduction of cell membrane permeability to sodium and potassium after fixation with aldehyde has been demonstrated (Fozzard and Dominguez, 1969). According to Bone and Denton (1971), after fixation with aldehydes cell membranes seem to be still impermeable to Na^+ and K^+. Studies of *Limnaea* eggs also show that ions are lost from the cell only very slowly after aldehyde fixation (Elbers, 1966). The suppression of Ca^{2+} uptake by isolated vesicles of the sarcoplasmic reticulum of rabbit skeletal muscle after aldehyde fixation is also known (Sommer and Hasselbach, 1967). This suppression of ion movement may be explained as follows.

There is evidence to indicate that aldehydes in concentrations usually employed for tissue fixation produce cell membrane depolarization by preventing repolarization after an action potential. The prolongation of the action potential as a result of aldehyde treatment has been demonstrated in cardiac Purkinje fibers (Fozzard and Dominguez, 1969), skeletal muscle, and crayfish nerve fibers (Shrager *et al.*, 1969). This depolarization is accompanied by increased membrane resistance to certain ions. Aldehydes may also act by causing a loss of specificity of certain ion conductance in favor of others. The reduction in total cell membrane conductance could also result from cross-links between proteins in the membrane introduced by the aldehyde.

Alterations in ion permeability and movement could change the electrical properties of the tissue. These alterations are likely to interfere with enzymatic activities. The suppression of extra ATPase activity in the isolated vesicles of sarcoplasmic reticulum of rabbit skeletal muscle after aldehyde treatment (Sommer and Hasselbach, 1967) could be due to alterations in the ion permeability of membranes. In this connection one cannot disregard the blocking of substrate and carrier sites. The net effect of the preceding physiological changes and others on the tissue structure are not known. It is apparent that studies on the structure–function correlation are needed. In particular, the effects of fixatives on enzymes involved in ion transport deserve consideration.

Osmolality

As stated earlier, it is generally recognized that the osmolality of the fixative exerts a significant influence on the quality of preservation of tissue fine structure. In the total osmolality of the fixative solution the concentration of the fixative vehicle (buffer salts and additives such as sucrose) seems to be more important than that of glutaraldehyde, although this aldehyde does exert considerable osmotic effect on certain tissues such as kidney. It has already been emphasized earlier that glutaraldehyde concentration can be changed over a wide range with relatively little change in the morphology of fine structure.

Strong evidence for the importance of the concentrations of substances in the fixative vehicle has been shown (Fahimi and Drochmans, 1965b; Maunsbach, 1966). As an example, liver and muscle tissues were best fixed with moderately hypertonic fixatives (400–500 mosmols) containing 2% glutaraldehyde; whereas strongly hypertonic fixatives (570 mosmols) and isotonic fixatives, both containing 2% glutaraldehyde, gave poor fixation in that the extraction of cellular constituents was considerable. Hampton (1965) reported that the concentration of sucrose in glutaraldehyde bears importantly on the morphology of fixed mouse Paneth cell granules. He demonstrated that these cells were preserved much better with glutaraldehyde containing 1% sucrose than with glutaraldehyde without the nonelectrolyte. Using the small-angle X-ray diffraction technique, Moretz et al. (1969) also indicated that the addition of sucrose to purified glutaraldehyde solution in phosphate-buffered saline is significant in obtaining a normal X-ray pattern of nerve.

Osmolarity of Glutaraldehyde

The presence of osmotically active impurities results in osmolarities generally much higher than the theoretical values. For example, osmolalities as high as 570 mosmols of a 3% solution of commercially supplied glutaraldehyde have been recorded (Maser et al., 1967), whereas the osmolality of a 3% solution of pure glutaraldehyde is only 300 mosmols. Since even the latter solution when buffered is hypertonic to a majority of biological fluids bathing the tissue, it is obvious that the buffered fixatives containing the former solution would be very hypertonic. The deleterious effect of hypertonic fixatives has already been discussed earlier. Moreover, the impurities may themselves affect the quality of fixation. Enzyme activity recovery, for instance, is inversely related to impurity content. Thus, to obtain optimum fixation and maximum retention of enzymes, glutaraldehyde should be ''pure'' and its osmolarity known. In other words, the quality of fixation can be controlled more effectively when the amount of the active component in glutaraldehyde is known.

Polymerization of glutaraldehyde, on the other hand, results in decreased osmolarities. When monomeric glutaraldehyde is polymerized to dimers or trimers, there is a decrease in the osmolarity. This is expected because the osmolarity of a solution is a measure of the total number of solute particles in solution per unit of solvent. When two or more individual monomeric molecules form a single larger dimer or trimer, the number of separate monomeric molecules is reduced and the osmolarity decreases. Osmolalities of aqueous 2% commercial glutaraldehyde solution from different hatches vary from 190 to 280 mosmols.

Average osmolalities of phosphate-buffered glutaraldehyde formulations of different concentrations follow. The commercially available glutaraldehyde was purified by charcoal treatment, and the osmolalities were measured with a commercial osmometer (Chambers *et al.*, 1968).

	Milliosmols	pH
Phosphate buffer (0.1 mol/l)	230	7.4
1.2% glutaraldehyde	370	7.2–7.3
2.3% glutaraldehyde	490	7.2–7.3
4.0% glutaraldehyde	685	7.1–7.3

A slightly hypertonic solution (400–450 mosmols) of buffered glutaraldehyde is recommended for general fixation. For mammalian cells and tissues the recommended formulation is 2% glutaraldehyde in 0.1 mol/l cacodylate buffer containing 0.1 mol/l sucrose (pH 7.2) with total and vehicle osmolalities of 500 and 300 mosmols, respectively.

As stated earlier, even after fixation with glutaraldehyde and other aldehydes the osmotic properties of membranes and other metabolic activities remain partly preserved; thus membranes remain sensitive to a change in the osmotic pressure of the washing and dehydration solutions. Cytological and biochemical studies lead one to question the completion of fixation at the molecular level by glutaraldehyde in intact tissues. Precautions must, therefore, be taken to minimize the differences between the osmolarity of washing solutions and that of the fixative. The osmolarity of washing solutions can be increased by adding either electrolytes or nonelectrolytes.

Temperature

Since the reaction between glutaraldehyde and cellular reactive groups is increased at high temperatures, the rate of penetration and fixation by glutaraldehyde is enhanced at room temperature. However, to minimize the extraction of cellular constituents due to autolysis, low temperatures are preferred. Fixation with glutaraldehyde at low temperatures also reduces the shrinkage of mitochondria, granularity of cytoplasm, and volume changes. For the preservation of very labile enzymes, fixation by immersion or vascular perfusion should be carried out in the cold. Low temperatures minimize the formation of artifacts especially in enzyme studies due to excessive polymerization of glutaraldehyde when it has been stored at subfreezing temperatures. Treatment with glutaraldehyde for 2 hr at 0–4°C is generally preferred for routine fixation of plant and animal tissues.

However, glutaraldehyde can be used in many tissues at temperatures ranging from 0 to 25°C with little apparent difference in the appearance of the fine structure. Sometimes the temperature of fixation can be critical. For example, labile structures such as certain microtubules may be lost because of rearrangement when glutaraladehyde is employed in the cold. Certain types of dense specimens may not allow adequate penetration by glutaraldehyde at cold temperatures. In such cases it is desirable to carry out fixation at room temperature. Alternatively, fixation can be accomplished at room temperature following a brief preliminary fixation in the cold. In vascular perfusion, glutaraldehyde should not be used below body temperature: otherwise it may cause vasoconstriction. (Temperature of fixation is also discussed on p. 60.)

Concentration

The process of protein cross-linking is affected by the concentration of the glutaraldehyde solution. Although changes in the concentration of the dialdehyde are always accompanied by changes in its osmolarity, this change within a reasonable limit may not affect the quality of tissue preservation. However, the quality of fixation of isolated cells is readily influenced by changes in the concentration of the fixative. Slow penetration by glutaraldehyde might prompt the investigator to use higher concentrations. This practice, however, is not desirable because the use of excessively high concentrations results in cell damage. The most desirable concentrations of glutaraldehyde solutions including the vehicle are hypertonic, but in these solutions the concentration of glutaraldehyde is usually hypotonic. Generally, high concentrations (e.g., 37%) cause an excessive shrinkage, whereas low concentrations (e.g., 0.5%) result in severe extraction of cell constituents from the tissue.

Although glutaraldehyde can be used safely in a relatively wide range of concentrations, 1.5–4.0% solutions are recommended for fixing plant and animal tissues. When it is necessary to fix relatively large tissue blocks, the peripheral zone (20–150 μm) of the block is best fixed with moderately low concentrations of aldehydes ($\sim$2%), whereas the core region requires higher concentrations ($\sim$8%) for satisfactory fixation.

Higher concentrations of glutaraldehyde have proved useful for obtaining superior fixation of certain tissues. A few examples are given. Lung tissue can be fixed with 8% aged, polymerized glutaraldehyde in 0.1 mol/l phosphate buffer (pH 7.2) for 2 hr. This fixative solution has an osmolality of $\sim$1000 mosmols. Nervous tissues, including CNS, can be fixed with 6% glutaraldehyde in 0.1 mol/l cacodylate buffer having an osmolality of 900 mosmols. As high as 19% glutaraldehyde has been used as an initial perfusate for preserving CNS prior to a conventional perfusate of a low concentration (Schultz and Case, 1970). This approach seems to minimize considerably the production of artifacts such as myelin figures and vacuolated mitochondria. Relatively high concentrations of glutaraldehyde (6%) seem to be desirable for fixing plant tissues. For example, *Avena* coleoptile is preserved better with 6% glutaraldehyde than with solutions of low concentrations.

It should be noted that high concentrations of glutaraldehyde may increase the surface of cristae membranes, but no such effect is seen on the membranes of rough endoplasmic reticulum. This would be consistent with the explanation that more metal ions are absorbed during staining by the inner membranes of mitochondria.

An interesting approach was utilized by Thornthwaite *et al.* (1978) to determine optimum glutaraldehyde concentration. They used electronic cell volume analysis to determine ideal concentration of glutaraldehyde for fixing nucleated mammalian cells. In this study high-resolution electronic cell volume spectra of glutaraldehyde-fixed cells were employed to determine the glutaraldehyde concentration that produced spectra most closely resembling those of the unfixed cells. The optimum concentrations of glutaraldehyde in 0.05 mol/l cacodylate buffer (pH 7.4) fixing mastocytoma tumor cells, human lymphocytes, and human granulocytes and monocytes were 3.8%, 4.9%, and 4%, respectively.

On the other hand, certain specimens are fixed better by using low concentrations of glutaraldehyde. Superior fixation of single cells is obtained with low concentrations of glutaraldehyde (0.2–1.0%). For example, preliminary fixation with a low concentration of glutaraldehyde (0.1%) followed by a second fixation with 3% glutaraldehyde seems to preserve platelets in suspension better; platelet fixation is thought to occur without coagulation of plasma proteins by using this approach. Relatively low concentrations of glutaraldehyde (1–2%) are useful for enzyme localization. One should be aware that too low a concentration of glutaraldehyde may increase the extent of artifacts, including the number of myelin figures in tissue specimens. It is interesting to note that similar artifacts may be introduced by purified glutaraldehyde.

pH

The importance of pH in fixation has been emphasized earlier. Among all the factors that influence the interaction of glutaraldehyde with proteins, pH is considered to be the most important in obtaining the maximum binding of aldehyde groups with proteins. An increase in pH generally results in an increase in the binding capacity of glutaraldehyde. For example, the maximum uptake of glutaraldehyde by collagen and cross-linking of collagen by glutaraldehyde occurs at pH 8.0 The guanidino groups of arginine may react only at pHs higher than 9.0. It has been shown that glutaraldehyde tans more rapidly at higher pH levels (Bowes and Cater, 1966) and that the amino groups of ovalbumin and bovine serum albumin react more rapidly with this dialdehyde with increasing hydrogen ion concentration (Habeeb and Hiramoto, 1968). The amount of precipitate formed between glutaraldehyde and noradrenalin increases at higher pH levels (Coupland and Hopwood, 1966).

A higher pH (8.0) seems to be more effective during prefixation of plant tissues with glutaraldehyde. Improved preservation of animal tissues by raising the pH from 7.0 to 8.0 has also been reported (Peracchia and Mittler, 1972a). However, at pHs higher than 7.5, glutaraldehyde tends to polymerize rapidly. This excessive polymerization may introduce artifacts. Therefore, fixation at higher pHs for routine electron microscopy cannot be recommended until more information on the mechanism of protein cross-linking by glutaraldehyde and the role of pH in it under fixation conditions becomes available.

Although a low pH in fixing biological specimens is not generally acceptable, some workers have obtained successful fixation of certain tissues at low pHs. For example, Moss (1966) obtained satisfactory preservation of a plant tissue at pH 4.0. He obtained the maximum binding of amino groups in anthers with 4% glutaraldehyde at pH 4.0. However, it was recommended that the fixation carried out at a low pH should be followed by a more alkaline glutaraldehyde fixation. Also, in the case of certain animal tissues, a lower pH (6.9–7.0) seems to produce better fixation (Busson-Mabillot, 1971). For routine electron microscopy, a pH of 6.9–7.5 can be used, depending on the type of the specimen. (For additional information on the role of pH in fixation, see p. 54.)

Fixation Method with Glutaraldehyde at Higher pHs

Specimens are fixed by immersion in 3–6% glutaraldehyde in a buffer (pH 7.0) for 1 hr at room temperature. The pH of the glutaraldehyde solution is then raised to 8.0 in three steps of ~0.3 pH unit each, at 30–60 min intervals by adding drops of 1 N NaOH while stirring. While approaching pH 8.0, 0.1 mol/l NaOH is used to avoid too rapid a change in the pH. The specimens are allowed to remain in glutaraldehyde at pH 8.0 for 30–60 min. For further details see Peracchia and Mittler (1972a).

Rate of Penetration

The rate of glutaraldehyde penetration is dependent primarily on the type of the tissue specimen, ambient temperature, and concentration. Penetration at room temperature is definitely faster than in the cold. Glutaraldehyde (2%) in a buffer (200 mosmols) penetrates into soft animal tissues (e.g., liver) ~0.7 mm in 3 hr at room temperature, while good fixation reaches to a depth of only ~0.5 mm in the same period of time. After 24 hr, glutaraldehyde penetrates to a depth of ~1.5 mm, while good fixation reaches to a depth of ~1.0 mm. According to Chambers *et al.* (1968), however, the maximum penetration of human liver by 4% glutaraldehyde in 24 hr at room temperature and in the cold was 4.5 mm and 2.5 mm, respectively. Hopwood (1967a, 1972) and Ericsson and Biberfeld (1967) have also presented data on the penetration rate of glutaraldehyde into rat liver. A mixture of glutaraldehyde (2%) and formaldehyde (2%) in the buffer (200 mosmols) penetrates human liver to depths of 2.0, 2.5, and 5.0 mm in 4, 12, and 24 hr, respectively (McDowell and Trump, 1976).

Even similar tissue types obtained from different species react differently to glutaraldehyde diffusion. Chambers *et al.* (1968) showed that 4% glutaraldehyde, for example, penetrated rabbit liver and human liver 1.5 mm and 3.0 mm, respectively, in 9 hr at room temperature. The rapid penetration into the human liver is presumably related to more efficient conduction of the fixative by its vessels.

It is apparent from the preceding data that to obtain uniform fixation of all cells in a tissue specimen within a period of 2 hr, the size of the specimen should not exceed 0.5 mm on a side. Prolonged fixation would not necessarily improve the quality of fixation because penetration and fixation should be completed prior to the onset of autolytic changes.

The depth of fixation is considered to be the same as the depth of the penetration by glutaraldehyde. The penetration of the dialdehyde into the tissue is indicated by a pale-yellow color (which is primarily due to the formation of Schiff's bases when the dialdehyde reacts with basic amino acids) and firm appearance. Since glutaraldehyde introduces Schiff positivity to the tissue fixed in it, the presence of aldehyde groups in the region showing pale-yellow color can be confirmed by its pink staining with Schiff's reagent.

The slow penetration by glutaraldehyde becomes a serious problem in the fixation of dense specimens (e.g., seeds) and plant cells having relatively impervious walls (e.g., yeast). This problem can, however, be solved by using glutaraldehyde in combination with rapid penetrants such as formaldehyde or acrolein. A mixture of glutaraldehyde (2%) and formaldehyde (2%) was employed to preserve grooves (cleftlike invaginations) on the surface of plasma membrane of yeast cells (Ghosh, 1971). These grooves did not preserve when the cells were fixed with glutaraldehyde alone.

The rate of penetration can also be increased by adding DMSO to the fixative, although its effects on enzymes and ultrastructure are not fully known. Glutaraldehyde solutions containing 2.0–10.0% DMSO are recommended. It has been demonstrated that DMSO enhances cell permeability of animal tissues (Gander and Moppert, 1969), plant tissues, and microorganisms. Excellent preservation of embryonic tissues has been obtained by employing a mixture of glutaraldehyde (3%), formaldehyde (2%), acrolein (1%), and DMSO (2.5%) (Kalt and Tandler, 1971). DMSO when added to certain incubation media also enhances staining of enzymes (Reiss, 1971). Also, the addition of DMSO to the fixative is known to preserve frozen tissues better (Etherton and Botham, 1970). The advantages as well as disadvantages of using DMSO for preserving both ultrastructure and enzymatic activity will be discussed later.

Shrinkage Caused by Aldehydes

Aldehyde fixative solutions used routinely for electron microscopy cause shrinkage of the cell. The degree of shrinkage in cytoplasm may be different than that in the nucleus. Glutaraldehyde causes more shrinkage in nuclei than that caused by OsO_4 or formaldehyde, but just the opposite is true in the case of cytoplasm. Thus, while comparing subcellular dimensions in fixed cells, the possibility of different degrees of change of dimensions induced in different parts of the same cell by the fixative cannot be ignored.

As stated earlier, aldehyde fixatives used routinely are hypertonic. Fixatives having a higher osmolarity generally cause greater shrinkage than that caused by fixatives with lower osmolarity. Various types of tissues show different degrees of shrinkage after fixation with glutaraldehyde. The wide range of shrinkage shown by the following examples is also due to differences in the preparatory procedures employed in different laboratories. Specimens may shrink ~5% in linear dimensions as compared with their size before fixation (Weibel and Knight, 1964). Rat brain tissue fixed with 4% glutaraldehyde and embedded in Epon showed a shrinkage of 9% (Hillman and Deutsch, 1978), whereas mouse ova shrank 8% compared with live cells (Konwiński et al., 1974). Glutaraldehyde (4%) caused a shrinkage of ~6% of rat liver in 18 hr at 4°C (Hopwood, 1967b), while 2% glutaraldehyde caused 5–10% shrinkage of calf erythrocytes (Carstensen et al., 1971).

Isolated cells usually shrink more as a result of fixation than intact tissues. Glutaraldehyde reduced the surface area of lymphocyte nuclei by 4–6% (Maul et *al.*, 1972) and caused shrinkage of nuclear pores (Willison and Rajaraman, 1977). Fixation with glutaraldehyde causes shrinkage of chloroplasts (Diers and Schieren, 1972). Shrinkage caused by aldehydes is usually neutralized by subsequent treatments such as postfixation in OsO_4.

Heated Glutaraldehyde

The use of heated glutaraldehyde is based on the concept that heating accelerates polymerization of glutaraldehyde, resulting in the formation of various polymers, especially α,β -unsaturated aldehydes. When glutaraldehyde solutions are heated above 90°C, their absorption peak intensifies at 235 nm, which is characteristic of the dimer. As discussed earlier, the dimer may be primarily responsible for cross-linking proteins, and thus preserving the fine structure. It should be noted that when heated solutions of glutaraldehyde are allowed to stand at room temperature, the intensity of absorption at 235 nm is decreased. This phenomenon, which has not as yet been explained, discourages any definite statement concerning the effect of temperature because fixation is usually carried out at room tempterature or at 4°C.

It has been shown that improved preservation of the fine structure in certain cases can be obtained by heating the glutaraldehyde during fixation (Tokin and Röhlich, 1965). Improved fixation of cerebral cortex has been obtained by heating the glutaraldehyde prior to fixation (Robertson and Schultz, 1970). Since monomeric glutaraldehyde penetrates faster, it is recommended that the heating be delayed for 20–30 min from the start of fixation (Peracchia and Mittler, 1972a). This delay is sufficient to allow the glutaraldehyde to penetrate into the tissue specimen prior to its gradual polymerization. It should be noted that using heated glutaraldehyde or heating the glutaraldehyde during fixation is not recommended for routine electron microscopy.

Fixation Method Employing Glutaraldehyde at Higher Temperatures

Specimens are fixed by immersion in 3–6% solution of glutaraldehyde in a buffer (pH 7.4) for ~30 min at room temperature. This is followed by warming the fixative solution containing the specimens on a hot plate with continuous stirring to 45°C within 5–10 min. The solution is allowed to cool at room temperature for 20–30 min. For further details the reader is referred to Peracchia and Mittler (1972a).

Limitations of Glutaraldehyde

Being an organic reagent, glutaraldehyde is unable to impart sufficient contrast or electron opacity to tissue to result in electron staining. Treatment of cells with glutaraldehyde monitored by light microscopy indicates that structural changes occur during fixation (Skaer and Whytock, 1976; Arborgh *et al.*, 1976).

A highly pleomorphic lysosomal system in chick cells is vesiculated by standard glutaraldehyde fixation (Buckley, 1973a,b). This system, however, can be stabilized in the presence of a relatively high concentration of calcium ions, but the best results are obtained with prior equilibration and fixation in the presence of calcium chloride (0.1 mol/l) and magnesium chloride (0.02 mol/l) plus sucrose (0.3 mol/l). A pleomorphic network that forms a dominant component of the living cytoplasm of plant hair cells is completely transformed into vesicles by glutaraldehyde (O'Brien *et al.*, 1973; Mersey and McCully, 1978). Glutaraldehyde causes clumping of chromatin and transforms nuclear sap into a coarse network.

Glutaraldehyde is incapable of rendering most lipids insoluble in organic solvents used during dehydration. Accordingly, glutaraldehyde-fixed tissues show cellular membranes as negative images. Longer durations of fixation with the dialdehyde may increase the production of artifactual myelin figures. The production of these figures in the CNS is probably related to delayed fixation by solutions of low concentrations of glutaraldehyde. The transformation of lipid complexes into myelin whorls in embryonic chick hepatocytes has been indicated (Curgy, 1968). It is assumed that this artifact results from the mobilization of complex lipids by glutaraldehyde followed by their molecular reorganization and staining with OsO_4. Since after the seventeenth day of embryonic development such artifacts are rarely seen, it is suggested that during the earlier stages of development, this artifact represents an abundance of available phospholipids or lipoproteins that are disorganized or transferred by glutaraldehyde.

It has been indicated that preparation of specimens for electron microscope autoradiography is accompanied by significant translocation and intercellular redistribution of radiolabeled saturated lipids, causing spurious labeling patterns (Poste *et al.*, 1978). Thus, the problem of redistribution and intercellular transfer of natural saturated phospholipids in the glutaraldehyde-fixed cells and tissues cannot be ignored, and the results of autoradiographic methods used for ultra-structural localization of lipids should be interpreted with caution.

Glutaraldehyde fixation in the cold may result in the loss or rearrangement of labile microtubules and a more dispersed pattern of ribosomes. Glutaraldehyde is unable to preserve viruses satisfactorily (e.g., wheat streak and tobacco mosaic viruses) in plant tissues.

The contraction of extracellular space in the nervous tissue during glutaral-dehyde fixation is a well-known phenomenon. Van Harreveld and Khattab (1968) have demonstrated that perfusion of cerebral cortex with glutaraldehyde caused an increase in the electrical impedance of the tissue, an accumulation of chloride into cellular elements, and a contraction of extracellular space. This transport of extracellular material is thought to be a consequence of an increase in the sodium permeability of the plasma membrane of cells that take up chloride and water during fixation with glutaraldehyde. Studies by Van Harreveld and

Fifkova (1972) indicated that glutamate is released from the intracellular into the extracellular compartment in thick retina during fixation with glutaraldehyde, and it was suggested that the action of glutamate on the plasma membrane is responsible in part for the contraction of extracellular space.

Caution is warranted in the use of glutaraldehyde for enzyme studies. Even though evidence has been presented indicating the activation of a nuclear acid phosphatase enzyme activity as a result of treatment with this dialdehyde (De Jong et al., 1967), generally glutaraldehyde is a powerful inhibitor of the activity of most enzymes and antigens. It may even inactivate horse liver alcohol dehydrogenase (Johnson, 1977), for which it acts as substrate. Glutaraldehyde also inhibits dehydrogenation of ethanol by this enzyme. Another type of artifact is formed because of the ability of glutaraldehyde to immobilize metabolites such as amino acids. Glutaraldehyde is thus capable of binding free amino acids to tissue constituents. This becomes significant in autoradiographic studies where amino acids labeled with radioisotopes are employed. Peters and Ashley (1967) have described a possible artifact caused by glutaraldehyde in the study of protein formation in the presence of labeled amino acids. On the other hand, this binding capability of glutaraldehyde can be utilized advantageously to immobilize diffusible compounds, including enzymes containing amino acid groups. In this connection it is known that this dialdehyde penetrates into the tissue before labeled amino acids are lost from the cell by diffusion.

Studies of the effect of glutaraldehyde on the levels of amino acids in rat brain indicated that, following perfusion fixation, significant increases in the levels of brain glutamic acid, alanine, valine, isoleucine, leucine, and tyrosin occurred compared with unfixed controls (Davis and Himwich, 1971). This is expected, since, through cross-linking, glutaraldehyde binds amino acids to proteins initially and then larger amounts can be extracted for the analyses. Some of the increase in the amino acid level is also attributed to the shrinkage of the tissue because the level of amino acids is based on wet tissue weight.

Fixation with glutaraldehyde may result in an apparent anomalous distribution of plasma membrane anionic sites (Grinnell et al., 1976). Two possible explanations to account for the glutaraldehyde effect are the following: (1) a direct binding may occur between the added polycationic ferritin and exposed aldehyde groups of glutaraldehyde bound to the cell surface, and (2) glutaraldehyde induces reorganization of the membrane, resulting in the appearance of new anionic sites.

Fixation with glutaraldehyde does not completely protect the specimen from the extraction of cellular materials (e.g., some proteins) during subsequent processing such as rinsing in a buffer. This loss is less at low temperatures. Some nuclear proteins, for instance, in glutaraldehyde-fixed (as well as unfixed) specimens undergo extraction by phosphate or veronal buffer. Different buffers extract different types and amounts of proteins.

Some of the problems mentioned earlier can be overcome by postfixation with OsO_4. Postfixation stabilizes the fine structure already maintained by glutaraldehyde so that it can withstand embedding in plastics. It is recognized that DNA of chromatin and general cytoplasmic details are preserved much better and more completely by double fixation than by either OsO_4 or glutaraldehyde. Postfixation with OsO_4 is an absolute requirement for electron cytochemical methods that are based on the "osmiophilic principle." In certain cases, postfixation can be carried out with $KMnO_4$ (Hayat, 1968a; deF. Webster, 1971).

Fixation with glutaraldehyde without subsequent treatment with OsO_4 is considered unsatisfactory for routine electron microscopy except for some cytochemical studies. For example, for a valid demonstration of acid phosphatase in nerve cell lipofuschin bodies; postfixation with OsO_4 should be avoided. Lipofuschin is osmiophilic, and thus it is difficult to separate lead phosphate precipitate from the reaction product of OsO_4 with the contents of lipofuschin granules (Brunk and Ericsson, 1972). In addition, longer periods of postfixation may be equivalent to an "acid rinse," which can dissolve lead phosphate precipitate (Desmet, 1962; Reale and Luciano, 1964).

Purification of Glutaraldehyde

The impurities present in glutaraldehyde can be removed by vacuum distillation, by filtering through activated charcoal, or by ion exchange. Using various methods, several workers (Clift and Cook, 1932; Fein and Harris, 1962; Fahimi and Drochmans, 1965a; Smith and Farquhar, 1966; Anderson, 1967; Hopwood, 1967b; Frigerio and Shaw, 1969) have carried out purification and determination of the concentration of glutaraldehyde with varying degrees of success. The simplest method is treatment with charcoal. However, this method yields a less pure glutaraldehyde, for charcoal is unable to remove certain impurities such as inorganic materials. When very high purity is desired, the method of choice is vacuum distillation, either at atmospheric pressure (Smith and Farquhar, 1966) or at reduced pressure (Anderson, 1967; Gillett and Gull, 1972). The concentration of glutaraldehyde in the distillate is determined with a recording spectrophotometer. Purified glutaraldehyde should show an absorption maximum in the ultraviolet at a wavelength of 280 nm; absorption at any other wavelength is caused by impurities. Since a direct linear relationship is found between concentration or osmolarity and the optical density at 280 nm, this parameter can serve to determine both the concentration and osmolarity of the distilled glutaraldehyde.

A very small and slow, but possibly significant, polymerization of purified glutaraldehyde occurs even under mildly alkaline conditions of fixation, and both animal and plant specimens catalyze the formation of polymers that absorb at 235 nm (Jones, 1974; Goff and Oster, 1974). Significant amounts of these polymers can be found in inadequately purified glutaraldehyde. Molin *et al.* (1978), indeed, found an index of 1.0–1.2 (235/280 nm) in glutaraldehyde purified by absorption with activated charcoal. Purified glutaraldehyde is polymerized faster at alkaline pH and at room temperature than at acidic pH and in the cold.

The Charcoal Method

Approximately 200 ml of commercial glutaraldehyde solution (50%) is added to 30 g of activated charcoal (Merck & Co., American Norit Co., or Fisher Sci.) in a large flask or beaker. The mixture is thoroughly shaken for ~1 hr at 4°C, and then vacuum filtered through Whatman No. 42 filter paper mounted in a Buchner funnel. The filtrate is remixed with 20% (w/v) fresh, activated charcoal and refiltered. The process is repeated at least twice; the number of washings depends on the amount of absorption at 235 nm, which shows the extent of impurities. The final yield of purified glutaraldehyde is ~20–30 ml.

It should be noted that each charcoal wash reduces both the volume of solution and its glutaraldehyde concentration. Starting with 200 ml of 25% glutaraldehyde and using 30 g of Norit Ex charcoal for each wash resulted in ~150 ml having concentration of ~22% after the first wash (Garrett *et al.,* 1972). After the second wash the volume was reduced to ~95 ml having a concentration of ~18%. Repeated washings with alkaline charcoal result in the elevation of pH of the glutaraldehyde solution. Starting with 25% glutaraldehyde with pH 3.1, after one, two, three, and four charcoal washes the pH was raised to 5.8, 6.9, 7.7, and 8.0, respectively (Trelstad, 1969). The resultant alkaline pH of the purified glutaraldehyde solution encourages its rapid polymerization. The rate of polymerization, however, can be reduced by slightly acidifying the solution by the addition of a few drops of HCl.

The Distillation Method

According to this simple method (Smith and Farquhar, 1966), the distillation is carried out at atmospheric pressure. The distillate is collected at ~100°C in 50 ml aliquots, which are monitored by measuring pH. Any sample showing a pH lower than 3.4 is discarded. Pure glutaraldehyde with a concentration of 8–12% is obtained.

Alternatively, a single-stage distillation under moderate vacuum (Anderson, 1967) yields a glutaraldehyde of equivalent purity. Approximately 250 ml of commercial glutaraldehyde are charged into a 500 ml Vigreaux distilling flask heated by an electrical heating mantle and connected to a Liebig condenser. The distillation is performed under vacuum at 15 mmHg. The temperature is raised to ~65°C and distillate is collected, which is a viscous, clear liquid. On interrupting the vacuum the distillate is immediately diluted with an equal volume of freshly boiled demineralized distilled water by slow addition of the latter (75°C) to the magnetically stirred distillate under a stream of nitrogen. It may be useful to remember that the boiling point of 25% aqueous solution of glutaraldehyde is ~101°C.

Determination of Glutaraldehyde Concentration

In spite of its wide application, the exact concentration of glutaraldehyde in the fixative solution is difficult to ascertain. The determination of changes in glutaraldehyde concentration during actual fixation process (cross-linking of proteins) is even more difficult. The availability of such information will result in understanding the cross-linking of proteins by glutaraldehyde and will help in the estimation of the remaining components (impurities) in the glutaraldehyde solution.

Several analytical procedures for monitoring changes in glutaraldehyde concentration during cross-linking of proteins have been reported. Fein and Harris (1962) determined the glutaraldehyde content in tanning solutions iodimetrically. This method requires at least 1 hr. Fahimi and Drochmans (1965a) analyzed glutaraldehyde by measuring its ultraviolet absorption at 280 nm. This procedure is difficult to apply to the analysis of fixative solutions, for it requires prior distillation, a sample free of impurities, and careful temperature control. Anderson (1967) analyzed glutaraldehyde by hydroxylamine titration; it requires ~2 hr and 11 steps. Other methods include those by Morel *et al.* (1971) and von Hesse (1973).

Frigerio and Shaw (1969) introduced a method based on the the formation of glutaraldehyde–bisulfite complex followed by iodometric titration of unreacted bisulfite. The measurement of the amount of unreacted bisulfite affords a simple, rapid, and precise procedure for the determination of glutaraldehyde concentration, in aqueous solutions or in fixative vehicles. This method is detailed in the following discussion.

One milliliter solution of 0.6–6% glutaraldehyde is transferred to a 125 ml glass-stopper Erlenmeyer flask. One milliliter of distilled water or a buffer is pipetted into a second similar flask. Approximately 20 ml of 0.25 mol/l sodium bisulfite ($NaHSO_3$) (prepared by dissolving 26 g/liter) are delivered to each flask. The reaction is allowed to continue for 5–10 min. The unreacted $NaHSO_3$ in the two flasks is titrated separately with standardized 0.1 mol/l triiodide (I_3^-). (Iodine solution is prepared by adding 25.78 g of resublimated iodine to 44 g of KI, and dissolving this mixture in ~100 ml of distilled water and diluting it to 1 liter.) The solution is turned yellow at first, but within 0.2 ml of the end point, it becomes colorless. The titration is continued until the solution again turns yellow and remains so for at least 3 min. The volume of I_3^- used is recorded. A carefully cleaned burette is necessary to deliver the iodine accurately because this solution tends to adhere to the walls.

Glutaraldehyde (GA) concentration is calculated by the following equation:

$$\% \, GA = \frac{(I) \, (V_2 - V_1) \, (100.12) \, (100)}{2S},$$

where I = the concentration of I_3^- expressed in moles/ml, V_2 = ml of iodine in the second flask containing water, V_1 = ml of iodine used in the first flask containing GA, 100.12 = mol. wt. of GA, and S = sample volume. If sample weight is used for S, the equation gives % w/w.

A simple spectrophotometric method for determining low concentrations of glutaraldehyde even in the presence of substances that absorb in the same wavelength region was introduced by Hajdu et al. (1975). The characteristic change of absorbance occurring during the reaction of glutaraldehyde with hydroxylamine, hydrazine, and methylamine allowed the development of this method. The consumption of glutaraldehyde during polymerization can be monitored by the method. Since this method is based on measuring absorbance differences, it is effective in the presence of substances (e.g., proteins) absorbing in the ultraviolet. It should be noted that at high initial absorbances, stray light effects (Mehler, 1954) should be considered.

Storage of Glutaraldehyde

As stated earlier, impurities increase spontaneously when glutaraldehyde is kept under ordinary conditions of storage. It was suggested that under these conditions, glutaraldehyde is oxidized to glutaric acid, which is the final oxidation product of this dialdehyde. Since barium glutarate is insoluble, some workers store glutaraldehyde over solid barium carbonate in order to neutralize the glutaric acid as it forms. This practice is undesirable since neutralization of the acid promotes further conversion of glutaraldehyde to an acid. More important, however, is that impurities that increase under ordinary conditions of storage show absorbance at 235 nm instead of at 207 nm, which is the characteristic of glutaric acid. The increase in absorbance at 235 nm is due to the formation of a polymer of glutaraldehyde rather than a product of oxidation. This increase is temperature dependent, only slightly affected by the presence of oxygen and unaffected by the presence of light (Gillett and Gull, 1972).

The two most important factors in the storage of glutaraldehyde are temperature and pH. The degree of polymerization is highest when solutions are stored at room or higher temperatures. In fact, the process of polymerization increases exponentially with the temperature (Rasmussen and Albrechtsen, 1974). However, polymerization is almost independent of temperature in the range of 1–25°C, and a linear relationship does not exist between temperature and the presence of polymeric glutaraldehyde above 0°C.

Glutaraldehyde polymerizes rapidly at high pHs. At pH 8.5 or over, for instance, polymerization occurs so rapidly, even at 4°C, that it is advisable not to mix the buffer and fixative until immediately prior to use (Rasmussen and Al-brechtsen, 1974). Approximately 50% polymerization of purified glutaraldehyde occurs at 4°C in 7 weeks at pH 6.5, in 3 weeks at pH 7.5, and in 6 days at pH 8.5. The desired pH can be obtained by the addition of a few drops of dilute solution of HCl. It should be noted that the rate of glutaraldehyde polymerization differs in various buffers at the same pH. Less polymerization occurs when glutaral-dehyde is stored in cacodylate buffer compared with that in phosphate buffer at the same pH (7.4).

Purified glutaraldehyde remains relatively stable for several months if stored at 4°C or below, provided the pH is lowered to ~5.0 (Trelstad, 1969). Purified glutaraldehyde can be stored for ~6 months at −14°C and for ~1 month at 4°C without significant polymerization. The most effective way to minimize the deterioration of purified glutaraldehyde is by storing it as an unbuffered, 10–25% solution at subfreezing temperatures (~−20°C). According to Rasmussen and Albrechtsen (1974), however, there is a sharp rise in polymerization on either side of neutrality. Probably, somewhat different storage conditions are required for purified and unpurified glutaraldehyde. There is no great advantage in storing glutaraldehyde in the dark or under inert gas, since commercial lots already contain sufficient acid to catalyze polymerization. However, the purified glutaraldehyde may be stored under oxygen-free conditions.

Glutaraldehyde-Containing Fixatives

Glutaraldehyde - Alcian Blue

Alcian blue is known to bind selectively to sulfate groups of mucosubstances (Hayat, 1975). It is used in combination with glutaraldehyde during fixation to obtain staining of cell coat and intercellular substances. Alcian blue–mucosubstances–glutaraldehyde complexes are osmiophilic, although individu-ally their components are not. This behavior of alcian blue is almost identical to that shown by ruthenium red. Unlike ruthenium red, alcian blue is not very toxic to living organisms. Amoeba continue to form new membranes and pseudopodia after exposure to alcian blue (Nachmias, 1963, 1968).

Benke and Zelander (1970) employed alcian blue during fixation with glutaraldehyde to demonstrate the cell coat in various types of cells. According to this procedure, tissue specimens are fixed with 4% glutaraldehyde containing 1% alcian blue for 1–18 hr. After a brief wash with a buffer, the specimens are postfixed with 2% OsO_4 for 1–4 hr. Both glutaraldehyde and OsO_4 solutions are buffered at pH 6.5. According to another procedure, perfusion fixation was employed to fix and stain gap junctions in mouse liver with 3% glutaraldehyde containing 0.1% alcian blue (Goodenough and Revel, 1971). Acid mucopolysac-charides have also been demonstrated by perfusing the tissue specimens intravitally with alcian blue (Ohkura, 1966). The electron-opaque aggregates range from 3.3 to 50 nm in diameter in the intercellular matrix. By adding alcian blue or cetylpyridinium chloride to glutaraldehyde and lanthanum to OsO_4, enhanced staining of the cell surface coat was obtained (Shea and Karnovsky, 1969; Shea, 1971). This procedure yields a higher contrast than that obtained by the addition of lanthanum alone.

Glutaraldehyde–alcian blue mixture seems to yield a superior preservation of pollen walls in comparison with that obtained with aldehydes followed by OsO_4 (Dunbar, 1978). Fixation with the mixture shows a weblike material filling the entire anther loculus from the inside of anther wall cells to the surface of pollen grains. The chemistry of alcian blue reaction with mucosubstances has been discussed by Hayat (1975). The manufacture of alcian blue dyes by ICI has ceased; astra blue 6GLL (Bayer U.K. Ltd., Dyestuffs Div., Bayer House, Manchester WA14 5PE) is a satisfactory replacement.

Glutaraldehyde – Digitonin

Ökrös (1968) introduced a modification of the Windaus (1910) digitonin reaction (which was originally carried out in ethanol) for investigating cholesterol-containing tissue specimens with the TEM. The use of this method was necessitated by the fact that since the reaction of glutaraldehyde with lipids in the tissue is slight, their retention is difficult. According to this method, a mixture of digitonin and an aldehyde is thought to be helpful in retaining and fine structural localization of free cholesterol and cholesterol esters.

It is known that certain saponins such as digitonin complex with most unesterified 3,β-hydroxysterols, including cholesterol, the principal free sterol present in vertebrate tissues. This reaction occurs on a 1:1 basis (Sperry, 1963) and results in the formation of a cholesterol–digitonide complex that is qualitatively insoluble in lipid solvents and survives embedding. This complex possesses strong osmiophilic properties. Glutaraldehyde does not seem to have any significant effect on the morphology or rate of cholesterol–digitonide complex formation. This dialdehyde is used to improve preservation of the fine structure. The complex is formed only if digitonin and free 3,β-hydroxysterol are present. It is not necessary that cholesterol be organized within a membrane or mixed with other lipids in order to complex with digitonin, for digitonin complexes with pure cholesterol (Elias et al., 1978). The reaction is inhibited by OsO_4 pretreatment, presumably because the heavy metal blocks the specific groups of 3,β-hydroxysterols that react with digitonin. As to the specificity of this reaction, all free 3,β-hydroxysterols react with digitonin; however, free cholesterol occurs in the highest concentration.

The repeated reports of increased retention of cholesterol in the tissue fixed in the presence of digitonin have prompted many workers to employ digitonin-containing fixatives for ultrastructural localization of free sterol in tissue. By using digitonin in combination with aldehydes, as much as 90% retention of free (unesterified) cholesterol in the tissue has been claimed (Scallen and Dietert, 1969; Darrah *et al.*, 1971; Parker and Odland, 1973). In these studies the retention of free cholesterol in the absence of digitonin was 3-8%. According to Mizuhira and Futaesaku (1971), excellent preservation of labeled cholesterol was obtained by employing digitonin in each of the fixation, washing, and dehydration solutions. Digitonin has been reported to be useful for autoradiographic demonstration of free cholesterol in myelin (Napolitano *et al.*, 1972) and in aorta (Albert and Rucker, 1975).

Fixation should be carried out at room temperature, dehydration should be limited to 70-95% ethanol, and infiltration should be accomplished with a 95% ethanol-Epon mixture. The preservation of cholesterol can be further improved by adding tannic acid along with digitonin to the glutaraldehyde solution.

Although a high quality of fine-structural detail can be maintained by using a mixture of glutaraldehyde and digitonin, a few morphological alterations appear in the tissue treated with the mixture (Fig. 3.1). For instance, the appearance of

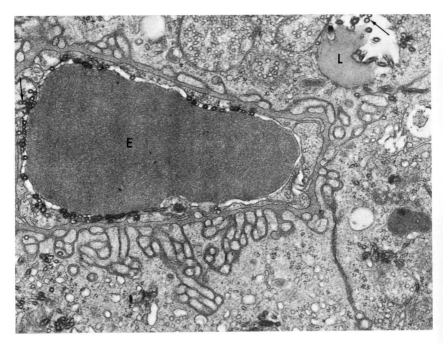

Fig. 3.1.   Rat adrenal gland, fixed in the presence of digitonin, shows artifactual crystals (arrows) containing digitonin-cholesterol complex. Lipid (L) and erythrocyte (E). ×17,000. (I. Ökrös, unpublished.)

elongate spicules, needles, lamellae, whorls, or myelin figures at certain loci in the tissue is not uncommon. According to one interpretation, cholesterol–digitonide complexes are visualized in such structures. It is thought that each structure is enclosed first by a unilaminar leaflet, and then variably cuffed by a membrane bilayer when phospholipid is present (Elias *et al.*, 1978). These workers speculate that digitonin–cholesterol complexes form the core of these structures and that the polar ends of cholesterol or digitonin molecules may react with OsO_4 to yield the narrow, outer electron-opaque layer (phosphatidylcholine bilayer). Another interpretation is that these structures represent lipids dislocated by the action of digitonin. Even though cholesterol is retained, it may be dislocated from its original position. It is difficult to differentiate between cholesterol in extramembrane precipitates and cholesterol in its original membrane site.

Evidence in support of specific binding of digitonin with free cholesterol *in situ* is lacking. Whorls and spicules can also be formed *in vitro* by the action of a digitonin-containing fixative on lecithin or cholesterol esters in the absence of free cholesterol (Frühling *et al.*, 1970; Vermeer *et al.*, 1978b). In fact, the retention of free cholesterol by the filter paper proved to be decreased by the addition of digitonin to the aldehyde fixative (Vermeer *et al.*, 1978a). These workers indicate that the solubility of cholesterol in water is increased by the presence of digitonin, whereas its solubility in ethanol is decreased when digitonin is present. Destructive action of digitonin on the outer membrane of mitochondria has been reported (Levy *et al.*, 1967).

Since digitonin is a potent detergent, capable of solubilizing membrane constituents, even when its application is preceded or accompanied by glutaraldehyde fixation, determination of precise quantitative retention of cholesterol is difficult. It is suggested that the usefulness of digitonin as a marker for the ultrastructural localization of free cholesterol is uncertain, for the structures such as spicules seem to be uninterpretable digitonin-induced artifacts.

The effect of digitonin on various cellular membranes is also uncertain. The differential effect of digitonin on the membranes of lysosomes and peroxisomes is well known. Approximately 1% digitonin causes a complete disruption of lysosomal membranes in the intact tissue, whereas it has little effect on the membranes of peroxisomes (Yokota and Fahimi, 1978). Similar effects of digitonin have been observed in liver cell fractions containing peroxisomes and lysosomes (de Duve and Baudhuin, 1966; Baudhuin, 1969). Peroxisomes show a certain degree of heterogeneity in their susceptibility to digitonin treatment, for under the influence of increased concentrations of digitonin, some peroxisomes remain intact, whereas others in the same cell exhibit a complete membrane disruption with catalase diffusion. The response of other cell organelles to different concentrations of digitonin remains to be established.

Method

Glutaraldehyde (2.5%)	5 ml
Formaldehyde (2%)	5 ml
Digitonin in buffer (0.2%)	5 ml

Paraformaldehyde powder and distilled water are heated to nearly boiling with stirring and a few drops of 1 *N* NaOH are added to clear the solution. The preceding mixture stays in good condition only for a few days; tissue specimens are fixed for ~2 hr at room temperature, followed by a prolonged wash in buffer and postfixation in buffered 1% OsO_4 (pH 7.2).

Glutaraldehyde - Formaldehyde

A mixture of glutaraldehyde and formaldehyde yields better preservation of the fine structure of a wide variety of specimens than that yielded by either aldehyde alone. The primary reason for this superiority is that formaldehyde, being a monoaldehyde, penetrates faster into the tissue specimen than does glutaraldehyde. Consequently, cellular structures are rapidly but temporarily stabilized by formaldehyde, which are subsequently fixed more permanently by the slow-penetrating glutaraldehyde. Fixation with the mixture results in increased depth of good fixation and decreased production of myelin figures.

Karnovsky (1965) originally suggested a mixture of 5% glutaraldehyde and 4% formaldehyde in 0.08 mol/l cacodylate buffer (pH 7.2) containing 0.05% (5 m*M*) $CaCl_2$. This formulation is extremely hypertonic, with an osmolality of 2010 mosmols. Presently, lower concentrations of glutaraldehyde (1–3%) and formaldehyde (0.5–2%) are in use, especially for vascular perfusion.

Method I. Glutaraldehyde (2.5%)–paraformaldehyde (2%)

Cacodylate buffer (0.2 mol/l)	25 ml
Paraformaldehyde (10%)	10 ml
Glutaraldehyde (25%)	5 ml
Distilled water to make	50 ml

The mixture is in 0.1 mol/l buffer. If necessary, the osmolarity can be adjusted by adding sucrose or NaCl. The pH may also need to be adjusted. This mixture or its modification is the most widely used general fixative for both plant and animal specimens.

Method II. Glutaraldehyde–formaldehyde

Glutaraldehyde (50%)	2 ml
Formaldehyde (40%)	10 ml
$NaH_2PO_4 \cdot H_2O$	1.16 g
NaOH	0.27 g
Distilled water	88 ml

This mixture gives easy and satisfactory preservation of specimens for both light and electron microscopy (McDowell and Trump, 1976).

Method III. Glutaraldehyde–formaldehyde
Glutaraldehyde   0.5–1.0%
Formaldehyde   4.0%
0.1 mol/l phosphate or cacodylate buffer (pH 7.0)
The preceding mixture containing a relatively high concentration of formaldehyde and low concentration of glutaraldehyde is effective in simultaneously localizing amines by the formation of fluorescent products and fixing CNS for electron microscopy (Furness *et al.*, 1978). By using this procedure, nerve cell bodies and terminal fields containing catecholamines can be located in Vibratome sections with the light microscope and, after further processing, can be viewed with the TEM. If Vibratome or cryostat sections are dried against glass slides, fluorescence intensity is increased and the sections can be permanently mounted. Tissue specimens can be stored in the mixture after perfusion and sections for fluorescence microscopy can be taken several weeks later. (The procedure of vascular perfusion of CNS with this mixture is presented on p. 228.)

A desirable feature of the fluorescence induced by this mixture is that if the tissue specimen is allowed to remain wet after the formation of the fluorophore, the reaction remains stable. If the specimen is dried, the fluorescence intensity is enhanced (Furness *et al.*, 1977), but if, after drying, the specimen is wetted, the fluorescence is abolished. These observations suggest that the fluorophore changes occur on drying (Furness *et al.*, 1978). It is possible that aqueous glutaraldehyde binds to and stabilizes the fluorophore and that, on drying, the bond with the dialdehyde breaks and a fluorophore similar to that induced by formaldehyde alone remains. The fluorophore induced by formaldehyde in freeze-dried specimens is known to be quenched by aqueous solutions (Corrodi and Jonsson, 1967). If the sections are dried to increase the fluorescence, the drying should be complete so that residual water does not destroy the fluorophore during storage.

This mixture may be useful for the localization of radio-labeled amines by the TEM autoradiography. The mixture is also useful for the subsequent localization of cholinesterase (Westrum and Broderson, 1976) and probably other enzymes.

Glutaraldehyde – Hydrogen Peroxide

According to Peracchia and Mittler (1972b), fixation with a mixture of glutaraldehyde and hydrogen peroxide (H_2O_2) results in better preservation than that obtained with conventional glutaraldehyde. This mixture has also been reported to better preserve enzymatic activity (Goldfischer *et al.*, 1971). The effectiveness of this mixture has been presumed to depend on the activity of reaction products of hydrogen peroxide and glutaraldehyde. Preliminary experiments with NMR spectroscopy suggest that the reaction products are primarily epoxides. The active compounds in the mixture may be α,β-epoxyaldehydes (Peracchia and Mittler, 1972b).

A reaction between hydrogen peroxide and double bonds of α,β-unsaturated aldehydes may give rise to epoxy groups. It has been stated earlier that α,β-unsaturated dimer is abundant in aqueous solutions of glutaraldehyde. A possible scheme for the synthesis of epoxy aldehydes follows (Peracchia and Mittler, 1972b):

$$
\begin{array}{c}
\text{CHO} \\
|\\
\text{CH—(CH}_2\text{)—C}=\text{CH—(CH}_2\text{)}_3\text{—CHO} \ + \ \text{H}_2\text{O}_2
\end{array}
$$

Unsaturated aldehyde and H_2O_2

$$
\begin{array}{c}
\text{CHO} \\
\text{CH—(CH}_2\text{)} \quad | \quad \text{H} \quad \text{(CH}_2\text{)}_3\text{—CHO} \\
\text{C——C} \\
\diagdown\text{O}\diagup \\
\ + \ \text{OH}^-
\end{array}
$$

Epoxy aldehydes

Epoxides are known to be unstable compounds that are capable of reacting with amino, imino, hydroxyl, and mercapto groups (March, 1968). Thus, the addition of epoxy groups to the unsaturated polymers of glutaraldehyde enhances their reactivity with tissue components. The improved preservation of polysaccharides may be due to high reactivity between epoxy and hydroxyl groups. In this reaction, β-hydroxyethers are probably formed.

The specimens fixed with glutaraldehyde–hydrogen peroxide mixture show decreased blackening during osmication. This may be due to a decrease in double bonds of glutaraldehyde–H_2O_2 as a result of epoxide synthesis. It is known that OsO_4 reacts with double bonds but not with epoxides.

An interesting explanation for the reported improved fixation with H_2O_2 may be that this reagent increases available O_2 during fixation. Since some biochemical evidence indicates that in glutaraldehyde and protein reaction, pyridine compounds are formed and that intermediate products in this reaction rapidly use O_2 under conditons of fixation, it is thought that respiration and fixation may compete for available O_2 during fixation. If O_2 is needed for irreversible cross-linking of proteins, methods that increase available O_2 should expedite fixation and improve the quality of tissue preservation (Johnson and Rash, 1980). Thus, procedures such as the use of H_2O_2, azide (respiratory poison), and aeration of fixative solutions may catalyze the rate of protein cross-linking with glutaraldehyde. The addition of 0.1% azide to glutaraldehyde has been reported to improve mitochondrial preservation in the deeper regions of tissue blocks

(Minassian and Huang, 1979). Anoxia (which sets in immediately after tissue slice is removed from the body) may retard the speed of protein cross-linking when fixation with glutaraldehyde is carried out by immersion. Further studies are needed to test the preceding suggestion.

Method
Cacodylate buffer (0.1 mol/l)   25 ml
Glutaraldehyde (25%)   5 ml
Hydrogen peroxide (15%)   5-25 drops

Drops of hydrogen peroxide are added to glutaraldehyde while continuously stirring. The final concentration of buffered glutaraldehyde ranges between 3 and 6%. Tissue specimens are fixed in the mixture for 1-2 hr at room temperature or for 3-4 hr at 4°C. During fixation, specimens attain a pink-orange color. If initially fixed at 4°C, the specimens are transferred to 3-5% glutaraldehyde (pH 7.4) for 1 hr at room temperature. After a thorough wash in the buffer, specimens are postfixed with 2% OsO_4 for 2 hr at room temperature. It is pointed out that hydrogen peroxide should not be mixed with formaldehyde, for explosive compounds may be synthesized.

Glutaraldehyde - Lead Acetate

Glutaraldehyde and lead acetate, $Pb(C_2H3O_2)_2 \cdot 3H_2O$, mixture is used to preserve soluble inorganic phosphate, which is lost when conventional fixation with glutaraldehyde is employed. Osmium tetroxide and uranyl acetate are not used, since they remove all the lead phosphate precipitate from the sections.

The preceding mixture was used to localize orthophosphate in the nucleolus in maize root-tip cells (Libanati and Tandler, 1969; Tandler and Solari, 1969). According to these authors, lead hydroxyapatite, $Pb_5(PO_4)_3OH$, is the chemical form of the precipitate formed by the glutaraldehyde-lead acetate fixative. The mixture has also been employed to show specific accumulation of inorganic phosphate adjacent to the plasma membrane and in the nucleolus of pancreatic B (beta) cells (Freinkel et al., 1978). Specimens are fixed with a mixture of glutaraldehyde (2%) and lead acetate (2%) in 0.1 mol/l cacodylate buffer (pH 7.0) for 3 hr at room temperature. The specimens are transferred to 4% lead acetate in 35% acetic acid at 4°C for 12 hr.

Glutaraldehyde - Malachite Green

Malachite green (Difco Lab., Detroit, Michigan) is a member of compounds known as diaminophenylmethanes. It has been used for staining bacterial spores and polysaccharides for light microscopy (Lillie, 1977). Malachite green in combination with glutaraldehyde was used to preserve certain lipid-containing granules in mammalian spermatozoa for electron microscopy (Teichman et al., 1972, 1974a). These granules were identified by thin-layer chromatography as fatty acids, phospholipids, glycolipids, and fatty aldehydes (Teichman *et al.*, 1974b). Fixation with glutaraldehyde alone fails to preserve these granules. Malachite green also enhances the electron opacity of other intracellular structures. Mammalian cardiac muscle, liver cells, and other tissues fixed with this mixture showed enhanced electron opacity of lipid-containing granules, ribosomes, and mitochondria (Pourcho *et al.*, 1978). This mixture has also been used for preserving electron opaque granules in bacteria (Kushnaryev *et al.*, 1980). Although this mixture stabilizes lipid-containing structures, the overall quality of ultrastructure preservation is relatively less satisfactory. The fixative mixture contains 2% glutaraldehyde and 0.1% malachite green.

Glutaraldehyde – Phosphotungstic Acid

Anionic sites in basement membranes and collagen fibrils have been demonstrated with cationic polyethyleneimine (PEI), which acts as a tracer particle for anionic sites (Schurer *et al.*, 1977, 1978). The tail vein of the animal (e.g., rat, 200 g body weight) is injected (0.2 ml of 0.5%) with PEI (mol. wt. 30,000–40,000) solution, adjusted to pH 7.3 with HCl, whose osmolality has been raised to 400 mosmols with NaCl. Approximately 15 min after injection, small tissue blocks are fixed by immersion in a mixture of glutaraldehyde (0.1%) and PTA (2%) for 1 hr at room temperature to obtain staining of the PEI particles and tissue preservation.

Alternatively, tissue blocks are immersed in a PEI solution (mol. wt. 1800, pH 7.3, 400 mosmols) for 30 min to obtain binding of PEI by all available anionic sites. The blocks are washed thoroughly in cacodylate buffer (pH 7.3, 400 mOsm) to remove excess of PEI. Phosphate buffer is undesirable, for it can precipitate PEI. Fixation is carried out as mentioned previously, followed by osmication in 1% OsO_4 in cacodylate buffer (pH 7.3, 220 mosmols) for 2 hr at 4°C. Sections are stained with 10% uranyl acetate in methanol and lead citrate for 2 and 1 min, respectively.

Glutaraldehyde – Trinitro Compounds

Aldehydes in combination with trinitro phenolic compounds (e.g., picric acid) have been employed to obtain improved fixation of certain tissues. Ito and Karnovsky (1968) further explored the usefulness of other trinitro compounds such as 2,4,6-trinitrocresol in fixation. These preservatives preserve smooth endoplasmic reticulum in the testicular interstitial cells and other steroid-secreting cells and peroxidase activity especially well, but certain types of microtubules are not consistently preserved. Fixative mixtures containing 2,4,6-trinitrocresol are superior to the formaldehyde–picric acid mixture.

Method
Buffer (0.2 mol/l) (pH 7.2)   45 ml
Glutaraldehyde (25%)   5 ml

Formaldehyde (4%)	50 ml
Trinitrocresol (2%)	50 ml

This fixative contains 2.5% glutaraldehyde, 2% formaldehyde, and 0.5% 2,4,6-trinitrocresol. Specimens are fixed for 1-2 hr at room temperature, washed with a buffer, and postfixed with buffered 1% OsO_4 at 4°C. The mixture can also be used for vascular perfusion. Trinitro compounds are potentially explosive.

Glutaraldehyde – Potassium Dichromate

That adrenomedullary cells are stained brown when treated with chromium salts was first demonstrated by Henle (1865). Later Wood and Barrnett (1964) showed that the chromaffin reaction is of value for the identification of biogenic amines at the submicroscopic level. For a review of the value of various methods used for ultrastructural demonstration of biogenic amines in terms of resolution, specificity, and sensitivity, the reader is referred to Bloom (1970).

The chromaffin reaction is useful for the identification of adrenalin and other aromatic compounds such as dihydric and polyhydric phenols, aminophenols, and polyamines. This reaction is unable to differentiate between various biogenic amines, since noradrenalin, dopamine, and serotonin also yield precipitates with chromium salts *in vitro* (Wood, 1966). However, differentiation between different amines can be achieved by using specific pharmacologic treatments, i.e., the synthesis of specific amines is blocked or inhibited by certain drugs. For example, parachlorophenylalanine inhibits tryptophan hydroxylase, which plays a key role in the synthesis of serotonin. Another approach to differentiate amines is by using formaldehyde, which eliminates the chromaffin reaction from the adrenal medulla, but not from enterochromaffin cells (Jaim-Etcheverry and Zieher, 1968).

Simultaneous aldehyde fixation and chromation is the best method for detecting amines in a variety of tissues (Fig. 3.2). Essential conditions in this method are the following: lower concentrations of glutaraldehyde (1%), slightly acidic pH (6.0) during fixation and prolonged incubation in 0.2 mol/l sodium chromate–potassium dichromate buffer prior to postfixation in OsO_4, short duration of fixation, and the use of sodium chromate–potassium dichromate as buffer. The slightly acidic pH is known to increase the redox potential of dichromate, thus favoring the formation of CrO_2. This method is elaborated later.

Mechanism of Reaction. Glutaraldehyde interacts with noradrenalin, resulting in the formation of an insoluble product capable of binding chromate. This reaction is thought to occur between the unsubstituted amino group of noradrenalin and the aldehyde groups of glutaraldehyde, leaving the dihydroxy groups free to react with dichromate ions (an oxidizing agent). Either the precipitated $Cr(OH)_3$ or the chromium-containing oxidation product of biogenic amine is responsible for the electron density. According to von Feuerstein and Geyer (1971), the electron density of the reaction product is primarily due to the

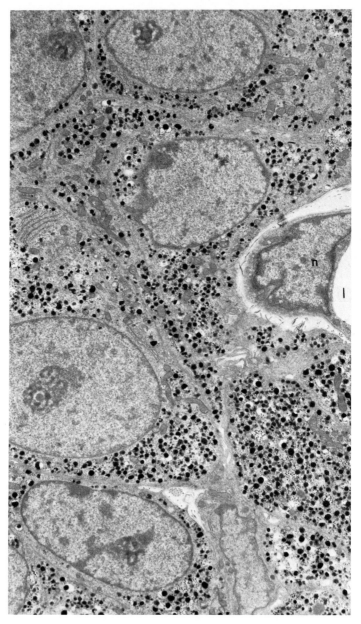

Fig. 3.2. Cat extraadrenal chromaffin cells fixed by perfusion with glutaraldehyde, followed by immersion fixation with a mixture of glutaraldehyde (3%) and potassium dichromate (2.5%). The cytoplasm contains many dense catecholamine granules. Note mitochondria, nuclei, lumen of blood vessel (1), endothelial cell nucleus (n), and rough endoplasmic reticulum. ×7395. (J. A. Mascorro, unpublished.)

formation of polynuclear complexes of the trivalent chromium with oxidation products of amines. An additional binding of Cr^{3+} ions to the matrix of storage granules is assumed, resulting in the simultaneous fixation of oxidation products. In fact, the presence of chromium in the reaction product has been confirmed by using X-ray microanalysis (Wood, 1975, 1976).

A schematic rationale for the glutaraldehyde–potassium dichromate method follows (Wood, 1977):

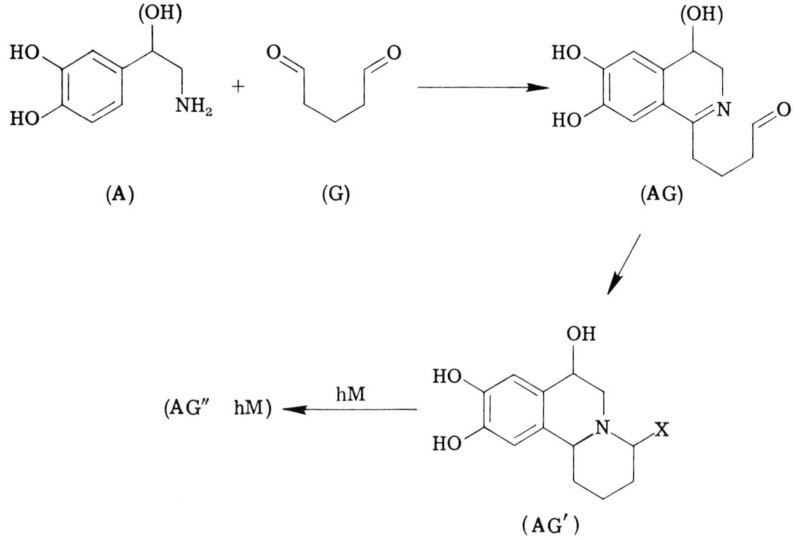

Glutaraldehyde (G) reacts with the biogenic amine (A) to form an intermediate Schiff's base (AG), which in turn forms an isoquinoline derivative (AG'). The addition of potassium dichromate to the AG' reaction results in the formation of AG"-hM (heavy metal) complex. This complex can be identified by electron microscopy.

Method I. Glutaraldehyde–potassium dichromate
This method is useful in demonstrating norepinephrine and argentaffin cells. Tissue specimens are fixed with 3% glutaraldehyde in cacodylate buffer (pH 7.2) for 4 hr followed by incubation in 2.5% potassium dichromate in cacodylate buffer (pH 4.1) for 4 hr.

Method II. Glutaraldehyde–formaldehyde–potassium dichromate
Glutaraldehyde   1%
Formaldehyde   0.4%
Sodium chromate–potassium dichromate (pH 7.2)   0.1 mol/l
Specimens are fixed in the preceding mixture either by immersion with agitation (1–10 min) or by vascular perfusion (5–15 min) at 4°C. After storing with

constant agitation in 0.2 mol/l sodium chromate-potassium dichromate buffer (pH 6.0) for 18 hr at 4°C, the specimens are postfixed with 2% OsO_4 in the same buffer (pH 7.2) for 1 hr at 4°C. Sections are stained with lead citrate, but not with uranyl acetate. This method is useful for localizing biogenic amines in a variety of tissues.

Glutaraldehyde - Potassium Ferricyanide - Osmium Tetroxide

Elbers *et al.* (1965) suggested the use of potassium ferricyanide, $K_3Fe(CN)_6$, during fixation for the preservation of phospholipids in the brain tissue. Dermer (1970) used this method to supposedly preserve pulmonary surfactants, which are preserved neither with gluraraldehyde nor with OsO_4. Subsequently, it was thought that labile lipids can be preserved by the addition of $K_3Fe(CN)_6$ to various fixatives.

The acellular layer lining the alveolar walls is a complex of lipids, proteins, carbohydrates, and inorganic substances. The major component of this material is the surfactant, dipalmitoyl lecithin, a highly saturated phospholipid. The method of tricomplex fixation was claimed to be effective in preserving the surfactants. It employs cations and anions, which form a link between the phospholipid amphoions, and the resulting electrostatic interactions provide cohesion between the phospholipid molecules. The essential components of the fixative are $Pb(NO_3)_2$ and $K_3Fe(CN)_6$; however, the former is the crucial component in this procedure, since de Bruijn (1973) has reported that excellent results were obtained without the addition of any cation.

Contrary to the claims mentioned previously and in the literature, amorphous precipitates formed on the alveolar surface of the lung, followed by treatment with tricomplex fixatives, are nonspecific and cannot be regarded to be related to surfactant phospholipids. It is possible that these precipitates may be artifacts, for it is not unlikely that the surface coat on the epithelial cells represent nonspecifically bound colloidal particles present in the fixative solution. The possibility that these precipitates are nonspecific has been strengthened by the evidence that identical precipitates are found even after lipid extraction with chloroform-methanol. Furthermore, identical precipitates have been found within the cytoplasm of isolated red blood cells (Gil, 1972).

Dermer (1970) fixed lung tissue by immersion in 2% glutaraldehyde in cacodylate buffer (pH 7.2), followed by in a mixture of $K_3Fe(CN)_6$ (0.05 mol/l) and $Pb(NO_3)_2$ (0.05 mol/l), and finally in 1% OsO_4. Adamson and Bowden (1970) applied this method to demonstrate purportedly saturated phospholipid in the lung. They used various methods, including intratracheal fixation and vascular perfusion with glutaraldehyde to preserve the tissue. The vascular perfusion used by Finlay-Jones and Papadimitriou (1972) seems to be the most reliable and is presented later. The advantage of this procedure is that by vascular perfusion the fixative comes in contact with the cell surface from the aqueous side and thus leaves the surfactant largely intact. In contrast, by immersion fixation the aerated lung tissue is forced to change from air to the aqueous phase. This treatment destroys the bimolecular leaflet of phospholipid oriented at the cell surface, which results in the formation of micelles that may be extracted.

Another approach to minimize the movement of surfactant is to have the lung collapsed, or its alveolar air evacuated and replaced with boiled linseed oil prior to tricomplex fixation. Linseed oil hardens during embedding and is unable to dissolve dipalmitoyl lecithin. By using this technique, lamellated osmiophilic bodies in the type II cells of the lung have been demonstrated (Pattle *et al.*, 1972).

The method of primary fixation with glutaraldehyde followed by a mixture of OsO_4 and $K_3Fe(CN)_6$ has been used to increase the visualization of lipids (de Bruijn and Den Breejen, 1975), to demonstrate membranous structure of liposomes, and to show membrane-bound particles (30–100 nm in diameter) in the subendothelial aortic space in experimental atheromatosis in rabbits (Vermeer *et al.*, 1978a) and in xanthomatous tissue (Vermeer *et al.*, 1978b).

Method I

$Pb(NO_3)_2$	0.83 g
$K_3Fe(CN)_6$	0.55 g
Distilled water to make	100 ml

Vascular perfusion is carried out with 2% glutaraldehyde followed by treatments with the preceding mixture and with 1% OsO_4.

Method II. Cells are fixed in a mixture of acrolein (1%) and glutaraldehyde (2.5%) in 0.067 mol/l cacodylate buffer (pH 7.4) containing 1 mM $CaCl_2$ for 24 hr at room temperature (Bluemink, 1972). Specimens are stored overnight in the same buffer and then postfixed in 1% OsO_4 in the same buffer, containing $K_3Fe(CN)_6$ (0.05 mol/l) and $CaCl_2$ (0.05 mol/l) for 3–4 hr in the dark at 4°C.

Method III. The trachea of the anesthetized animal (e.g., rabbit) is exposed and clamped and the thorax is opened. Vascular perfusion into the right ventricle is carried out with 3% glutaraldehyde buffered to pH 7.4 with 0.1 mol/l cacodylate and HNO_3 at 4°C. After an initial influx of the perfusate, the left atrium is opened and 1 ml of glutaraldehyde per gram body weight is injected slowly. The heart, lungs, and trachea are dissected from the animal and immersed in fresh glutaraldehyde for 1 hr at 4°C.

The lungs are then filled to full expansion by injecting a mixture of $Pb(NO_3)_2$ and $K_3Fe(CN)_6$ given previously in Method I. The trachea is clamped and the whole lung incubated in the mixture for 30 min at 37°C. Small pieces of the tissue are excised, washed in the buffer, and postfixed in buffered 1% OsO_4 for 1 hr. The success of this method depends on satisfactory fixation of the tissue. Therefore, tissue blocks for sectioning should be chosen from those areas of the lung that have been fully permeated by the salt mixture.

Glutaraldehyde - Potassium Ferrocyanide - Osmium Tetroxide

Primary fixation with glutaraldehyde followed by fixation with a mixture of OsO_4 and potassium ferrocyanide has proved useful for the preservation of the male reproductive system, including Leydig cells; steroid-secreting cells are especially well fixed (Russell and Burguet, 1977). The procedure eliminates dense cytoplasmic matrix and improves the preservation of membranes, including those of Golgi and smooth endoplasmic reticulum, which appear tubular instead of vesicular (Figs. 3.3, 3.4). Microfilaments and glycogen are also well preserved. Karnovsky (1971) indicated that the fixative mixture substantially enhanced the contrast of the outer surfaces of membranes. This mixture is recommended for revealing the fine structure of the myelin sheath (Fig. 3.5).

Method. Fixation is carried out by vascular perfusion for ~20 min with 5% glutaraldehyde in 0.05 mol/l cacodylate buffer (pH 7.4) through the abdominal aorta by a retrograde method (Vitale *et al.,* 1973) or through the testicular artery. The testes are removed, cut into small pieces (1 mm), and washed overnight in the buffer. Specimens are postfixed in a freshly prepared mixture of 1.5% OsO_4 and 2.5% potassium ferrocyanide for 2 hr at room temperature. For the preservation of Leydig cells, the reader is referred to Russell and Burget (1977).

Glutaraldehyde - Ruthenium Red

The use of ruthenium red as a stain for cell surface material and its staining mechanism have been discussed by Hayat (1975). Ruthenium red, being a strong oxidant, probably acts not only as a stain but also as a fixative along with OsO_4. When ruthenium red is applied during fixation, membranes and myofilaments are well preserved and sharply defined without poststaining. Alternatively, ruthenium red and OsO_4 together may form RuO_4, which can react both with some of the more polar lipids and with proteins, glycogen, and common oligosaccharides. Osmium tetroxide is thought to be reduced catalytically to osmium blacks by ruthenium red bound to the tissue. For general purposes the following method of fixation and staining primarily for acid mucopolysaccharides is recommended.

Solution A
Aqueous glutaraldehyde (4%)	5 ml
Cacodylate buffer (0.2 mol/l, pH 7.3)	5 ml
Ruthenium red stock solution (1,500 ppm in water)	5 ml

Solution B
Aqueous OsO_4 (5%)	5 ml
Cacodylate buffer (0.2 mol/l, pH 7.3)	5 ml
Ruthenium red stock solution (1,500 ppm in water)	5 ml

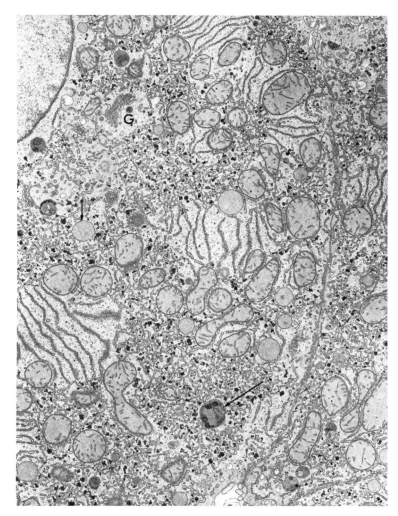

Fig. 3.3.   Rat liver (fasted 18 hr prior to sacrifice) was fixed by perfusion with a mixture of 2.7% glutaraldehyde and 0.8% formaldehyde in 0.2 mol/l sodium bicarbonate buffer (pH 7.4, 770 mosmols). Small pieces from the median lobe were further fixed with the same fixative for 2 hr at room temperature. After a wash in the buffer overnight, the specimens were postfixed with a mixture of 1% OsO_4 and 1.5% potassium ferrocyanide. Note Golgi (G), lysosome (long arrow), microbody (short arrow), tubular smooth endoplasmic reticulum, and glycogen (dark granules). ×11,050. (D. L. Schmucker, D. L. Mooney, and A. L. Jones, unpublished.)

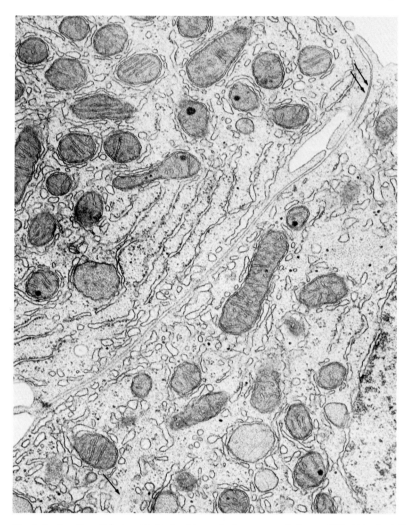

Fig. 3.4. Leydig cell from rat testis fixed by perfusion as presented on p. 254. After the initial fixation in glutaraldehyde and an overnight wash in the buffer, tissue slices were postfixed with a freshly prepared mixture of 1.5% OsO$_4$ and 2.5% potassium ferrocyanide for 2 hr at room temperature. This fixation procedure yields optimal ultrastructural preservation of Leydig cells. Note that the smooth endoplasmic reticulum appears as an extensive network of interconnected tubules of uniform diameter. Note mitochondria, peroxisomes, nucleus, rought endoplasmic reticulum, microtubules (single arrow), and a gap junction (double arrow) between the two cells. ×34,850. (Russell and Burguet, 1977.)

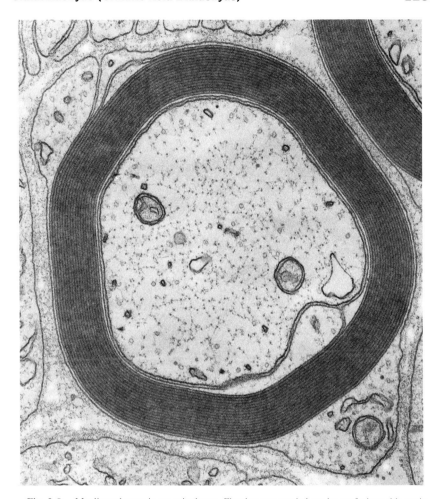

Fig. 3.5.   Myelinated axon in rat spinal root. Fixation was carried out by perfusion with a mixture of 3% glutaraldehyde, 3% formaldehyde, and 0.1% picric acid in 0.1 mol/l cacodylate buffer (pH 7.4). Small pieces of the tissue were further fixed in the same fixative solution overnight. After washing in the same buffer for 20 min, specimens were postfixed with a mixture of 1% OsO_4 and 1.5% potassium ferrocyanide in the same buffer for 1–2 hr at room temperature. This was followed by a rinse with 0.1 mol/l maleate buffer (pH 6.0) for 20 min and *en bloc* staining with 1% uranyl acetate in the same buffer for 1 hr. This seems to be the best method for preserving the structure of the adult myelin sheath. ×41,250. (Coggeshall, 1979.)

Tissue specimens are fixed and stained with solution A for 1 hr at room temperature. After rinsing in the cacodylate buffer, the specimens are postfixed with solution B for 3 hr at room temperature. Poststaining with uranyl acetate and lead acetate is not necessary. The application of ruthenium red during fixation may cause artifacts such as myelinlike structures (Hayat, 1975). Poor diffusion of this dye into certain tissues can be overcome by employing it in combination with glutaraldehyde via vascular perfusion (Hayat, 1975).

Glutaraldehyde - Spermidine Phosphate

The preservation of fine structure in amoebas is difficult by conventional fixation procedures. A mixture of glutaraldehyde and the membrane-stabilizing spermidine phosphate, N-(3-aminopropyl-)-1,4-butan diamine phosphate (Prescott *et al.*, 1966), is effective in preserving thick and thin filaments within the ectoplasm of amoebas (Hauser, 1978). Diamine spermidine reacts with glutaraldehyde to form typical addition products of the configuration R_1—$CH{=}N{-}R_2$ called Schiff's bases. The Schiff's base with one reactive carbonyl group formed in the initial step is the actual fixative.

The fixative solution consists of 2×10^{-2} M spermidine phosphate in 0.05 mol/l collidine buffer (pH 7.8), to which is added 25% glutaraldehyde with continuous stirring to give a final concentration of 2.5×10^{-2} M. Spermidine phosphate dissolves in collidine more easily at 40–50°C. Primary amines react with glutaraldehyde at room temperature within 2 min, resulting in a yellowish reaction product, which turns deep orange after some time.

Specimens are fixed for 10 min in a freshly prepared mixture (yellow) of glutaraldehyde and spermidine phosphate. After centrifugation, the specimens are fixed further in 2.5% glutaraldehyde in collidine buffer for 10 min (Hauser, 1978). Postfixation is carried out in 1% OsO_4 for 30 min. In all buffers, bivalent cations should be avoided because a precipitate is easily formed in the presence of phosphate complexes. Thin sections are poststained with lead followed by uranyl acetate.

Glutaraldehyde - Tannic Acid

Commercially available tannic acids are complex mixtures of polyphenolic compounds with a galloylated glucose structure and vary considerably in molecular weights (1000–1700) (Mallinckrodt Inc., St. Louis, Mo.). The chemistry of tannic acid is known primarily because of its use in the leather industry, which is interested in its reactions with proteins, especially with collagen (Gustavson, 1949; Lollar, 1958; Haslam, 1966). It is assumed that tannins react with peptide bonds and amine and amide residues present in polar amino acid side chains.

Mizuhira and Futaesaku (1971) introduced tannic acid as a supplementary fixative for electron microscopy. They suggested that it might fix soluble proteins by forming complexes with proteins/polypeptides and heavy metals, yielding a precipitate of high electron density. The precipitate is insoluble in dehydration solvents. They suggested three possible modes of action of tannic acid: (1) the pyrogallol groups of tannic acid may form hydrogen bonds with proteins at the sites of peptide bonds, (2) tannic acid may carry a negative charge in aqueous solution that would bind with proteins that are positive under acidic conditions, and (3) tannic acid–protein molecules bind with OsO_4. The pH dependency of the reaction product suggests an ionic interaction. The instability of the reaction product under hydrophobic conditions suggests that hydrogen bonding involving the multiple hydroxyl groups of tannic acid might be a contributing factor in its formation and maintenance in an aqueous environment (Kalina and Pease, 1977a). Presumably, tannic acid cross-links a variety of proteins (especially in elastic fibers) in competition with glutaraldehyde; these proteins then have free aromatic hydroxyl groups that can react with heavy metals such as OsO_4. Tannic acid reacts preferentially with NH_3^+ groups of proteins.

Since tannic acid enhances the contrast of various intra- and extracellular structures, including cytomembranes (Tilney *et al.*, 1973b; Rodewald and Karnovsky, 1974; Shienvold and Kelly, 1974; Simionescu and Simionescu, 1976a,b), the previously mentioned reaction mechanisms do not directly explain the increased contrast shown by cytomembranes after treatment with this acid. A partial explanation is that tannic acid is not a true fixative but is a multivalent agent, acting primarily as a mordant between osmicated cellular structures and lead stains (Simionescu and Simionescu, 1976a,b; Wagner, 1976; Kalina and Pease, 1977b). In this scheme, osmication is a prerequisite for the increased contrast. A carboxyl group and at least one hydroxyl group on the tannic acid are thought to be the minimal functional components necessary for the mordanting effect (Simionescu and Simionescu, 1976b). However, the ability of tannic acid to stabilize certain cellular structures and thus minimize their extraction cannot be overlooked. This reagent, for example, seems to prevent or delay microtubular disruption caused by postfixation with OsO_4. Also, microfilament stability is increased in the presence of tannic acid. *In vivo* stabilization of microfilaments with tannic acid is also likely. Tannic acid is known to preserve microtubule-associated microfilaments (Seagull and Heath, 1979).

Studies on the interaction between tannic acid and saturated phospholipids and type II pneumocytes of lung tissue (containing a high proportion of saturated lipids) indicate that the acid reacts with the choline base of synthetic phosphatidyl cholines and sphingomyelin to form complexes that can be stabilized by osmication (Kalina and Pease, 1977a,b). Staining with lead results in additional contrast of polar regions in the ordered lamellar structures. The increased contrast exhibited by these phospholipids seems to be limited to the hydrophilic layers corresponding to the polar regions of the phospholipid molecules. However, active groups of proteins may also play a part in the increased contrast of natural membranes after treatment with tannic acid (Fig. 3.6). The importance of the use of tannic acid in enhancing the contrast of saturated phosphatidyl cholines is apparent when one considers that both glutaraldehyde and OsO_4 do not react with saturated phospholipids and that phosphatidyl cholines are a major component of most mammalian membranes (often over 50% of the total phospholipids present) (van Deenen, 1966). It must be emphasized that presently the knowledge on the reactions of tannic acid with active cellular groups is rather limited.

Tannic acid can be used either prior to or after osmication, depending on the objective of the study. Treatment with tannic acid prior to osmication is preferred for optimal preservation of ordered structures. On the other hand, treatment with tannic acid after osmication is most effective in favoring high-density lead staining of cellular structures such as secretory bodies in type II pneumocytes of lung tissue (Kalina and Pease, 1977a). A mixture of glutaraldehyde and tannic acid imparts to cell membranes a negative image similar to that obtained when specimens are fixed in glutaraldehyde without postfixation with OsO_4. Tannic acid interferes in the reaction between OsO_4 and membrane components.

Glutaraldehyde–tannic acid is effective in preserving intercellular glycosaminoglycans (Singley and Solursh, 1980). Both hyaluronic acid and chondroitin sulfate can be precipitated by this method. This reagent has been used as a mordant for heavy metal staining of mucins and other complex carbohydrates (Sannes et al., 1978). The globular subunit structure of microtubules (Tilney et al., 1973b) and microfilaments alone or in association with microtubules can be easily visualized by using this method. This mixture has also been used to negatively stain F-actin of muscle and actinlike microfilaments of nonmuscle in situ (LaFountain et al., 1977). Tannic acid does not penetrate deep into the tissue nor does it diffuse through unfixed membranes. However, when cells are treated with tannic acid following a brief fixation with OsO_4, the acid penetrates the membrane and both inner and outer leaflets and surfaces as well as intracellular membranes show increased electron density. The small molecular weight of tannic acid makes it a desirable tracer for interfaces of adjoining cells, especially for detecting gaps in cell junctions.

Tannic acid–glutaraldehyde mixture is effective in preserving the components of the pollen wall, which are not visible when the specimens are fixed with aldehydes followed by osmication (Dunbar, 1978). The walls of pollen grains fixed with this mixture show globular subunits that are lost with conventional double fixation.

A wide range of tannic acid concentrations has been employed for different tissues. Human skin was fixed with 3% glutaraldehyde in Millonig buffer (pH 7.4) containing 0.25% tannic acid (Cotta-Pereira et al., 1976). The fixative

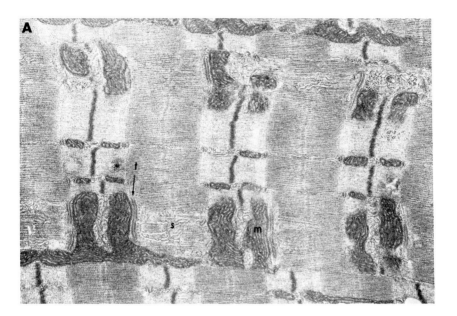

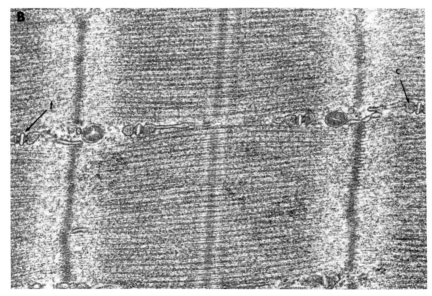

Fig. 3.6. Longitudinal section of rabbit skeletal muscle. Small cubes (0.4 mm on a side) of the tissue were fixed with 2.5% glutaraldehyde in 0.1 mol/l cacodylate buffer (pH 7.2) containing 1% tannic acid for 2 hr at 4°C and postfixed with 1% OsO₄ in 0.1 mol/l veronal acetate buffer (pH 7.2) containing 2.4 mM CaCl₂ and 0.06 mol/l NaCl for 2 hr at 4°C. *En bloc* staining was carried out with 0.5% uranyl acetate in veronal buffer (pH 6.0) for 2 hr. Tannic acid was penetrated well into the tissue (lower part). The fixative used for the upper part did not contain tannic acid. Note sarcoplasmic reticulum (s), transverse tubules (t) between terminal cisternae (c), mitochondria (m), and bands. A, ×14,063; B, ×45,000. (Saito *et al.,* 1978.)

enabled a distinction among oxytalan, elaunin, and elastic fibers. Rabbit skeletal muscle was fixed with 2.5% glutaraldehyde in 0.2 mol/l cacodylate buffer (pH 7.2) containing 8% tannic acid (Bonilla, 1977). This study revealed prominent staining of the junctional T-system, cell surface, and lamellar structures at the cell periphery. Mouse liver was fixed with 2.5% glutaraldehyde in 0.2 mol/l cacodylate buffer (pH 7.2) containing 4% tannic acid for delineating gap and tight junctions (van Deurs, 1975). Megakaryocytes and platelets were fixed with 2.5% glutaraldehyde in 0.1 mol/l cacodylate buffer (pH 7.4) containing 8% tannic acid and 0.015 mol/l $CaCl_2$. The mixture had a pH of 5.8. The cells were pelleted and the solution was replaced by 1% OsO_4 in the buffer for 30 min at 4°C (Fedorko and Levine, 1976). This method stained surface membranes, the demarcation membrane system in megakaryocytes, and the open canalicular system in platelets.

Rat testes were perfused with 4% glutaraldehyde in 0.05 mol/l phosphate buffer (pH 7.2) containing 2% tannic acid for 15 min for preserving and promoting the staining of protein constituents of the extracellular fluid and for delineating the extracellular compartment and changes in capillary permeability under experimental conditions (Aoki *et al.*, 1976).

Specimens fixed with tannic acid–glutaraldehyde should be washed in the buffer prior to postfixation with OsO_4, for tannic acid readily reacts with OsO_4. The fixative should be prepared immediately before use, since a precipitate is formed in a fairly short time. Because of the acidity of tannic acid, the pH of the fixative mixture is drastically lowered, which can be prevented by using 0.2 mol/l buffer. The maximum precipitation occurs at ~pH 7.4. The extent of reaction is substantially dependent on the concentration of tannic acid. For example, the diameter of microfilaments preserved with tannic acid tends to increase in proportion to the concentration of tannic acid used. It should be noted, however, that increased concentration of tannic acid will result in intense staining of the cytoplasm, which may obscure microfilaments. Therefore, the need of optimal concentration of tannic acid is apparent.

Method I. Glutaraldehyde (2.5%)–tannic acid (4%)
    Glutaraldehyde (5%)   10 ml
    Tannic acid (8%)   10 ml
The pH may have to be adjusted by adding dilute solution of NaOH. Specimens are postfixed with 1% OsO_4 (Fig. 3.6B).

Method II. Glutaraldehyde (2.5%)–paraformaldehyde (2%)–tannic acid (4%)
    Glutaraldehyde (5%)   20 ml
    Paraformaldehyde (16%)   10 ml
    Tannic acid (16%)   10 ml
    Sucrose   0.05 mol/l
All solutions are prepared in a buffer. Specimens are postfixed in 1% OsO_4.

Glutaraldehyde - Uranyl Acetate

Glutaraldehyde and uranyl acetate mixture is effective in preserving structures susceptible to easily losing their DNA content, for it gels DNA in minutes. Therefore, it was introduced for preserving bacteria containing intracellular phage (Séchaud and Kellenberger, 1972). (For details see p. 390.)

FORMALDEHYDE

$$H-C\underset{H}{\overset{O}{\diagdown}}$$

Formaldehyde is a colorless gas. It is easily soluble in water, and is available commercially as formalin (37–40%). It contains a small amount of formic acid (<0.05%) and a considerable amount of methanol (6–15%). The methanol in formalin hinders polymerization by breaking down high-molecular weight oligomers of polymethylene glycols, forming hemiacetals. Because the latter are more soluble than the former, the precipitate formation is prevented:

$$HOCH_2OH + CH_3OH \leftrightharpoons HOCH_2OCH_3 + H_2O$$

$$HOCH_2OCH_2OH + CH_3OH \leftrightharpoons HOCH_2OCH_2OCH_3 + H_2O$$

Another mechanism may play an important role in preventing paraformaldehyde precipitation. According to this mechanism, methanol stabilizes formalin solution by inducing (1) depolymerization, which causes a decrease in the concentration of the higher and less soluble homologues, and (2) formation of more soluble products (hemiacetals) (Dankelman and Daemen, 1976).

Formaldehyde, a monoaldehyde, is the simplest member of the aldehyde fixatives. It reacts in an aqueous solution as methylene glycol (A), and in acid conditions it contains higher concentrations of the more reactive electrophile (B).

$$\underset{H}{\overset{H}{\diagdown}}C\underset{OH}{\overset{OH}{\diagup}} \qquad \underset{H}{\overset{H}{\diagdown}}\overset{+}{C}\underset{}{\overset{OH}{\diagup}} \longleftrightarrow \underset{H}{\overset{H}{\diagdown}}C{=}\overset{+}{OH}$$

(A) (B)

Formalin consists of free formaldehyde, methylene glycol, and polyoxymethylene glycols, and in very small concentrations methylal, methyl-formate, trioxane, and acetals of the polyoxymethylene glycols (Walker, 1964). Monomeric formaldehyde probably exists as $HOCH_2OH$ in solution. It has a strong tendency to polymerize into a dimer, trimer, etc., with the general formula $HO(CH_2O)_n H$. Only a small part (11%) of the formaldehyde is monomeric in formalin; when formalin is diluted to 2%, the monomer predominates. The formation of formaldehyde oligomers is as follows:

$$CH_2O + H_2O \rightleftharpoons HOCH_2OH$$

$$HOCH_2OH + (n - 1)HCHO \rightleftharpoons HO(CH_2O)_n H$$

The concentration of formaldehyde in the solution determines the maximum value of n and the distribution of these oligomers. The higher the value of n, the less will be the solubility of oligomers in water. The solubility of the oligomers in water also decreases at low storage temperatures.

Since formalin contains various impurities, formaldehyde produced by the dissociation of paraformaldehyde powder is more efficient as a fixative. Formaldehyde is generally not recommended for preserving ultrastructure except in special cases or in combination with glutaraldehyde or with other fixatives. This monoaldehyde, for example, has proved useful for fixing very dense tissues such as seeds, which are not penetrated easily by glutaraldehyde. In this connection it has been used alone or mixed with glutaraldehyde. Since chromic acid reacts with both nucleic acids and proteins, this reagent in combination with formaldehyde has proved effective in stabilizing nucleic acids of phages and viruses, and their integrity for the most part is preserved during dehydration and embedding (Langenberg and Sharpee, 1978).

Formaldehyde (4%) is used in prefixing large slices of surgically removed tissues before they are cut into small pieces and fixed with glutaraldehyde; a prefixation for 30 min at 4°C is normally used. However, caution is warranted in interpreting the ultrastructure of biopsy and surgical specimens fixed by immersion in formaldehyde, for they contain a variety of fixation artifacts that can be easily confused with subcellular abnormalities caused by a disease. A case in point is the presence of septate junctions between digestive organelles in phagocytic cells in human malekoplakia biopsy specimens fixed with formaldehyde; such junctions are thought to be fixation artifacts (Nistal *et al.*, 1978). The junctions are absent when specimens are fixed with a mixture of glutaraldehyde and formaldehyde. Furthermore, such septate junctions occur in both normal and pathological biopsies fixed with formaldehyde. Formaldehyde fixation causes a general swelling and distortion of cytoplasmic organelles.

Reaction with Proteins

The reactions of formaldehyde with proteins are numerous and well understood. Formaldehyde is thought to cause cross-linking of peptide chains, and the reactive functional groups so far identified include amino, amido, guanidino, thiol, phenolic, imidazolyl, and indolyl. The participation of lysine in the cross-linking reaction has been confirmed (Caldwell and Milligan, 1972).

In the reaction of formaldehyde with proteins, the first step involves the free amino groups with the formation of amino methylol groups, which then condense with other functional groups such as phenol, imidazole, and indole to form methylene bridges (—CH$_2$—). The reactions of formaldehyde with proteins follow (Lojda, 1965):

1. Addition:

(a) RH + H—C$\overset{O}{\underset{H}{}}$ ⇌ R—CH$_2$OH (with groups —NH$_2$, ═NH—CONH—)

with—OH ⟶ Hemiacetal
with—SH ⟶ Hemithioacetal

(b) $\underset{}{>}$CH + H—C$\overset{O}{\underset{H}{}}$ ⟶ $>$C—CH$_2$OH

2. Condensation:

(a) —NH$_2$ + H—C$\overset{O}{\underset{H}{}}$ ⇌ —N═CH$_2$ + H$_2$O

(b) ⟨O⟩—OH + CH$_2$(OH)$_2$ + HO—⟨O⟩ ⟶ ⟨O⟩—O—$\overset{H}{\underset{H}{C}}$—O—⟨O⟩ + 2H$_2$O

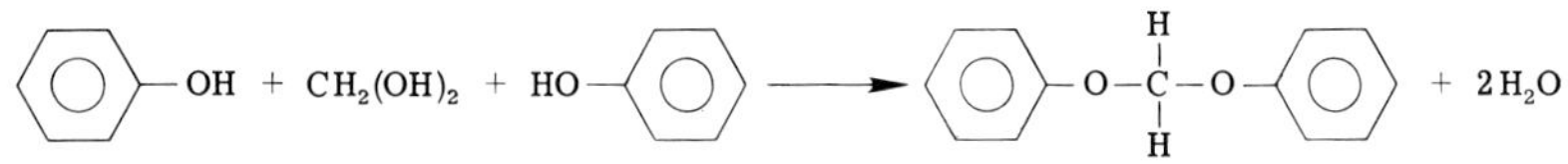

(Phenoplasts)

(c) R—CH$_2$OH + RH ⇌ R′—CH$_2$—R′ + H$_2$O

(Formation of
methylene bridges)

Formaldehyde reacts readily with compounds containing an active hydrogen atom and forms additive compounds such as hydroxymethyl. The addition compounds are formed freely with amino, imino, and peptide groups. The preceding reactions show, for example, that with hydroxyl and sulfhydryl groups the addition compounds formed are hemiacetal and hemithioacetal, respectively. The addition compounds in turn react (condense) with other compounds containing an active hydrogen atom, which results in the formation of methylene bridges (—CH$_2$—). The occurrence of these bridges is considered responsible for the fixation of proteins by formaldehyde under conditions appropriate for electron microscopy.

Reactions with wool keratin indicate the types of compound formed and were identified to be δ-N-hydroxymethylglutamine (I) and ε-N, ε-N'-methylenedilysine (II) (Caldwell and Milligan, 1972):

$$\underset{HOCH_2NHCCH_2CH_2CHCO_2H}{\overset{O\qquad NH_2}{\overset{\|\qquad |}{}}} \quad (I)$$

$$\underset{CH_2(NHCH_2CH_2CH_2CH_2CHCO_2H)_2}{\overset{NH_2}{\overset{|}{}}} \quad (II)$$

The formation of product (II) involves addition of a lysinyl residue side chain amino group to formaldehyde, which is followed by condensation of the resulting *N*-methylol derivative with another lysinyl residue side chain. This product represents a cross-link.

The formaldehyde condensation method of Falck and Hillarp has demonstrated that N-terminal tryptophan dipeptides and tryptophan-containing polypeptides and proteins react with the monoaldehyde giving strong fluorescence (Håkanson and Sundler, 1971). Probably all peptides and proteins with NH_2-terminal tryptophan residues are capable of forming fluorophores on formaldehyde condensation. Strong formaldehyde-induced fluorescence in tissues with low levels of known monoamines has also been documented over the years (Falck *et al.*, (1962).

Considerable time and effort has been spent attempting to identify the reaction sites in the cross-linking phenomenon in model compounds. Formaldehyde residues, for instance, are found between aryl carbon and nitrogen atoms:

$$R\ \overset{*}{C}H_2NHCH_2CO_2H$$
$$R\ \overset{*}{C}H_2NHCHCH_3CO_2H$$

R =

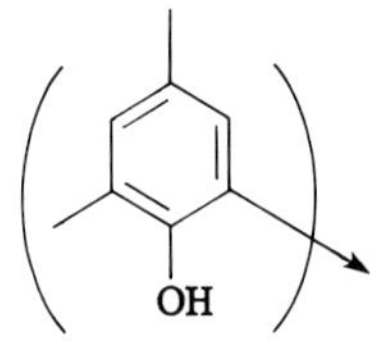

or between two nitrogen atoms:

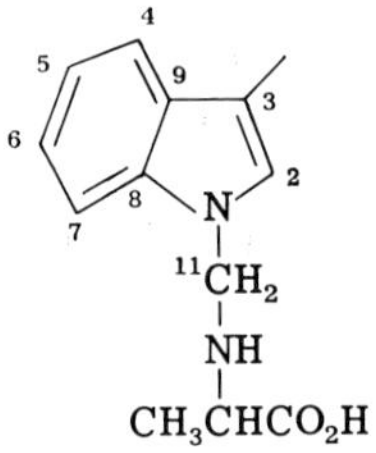

The cross-linking reaction involving tyrosine and lysine and formaldehyde is represented by the following scheme (Dewar *et al.*, 1975):

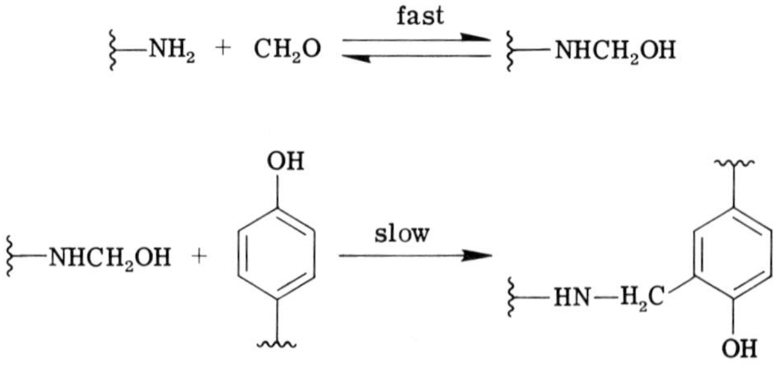

Dewar *et al.* (1975) have identified the site of cross-linking of amino acids by formaldehyde in peptide model compounds.

It should be noted that the majority of the preceding reactions are reversible, and formaldehyde for the most part is removable by washing with water. The reactions of formaldehyde with proteins are influenced by several factors including the concentration of the fixative solution, temperature, pH, and duration of fixation. In general, higher levels of these factors result in an increased binding of formaldehyde. The maximum binding seems to occur at pH 7.5–8.0. By using interference microscopy, it was calculated that after 2 hr fixation with formalin the bound formaldehyde constituted 3.6% of the dry mass of isolated nuclei (Abramczuk, 1972). At higher pHs, formaldehyde transforms collagen fibrils into a gel that resists degradation *in vivo* as well as *in vitro* by collagenase (Harris and Farrell, 1972). The presence of three to eight cross-links per molecule of collagen is sufficient to retard collagenolysis.

The addition of acrolein to formaldehyde minimizes the extraction of proteins from the tissue. According to Artvinli (1975), bicarbonate–formaldehyde decreases the solubility of tissue proteins and preserves glycogen better than glutaraldehyde. It has been suggested that the superiority of this fixative is due to the presence of bicarbonate ions, which penetrate cells readily and thus act as a buffer during the liberation of hydrogen ions resulting from the interaction of formaldehyde with free amino groups of proteins. In the presence of bicarbonate, formaldehyde reactions with proteins become irreversible. This results in a continuous decrease in the number of formaldehyde molecules in the tissue through their binding to proteins. This phenomenon may maintain the difference between formaldehyde concentration at the site of actual fixation in the tissue and fixative solution, thus preventing the deceleration of the diffusion rate (Artvinli, 1975).

Reaction with Lipids

Although formaldehyde-fixed tissues fail to show lipids after dehydration and embedding, formaldehyde is capable of changing to some extent the physical and chemical properties of lipids. Available evidence indicates that formaldehyde reacts at least with unsaturated fatty acids in tissues during fixation and the site of reaction is double bonds (Wolman and Greco, 1952; Wolman, 1955; Jones and Gresham, 1966). Jones (1969b) isolated and characterized the new products formed in the reaction of formaldehyde with pure unsaturated fatty acids under conditions identical with those of fixation. It is also known that carbonyl groups introduced by formaldehyde during fixation are demonstrable by Schiff's reagent. However, since formaldehyde-fixed specimens fail to show lipids after dehydration, it is thought that this monoaldehyde is the fixative of choice when lipid extraction is desired. After fixation with formaldehyde, the lipid-depleted membranes consist largely of protein.

Reaction with Nucleic Acids

Formaldehyde is used extensively in structural and functional studies of nucleic acids and nucleoproteins as an agent for causing denaturation as well as for preventing renaturation of these biomolecules. The most important features of formaldehyde interaction with nucleic acids and nucleoproteins are (1) it reacts with both the proteins and nucleic acids without destroying polypeptide or polynucleotide chains, (2) it modifies the bases and forms cross-links in nucleic acids, (3) it preserves the conformation of nucleoproteins, and (4) its small molecule penetrates through the protein shell into nucleic acids. The use of formaldehyde as a probe for nucleic acid structure has provided significant information on the reaction mechanism of the monoaldehyde with DNA and RNA. Formaldehyde has been used extensively as a probe for determining the secondary structure of DNA as well as understanding the mechanism of DNA unwinding (Vologodskii and Frank-Kamenetskii, 1975; Stevens *et al.*, 1977).

It has been shown that formaldehyde reacts with the amino groups of DNA nucleotides and that the reaction with formaldehyde proceeds much more rapidly with free nucleotides or denatured DNA than with native DNA (Stollar and Grossman, 1962). The studies of native calf thymus DNA demonstrated that the overall reaction can be formulated as an equilibrium conformational "opening" step, followed by a slow chemical reaction of formaldehyde with nucleotide amino groups normally involved in interchain hydrogen bonding (von Hippel and Wong, 1971).

Li (1972) has shown that histones are fixed on DNA in nucleohistones if the latter are treated with formaldehyde (1%) at 0°C for 24 hr; probably covalent bonds are involved in the fixation of histones on DNA. As a result of this reaction, methylene bridges occur on the two neighboring amino groups of basic residues as well as on the three aminated nucleotide bases. After formaldehyde fixation, histones are no longer acid- or salt-dissociable from DNA.

When cells are pretreated with formaldehyde that is then removed, DNA resistance to denaturation (e.g., thermal denaturation) increases, presumably because of chromatin cross-linking (Traganos *et al.*, 1975). The reaction of formaldehyde with DNA, under denaturation conditions, results in a small portion of the monoaldehyde being bound in the secondary structure of DNA (Beer and Thomas, 1961; Inman, 1966). After reaction with formaldehyde, DNA is not capable of renaturation. However, formaldehyde itself is a denaturing agent of DNA. The denaturation effect of formaldehyde on chromatin may be caused by histone–DNA or histone–histone cross-linking. It has been suggested that formaldehyde fixation causes redistribution of histones among DNA molecules (Polacow *et al.*, 1976).

Formaldehyde is widely used for the stabilization of viruses. This monoaldehyde has been used to produce RNA–protein complexes in rodlike plant viruses. Such complexes are resistant to the action of high temperatures, detergents, and mercaptoethanol (Mazhul *et al.*, 1978). Formaldehyde also cross-links DNA and proteins.

Ultraviolet absorption spectra of formaldehyde-fixed plant tissue specimens indicate that a major portion of the DNA is retained (Saurer, 1969). That formaldehyde inactivation of phage TL *E. coli* is caused by reaction with the DNA has been demonstrated (Sauerbier, 1960). It has been shown that fixation with formaldehyde at room temperature blocks DNAse action, but fixation at 0°C produces little effect on DNAse action (Swift, 1966). It has been suggested that formaldehyde fixation at room temperature partially inhibits but does not prevent DNAse action. The resistance of formaldehyde-fixed DNA to extraction by both DNAse and mineral acids in these studies may be due to denaturation of the DNA molecule.

It is thought that denaturation is initiated at AT-rich regions in the interior of the DNA molecule. Overall denaturation rate increases with increasing pH and temperature. Interchain hydrogen bonds break prior to reaction (McGhee and von Hippel, 1977). Denaturation involves adduct formation with the functional groups of both thymine and adenine. The thymine reaction seems to be reversible, whereas the adenine reaction is effectively irreversible. It has been suggested that the bases unstack prior to reaction with formaldehyde (McGhee and von Hippel, 1977).

According to Rosenfeld *et al.* (1970) the complete kinetics of the reaction with duplex DNA is sigmoidal; on the other hand, the kinetics of the RNA (e.g., tRNA) is pseudo first order at room temperature. This can be explained by the fact that some of the bases in these molecules are not hydrogen bonded. The implication is that open regions react with formaldehyde rapidly, whereas closed (base-paired) regions react slowly or not at all. The formaldehyde reaction has been used in electron microscopy to demonstrate regions of differing stability in the DNA helix (Beer and Thomas, 1961; Inman, 1967).

Studies by other workers, however, suggest that the majority of the RNAs react with formaldehyde, whereas native DNA does not at room temperature. This may be due to the inaccessibility of the bases in the double helical structure. It is expected, nevertheless, that the formaldehyde reaction will be promoted by elevated temperatures and solvent conditions that disrupt the secondary structure. It has been known for some time that the native DNA structure is not infinitely stable with respect to interaction with formaldehyde.

Generally, it is thought that the major reaction of formaldehyde with nucleic acids is largely reversible. It is well known that the binding of formaldehyde with amino groups of the bases is reversible. Eyring and Ofengand (1967) indicated, for instance, that immediately after the reaction of formaldehyde with nucleosides, hydroxymethylation occurred that was reversible. Similarly, the reaction of formaldehyde with polynucleotides was completely reversible (Haselkorn and Doty, 1961). The reaction occurs in two steps for helical polynucleotides: the first is the denaturation of the helix, and the second is the formaldehyde addition to the amino groups freed by the denaturation. On the other hand, Lewin (1966) found that formaldehyde reacts not only with basic amino groups of adenine, cytosine, and guanine, but also with acidic imino groups of thymine and guanine. Imidazole ring nitrogens of histidine residues also react with formaldehyde to form N-hydroxymethyl derivatives (Martin *et al.*, 1975).

It should be noted that under certain conditions the reaction between formaldehyde and nucleic acids can be irreversible. Collins and Guild (1969) showed that, at least, one reaction of formaldehyde with DNA occurs at 100°C at pH 8.0, which is irreversible at 20-37°C. Other studies have also suggested that formaldehyde may form stable methylene bridges ($-NH-CH_2-NH-$) between nucleotides (Staehelin, 1958; Feldman, 1962; Alderson, 1964). Finally, it may be said that the binding of formaldehyde with nucleic acids ranges from easily reversible to only partially reversible even after prolonged dialysis.

Reaction with Carbohydrates

Formaldehyde does not preserve soluble polysaccharides but prevents the extraction of glycogen, provided the duration of fixation is not too long. Quantitative studies indicate that glycogen is preserved in the rat brain perfused with formaldehyde (Guth and Watson, 1968). During initial ischemia, glycogen is decreased to ~60%, but after the arrival of the aldehyde fixative, glycogen remains more or less constant. Acid mucopolysaccharides are not preserved by formaldehyde unless they are bound to proteins. Formaldehyde, on the other hand, is very effective in fixing mucoproteins.

Preparation of Paraformaldehyde

Formaldehyde generated from paraformaldehyde powder is the most effective monoaldehyde for preserving enzyme activity; it is preferred over formalin (as it is commercially available) because the latter contains undetermined amounts of impurities such as methanol and formic acid. Paraformaldehyde polymer is slowly and partially soluble in water at 60°C. The addition of 1.0 mol/l NaOH accelerates and completes dissolution of the powder. When paraformaldehyde is dissolved in the alkaline form of the buffer containing disodium phosphate or sodium cacodylate, it may be cleared without the addition of NaOH. Water produces a partial hydrolysis of the polymer, which is completed by the presence of a small amount of alkali. Solutions should be prepared immediately prior to use.

Formaldehyde Fixatives

Method I. Formaldehyde–picric acid. Stefanini *et al.* (1967) introduced a modification of Bouin's fluid, which contains 0.02% picric acid and 2% paraformaldehyde (PAF) buffered to pH 7.2 with 0.1 mol/1 phosphate. This fixative has an osmolality of 895 mosmols and a pH of 7.3. It can be prepared in large quantities and stored at room temperature for months without deterioration. The fixative penetrates faster than conventional fixatives and is said to be affected only a little by tissue fluids and plasma proteins.

The preceding formulation has proved especially useful for fixing human and rabbit ejaculated spermatozoa. The first two drops of the ejaculate are collected directly into 20 ml of PAF at room temperature. Proteinaceous seminal plasma coagulates rapidly into clumps of different sizes. These clumps are discarded and the remaining nearly pure cellular suspension is centrifuged for 10 min at 1000 rpm. The supernatant is discarded and the pellet of highly concentrated spermatozoa is washed for 15 min in the phosphate buffer. The total duration of fixation in PAF is 25 min. The pellet is then postfixed for 15 min in 10 ml of 1% buffered OsO_4 and rapidly dehydrated. The pellet is broken into smaller sizes and embedded according to standard procedures.

The fixative has also proved useful for fixing whole small human embryos, rabbit embryos, and whole kidneys of adult rabbits and rats (Zamboni and De Martino, 1967) and for preserving antigenicity and ultrastructure in anterior pituitary glands of rats (Kawarai and Nakane, 1970).

Method II. Formaldehyde–chromic acid. This formulation has proven effective in preserving mosaic virus in plant tissues. A mixture of formaldehyde (0.2%) and chromic acid (0.1%) in cacodylate buffer (pH 7.2) is used to fix virus-infected leaf tissue for 16–24 hr at 4°C. Initial fixation is carried out under vacuum until the blocks sink to the bottom of the vial; vacuum should be terminated to check whether the blocks will sink. After a thorough wash (four washes of 15 min each) in the cold buffer, the blocks are postfixed with 1% OsO_4. Although tissue blocks may remain green, OsO_4 reaction with membranes has been completed. The results are shown in Fig. 3.7; chloroplast ribosomes and a virus bundle are visible.

ACROLEIN

Acrolein (CH_2=CH—CHO), a highly reactive, volatile liquid, which owes its common name to its acrid odor when formed by the scorching of oils and fats. It is an olefinic aldehyde with three carbon atoms and conjugated double bonds:

$$H_2C=CH-C\overset{\displaystyle O}{\underset{\displaystyle H}{\diagdown}}$$

Essentially, it is an α,β-unsaturated carbonyl derivative and is the simplest in the series. Acrolein can be prepared by contacting propylene and oxygen in the vapor

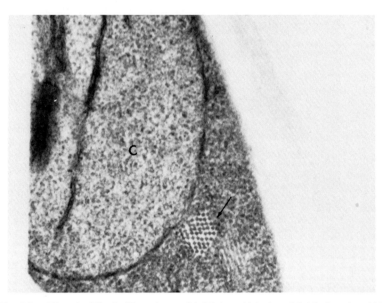

Fig. 3.7. Wheat leaf fixed with a mixture of 0.2% formaldehyde and 0.1% chromic acid in 0.1 mol/l cacodylate buffer (pH 7.2) for 16–24 hr at 4°C, followed by four washes (15 min each) in the buffer and postfixation with 1% OsO₄ in the same buffer. Note the presence of ribosomes in the chloroplast (C); ribosomes are generally absent in chloroplasts of conventionally fixed tissues. The arrow points to the Agropyron mosaic virus bundle in the cytoplasm. (W. G. Langenberg, unpublished.)

phase in the presence of a suitable catalyst such as bismuth molybdate, but the main source of acrolein is the dehydration of glycerol at high temperatures. It has a molecular weight of 56.062 and a viscosity of 0.393 centistokes at 20°C. It is freely soluble in water and absorbs intensely in the neighborhood of 211 nm in aqueous solutions. Acrolein undergoes most reactions of ethylenic compounds, including addition of halogens and hydrogen halides at the ethylenic bond. It has a propensity to react with substances that bear the sulfhydryl group, or thiols.

Luft (1959) introduced acrolein as a primary fixative for electron microscopy. Because it penetrates and reacts faster than most other fixatives and causes little shrinkage, it was originally recommended as an alternative to OsO₄ as a primary fixative. Since the rate of glutarldehyde penetration is relatively slow (~0.4 mm in rat liver in 1 hr), the core of the tissue block (1 mm³) is not well fixed. Therefore, acrolein, which penetrates much faster (~1 mm in rat liver in 1 hr) than glutaraldehyde or formaldehyde, is useful where fixative penetration is a problem.

The problem of fixative penetration is encountered with specimens that are large, dense, or covered by impermeable substances such as waxes and chitin.

Thus, it is effective in the fixation of large tissue blocks that, for practical reasons, cannot be dissected into smaller pieces. Acrolein can also be utilized in studies where only surface layers need to be examined, which is true in most of the studies carried out by the SEM. The surface layers of large tissue blocks can be fixed by a short exposure ($\sim$20 min) to acrolein. Homogenous tissues such as muscle often do not require overall fixation. Because of its volatility, acrolein vapor is ideal for the fixation of mineralized tissues such as bone. Since aqueous solutions induce alterations in the chemical and structural arrangement of the mineralized phase of these tissues and in the relationships between the mineral phase and certain of its intracellular and extracellular components (Landis *et al.*, 1980), the advantage of anhydrous fixation of calcified tissues is apparent. A brief exposure ($\sim$15 min at room temperature) to undiluted acrolein vapors is sufficient for adequate fixation. Dehydration can be carried out with anhydrous Cellosolve for 12 hr.

Acrolein has been used for fixing yeast (Schwab *et al.*, 1970) and mammalian embryos at early stages (Saito *et al.*, 1973). Satisfactory preservation of algae, fungi, and higher plants was obtained by using a mixture of acrolein and glutaraldehyde (1:1) and then postfixing with $KMnO_4$ (Hayat, 1968a,b). Mayor and Jordan (1963) obtained both the structural preservation of viruses and specific antigenic properties through acrolein fixation. Acrolein in combination with glutaraldehyde, formaldehyde, and DMSO has proved useful in the preservation of amphibian unfertilized eggs and fertilized embryos (Kalt, 1971; Kalt and Tandler, 1971; Singal and Sanders, 1974). A mixture of acrolein and glutaraldehyde proved useful for preserving eggs of *Ambystoma* (Bluemink, 1970). Satisfactory preservation of seeds during the late developmental and early germinative stages was obtained by a mixture of acrolein, glutaraldehyde, and formaldehyde followed by osmication (Mollenhauer and Totten, 1971). Frog skeletal muscle fixed with 2% acrolein in Ringer and postfixed in 1% OsO_4 in Ringer showed a shrinkage of only $\sim$10.4% (Davey, 1972; Birks and Davey, 1972).

Studies by Sandborn (1966) showed that a mixture of acrolein and glutaraldehyde gave a better preservation of cytoplasmic microtubules than that obtained with either aldehyde alone. Schultz and Case (1968), on the other hand, demonstrated that neuronal microtubules were lost when the tissue was fixed in acrolein alone by vascular perfusion even though other morphological features were preserved satisfactorily. It is accepted now that glutaraldehyde alone at room temperature should be used for the preservation of cytoplasmic microtubules.

The use of acrolein in enzyme cytochemistry has been examined (Sabatini *et al.*, 1963; Flitney, 1966; von Feustel and Geyer, 1966; Saito and Keino, 1976) and the results lack consensus. However, the extreme reactivity of acrolein severely restricts its use in enzyme cytochemistry. Acrolein is recommended only when glutaraldehyde or formaldehyde is unsuitable.

Acrolein fixatives are prepared usually without purifying commercially available lots. The fixative solutions are usually prepared with 10% acrolein and are hypotonic. Osmolarity can be raised by adding calcium or magnesium chloride, but not sucrose. Specimens fixed with acrolein are usually postfixed in OsO_4. The preparation and use of various acrolein fixatives are given later.

Reaction with Proteins

Among all the tested monoaldehydes, acrolein is probably the most reactive. It shows a high specificity for proteins and is thought to form linkages with amino groups and yields initially

$$CH_2{=}CH{-}CH{-}\overset{|}{\underset{|}{N}}{-}$$
$$OH$$

which then is rapidly polymerized (van Winkle, 1962). Acrolein reacts rapidly at room temperature with sulfhydryl groups through its ethylene linkage. Reactions of this type lead to tissue-bound aldehyde groups. The reaction leading to a protein-bound aldehyde group, however, is due to a reaction not only of sulfhydryl groups but also of certain other groups present in proteins. It has been suggested, for instance, that the double bond of acrolein reacts with SH, aliphatic NH_2, and NH, and imidazole groups (van Duijn, 1961; Jones, 1969a). The carbonyl groups introduced by acrolein can be demonstrated with Schiff's reagent. Fixation with acrolein has been employed to bring out protein by staining the free aldehyde group of the protein-bound acrolein with Schiff's reagent or silver nitrate methenamine (van Duijn, 1961; Marinozzi, 1963).

A detailed study of the effect of various fixatives on fluorogenic detection of primary amines in plant tissues demonstrated the usefulness of acrolein (Bruin *et al.*, 1976). This study suggests that after fixation with acrolein, primary amino groups remain available and react with fluorescamine, causing a strong fluorescence of the tissue. This suggestion is somewhat inconsistent with the previously mentioned reaction between acrolein and several groups in protein. A possible explanation is that "hidden or buried" amino groups do not react with acrolein, perhaps because of steric hindrance. In other words, these groups are somehow protected by acrolein.

Reaction with Lipids

Some information is available regarding the effect of acrolein on the extraction of tissue lipids. Published data on the effects of acrolein and formaldehyde on the changes of lipid constituents of rat brain indicate that the former is much more rapid in action and causes less extraction of lipids than the latter (Norton *et al.*, 1962). Dog erythrocytes fixed with pure acrolein having a concentration of up to 3% show less than 5% loss in cell phospholipids; whereas 10% acrolein and acrolein containing hydroquinone stabilizer cause a much higher loss of phospholipids (Carstensen *et al.*, 1971).

The preceding and other studies imply that acrolein reacts readily with lipids, especially with membrane phospholipids. The resulting new, active site on the phospholipid may achieve cross-linking with nearby protein (Carstensen *et al.*, 1971). Acrolein is thought to react with fatty acids (Hall and Stern, 1955) forming esters of β-hydroxypropionaldehyde:

$$R—COOH \xrightarrow{\text{CH}_2=\text{CH}—\text{CHO}} R—COO—CH_2CH_2CHO$$

which is capable of reacting with proteins. Linoleic acid reacts much more rapidly than oleic acid with acrolein (Jones, 1972).

The preliminary studies on the reaction of acrolein with unsaturated fatty acids indicate the possibility of the production of compounds having the following structure (Jones, 1972); such compounds could react further with more acrolein.

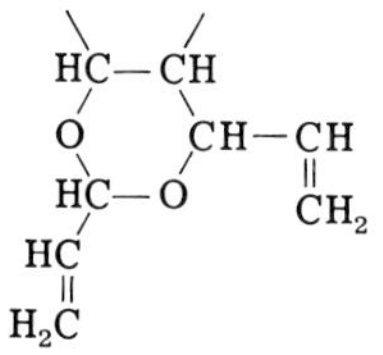

$$
\begin{array}{c}
\diagdown \quad \diagup \\
HC—CH \\
\diagup \quad \diagdown \\
O \quad\quad CH—CH \\
\diagdown \quad \diagup \quad\quad \| \\
HC—O \quad\; CH_2 \\
\diagup \\
HC \\
\| \\
H_2C
\end{array}
$$

It is likely that at lower concentrations of acrolein, after the initial reaction, the resulting active sites on phospholipids undergo a reaction with an adjoining site of protein, which results in the cross-linking. At higher concentrations, the new reactive sites react with a second molecule of acrolein instead of protein, which results in the formation of derivatives that are extractable (Carstensen *et al.*, 1971; Jones, 1969a; Schmid and Takahashi, 1968; von Feustel and Geyer, 1966). This assumption could be a possible explanation for the increased extraction of phospholipids after fixation with concentrated solutions of acrolein.

Precaution in the Handling of Acrolein

Acrolein has a strong tendency to polymerize on exposure to light, air, and certain chemicals with generation of heat. The containers for storing this reagent must be cleaned thoroughly before use, and special care should be taken to prevent contamination since even traces of contaminants can initiate polymerization. Acrolein available commercially contains an oxidation inhibitor (usually 0.1% hydroquinone), and thus treated it can be stored in a cool place for several months without significant polymerization. The inhibitor can be removed by distillation (Carstensen *et al.*, 1971); however, purification of acrolein is not necessary. For routine use the reagent should be stored in small brown bottles in a cool place and must be kept tightly stoppered. Acrolein is considered contaminated if its solution is turbid or if a 10% solution in tap water shows pH below 6.4. Such acrolein must be redistilled before use. The method of distillation is explained in detail by Albin (1962).

Acrolein is a hazardous chemical because of its flammability and extreme reactivity. It is highly toxic through vapor and oral routes of exposure, is irritating to respiratory and ocular mucosa, and induces uncontrolled weeping. Acrolein is moderately toxic through skin absorption and is a strong skin irritant even at low concentrations. Although acrolein is highly toxic by the vapor route, the sensory response to very low vapor concentration gives adequate warning. The physiological perception of the presence of acrolein begins at 1 ppm, at which concentration an irritating effect on the eyes and nasal mucosa is felt. Thus, there is little risk of acute intoxication because, as stated earlier, the lachrymatory effect compels one to leave the polluted area. It does not have toxicologic effects from repeated exposure to low tolerated concentrations. Laboratory safety manuals, however, quote a threshold level between 0.05 and 0.1 ppm of air. Above this level acrolein sensitizes the skin and respiratory tract and causes bronchitis and pneumonia in experimental animals. In the case of human beings, when there is contact with acrolein, cutaneous or mucosal local injury may be observed. Human and animal toxicity caused by exposure to acrolein has been discussed by Izard and Libermann (1978).

It is strongly recommended that gloves and fume hoods be used when handling this chemical. In the preparation of solutions, acrolein should be slowly added with stirring rather than the reverse. Any acrolein that contacts the skin should be removed immediately by washing with soap and water. Waste acrolein should be disposed of by pouring into 10% soldium bisulfite solution, which acts as a neutralizer.

Acrolein Fixatives

As stated earlier, acrolein is highly volatile and toxic. Therefore, its solutions should be prepared under a fume hood. The following formulations are recommended.

Method I.   10% Acrolein in 0.1 mol/l buffer

Buffer (0.2 mol/l)	50 ml
Acrolein	10 ml
Distilled water to make	100 ml

Method II.   1% Acrolein–2.5% glutaraldehyde in 0.1 mol/l buffer

Buffer (0.2 mol/l)	50 ml
Acrolein	1 ml
Glutaraldehyde (25%)	10 ml
Distilled water to make	100 ml

The mixture is a general-purpose fixative and can be used for tough specimens that are not well fixed with glutaraldehyde alone. Fixation can be carried out at room temperature or at 4°C depending on the specimen type.

Method III. Acrolein–glutaraldehyde

Glutaraldehyde (25%)	12 ml
Acrolein	3 ml
Millonig's phosphate buffer	85 ml

Specimens are fixed for 1–3 hr; fixation is started in the cold and allowed to continue at room temperature. After a thorough rinse, specimens are postfixed in 2% OsO_4 in the same buffer for 2 hr in the cold. Alternatively, postfixation can be accomplished in 0.6% $KMnO_4$ for 30 min in the cold. The method is effective in fixing fungal cells (Dawes, 1969, 1979).

Method IV. 1.5% Acrolein–3% glutaraldehyde–1.5% paraformaldehyde

Buffer (0.2 mol/l)	5 ml
Acrolein (10%)	3 ml
Glutaraldehyde (10%)	6 ml
Paraformaldehyde (6%)	3 ml
Distilled water	1 ml

The mixture should be prepared immediately prior to use. The three aldehydes combine the three most important requirements for achieving successful fixation and they are preservation of general structure (glutaraldehyde) rapid penetration (acrolein), and rapid protein stabilization (paraformaldehyde). This mixture is especially effective in preserving dense animal and plant specimens (Hayat, 1980).

Method V. Acrolein–glutaraldehyde–paraformaldehyde

Acrolein	1.5%
Glutaraldehyde	2.0%
Paraformaldehyde	1.5%
Uranyl acetate	2.0%

The mixture is prepared in collidine buffer (pH 6.8) immediately prior to use. Specimens are postfixed with 2% OsO_4 in collidine buffer (pH 7.2). This mixture is effective in preserving highly hydrated, especially delicate plant cells and embryoids growing in culture for scanning electron microscopy (Homès, 1974).

Method VI. Acrolein–glutaraldehyde–formaldehyde

Distilled water	25 ml
Paraformaldehyde	2 g
Glutaraldehyde (25%)	5 ml
Acrolein (10%)	1 ml
Phosphate buffer (0.2 mol/l) to make	50 ml

The concentrations of formaldehyde, acrolein, and glutaraldehyde are 4%, 2%, and 2.5%, respectively. The pH is ~7.4 and osmolality varies from 1270 to 1350 mosmols. The grainy appearance of membranes is abolished by *en bloc* staining with uranyl acetate. This mixture was used by Rodriguez (1969) for vascular perfusion of the CNS.

Method VII.   1% Acrolein–3% glutaraldehyde–2% paraformaldehyde–2.5% DMSO

Paraformaldehyde (2%) in a buffer	44 ml
Acrolein	0.5 ml
Glutaraldehyde (25%)	5.0 ml
DMSO	1.25 ml
CaCl$_2$	0.008 g

The final concentration of CaCl$_2$ is 0.001 mol/l. The procedure has proved useful for fixing amphibian embryos (Kalt and Tandler, 1971).

Method VIII.   Acrolein–glutaraldehyde–paraformaldehyde

Glutaraldehyde (10%)	6 ml
Acrolein (10%)	3 ml
Paraformaldehyde (6%)	5 ml
Buffer (0.2 mol/l)	5 ml
Distilled water	1 ml

This mixture has proved especially effective for fixing dense tissues such as seeds rich in protein bodies and lipids (Mollenhauer and Totten, 1971). The three aldehydes are used in a mixture that should be prepared immediately prior to use.

Method IX.   Acrolein–glutaraldehyde–KMnO$_4$

Glutaraldehyde (3%)	5 ml
Acrolein (3%)	5 ml

The preceding mixture is prepared in 0.2 mol/l collidine buffer (pH 7.2). After fixation in the mixture and washing with the buffer, specimens are postfixed with aqueous 0.4% KMnO$_4$ for ~20 min at 4°C. The method is effective in fixing algae, fungi, and tissues of higher plants (Hayat, 1968a,b).

Method X.   Acrolein (2%)–glutaraldehyde (2%)–DMSO (50%)

Aqueous DMSO (50%)	42 ml
Glutaraldehyde (25%)	4 ml
Acrolein (25%)	4 ml

The preceding mixture is effective in fixing cells in culture (e.g., yeast) that resist penetration by glutaraldehyde. After primary fixation in this mixture, cells are postfixed in OsO$_4$. For additional details the reader is referred to Schwab *et al.* (1970).

POLYALDEHYDE

Low-molecular-weight aldehydes penetrate cells rather easily, extensively cross-link cell components, and are lethal. A high molecular-weight fixative, supravital polyaldehyde, has been introduced, that is thought not to enter the cell, is specific for external surfaces of plasma membranes, and allows continued metabolic activity of the cell (Phillips *et al.*, 1977). The fixative has been used for ultrastructural mapping with ferritin and for studying structural and functional elements of the cell surface.

The fixative is derived from a branched polyglucoside (dextran, mol. wt. 40,000). On exposure to periodate, many vicinal hydroxyls seem to be oxidized to aldehydes without disrupting the long-chain, branched polysaccharide. The oxidation of a large number of vicinyl hydroxyl groups should result in a molecule having multiple free reactive aldehyde groups, which probably react with susceptible groups (especially NH_2) in cell surface proteins, glycoproteins, glycolipids, and polysaccharides. The preparation of the fixative is as follows (Phillips *et al.*, 1977):

A flask containing 40 ml of 0.125 mol/l sodium acetate (adjusted to pH 5.5 with HCl) is equilibrated by stirring with nitrogen gas for 30 min at room temperature (23°C). Two grams of sodium metaperiodate are added with continuous stirring. Exposure to light is prevented by covering the flask with a foil. This is followed by adding 10 cm³ of 10% dextran, and the reaction is allowed to continue for 3 hr at room temperature in the presence of nitrogen. The solution is dialyzed against Dulbecco's PBS (pH 7.2), which had been equilibrated with nitrogen and then cooled to 4°C. Several changes of the dialysis fluid are made at 12–24 hr intervals. The polyaldehyde solution (2%) obtained is stored in a sealed container under nitrogen at 4°C.

ROLE OF ALDEHYDES IN QUANTITATIVE ELECTRON MICROSCOPY

The validity of quantitative and stereological data depends on the faithful preservation of not only cell and tissue dimensions but also relationships between cells and among cell organelles. Moreover, there should not be a substantial alteration in the quantity of intracellular substances such as amino acids and substrates. Drastic changes in the activity of cellular components such as enzymes are also undesirable. To accomplish this preservation, fixation parameters have to be optimal as well as standardized. The need to use altogether different techniques to prepare similar specimens for biochemical analyses and for untrastructural studies has hampered the achievement of more complete and cohesive quantitative data. Techniques such as microwave irradiation, rapid freezing, or immersion in Freon, used for biochemical analyses, are unsuitable for preserving ultrastructural details.

Although chemical fixation has limitations in the preservation of labile biochemical constituents, certain intracellular components can be quantitated in specimens fixed with aldehydes. Chemical fixative can allow quantitation of certain macromolecules as well as of components of low molecular weight. Such studies have been carried out only in a few tissue types. Glycogen in the rat brain can be preserved and quantitated after a careful vascular perfusion with formaldehyde (Guth and Watson, 1968). Studies with brain tissue indicated that glycogen levels decreased by ~60% during the initial operative procedure, but remained constant after fixation with aldehydes (Conger et al., 1978). In this study glycogen content declined ~90% in tissue that was not fixed with an aldehyde. In the same study several amino acids (e.g., glutamate and aspartate) were measured reliably in the tissue fixed by vascular perfusion. Cellular substances such as lactate, pyruvate, and alanine showed an increase.

Studies with norepinephrine and epinephrine suggest that primary amines are preserved in specimens fixed with glutaraldehyde, whereas secondary amines are extracted (Coupland and Hopwood, 1966; Hopwood, 1968a). Purines, phosphate esters, carbohydrates, and indole derivatives are extracted from brain after fixation with formaldehyde (Schneider and Schneider, 1967; Hopwood, 1969b). A number of metabolites (e.g., glucose, glucose-6-phosphate, lactate, ATP, and phosphocreatine) in the brain change rapidly when the blood supply is stopped (Lowry et al., 1964; Conger et al., 1978). However, vascular perfusion with aldehydes terminates the changes and decline in the level of some of these metabolites.

The preceding data indicate that various intracellular substances react quite differently to conditions before, during, and after fixation. For example, various amino acids may remain constant, or increase or decrease in level. Very little is known about the complex factors that govern these changes. A decline in the level of certain free amino acids may be due to their binding to the tissue as a result of cross-linking introduced by glutaraldehyde. A rapid reduction in ATP may facilitate the maintenance of certain metabolites since energy is needed for their metabolism. Conger et al. (1978) have discussed some of the factors involved in the changes observed in the specimen fixed with aldehydes.

TISSUE STORAGE

The most desirable practice in the preparation of specimens for electron microscopy is to fix and embed them immediately after their collection. Loss of cellular materials during their processing may increase when specimens have been stored prior to fixation. However, because of unavoidable circumstances, it is sometimes not possible to process specimens immediately after they have been collected or excised from the source, so they must be stored for a period of time. The necessity of storage may arise, for instance, in the case of surgical or biopsy specimens, where the facilities needed to process the specimen for electron microscopy are not at hand. In some cases electron microscopy of such specimens may have to await results obtained by other means.

The size of the tissue specimen and the objective of the study primarily determine the type of the fixative solution to be used. Specimens of very small size can be stored, especially for relatively short periods of time, in the aldehyde fixative solutions used routinely for electron microscopy. Medium-sized (~3 mm) tissue specimens can be stored for up to 12 months in a mixture of formaldehyde (4%) and glutaraldehyde (1%) in a buffer (pH 7.4) having an osmolality of 200 mosmols. Large (2–3 cm) tissue blocks require fixatives having the maximum penetration power. Such fixatives can be developed by using acrolein or formaldehyde to which may be added trinitro compounds. One such formulation is as follows:

2% formaldehyde
2% glutaraldehyde
0.1 mol/l phosphate buffer (pH 7.4)
0.2% trinitrocresol
1.5 mM calcium chloride

Another fixative for large specimens consists of 10% paraformaldehyde in 0.2 mol/l collidine buffer, which is supposed to extract tissue components and thus facilitate the penetration (Winborn and Seelig, 1970).

Storage of aldehyde-fixed specimens in a buffer is not recommended. It has been shown that storage of glutaraldehyde-fixed sections of rat liver in various buffers (phosphate, cacodylate, and Tris) results in progressive deterioration of peroxysome structure and diffusion of catalase into the adjacent cytoplasm (Fahimi, 1974). Adverse effects (morphological damage and diffusion of cellular materials) of storing aldehyde-fixed specimens in a buffer have not been adequately explored.

Osmium Tetroxide

Infrared absorption spectra (Woodward and Roberts, 1956; Dodd, 1959) and X-ray diffraction (Ueki *et al.*, 1965; Seip and Stølevik, 1966) studies indicate that the OsO_4 molecule is tetrahedral (see p. 178 for its structure and charges on the atoms), perfectly symmetrical, and therefore, as a whole, nonpolar. The nonpolarity of OsO_4 facilitates its penetration into charged surfaces of tissues, cells, and organelles, which is in part responsible for its effectiveness as a fixative. Precise calculations reveal that the Os—O bond length is 1.711 Å ± 0.003 and that the O—O distance is 2.795 Å (Seip and Stølevik, 1966).

The term "osmic acid" is a misnomer because this reagent is a electrolyte and forms no salts. Metallic osmium occurs naturally in close association with iridium and is innocuous, but OsO_4 slowly formed on exposure of the spongy metal to air is responsible for its toxicity. Osmium tetroxide has a molecular weight of 254.2, melts at 41°C, and boils at 131°C. It exists as faintly yellow monoclinic crystals having an acrid chlorinelike odor. The reagent dissolves in water rather slowly; solubility in water is ~7.24% at 25°C. It dissolves in neutral distilled water without change of pH. It also dissolves in benzene, paraffin oil, carbon tetrachloride (CCl_4), and saturated lipids. Osmium textroxide is 518 times more soluble in CCl_4 than in water. Osmium tetroxide has an absorption peak close to 250 nm and volatilizes readily at room temperature; its vapor pressure is given by $\log p = 10,000/4.57T + 5.49$. It is convenient to remember that 1% solution of OsO_4 in distilled water is 0.04 mol/l. Since aqueous solutions of OsO_4 are hypotonic, it is necessary to increase the osmolarity of the fixing solution by adding electrolytes or nonelectrolytes.

Osmium tetroxide acts not only as a fixative but also as an electron stain, and this is its major advantage over most other known fixatives. Reduced osmium imparts high contrast to the osmiophilic structures in the specimen. It also acts as a mordant; OsO_4-dependent enhancement of lead staining is well documented.

Osmium tetroxide is a noncoagulant type of fixative; that is, it is able to stabilize some proteins by transforming them into clear gels without destroying many of the structural features. Tissue proteins stabilized by this reagent are not coagulated by alcohols during dehydration. The most important application of OsO_4, however, is its use as a postfixative, for it preserves many lipids and imparts electron density to cell components. Osmium tetroxide fixation has been found to reverse the sign of birefrigence of pancreatic acinar cells from positive to negative (Munger, 1958).

In the early stages of electron microscopy, OsO_4 was the most widely used fixative for preserving biological specimens. The use of buffered, slightly alkaline, and hypertonic OsO_4 results in better preservation of cellular details than that obtained from acid fixation. The quality of tissue preservation obtained through fixation with glutaraldehyde followed by OsO_4 has not been surpassed by any other fixation combination. The fixative has also proved very effective in freeze-substituted tissues.

The main disadvantages of OsO_4 are its slow rate of penetration into most tissues and its inability to cross-link most proteins. As a result, the fine structure may be changed considerably prior to the completion of fixation when it is used as a primary fixative. Even when tissue blocks range in size from 0.5 to 1 mm^3, the core of the block may remain incompletely fixed. Consequently, OsO_4 is not used as a primary fixture in routine electron microscopy (see p. 175 for its use as a prefixative). On the other hand, the slow rate of penetration by OsO_4 during postfixation is not detrimental, because cell structure has already been partially stabilized by an aldehyde.

Some evidence indicates that postfixation with OsO_4 is undesirable when studying the cytochemistry of certain cellular proteins and membrane glycoproteins. A substantial amount of cellular protein in glutaraldehyde-fixed tissue is extracted following postfixation with OsO_4. It has been shown that postfixation with OsO_4 results in increased solubility of cell membrane proteins and in altered morphology (McMillan and Luftig, 1973). The ultrastructure of cell surface glycoprotein, seen in freeze-etch preparations, differs in tissue fixed with glutaraldehyde alone compared with that of tissue exposed to an aldehyde followed by OsO_4 (Ito, 1974). Carbohydrate-containing moieties of macromolecules are also vulnerable to damage when tissue blocks are exposed to OsO_4. Postfixation with OsO_4 diminishes periodic acid–thiocarbohydrazide–silver proteinate reactivity.

The knowledge of the extent to which distortion or even destruction of various tissue components proceeds during fixation with OsO_4 is insufficient. Information of this nature is extemely helpful in assessing the optimum duration of fixation to study a specific tissue type. In the absence of this kind of data, one has no choice but to find the most desirable duration of fixation through trial and error. The work done in this area deals primarily with the binding capacity of OsO_4 with lipids, which is greater than that with proteins and carbohydrates. This subject is discussed in detail later.

Osmium tetroxide should not be used for preservation of tissue prior to incubation for preserving enzymatic activity, for it is a potent inactivator of enzymes. It has been demonstrated, for instance, that OsO_4 is fully capable of inactivating malt diastase as well as pancreatic amylase (Sasse, 1965). The inhibition of enzyme is thought to be caused by the heavy metal atom of OsO_4 as is true in the case of other heavy metal compounds, and not due to the oxidative effect of OsO_4. Although in general the activity of digestive enzymes following fixation with OsO_4 seems to be reduced, there are exceptions. Fixation with OsO_4 does not have any apparent inhibitory effect, for instance, on the activity of elastase applied to thin sections for the specific digestion of elastin (Kadar et al., 1971). Osmium tetroxide is routinely used as a postfixative after incubation for enzymatic acitivity.

In summary, the wide usefulness and applicability of OsO_4 are based on the following three major characteristics of this metal:

1. Osmium can exist in nine oxidation states (0-VIII), five of which (0, II, III, IV, and VIII) possess reasonable stability and have a wide chemistry. An important consequence of this is that many pathways exist for the reaction of osmium with a great number of substrates.

2. Osmium tetroxide is soluble in both polar (aqueous) and nonpolar (hydrocarbon) media. This property is of importance for the penetration of OsO_4 into hydrophobic regions of tissue and cell and their subsequent fixation.

3. Osmium is electron opaque. This advantage obviously applies only to electron microscopy, but the brown or black color of many osmium reaction products not only is useful in light microscopic study but also aids in selecting specimens with the light microscope for electron microscopy.

Reaction with Unsaturated Lipids

For any understanding of the pheonomenon of fixation, including the appearance of lipids in the electron microscope, one must know the reaction products of OsO_4 with lipid compounds and the sites of reaction involved. Only such information would allow one to interpret electron micrographs in terms of the molecular structure of cell components.

There is ample evidence indicating that fixation with OsO_4 results in at least partial retention of lipids. It has been demonstrated, for instance, that fixation with OsO_4 is indispensable for subsequent demonstration of lipid droplets with

Sudan dyes in blood leukocytes (Coimbra and Lopes-Vaz, 1971). Since sudanophilia is not altered by removing the reduced osmium, the staining results are due to immobilization of finely dispersed free lipids rather than due to an interaction between Sudan dyes and osmium.

Since saturated fatty acids are not altered chemically by OsO_4, it is most likely that the unsaturated fatty acids are preferentially involved in the fixation process. Moreover, the facts that both oleic acid and olein (which contain double bonds) reduce OsO_4 whereas palmitic and stearic acids and their triglycerides (which do not contain double bonds) do not reduce OsO_4 lead to the conclusion that OsO_4 oxidizes olefinic double bonds. It is also known that hydrogenation and bromination of the double bonds prevent the fixation of phospholipids. Procedures (such as those using dilute permanganate or periodate) that oxidize the double bonds in the lipid bilayer prior to exposure to OsO_4 block the initial staining reaction with OsO_4. The use of an antioxidant (pyrogallol) that protects the double bonds results in strong staining of the membrane with OsO_4. Thus, the presence of a double bond in the lipid molecule is a prerequisite for reaction with OsO_4 under the conditions of preparative electron microscopy. Osmium tetroxide may also react with one or more of the highly unsaturated precursors of cholesterol, especially squalene with six double bonds or farnesol pyrophosphate with three double bonds (Wollman, 1972; Griffiths and Beck, 1977).

Indirect evidence supporting the interaction between OsO_4 and unsaturated lipids was obtained by studying frozen, fractured membranes. Studies of the effects of OsO_4 on the fracture properties of liposomes and *Mycoplasma laidlawii* cell membranes containing lipids of various degrees of saturation indicate that the degree of decrease in the membrane fracture faces was related to the increased number of double bonds and their position in the fatty acid chain (James and Branton, 1971). This decrease in the membrane fracture faces could be accounted for by the formation of strong, covalently bonded diesters or polymers across the bilayer.

The studies by Criegee (1936, 1938) and Criegee *et al.* (1942) resulted in the realization that oxidation of a double bond by OsO_4 would lead to the formation of monoesters in the following manner:

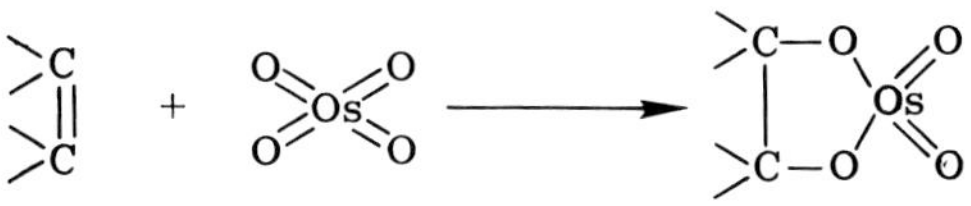

The monoester is not very stable and is easily hydrolyzed to form a diol and the osmate ion, $OsO_2(OH)_4^{2-}$. The diol can then react with an adjacent monoester to form a stable diester as follows:

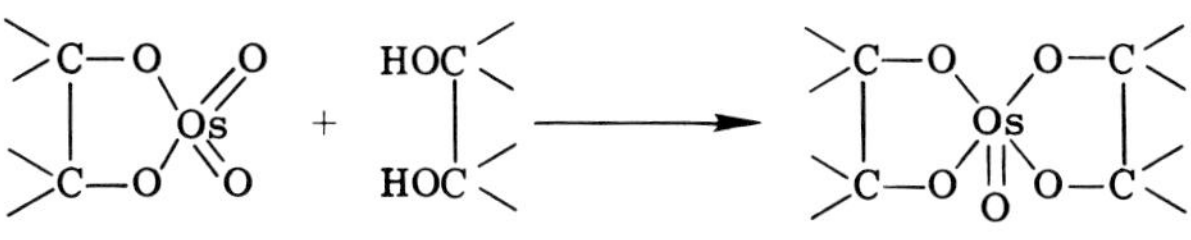

Since this diester results from the cross-linking of two molecules of olefin, the preceding reaction scheme seems to be a satisfactory explanation of lipid fixation by OsO_4. Hydrolysis and reduction of initially formed osmate esters are important processes in fixation and staining.

However, it is pointed out that the complex (the monomeric unliganded ester) postulated by Criegee is a reasonable intermediate, but there is no direct evidence for its existence. The complexes isolated by Criegee are actually dimers. A more feasible general reaction is as follows:

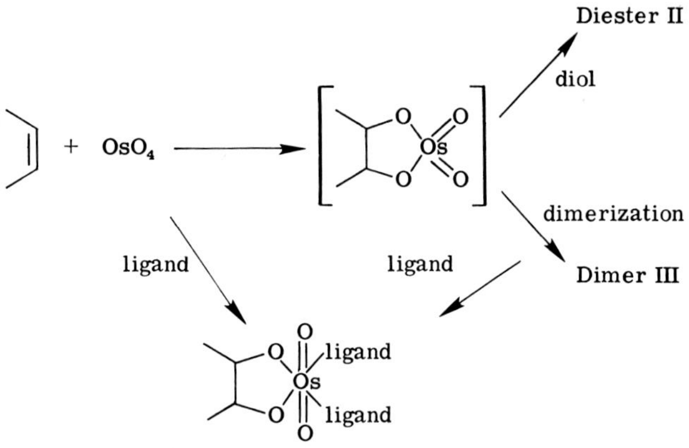

According to Wigglesworth (1957), the monoester may also react directly with a double bond to accomplish cross-linking as follows:

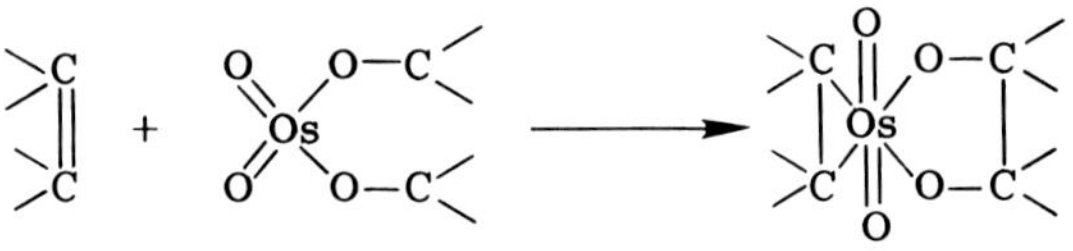

He also suggested that OsO_4 may form monoester linkages in the following manner:

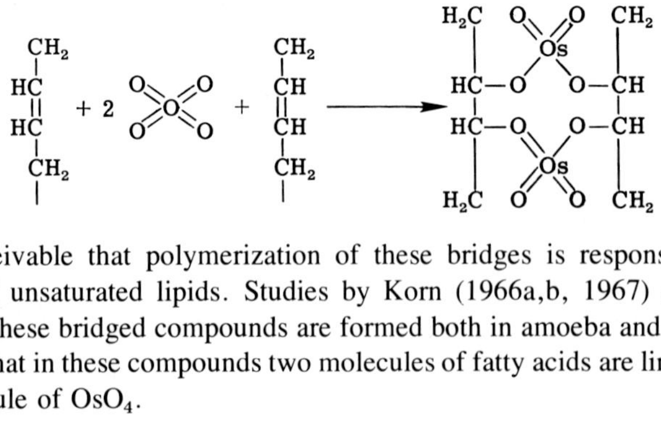

It is conceivable that polymerization of these bridges is responsible for the fixation of unsaturated lipids. Studies by Korn (1966a,b, 1967) conclusively show that these bridged compounds are formed both in amoeba and *in vitro*. He indicated that in these compounds two molecules of fatty acids are linked through one molecule of OsO_4.

As stated earlier, OsO_4 oxidizes olefins to vicinal diglycols through the intermediate formation of cyclic osmate esters. The intermediate product of this reaction is hard to identify. However, at least three types of esters have been proposed and they are (I) monoesters (Riemersma, 1968), (II) diesters (Korn, 1967), and (III) dimeric monoesters between two unsaturated lipids (Collin et al., 1973).

(I)   (II)   (III)

Monoesters, if formed, are not particularly stable in water. The structures of both the diester, II, and the dimer, III, have been established by X-ray studies carried out by Collin et al. (1973, 1974). In the diesters two molecules of fatty acid are linked through one molecule of osmic acid. The formation of diesters or monoesters depends on the disposition of C—C bonds in the unfixed tissue, since in the former these bonds are expected to be 0.51 nm from each other and in the latter ~0.77 nm apart (Collin et al., 1974).

It should be noted that these molecules are certainly pentacoordinate; the hexacoordinate octahedral esters are the liganded ones referred to previously. These structures have also been established by X-ray studies (Neidle and Stuart, 1976).

Agreement is lacking as to whether monoesters or diesters play a major part in the reaction of OsO_4 with unsaturated lipids. It is likely that various types of osmic ester intermediates are obtained during fixation subject to the reaction conditions. During fixation, in addition to the expected Os(VI) and Os(IV) species, Os(III) oxidation state may be present in the fixed tissue (White et al., 1976, 1980). Subsequent hydrolysis and further reduction may yield the latter species. Behrman (1980a,b) has questioned some of the evidence used by White et al. (1976) to support the idea that the oxidation state of osmium in fixed tissues is predominantly Os(IV). Probably a mixture of these complexes is present in the specimens. Os(VI) and Os(IV) could be present as osmate esters and osmium dioxide, respectively, and Os(III) could be present as oxo or amino complex(es). Substantial quantities of Os(II) species may also be present in the tissue (see Battistoni et al., 1977; Casciani and Behrman, 1978).

The reaction mechanisms discussed earlier could be expected to result in the attachment of OsO_4 inside the apolar layers rather than at the polar regions. However, overwhelming evidence indicates that the buildup of reduced osmium $(OsO_2 \cdot nH_2O)^-$ is greater at the polar end than in the apolar regions. This does not mean that the polar end is the primary site of reaction, although in one phospholipid, phosphatidyl serine, a direct reaction of OsO_4 with the hydrophilic group has been demonstrated. Indeed, the C=C bonds are the primary site of reaction with OsO_4, while secondary reactions may involve the polar ends of the phospholipid molecules. Phospholipids show a complete disappearance of the C=C absorption band after reaction with OsO_4. One molecule of OsO_4 reacts with one double bond. For every diester bond formed, one osmium atom in the form of a lower oxide is produced.

Trilaminar Appearance of Membranes

The fundamental problem in explaining the mechanism of interaction between OsO_4 and unsaturated lipids is to reconcile the trilaminar appearance of the membrane having a width of ~8 nm with its lipid bilayer structure. At least three schemes have been proposed to explain the appearance of the membrane:

1. Osmium oxides migrate from the hydrophobic region and are subsequently deposited at the hydrophilic interface. The preponderance of Os(IV) and Os(III) in fixed membranes (White *et al.*, 1976) favors the migration scheme. Since stained membranes contain considerable amounts of lower-valent, polar osmium adducts, it is quite likely that the final osmium products are deposited at the polar ends.

Indeed, if there are C=C present in the membrane, these are almost certainly the initial site of reaction. The reaction of OsO_4 with olefins is extremely rapid and highly exothermic, indicating a low energy of activation (9.5 kcal/mol) (Subbaraman *et al.*, 1972) and large free-energy decrease. [White *et al.* (1976) have observed the solid phase reaction of OsO_4 with cholesterol at $-80°C$.] It is likely that at least some of these initially formed products would hydrolyze and that the osmium-containing hydrolysis products [e.g., $OsO_2(OH)_4{}^{-2}$] could then diffuse away; they then might be bound by coordinating ligands (N, O, S) occurring in polar head groups and surface membrane proteins. Such bound osmium could serve as initiation sites for the formation of oxo-bridged osmium dimers and polymers. Thus, some osmium would remain in the hydrophobic interior of the bilayer, but much more would be found at the hydrophilic interface producing the observed contrast.

2. Osmium oxides are moved to the hydrophilic interface by bending of the lipid osmate esters (Riemersma, 1963; Stoeckenius and Mahr, 1965). However, marked bending of the hydrocarbon chain of a lipid osmate ester would seem, *a priori*, to be unlikely on thermodynamic grounds. Furthermore, this mechanism would require such bending to take place far more often than cross-linking via Os(VI) diesters or dimeric monoesters to account for the observed contrast in the stained membrane.

3. Unsaturated lipids may be cross-linked via covalent bonds that do not directly involve osmium (D. L. White, personal communication, 1978). For example, the osmate ester initially formed by the reaction of OsO_4 with a C=C could undergo homolytic C—O or O—Os bond scission to yield a free-radical intermediate. This radical could then attack the double bond on an adjacent

hydrocarbon chain. The osmium would act as a free-radical chain initiator. Anionic or cationic mechanisms can similarly be postulated. Such propagation mechanisms could explain the lack of electron density in the center of OsO_4-stained membranes.

It has been suggested that polar groups as well as fatty acid double bonds take part in the reaction with OsO_4 (Khan *et al.,* 1961; Riemersma, 1968). Also, it has been proposed that reduced osmium behaves as an anionic dye, and migrates from its site of formation at double (ethylene) bonds by interacting with cationic groups of phosphatide molecules toward the polar groups at the lipid–protein interface of membranes (Riemersma and Booij, 1962; Riemersma, 1963, 1970). An osmate anion bound to a lipid alkyl chain (i.e., an osmic acid monoester group) is shifted into the lipid polar group region by virtue of electrostatic interactions with the quarternary ammonium ions present. This proposed migration and ultimate deposition of osmium at the lipid–protein interface might explain the trilaminar appearance of membranes. Further reduction of Os(IV) esters to osmium dioxide hydrate is brought about by dehydration solvents (ethanol) and by reducing materials present in the tissue. Osmium dioxide is expected to be one of the final osmium-containing products, and the fact that it does not diffuse is probably a major factor in obtaining reproducible morphology.

Some evidence has been presented showing that osmates or osmium dioxide are likely to migrate toward the cations of phospholipid ''head'' groups (Adams *et al.,* 1967). It was demonstrated that the black reaction product of OsO_4 with the olefinic bonds in propylene groups in polythene was not removed by other quaternary cationic compounds such as 20% aqueous cetyltrimethylammonium bromide or 2% alcian blue. A series of experiments by Korn (1966a,b, 1967) indicated that no reaction occurred with polar groups and that osmium dioxide accumulated in test tube reactions. On this basis, it was speculated that in membrane systems the deposition of osmium dioxide hydrate at the aqueous interface does not involve interactions with lipid polar groups.

It has been shown that when tissue sections are briefly exposed (20 min) to OsO_4 vapors, membranes generally become uniformly dense (Pease, 1973b). As the section is treated for a longer period of time (24 hr) to OsO_4 vapors, the relative density of the polar regions of the membrane begins to build up. This evidence indirectly supports the migration concept.

Additionally, the trilaminar appearance can be explained by assuming that the other heavy metals used as poststains react with the hydrophilic groups of phospholipid molecules and that these metals also react with osmium. Uranyl and lead salts, when applied with or without posttreatment with OsO_4, tend to be deposited in polar regions. Moreover, any concrete evidence to support an alternative theory is lacking. From the preceding discussion it is apparent that additional information is required to elucidate the possible role played by polar groups of lipids and structual proteins intimately associated with lipids in the trilaminar appearance of biological membranes fixed with OsO_4. The trilaminar appearance of the complex and varied mosaic structure of the membrane is difficult to explain on the basis of the tentative mechanisms discussed earlier.

Reaction with Saturated Lipids

It is interesting to note that although saturated lipids do not react with OsO_4 at room temperature, some of these lipids can react with OsO_4 at higher temperatures. Chapman and Fluck (1966), for instance, demonstrated that fully saturated phosphatidylethanolamines react with OsO_4 at an elevated temperature of ~60°C. One explanation of this temperature-sensitive reaction is the physical condition of the saturated lipid at the time of reaction. This becomes clear when one considers that saturated lipids are in a crystalline condition at room temperature, whereas unsaturated lipids are in a liquid crystalline condition, and that OsO_4 reacts with the latter because hydrocarbon chains are "liquidlike" in structure in the liquid crystalline phase. Since fully saturated lipids such as phosphatidylethanolamines can react with OsO_4 at 60°C, it is assumed that the hydrocarbon chains melt and allow the water to penetrate into the crystalline structure of this lipid, which results in the orientation of polar groups of the lipid toward the aqueous phase and reaction with OsO_4.

The other explanation is that, in addition to the facile reaction with C=C, OsO_4 is known to react under certain conditions with other functional groups, in particular, hydroxyl and amino groups. Thus, the reaction with saturated lipid polar groups is not surprising, especially at elevated temperatures. In this scheme the physical condition of the lipid may have little bearing on its reaction.

Since fully saturated phosphatidylethanolamine does not contain double bonds, OsO_4 must react with the polar group. However, the nature of this reaction seems to be different from the one involved with unsaturated lipids; probably the amino group of this ethanolamine is involved. Most of the saturated lipids (e.g., lecithin) cannot be made to react with OsO_4 even at higher temperatures. Apparently, further experimentation is needed to clarify the nature of the difference between the reaction of OsO_4 with unsaturated lipids at room temperature and with fully saturated phosphatidylethanolamines at higher temperatures.

Experiments on artificial multilayers demonstrated the fixation effect of OsO_4 on fully saturated fatty acid molecules without a discernible effect on the contrast of these structures (Schidlovsky, 1965). Synthetic as well as natural dipalmitoyl lecithin can be stained by fixation in OsO_4 followed by dehydration without an intermediate washing. Prolonged washing after fixation with OsO_4 eliminates the possibility of interaction between reduced osmium, dehydration solvents, and saturated dipalmitoyl lecithin. Secondary osmication of the washed lung with a mixture of OsO_4 and the dehydration solvent (e.g., ethanol) has also been employed to obtain the staining of lung dipalmitoyl lecithin (Kaibara and Kickkawa, 1971). However, since the surface of film of mammalian lung contains other saturated and unsaturated lipids than dipalmitoyl lecithin, the possible role played by these lipids in the osmiophilia cannot be disregarded.

It is known that OsO_4 is soluble in most lipids and can be readily taken up in an unreduced form by fully saturated lipids. This unreduced osmium can be reduced by organic solvents during dehydration, a phenomenon called "secondary blackening." Thus, the reaction of OsO_4 with the double bonds of unsaturated lipids is not solely responsible for the production of black deposits. Apparently then, structures that appear osmiophilic in the fixed tissue do not necessarily indicate the presence of unsaturated lipids or chemical reaction with the double bond. Schidlovsky (1965) emphasized the role played by liquid fixation and solvent dehydration in the distribution of electron-scattering properties in the tissue. The exact mechanism responsible for "secondary blackening" is not known.

Reaction with Proteins

Much of the information of the reaction of OsO_4 with proteins has been obtained through studies of the addition of OsO_4 to protein solutions or films. The results of these types of studies can be applied, with some reservations, to the actual situation existing in cells due to the fact that the protein concentrations forming gels are not vastly different from the overall concentrations of proteins in cells. These studies have demonstrated that weak solutions of OsO_4 can form gels with proteins such as albumin, globulin, and fibrinogen (Fischer, 1899; Brand *et al.*, 1944; Baker, 1950; Porter and Kallman, 1953; Millonig and Marinozzi, 1968); however, these proteins differ in their reactivity with OsO_4. For example, fibrinogen and globulin react with OsO_4 at lower concentrations and at much more rapid rates than does albumin. This difference in reactivity is claimed to be related to the amount of tryptophan present (Porter and Kallman, 1953). Since OsO_4 is an additive fixative, it probably reacts at the double bonds of tryptophan. (Tryptophan, however, is quite readily oxidized at the pyrrole ring.)

Millonig (1966) indicated that 10% bovine serum albumin was easily gelled by 1.3% OsO_4 within 3 hr at 20°C. Isolated human albumin managed to bind a large amount of osmium (0.4 mg/mg protein) after a 4 hr reaction (Hayes *et al.*, 1963), and the albumin solution turned dark brown after 3 days of treatment with OsO_4 (Adams, 1960). Studies of models of bovine serum albumin and bovine γ-globulin by gel filtration and polyacrylamide gel electrophoresis also showed that OsO_4 had moderate cross-linking ability toward these proteins (Hopwood, 1970). It was found that OsO_4 caused a greater increase in viscosity of these proteins than would be expected to result from denaturation. Generally, these gels are free from coarse coagulation, which indicates an extremely fine micellar bonding. It is known that these gels are not coagulated by alcohols and heat.

That OsO_4 reacts with various amino acids, peptides, and proteins under the conditions of preparative electron microscopy has been demonstrated (Adams, 1960; Hake, 1965; Burkl and Schiechl, 1968; Lisak et al., 1976). Osmium tetroxide seems to react readily at alkaline pH with cystein and methionine; moderately with tryptophan, histidine, proline, and arginine; and only slightly with lysine, asparagine, and glutamine. Sulfur-containing amino acids appear to be oxidized to their sulfone derivatives. Hake (1965) showed the formation of carboxylic acids through oxidation of α-amino acids. Studies of the reaction between OsO_4 and proteins of red blood cell membranes indicated that the amino acids of these proteins differed in their ability to reduce OsO_4 (Eddy and Johns, 1965). Available evidence indicates that oxidative deamination of tissue proteins by OsO_4 is not uncommon. Oxidative deamination is accompanied by the evolution of ammonia, which probably originates from side-chain animo groups. According to Needles (1967), deaminated proteins have little ability to form cross-links.

By employing the OTAN (OsO_4-α-naphthylamine) reaction, which detects osmium bound in the tissue, Elleder and Lojda (1968a,b) suggested that certain protein-rich tissue components are able to bind osmium. Furthermore, since neurosecretory substances do not contain a large amount of lipid but are rich in protein-bound sulfhydryl groups, the conspicuous staining of neurosecretory cells with OsO_4 and ethyl gallate may be due to the presence of reactive sulfhydryl (—SH) groups (Wigglesworth, 1964). It has been suggested that the increase in the osmiophilia of axonal membranes of crayfish as a result of electrical stimulation is due to the unmasking of —SH groups in membrane proteins and their reaction with OsO_4 (Peracchia and Robertson, 1971). When the —SH groups are blocked by maleimide or N-ethylmaleimide before fixation with OsO_4, the increase in osmiophilia does not appear. The reduction of OsO_4 by —SH groups in a free system was shown by Adams (1960) and Hanker et al. (1964). The reduction potential of —SH groups is unquestionably sufficient to reduce OsO_4 very rapidly even though some workers believe that indisputable evidence for the reduction of OsO_4 by tissue —SH groups is lacking. In support of the latter view, which is not widely held, it has been assumed that —SH groups are located relatively far apart in the protein chain and may often be masked. Also, the reduction of OsO_4 by —SH groups in the tissue has not been directly demonstrated. Schneider and Schneider (1967) indicated that OsO_4 had only a slight denaturing effect on proteins as assessed electrophoretically.

From the foregoing discussion it is evident that OsO_4 does interact with certain proteins and a small amount of the fixative certainly reacts and blocks sulfhydryl, disulfide, phenolic, hydroxyl, carboxyl, amino, and certain heterocyclic groups.

These groups differ in the extent of their reactivity with OsO_4; for instance, —SH groups are far more reactive than are disulfide bonds.

Both soluble and membrane proteins undergo alterations in their secondary structure when treated with OsO_4 (Lenard and Singer, 1968). Approximately 40–60% of the α-helical content of proteins in red blood cell membranes was lost after treatment with 2% OsO_4 in phosphate buffer (pH 7.5) for 30 min at 4°C; such a loss was increased to ~70% when the specimen was pretreated with glutaraldehyde. Mitochondrial proteins are also partially aggregated after treatment with OsO_4 (Wood, 1973). Osmium tetroxide probably causes alterations in both the primary and secondary structures of proteins, resulting in their extraction. The isoelectric point of proteins is lowered following treatment with OsO_4, which indicates the disappearance of basic groups. After reaction with proteins, some of the OsO_4 is converted into a nonvolatile form (Hayes *et al.*, 1963).

Although the appearance of a black or brown color is an unequivocal indication of the reactivity of OsO_4 and the accumulation of lower oxides of osmium, the absence of color does not necessarily mean that the cellular structure is nonreactive with OsO_4. The absence of a detectable change in color may be due to the fact that only small amounts of lower oxides form, or that osmium is bound in a higher valence state (Elleder and Lojda, 1968a), or that the lower oxides migrate to other sites after the reaction. Thus, a change in color alone should not be used as the only criterion for the chemical reaction between OsO_4 and proteins. It is probable that certain amino acids react with OsO_4 without reducing it to lower oxides of osmium. The cross-linking of certain proteins would thus be accomplished without an apparent change in their color.

Osmium tetroxide may still preserve proteins although it stains them weakly or not at all. It is emphasized that even if OsO_4 is responsible for introducing relatively few cross-links in proteins, it may still contribute significantly to the preservation of fine structure. Finally, it may be said that while the relative increase in electron opacity of proteins by OsO_4 is rather small, too little is known of the efficacy of OsO_4 as a protein fixative. In fact, some workers (Adams *et al.*, 1967; Adams and Bayliss, 1968) strongly question the reactivity of OsO_4 with nonlipidic substances, especially proteins present in the tissue. They suggest that the reported binding of osmium to protein structures may be due to the protein-bound lipids. The correct interpretation of electron micrographs is dependent on the availability of additional information on the interaction between OsO_4 and tissue proteins under fixation conditions.

Reaction with Lipoproteins

In order to understand the mechanism of fixation, it is imperative to study the nature and extent of reaction of OsO_4 not only with lipids and proteins but also with lipoprotein complexes, which make up most of the membranes. Studies of lung lipoprotein myelinics treated with OsO_4 showed that the metal reacted readily with the lipoprotein and that the primary site of osmium deposition was the aqueous phase (Dreher *et al.*, 1967). Although protein reacted readily with OsO_4, the reduction did not destroy the lipoprotein monolayers even though the structural properties apparently were altered. After OsO_4 treatment the hydrophobic layer was doubled in thickness whereas the hydrophilic layer remained unchanged in thickness. The variations observed in the thickness of cell membranes appear to be related to the number of double bonds near the aqueous phase, the extent of penetration of hydrophilic protein into the hydrophobic lipid layer, and the relative amounts of lipid and protein present (Dreher *et al.*, 1967). This is a significant step toward explaining the variations in thickness of different membranes.

Osmium tetroxide may fix the relative positions of amphipathic proteins in membranes and immobilize molecular motion in the lipid bilayer (Jost *et al.*, 1973). It may build bridges between the aliphatic chains of lipids and the peptide bonds of certain membrane proteins (Litman and Barrnett, 1972; Nermut and Ward, 1974). Osmium tetroxide is known to interfere with membrane cleavage (Nermut and Ward, 1974) and membrane fusion (Poste and Papahadjopoulos, 1976). The exact role played by protein in the binding of osmium in lipoprotein complexes is rather difficult to assess. It has been reported that as the protein percentage increases in the lipoprotein molecule, the osmium uptake correspondingly decreases (Hayes *et al.*, 1963). The functional groups in proteins and lipids that react with OsO_4 have been discussed earlier.

Reaction with Nucleic Acids

The information on the interaction of OsO_4 with nucleic acids has important biochemical implications. Such information may be obtained by various methods, including X-ray crystallography (Rosa and Sigler, 1974), direct visualization of base sequences with the TEM (Whiting and Ottensmeyer, 1972), and correct interpretation of cell structures treated with this metal. The reaction of OsO_4 with nucleic acid components, both in the presence and in the absence of ligands (e.g., pyridine, ammonia, cyanide, and thiocyanate) has been studied. Osmium tetroxide in the presence of pyridine reacts with the pyrimidine moieties (thymine, uracil, and cytosine) in polynucleotides (Chang *et al.*, 1977), whereas adenosine and guanosine under similar conditions are not oxidized (Burton and Riley, 1966; Burton, 1967). Preliminary data indicate that thymine is attacked approximately 10 times more rapidly than is uracil. Osmium tetroxide–pyridine complex reacts with isopentenyl adenine approximately 4600 times faster than with thymidine (Ragazzo and Behrman, 1976).

The usual reactive sites in nucleic acids are the 2,3-glycol moiety in a terminal ribose group (Daniel and Behrman, 1975; Conn *et al.*, 1974) for the Os(VI) reaction, and the 5,6 double bond of uracil and the thymine residues for the Os(VIII) system. Osmium(VIII) reagents also oxidize thio bases such as 4-thiouridine and react rapidly with the isopentenyladenine group. Similar reactions occur at 0°C with denatured DNA, but not with double-stranded DNA (Beer *et al.*, 1966).

Using higher concentrations of OsO_4 and higher temperatures (23°C), Beer *et al.* (1966) demonstrated that OsO_4 in buffer solution that did not contain a ligand reacted predominantly with the thymine base of denatured DNA. They proposed that the base is converted to 4,5-dihydroxythymine, and that 1 mole of osmium reacted with 1 mole of the nucleotide. In the presence of ligands such as pyridine (which contains tertiary nitrogen), stable osmate ester derivatives are formed (Daniel and Behrman, 1976). Exposure of yeast tRNA crystals to a mixture of Os(VI) and pyridine produced a derivative containing approximately one atom of osmium and two molecules of tRNA (Rosa and Sigler, 1974). The adduct is stable and cannot be reversed by treatments known to disrupt the secondary and tertiary structure of the tRNA molecule.

Subbaraman *et al.* (1971) reported reactions between OsO_4–pyridine complexes and thymidine residues in aqueous solutions, which resulted in the formation of stable bis(pyridine) osmate esters. A proposed structure of the reaction product follows. Although their exact geometry is not known, the products could be osmyl complexes with linear $O=Os=O$ (Griffith, 1967).

It is clear from the available data that OsO_4 will readily add across the 5,6 double bond of a pyrimidine base to form a hydrolytically labile adduct that, in the presence of pyridine, forms a stable bispyridyl osmate ester. [Phosphate ions increase the rate of reaction between the OsO_4–pyridine complex and the olefinic group in thymine derivatives; the rate increase is thought to be due to an ionic strength effect (Clark and Behrman, 1976).]

Similarly, DNA can be converted to a new product which remains largely unbroken and linear but in which the majority of the thymine residues are converted to an addition product containing one osmium atom and two cyanide atoms (Di Giamberardino *et al.*, 1969). A single atom of osmium coupled to thymidine residues does not provide adequate contrast to visualize the base clearly. Cyanide may provide additional electron scattering groups for the ester by binding additional negative charges. One can conclude from the preceding and other biochemical studies that although nucleic acids in the tissue seem to be inert to OsO_4, the previously mentioned preferential reactions are important in the development of methods (base-specific markers) for electron microscopy of the base sequence in DNA and RNA.

It is useful to present a brief discussion on the effects of OsO_4 on the nucleic acids in the tissue. After fixation in OsO_4, coalescence of DNA fibers into coarse aggregates during alcohol dehydration is a common phenomenon. This type of clumping of intramitochondrial DNA fibers has been demonstrated in a wide variety of species. The clumping or condensation of DNA can be prevented under certain conditions of fixation. Bacterial nuclei can be stabilized by employing OsO_4 in the presence of Ca^{2+} and amino acids (Kellenberger *et al.*, 1958). The gelation of a solution of DNA or nucleohistones by OsO_4 in the presence of Ca^{2+} and tryptophane has been reported (Schreil, 1964). It has been shown that Ca^{2+} increased the stability of the double helix of DNA (Bach and Miller, 1967).

The clumping of DNA can also be prevented by prefixation with $KMnO_4$ or by postfixation with uranyl acetate prior to dehydration. The stabilizing effects of these methods becomes quite clear when one considers that intramitochondrial DNA in a clumped state shows a thickness of up to 25 nm, whereas in a stabilized state the thickness ranges from 1.5 to 5.0 nm. The gelation of DNA by the preceding procedures prevents the damaging effects of alcohol dehydration and results in little shrinkage or formation of coarse aggregates. Apparently, the formation of gel prior to dehydration is necessary in order to preserve the fine fibrillar structure of DNA.

The gelation of DNA produced by the various methods discussed above appears to be of a varied nature. According to Schreil (1964), the gelation produced by OsO_4 in the presence of Ca^{2+} and tryptophane gives the most accurate preservation of the *in vivo* disposition of DNA. Although it is known that uranyl acetate interacts with DNA and that $KMnO_4$ fixation retains DNA, the exact role played by the amino acid tryptophane in the stabilization of DNA is not clear. Whether tryptophane interacts with OsO_4 and/or with DNA is not known.

Ribonucleoproteins and/or RNA have been considered to be responsible for the staining of nucleoli with OsO_4 at the light microscope level (Battaglia and Maggini, 1968; Stockert and Colman, 1974). However, it is possible that the preservation and staining of nuclei do not depend solely on the interaction between OsO_4 and nucleic acids. It is known that nucleic acids in most eukaryotic organisms are chemically linked with basic proteins, and it has been suggested that a high content of arginine and lysine in the nuclear histones may account in part for staining of the nucleus (Wigglesworth, 1964). Furthermore, the presence of unsaturated lipids in the nucleus cannot be ruled out. It is known that chromatin contains lipids of various kinds. Thus, the staining of nuclei by OsO_4 may also be due to the interaction of unsaturated lipids with this fixative. Staining methods using OsO_4 have been presented by Hayat (1975). Single osmium atoms attached selectively to uracil moieties in polyuridylic acid and to thymine in DNA have been visualized using dark field transmission electron microscopy

(Whiting and Ottensmeyer, 1972). Similar atoms in stained polynucleotides have been viewed with the STEM (Cole *et al.*, 1977).

Reaction with Carbohydrates

The available evidence indicates that OsO_4 does not interact with most of the pentose or hexose sugars or their polymers, although there are a few exceptions. The possiblity of a slow oxidation of sucrose to oxalic acid by OsO_4 was suggested by Bahr (1954). Glycogen solutions are known to blacken after a prolonged treatment with OsO_4 at 50°C. Glycogen in the sections of OsO_4-fixed tissue has been shown to reduce silver nitrate–methenamine solution even in the absence of a preliminary oxidation with periodic acid (Millonig and Marinozzi, 1968). This reaction indicates that OsO_4 oxidizes glycogen with the aldehyde groups.

Osmium tetroxide probably produces vicinal hydroxyl groups by oxidation, which are in turn split to form aldehydes by, for instance, periodate oxidation (McManus and Mowry, 1958). Also, OsO_4 is reduced by aldehyde thiosemicarbazones formed by condensation of thiosemicarbazide with the aldehydes that are formed by periodate oxidation of polysaccharide 1,2-glycol groups (Hanker *et al.*, 1964; Seligman *et al.*, 1965). An attempt has been made to explain the chemistry of the putative linking of free hydroxyl groups in glucose by OsO_4 (Luzardo-Baptista, 1972). It has also been shown that OsO_4 reacts with solutions of pure amino sugars, producing black droplets (Wolman, 1957).

The preceding data indicate the difficulty in assessing accurately the degree of interaction of OsO_4 with carbohydrates. It must be mentioned, however, that most carbohydrates in OsO_4-fixed tissue are extracted during rinsing and dehydration. Neutral polysaccharides such as glycogen are relatively less readily water soluble and thus show less leaching. Since glycogen shows little increase in electron opacity after fixation with OsO_4, it is apparent that the interaction of this fixative with glycogen does not involve binding of the lower oxides of osmium to the latter. The electron opacity of glycogen can, however, be enhanced selectively by adding 0.05 mol/l $K_3Fe(III)(CN)_6$ to 1% OsO_4 during postfixation (de Bruijn, 1968). The electron opacity of lipid droplets and membranes is also increased by this method. If the objective of the study is to differentiate between ribosomes and glycogen, sections should not be poststained.

The mechanism involved in the glycogen staining reaction with OsO_4–$K_3Fe(CN)_6$ or $K_2OsO_2(OH)_4$–$K_4Fe(CN)_6$ mixture has been partly elucidated (de Bruijn and Den Breejen, 1976). It is thought that the $Os(VI)$–$Fe(II)$ complex reacts selectively with unchanged diols in tissue glycogen and that this complex is potentially able to stain endogenous tissue aldehydes and carboxyl groups as well as those introduced by oxidizing agents. Hydrogen peroxide and OsO_4 are potentially able to introduce carboxyl groups in glycogen.

The possibility that OsO_4 oxidizes certain carbohydrates without the formation of a black precipitate cannot be ruled out. In case OsO_4 interacts with carbohydrates, the fixation capacity of the reagent would be altered. Such interaction may reduce the effective concentration of OsO_4 in a given volume of the fixative mixture. A reduction in the concentration of OsO_4 would result in a slower rate of penetration and fixation.

Reaction with Phenolic Compounds

The discoverer of osmium, Smithson Tennant (1805), aptly pointed out the formation of a deep blue color when OsO_4 reacts with "infusion of galls." The first suggestion that OsO_4 reacts with phenol-rich regions in plant cells was made by Schultze and Rudneff as early as 1865. In recent years, the presence of phenol-containing regions in cells of higher (e.g., Baur and Walkinshaw, 1974; Mueller and Beckman, 1976) and lower plants (e.g., Fulcher and McCully, 1971; Evans and Holligan, 1972) and animals (e.g., Tranzer et al., 1972) has been reported. Sufficient evidence is now available indicating that phenol-containing regions of cells are osmiophilic. These regions show electron opacity with OsO_4 alone, without poststaining with uranyl acetate and lead citrate.

When phenolic-containing cells are fixed with glutaraldehyde followed by OsO_4, phenolics leach from the vacuoles into the cytoplasm where they subsequently react with OsO_4. This results in a dense, osmiophilic cytoplasm, the details of which are obscured. Leaching of the phenolics from the vacuoles can be minimized by prefixation with low concentrations of glutaraldehyde (Mueller and Beckman, 1974). The most effective approach to prevent leaching of phenolics is to add caffeine, nicotine, or cinchonine to glutaraldehyde during prefixation (Mueller and Rodehorst, 1977). These alkaloids react with phenolics in the vacuoles and prevent their leaching (Fig. 4.1). Caffeine causes the phenolic material to condense into globules, while the other two alkaloids precipitate the phenolics into amorphous masses against the tonoplast. Caffeine has been used to precipitate tannins *in vivo* for study with the light microscope (Flasch, 1955); this treatment does not seem to cause any apparent injury to the cell. Caffeine is most effective at concentrations of 0.1 to 1.0%, whereas nicotine and cinchonine should be used at a concentration between 0.05 and 0.1%.

Mechanism of Reaction

Under experimental conditions similar to those of conventional tissue fixation, OsO_4 reacts rapidly with phenols containing o-dihydroxy groups and yields electron-dense, stable-chelate complexes (Nielson and Griffith, 1978). The presence of two o-hydroxy groups in an aromatic ring of the phenol is a prerequisite for this reaction.

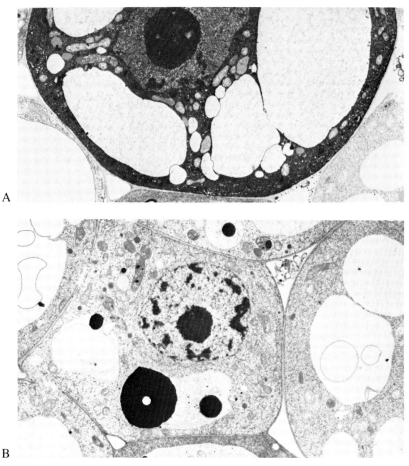

Fig. 4.1. Endodermal cells of 3-day-old cotton roots (3 mm from the root tip). A, fixed with glutaraldehyde followed by OsO₄; the cell shows osmiophilic cytoplasm and nucleus by leaching of phenolics. B, fixed with glutaraldehyde containing caffeine, followed by OsO₄; phenolics are condensed. ×4200. (W. C. Mueller, unpublished.)

Osmium tetroxide reacts with these phenolic compounds in water or aqueous acetone medium and yields stable precipitates with the following basic polymeric structure:

$$\left[\left(R - \underset{}{\overset{}{\bigcirc}} \begin{array}{c} O \\ \\ O \end{array} \right)_2 OsO \right]_n \cdot n\,H_2O$$

In phosphate or cacodylate buffer (pH 6.8), OsO_4 reacts with o-dihydroxy phenolic compounds, yielding stable precipitates with the following structure:

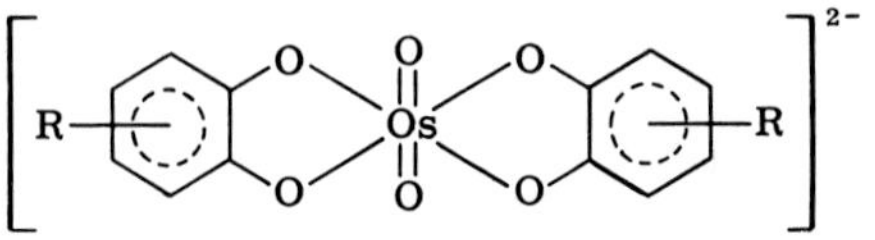

The presence of such complexes may also explain the increased contrast obtained by treating the tissue specimens with o-dihydroxy phenolic species such as tannic acid or D-catechin (Futaesaku and Mizuhira, 1974) prior to postfixation with OsO_4.

Method. Root specimens are fixed in 2.5% glutaraldehyde containing 0.5% caffeine in 0.05 mol/l phosphate buffer (pH 6.8) for 2 hr at room temperature. The specimens are washed in the buffer containing caffeine for 1 hr, followed by postfixation in 1% OsO_4 without caffeine for 1 hr. Thin sections are stained with aqueous uranyl acetate (2%) for 5 min followed by lead citrate for 5 min. Figure 4.1 shows the results of this technique.

LOSS OF LIPIDS

It has been assumed in the past that most lipids are preserved during fixation by OsO_4. One of the reasons for this misconception is that the total amount of masked lipids present in the living cell has been underestimated. It has been shown that the insoluble fragment of protoplasm in living cells contains as much as 45% lipids (Smith et al., 1957). It is recognized that even chromatin contains a high proportion of lipids. The fact that fixatives penetrate many times more slowly in tissues than into protein gels containing equal amounts of protein also indicates the presence of large amounts of lipid in the tissues. It is accepted that one of the major barriers to fixative (other than OsO_4) penetration is the lipid component of cytoplasmic membranes.

Another reason for the misunderstanding is that the morphology of some organelles changes little even though a considerable amount of the lipid component is lost. For example, although lipids are necessary for electron transfer in mitochondria, the characteristic morphology of mitochondria can persist even after extraction of phospholipids with aqueous acetone. A detailed study of the fine structure of bovine heart mitochondria after lipid extraction showed that the outer membrane was lost only when more than 80% of the lipid was removed, and in the remaining membranes the characteristic triple-layered appearance of the membrane was preserved even when more than 95% of the lipid was extracted (Fleischer et al., 1967). Similar results were obtained with myelin figures

(Napolitano *et al.*, 1967). These results may be explained on the basis of OsO_4 reaction with residual lipids and/or with proteins or even with some carbohydrates.

Studies of fixed chloroplasts showed that after 20% of the glycolipids and 40% of the chlorophyll were extracted, the characteristic structural features were retained, but when most of the lipids were extracted with chloroform–methanol, structural features could not be demonstrated (Ongun *et al.*, 1968). It is apparent that in chloroplasts some lipid fixation is necessary for retention of ultrastructure. It follows from these considerations that the absence of any apparent change in the morphology of cell components cannot be indicative of the preservation of lipids, and that various organelles differ in their requirement of the amount of fixed lipids necessary to maintain their ultrastructure.

The visualization of electron-opaque masses in sections of tissues fixed with OsO_4 probably does not accurately represent lipids present in the living cells. It is conceivable that electron-opaque masses conceal the "fixed" lipids. In fact, it has been reported that empty spaces are left behind after the removal of osmium by oxidation from the electron-opaque osmium reaction products (Casley-Smith, 1967). It is also worth considering that dehydration and even embedding may affect the final relative concentration of reduced osmium in the tissue. That reduced osmium deposits are soluble in xylene was indicated by Chou (quoted by Baker, 1958) and by Marinozzi (1963), although no experimental data are available on the effect of ethanol or acetone on reduced osmium.

X-ray diffraction studies of frog sciatic nerves indicate that fixation by OsO_4 is not enough to prevent extraction of cholesterol during dehydration (Moretz *et al.*, 1969). Bahr (1955) and Dallam (1957) also pointed out a considerable loss of unsaturated lipids by ethanol extraction from tissues fixed with OsO_4. Although fixation with glutaraldehyde followed by OsO_4 seems to minimize lipid loss, especially that of neutral lipids, a considerable amount of lipid is lost during dehydration. It should be noted that only minute amounts of lipids are extracted during fixation; the major loss occurs during dehydration.

After double fixation, the lipid losses during dehydration were 25% in the amoeba (Korn and Weisman, 1966), 7.2% in rat liver (Ashworth *et al.*, 1966), 39% in tritiated choline-labeled lung phospholipids (Morgan and Huber, 1967), and 9.5% in triolein-labeled macrophages (Cope and Williams, 1968). The higher lipid loss from lungs may be due to the higher content of saturated fatty acids in this organ than in amoeba. The lipids extracted in the amoeba at room temperature were mostly neutral lipids. According to another study the total loss of neutral lipids and phospholipids in intestine amounted to ~16% (Dermer, 1968). According to Korn and Weisman (1966), phospholipids are better preserved than are neutral lipids, whereas Dermer (1968) indicates that both phospholipids and neutral lipids are retained to the same extent.

On the other hand, retention of as much as 70–78% of liver [^{3}H]lipid was reported by Stein and Stein (1967), and ~90% of the incorporated [14C]linolenic acid was found in the fixed and dehydrated small intestine of hamster (E. W. Strauss, 1967, see Saunders *et al.*, 1968). Saunders *et al.* (1968) demonstrated that in fixed and dehydrated rat jejunal mucosa incubated *in vivo* with saturated and unsaturated lipids, less than 50% of [14C]palmitate or [14C]cholesterol but more than 85% oleate and [14C]linoleate were retained. It is relevant to this point that only 0.7–7.2% of total lipids of rat liver is lost after fixation with OsO$_4$. Ongun *et al.* (1968) found that OsO$_4$ can completely fix glycolipids, phosphatidyl glycerol, and phosphatidyl choline in isolated spinach chloroplasts. Stein and Stein (1971) have reviewed the phenomenon of lipid loss from tissues prepared for electron microscopy.

Although the major lipid loss occurs during dehydration, some lipids are extracted by OsO$_4$ itself. Proteins diffuse out in OsO$_4$-fixed cells unless prefixed with an aldehyde. These two events cause cell membranes to become freely permeable to small ions and molecules. This is the reason that the osmolarity of OsO$_4$ vehicle is not very important.

The varying results discussed here suggest that the degree of extraction is dependent not only on the type of intracellular lipids but also on the type of tissue involved. Moreover, the degree of extraction of indigenous tissue lipid and that of absorbed lipid is not similar. It also appears that lipid loss is not related entirely to the degree of saturation of fatty acids present in the lipid fraction of a cell. The details of the loss of free lipids during embedding have been discussed by Hayat (1981).

Dehydration at 4°C or a partial dehydration procedure reduces lipid loss. In the partial dehydration procedure the steps of 100% ethanol and propylene oxide are eliminated, and the final dehydration is completed by using Epon monomer (see Idelman, 1964, 1965). The retention of free fatty acids can be improved by adding CaCl$_2$ to the fixative (Mitchell, 1969; Strauss and Arabian, 1969); CaCl$_2$ forms highly insoluble salt with free fatty acids. Cholesterol can be preserved in the tissue by adding digitonin in aqueous solution of glutaraldehyde and OsO$_4$ fixatives (see also p. 107). By using the partial dehydration procedure, a loss of radioactivity of only 11.2% was detected when intracellular triglycerides had been labeled by injection of [14C]palmitic acid to rats (Stein and Stein, 1967). According to Stein *et al.* (1969), free or esterified cholesterol is better preserved in the liver after intravenous injection of labeled chylomicrons by using Aquon, which is a highly water-miscible derivative of Epon (Gibbons, 1959; Hayat, 1981). In this study the resin was used both for dehydration and for embedding.

LOSS OF PROTEINS

The biphasic effect (gelation and then extraction) of OsO$_4$ on tissue constituents is well known. It is recognized that tissues fixed with OsO$_4$ lose proteins during both fixation and dehydration. For example, mitochondria in rat liver tissue lost ~22% of their proteins during fixation in OsO$_4$ and a further 12% during dehydration (Dallam, 1957). The loss of proteins was also demonstrated by Luft and Wood (1963), who labeled tissue protein in rat with [^{35}S]methionine and showed, under the conditions of preparative electron microscopy, that ~8% of the label was lost during fixation in OsO$_4$ and an additional 4% during dehydration. Glycoproteins labeled with [^{14}C]fucose showed a loss of 3% during glutaraldehyde fixation, 1% during postfixation with OsO$_4$, and 2% during dehydration (Sturgess et al., 1978).

Extraction of proteins occurs not only when OsO$_4$ is used as a primary fixative, but also when it follows prefixation with glutaraldehyde. This has been demonstrated in erythrocyte membranes (McMillan and Luftig, 1973; Parish, 1975). According to Lenard and Singer (1968), conformational changes of proteins are more severe after glutaraldehyde fixation followed by OsO$_4$ than with either fixative alone. It seems that prefixation with glutaraldehyde does not prevent leaching of some proteins, especially when specimens are postfixed with OsO$_4$.

The degree of extraction is primarily dependent on the duration of fixation and dehydration, type of buffer employed, and the type of proteins involved. The role played by various buffers in protein extraction has been discussed earlier. The difference in the extraction of different proteins was indicated by Bahr (1955), who demonstrated that in rat liver the amount of extracted protein exceeded 50% of its dry weight after a fixation of 4 hr with OsO$_4$, whereas muscle showed an extraction of less than 50% of its dry weight after a similar treatment. He also showed that liver, muscle, and tendon of rat exhibited an increase in the extractability of proteins with time, whereas skin did not show such an increase. The extraction effect of OsO$_4$ has been utilized advantageously for studying the ultrastructure of aortic elastica (Cliff, 1971). Rat aorta fixed for periods up to 72 hr with OsO$_4$ revealed an underlying fibrillar structure due to the differential extraction of the amorphous matrix material.

Prolonged fixation by OsO$_4$ generally results in progressive destruction of cellular proteins, which leads to increased extractability during dehydration. Although the exact mechanism of extraction of proteinaceous substances as a result of prolonged fixation with OsO$_4$ is not known, discussion of a few possible mechanisms follows.

It has been suggested that the initial, partial denaturation of proteins is followed by further oxidation of certain proteins leading to the production of soluble end products that are capable of being washed out of the cells after long fixation (Porter and Kallman, 1953; Wolman, 1955). Another possible explanation of the loss of proteins is that OsO$_4$ like other oxidizing agents causes cleavage of certain protein molecules. Several workers have shown that oxidizing agents can cause cleavage of proteins through disulfide bridges in histidine, tryptophan, or tyrosine, or through other groups (Sanger, 1949; Witkop, 1961; Joly, 1965). The cleavage of protein molecules could result in the loss of peptide fragments especially with OsO_4 since it possesses low cross-linking ability. Studies by Hopwood (1969a), utilizing Sephadex G-50 separation, demonstrated that OsO_4 causes oxidative cleavage of proteins, although much less than that caused by $KMnO_4$.

An explanation of the loss of protein in zymogen granules was suggested by Amsterdam and Schramm (1966). Zymogen granules isolated from rat parotid and pancreas lost a large part of their protein content within a few minutes after treatment with OsO_4. Since these granules are quite stable without OsO_4 treatment, it was assumed that the fixative reacted with the granule membrane and increased its permeability without cross-linking the proteins inside the granules. As a result, the soluble proteins diffused out of the granules.

Studies of the effect of OsO_4 on the conformation of protein molecules by means of circular dichroism measurements in the spectral region of the peptide absorption band indicated that OsO_4 fixation either alone or preceded by prefixation with glutaraldehyde caused the loss of helical structure of the proteins (Lenard and Singer, 1968). It was also indicated that one-quarter to one-third of the membrane protein is in a helical conformation and the remainder is in the random-coil form. The importance of these studies cannot be overemphasized in view of the role played by the three-dimensional folding (the conformation) of the individual protein chains in determining the structure of cell components.

SWELLING CAUSED BY OSMIUM TETROXIDE

Osmium tetroxide hardens tissue slightly and generally causes some gross swelling during fixation. Isolated cells usually show swelling even in isoosmotic solutions of OsO_4. Animal tissues exhibit a marked and rapid swelling when fixed with OsO_4. In the case of liver tissue, for example, as much as 30% swelling was found after 4 hr fixation with OsO_4, and at least half of this value was reached after only 15 min in the fixative (Bahr *et al.*, 1957). Comparative studies by these workers indicated that animal tissues such as brain, spleen, kidney, muscle, and liver do not differ significantly in their fundamental pattern of swelling during fixation. This swelling is nearly neutralized by the shrinking action of the dehydration solvents. Further shrinking of the specimen occurs during infiltration by the embedding medium.

Sea urchin eggs showed a swelling of 3.2% after fixation with 1% OsO_4 in seawater (Kushida, 1962). This swelling was temporary because the cells underwent 12.2% shrinkage in ethanol and an additional 1.7–21.1% shrinkage in standard embedding media including epoxy and polyester resins. *Limnaea* eggs fixed in 1% OsO_4 solution in distilled water at pH 6 showed a swelling of $\sim$ 25%, while when fixed in a 2% isotonic solution of OsO_4 a volume swelling of $\sim$10% was still found (Elbers, 1966).

Fixation temperature seems to exert little influence on the extent of volume changes, and swelling is a consequence of the chemical action of OsO_4. An increase in the volume of the tissue during fixation is generally closely followed by an increase in its weight. The increase in the specific weight of the tissue appears to be primarily an expression of the binding of osmium in the tissue during fixation. This increase, however, reverses during dehydration and finally increases during infiltration by embedding resins.

Specimen swelling during primary fixation in OsO_4 can be prevented by adding $CaCl_2$ or NaCl to the fixative vehicle, and during secondary fixation the swelling can be avoided by adding either an electrolyte (NaCl) or a nonelectrolyte (glucose or sucrose). The addition of $CaCl_2$ causes the cross-linking of negatively charged proteins, resulting in a reduced osmotic pressure within the cell. The addition of nonelectrolytes to OsO_4 solution, during primary fixation, results in less than satisfactory fixation of albumin in model experiments (Millonig, 1964; Millonig and Marinozzi, 1968) and loss of cellular materials. Nonelectrolytes should not be added to prevent the swelling of isolated cells, for these materials do not prevent the flow of water into the cells. Sea urchin eggs show more swelling in the OsO_4 containing sucrose than in OsO_4 without sucrose (Millonig and Marinozzi, 1968). To prevent swelling, a final concentration of 1–3 mM of $CaCl_2$ in OsO_4 solution is recommended. Caution is warranted, for $CaCl_2$ may cause a granular precipitation of cell proteins and phosphates present in the buffer (also see pp. 33 and 45).

PARAMETERS OF FIXATION

Concentration of Osmium Tetroxide

Osmium tetroxide is most effective when used in optimal concentration; higher concentrations can cause oxidative cleavage of protein molecules, which would result in the loss of peptide fragments (Hopwood, 1969a). Presently, the most commonly used concentration of OsO_4 ranges from 1 to 2% in a buffer. Osmium tetroxide solutions less than 1% may be desirable for cytochemical and certain morphological studies. Concentrations ranging from 0.2 to 0.5% are desirable for particulate specimens. Some ciliates have been postfixed satisfactorily with 0.1% OsO_4 (Shigenaka *et al.*, 1973).

Temperature of Fixation

It has already been pointed out earlier that various cell components differ with regard to their appearance as a result of changes in fixation temperature, which can cause both qualitative and quantitative alterations in tissue. In some cell types, microtubules are lost when cells are fixed in cold OsO_4 presumably because of its slow rate of penetration. In addition, fixation reactions are considerably slower at lower temperatures. Nevertheless, fixation should generally be carried out at ~4°C, even though OsO_4 is a slow penetrant, since autolytic activity is reduced at low temperatures. Furthermore, low temperatures reduce leaching or extraction of cell constituents during fixation. Also, if a relatively long exposure of the tissue to OsO_4 is necessary, it can be accomplished with less damage under cold than under warm temperatures. Fixation in OsO_4 at low temperatures has proved valuable in improving the uniformity of the quality of preservation. Fixation at higher temperatures may cause shrinkage of mitochondria and increased granularity of the cytosome. Alternatively, a preliminary brief fixation by OsO_4 can be carried out in the cold followed by fixation at room temperature. Perhaps the quality of ultrastructure preservation, with a few exceptions, is not significantly affected by varying the temperature, provided the specimen size is very small and duration of fixation is not very long.

Rate of Penetration

A knowledge of the rate of penetration and the amount of osmium uptake by the tissue is important in order to interpret electron micrographs correctly. This information may indicate to what extent an externally introduced heavy metal participates in the morphological appearance of a micrograph. In other words, the amount of osmium uptake by the tissue in a given period of time greatly influences the contrast and general appearance of the electron micrograph. Measurements of the rate of penetration are also important for determining the optimal duration of tissue fixation. The rate of penetration is controlled primarily by the diffusion gradient at the penetration front, and the diffusion gradient, in turn, is influenced by several factors, some of which are discussed here.

The speed of OsO_4 penetration is partly dependent on the tissue density when solutions of equal concentrations are used. Generally, the higher the tissue density, the slower will be the speed of penetration. The concentration of OsO_4 in the fixative solution is another factor that influences the rate of diffusion. The rate of penetration generally increases with an increase in the concentration of OsO_4; however, this relationship is not linear. Higher temperatures and fixation by perfusion also accelerate penetration by OsO_4.

The addition of fixation vehicles to balance the OsO_4 solution osmotically with the cell interior is another important factor that significantly influences the rate of penetration. The addition of vehicles results in a slower rate of penetration, since OsO_4 in water alone penetrates most rapidly. The net decrease in the rate of penetration obviously depends on the types and quantities of salts employed. In general, the more osmotically balanced the fixation mixture, the less the swelling of the tissue, and also the slower the rate of penetration. Generally, the addition of electrolytes, such as NaCl, results in a decreased rate of penetration, but this decrease is less than that caused by the addition of nonelectrolytes such as sucrose. According to Hagström and Bahr (1960), even the addition of 0.15 M sucrose to 1% OsO_4 solution results in as much as a 40% reduction in the penetration depth in rat liver tissue, whereas the addition of veronal acetate buffer does not affect the rate of penetration significantly. It is advisable, therefore, to increase the duration of fixation and the concentration of OsO_4 in the presence of salts in the fixation mixture.

On the other hand, since the diffusion gradient decreases with the duration of diffusion, the rate of OsO_4 penetration is expected to decrease with the continuance of fixation irrespective of the density or physiological condition of the tissue. The diffusion rate of OsO_4 into gelatin–albumin gel decreases with time, suggesting the formation of a barrier. This barrier may also hinder the penetration by solvents and embedding media. However, the rate of penetration throughout the duration of fixation is controlled by many factors. In most tissues the rate of penetration varies during the total duration of fixation. For example, the uptake of OsO_4 by rat tail tendons is slower in the beginning than in the later stages of fixation. This initial slow rate is probably due to the tightly packed collagen fibrils, diffusion barriers, and slow rate of chemical reaction between the cellular materials and the fixative. In contrast, some tissues, such as fat, exhibit a quicker uptake in the beginning. In this case only a thin outer layer reacts quickly with OsO_4, which then prevents a deeper penetration by the fixative.

The size of the tissue specimen is apparently also responsible for variation in the rate of penetration, which would be different at different depths from the surface of the specimen. It is obvious, therefore, that the smaller the size of the specimen, the more uniform will be the rate of penetration. Since deeper layers of relatively large specimens will be penetrated rather slowly, the cells of these layers will be exposed to low concentrations of OsO_4. Thus, only the outer layers of a tissue block are well fixed by the slowly penetrating OsO_4. It is generally recommended, therefore, to use only the peripheral layers for examination unless these have been injured mechanically.

Unlike most polar oxidizing agents, OsO_4 is able to penetrate both hydophilic and hydrophobic lipids (Adams, 1958). The reagent penetrates tissues very slowly but reacts rapidly. It is excessively superficial in its action, so that specimens larger than 0.5 mm or so in diameter are often not fixed uniformly. For this reason the tissue specimen to be fixed should be cut into small pieces (less than 1mm³) or, alternatively, sectioned with an automatic sectioner.

In most types of cells and tissues the maximum depth of OsO_4 penetration is reached within 1 hr. Osmium tetroxide (1%) stops all respiration completely after 40 min of incubation of tissue blocks (Burkl and Schiechl, 1968). The rate of penetration is calculated to be ~800 μm deep during the first hour of fixation.

After this period the rate is progressively slowed because the fixed outer layers of cells resist deeper penetration of the fixative. This slow penetration is also due to a progressive decrease in the concentration of OsO_4 fixative. It is apparent, therefore, that the fixation of deeper layers of a 0.5 mm specimen requires a long time. In conventional fixation the core of the specimen is not as well fixed as the outer layers. Osmium tetroxide penetrates slightly slower than does glutaraldehyde.

Slow penetration of OsO_4 might tempt the investigator to use higher concentrations of the fixative. This is not practical because, as stated earlier, OsO_4 is poorly soluble in water. The rate of penetration, however, can be increased by using OsO_4 in combination with potassium dichromate (Weissenfels, 1960).

The total uptake of osmium, per unit weight of tissue, from the fixing fluid differs depending primarily on the tissue type. For example, pancreas, liver, and kidney tissues contain about three times as much osmium as muscle and skin tissues after an equivalent time of fixation (Bahr, 1955). This higher uptake by the former tissues is probably due to the fact that these tissues are richer in lipoprotein membranes. The uptake of osmium per unit membrane is probably the same in the two groups of tissues. The accumulated reduced osmium may account for as much as 46% of the tissue dry weight after rat heart has been fixed by perfusion with OsO_4 (Krames and Page, 1968).

Duration of Fixation

The duration of fixation is intimately linked to the fixation temperature in all respects. As a general rule, the most desirable duration of fixation is a compromise between the two simultaneous effects of the fixative: (1) fixation and (2) extraction of cellular materials. In other words, length of fixation should be determined on the basis of achieving the best possible fixation and the least possible extraction and alteration of tissue components. Leaching or extraction of cellular materials is undesirable except in some cases where increased image contrast of unextracted elements is the primary objective. Prolonged treatment with OsO_4 is known to increase the contrast of cystine-rich proteins such as keratin. Since OsO_4 is unable to make all cellular constituents insoluble in water, prolonged fixation causes extraction of cell constituents, especially of proteinaceous substances. Therefore, if possible, a short fixation time should be employed.

It is difficult to recommend a definite duration of fixation because the rate of OsO_4 uptake varies in different types of tissues. Thus, each type of tissue has its own specific requirement in terms of an ideal duration of fixation. The size of the tissue specimen and the type of buffer employed also influence the optimal duration of fixation. Other factors that influence the fixation time include the concentration of OsO_4 and the concentration of organic matter in the cell. It is known that cells containing very low concentrations of organic matter are difficult to fix with OsO_4. It is admitted that the optimal duration of fixation for most tissues is yet to be determined.

For isolated cells and particulate specimens, a few minutes of fixation may be satisfactory. For dense tissue blocks, $\frac{1}{2}$ to 2 hr may be needed. Ordinarily, a 15 min to 2 hr fixation time is ample. Osmium tetroxide has been employed as a postfixative for as long as 12 hr at room temperature to obtain preservation and staining of early embryonic tissues (Kalt and Tandler, 1971). Fixation at cold temperatures requires longer fixation times. For most purposes a 1% OsO_4 solution having a pH of 7.2–7.4 with an osmolality of 300 mosmols is recommended. A better preservation of nuclear structures and spindle fibers is obtained when OsO_4 is employed in a slightly acid medium.

OSMIUM TETROXIDE AS A PRIMARY FIXATIVE

In routine electron microscopy, prefixation with OsO_4, instead of glutaraldehyde, is not desirable. However, certain cellular structures are better preserved when specimens are prefixed with OsO_4. When cells are prefixed with OsO_4, their membranes become completely permeable within a few minutes, whereas glutaraldehyde-fixed cells on exposure to OsO_4 lose their total impermeability relatively slowly. It is likely that cytoplasmic movements continue during the first few minutes of fixation with glutaraldehyde resulting in the damage of certain delicate cell structures. Such structures can be preserved better by using OsO_4, which stops cell metabolism instantaneously when in direct contact. Examples of such structures are mucilage strands, cilia, flagella, and hairs. All of these structures are extremely delicate. Also, an increase in the number of coated vesicles and reticulosomes are seen in mossy endings of rat cerebellum when the tissue is prefixed with OsO_4 (Paula-Barbosa and Gray, 1974).

The cytoplasmic rearrangement, which occurs during glutaraldehyde fixation, can be minimized in some cases by brief prefixation (~ 1 min) with either OsO_4 or a mixture of OsO_4 and $K_2Cr_2O_7$ (Wohlfarth-Bottermann, 1957) prior to fixation with glutaraldehyde. This type of pretreatment yields satisfactory preservation of certain microtubules. However, it should be noted that generally while ciliary microtubules are well preserved by prefixation with OsO_4, cytoplasmic microtubules are better preserved when prefixed with glutaraldehyde at room temperature.

OSMIUM TETROXIDE AS A VAPOR FIXATIVE

Some of the undesirable effects of sequential double fixation may be eliminated by exposing the specimens to fixative vapors, especially of OsO_4, prior to primary fixation by immersion. It has been indicated that this procedure results in better overall preservation of cell components, particularly that of lipids, polyribosomes, and mitochondrial matrix (Hayat and Giaquinta, 1970). This approach provides an effective means for immobilization of certain diffusible substances, since they may form polymers with OsO_4 prior to the undesirable action of foreign liquids. Thus, the damaging effects of these liquids is obviated during the early stages of fixation.

Only surface layers of a tissue block are fixed by vapors. However, monolayers of cells and cell fractions can be easily fixed with vapors of OsO_4 or aldehydes. The monolayer along with its support and the fixative in an open vial are placed in a closed container, such as a Petri dish, and allowed to remain at room temperature for several minutes.

Since aqueous fixation causes extensive displacement and loss of elements, vapor fixation is useful for specimens prepared for X-ray microanalysis. Fixation with OsO_4 vapors has proved useful for freeze-dried tissues. These tissues are placed in a desiccator containing an open 0.25 g vial of OsO_4 crystals. Osmication seems to be completed in ~ 8 hr. The tissue turns grayish with these vapors, and turns black only after introduction of a fluid of some sort (Coulter and Terracio, 1977). Lim *et al.* (1976) fixed freeze-stopped tissue in a chamber consisting of a capped plastic centrifuge tube containing an open ampule of OsO_4 crystals. The inside of the chamber was maintained at $-90°C$ at all times before and after fixation. After an overnight fixation, the specimens were gradually brought to $4°C$ in a refrigerator. Coulter and Terracio (1977) have published the details of a vapor fixation chamber.

REMOVAL OF BOUND OSMIUM FROM SECTIONS

The specificity of certain poststaining techniques is diminished by the bound osmium in tissues that have received prior treatment with OsO_4. Methods are available for removing this osmium from ultrathin sections. Unmounted, ultrathin sections are floated on 10% aqueous solution of periodic acid for 20–30 min at room temperature. Alternatively, sections are exposed either to a saturated solution of potassium periodate for 30–36 min or to a 15% solution of hydrogen peroxide for 10–15 min at room temperature. The previously mentioned treatments may themselves selectively extract cellular substances. An example of such an extraction is the leaching of sulfur-rich keratohyalin granules from sections containing keratinocytes.

OSMIUM BLACKS

Osmium blacks include the free metal oxides of osmium such as $OsO_2 \cdot nH_2O$ (Riemersma, 1963, 1968, 1970; Hanker *et al.,* 1964; Korn, 1967; Adams *et al.,*

1967; Dreher *et al.*, 1967), and coordination polymers of Os(IV) (Hanker *et al.*, 1967) as well as cyclic osmate esters (Criegee, 1936). When OsO_4 is reduced by the unsaturated lipid components of tissue, osmium blacks are formed. Since OsO_4 is soluble in lipids, its reduction by alcohol results in further blackening during dehydration. The excessive blackening, however, can be minimized by fixing the tissue in darkness. The nature and amount of osmium blacks formed are dependent on the concentration of OsO_4 present in the fixative as well as the type of reducing agent.

The formation of osmium blacks (cyclic osmate esters) was first hypothesized by Criegee (1936). This hypothesis was strengthened by the work of Becker (1959), Holt and Hicks (1966), and Korn (1966a). Studies by Hanker *et al.* (1967) demonstrated that osmium blacks are not a single, uniform substance, but vary considerably depending on the reductant, which may even be a sulfur-containing organic substance. Infrared and elemental analyses made by Hanker *et al.* (1967) and Seligman *et al.* (1968) indicated that osmium blacks could be prepared that are coordination polymers of osmium with organic sulfur ligands and include hydrated osmium oxides.

Osmium blacks are amorphous and generally insoluble in tissue constituents. Because of their electron-scattering properties, the atoms of coordination compounds of osmium remaining in the tissue contribute to the formation of contrast in electron micrographs. Because these compounds and polymers of osmium are insoluble in water, in the organic solvents used for dehydration, and in the epoxy and acrylic monomers used in the preparation of ultrathin sections, they are extremely useful for the ultrastructural demonstration of enzyme activity. The demonstration of oxidoreductases by utilizing DAB and OsO_4 producing osmium blacks is well known. Hanker (1975) has explained how the oxidation of DAB and interaction with OsO_4 result in the formation of osmium blacks. The disubstituted indigo dyes are also amenable to osmication, resulting in the production of osmium blacks for localizing 5'-nucleotide phosphodiesterase activity. In this reaction, the olefinic linkage of the indigo molecule is probably involved.

Besides their use in localizing enzyme activity, osmium black end products have been used in a periodic acid–*p*-fluorophenylhydrazine reaction for localizing mucosubstances (Bradbury and Stoward, 1967) and for enhancing the contrast of osmicated lipid-containing membranes by stepwise treatment with osmiophilic thiocarbohydrazide, rinsing, and posttreatment with OsO_4. This latter reaction results in the bridging of osmium to osmium through thiocarbohydrazide (Hanker *et al.*, 1966; Seligman *et al.*, 1966).

OSMETH

Osmeth ($C_6H_{12}N_4 \cdot 2OsO_4$) is one of several compounds of osmium with ammonia and various heterocyclic nitrogen compounds. This compound is a molecular addition complex of hexamethylenetetramine (methenamine) with OsO_4 (see its structure below). The osmeth molecule differs from OsO_4 in that it is less symmetrical and polar. Thus, osmeth penetrates membranes more slowly than OsO_4. The advantages are that it is safer to handle than OsO_4, and its solutions do not blacken as much as OsO_4 solutions; this may be due to the fact that they are less concentrated with respect to OsO_4. The latter advantage could minimize spurious osmium black deposits as a result of postfixation with OsO_4. Hanker *et al.* (1976) have suggested the use of osmeth as a substitute for OsO_4 for cytochemical reactions involving osmiophilic end products using relatively thin (25–50 μm) chopper sections or isolated cells.

Osmeth is soluble only with difficulty in cold water, but can be dissolved in dimethylformamide and then diluted to the desired concentration with distilled water. Solutions of 0.1–0.25% concentration seem to yield adequate fixation and staining of most tissues. A 0.25% osmeth solution is prepared by dissolving 25 mg of this compound in 1 ml of dimethylformamide by stirring. The solution is brought to a total volume of 10 ml by adding distilled water with continuous stirring. This solution can be stored in a tightly closed vial in the cold.

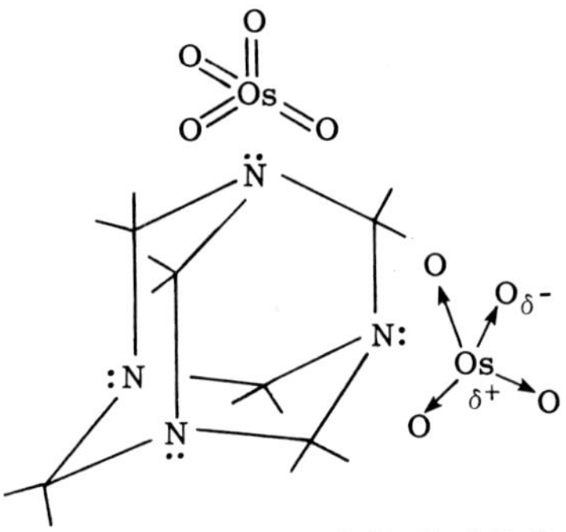

$$C_6H_{12}N \cdot 2\,OsO_4$$

PREPARATIONS FOR AND PRECAUTIONS IN THE HANDLING OF OSMIUM TETROXIDE

Osmium tetroxide is dangerous to handle because of its toxicity and vapor pressure; the solid has a vapor pressure of 11 mm at 25°C (Griffith, 1965). Its fumes are injurious to the nose, eyes, and throat. Hands or any other part of the body must not be exposed to this reagent. This reagent is a strong oxidizing agent and is readily reduced by organic matter and exposure to light. Even the smallest amount of organic matter may reduce it to the hydrated dioxide, which is worthless as a fixative. However, reduction can be avoided by the complete exclusion of dust and organic matter and by use of a brown glass bottle. According to Baker

(1958), reduction by light can be prevented by the addition of strong oxidizing agents such as potassium dichromate; however, it is better to add a few crystals of $MgCl_2$ to slow the reduction. If OsO_4 solution shows violet to light brown coloration because of the formation of osmium dioxides, it can be regenerated by adding a drop of hydrogen peroxide. Extreme care should be taken to avoid contamination by dust particles and exposure to light.

Osmium tetroxide must be handled in a fume hood, and because the reagent is rather expensive, unstable, and volatile, solutions should be prepared with utmost care. The first step in the preparation of its aqueous solution is to remove the label (after reading it!) from the glass ampule containing the OsO_4 crystals. (Osmium tetroxide is also supplied as an aqueous solution in glass ampules.) The glass ampule, a glass-stoppered bottle, and a heavy glass rod are carefully cleaned with concentrated nitric acid (to remove all the organic matter), and then thoroughly washed with distilled water to eliminate all traces of the acid.

The bottle and the rod should be dried in an oven, and they should never be wiped with a paper or cloth towel because these materials invariably leave behind some lint that would reduce the solution to hydrated dioxide. A measured amount of distilled water, buffer, or another vehicle is added to the bottle, and the glass ampule, after scoring its neck with a file (ampules with prescored necks are also available), is gently placed into the same bottle. After the glass ampule has been broken with a heavy glass rod, the bottle is quickly stoppered and is shaken vigorously. Several hours are required to dissolve OsO_4 crystals completely in the vehicle (the solution can be prepared more rapidly by gentle heating on a steam bath over a magnetic stirrer or in a few minutes by using a sonicator). This solution should be protected from exposure to light and contamination caused by organic matter and laboratory dust.

It is emphasized that since OsO_4 is extremely volatile and its solutions rapidly decrease in concentration, solutions should be prepared in small quantities and stored in a tapered flask fitted with a glass stopper and Teflon sleeve or Teflon tape. It is emphasized that the use of ground glass stoppers is not recommended, for they do not prevent decrease in the concentration of OsO_4 even when maintained at 4°C. The only effective way to keep OsO_4 solutions is by using Teflon liners on glass stoppers. The flask must be tightly stoppered, wrapped in aluminum foil, and stored in a refrigerator. The solution is thought to be stable for several months under the previously mentioned conditions of storage. Alternatively, it may be stored in a ground-glass-stoppered bottle, but it is less stable under these conditions.

Aqueous soltuions of OsO_4 (2%) can be stored in small amounts in vials in a freezer, and can be used, after thawing and diluting with a buffer solution, when needed. These solutions can be kept in the freezer for at least several months without any adverse effect. Such practice can be helpful when the time available on a certain day is limited.

Studies on the volatility of OsO_4 solutions indicate that the strength of a 2.22% aqueous solution of OsO_4 drops to 2.15% in 24 hr when stored in an ordinary glass-stoppered bottle at 0°C; more dilute solutions tend to deteriorate more rapidly (Frigerio and Nebel, 1962). Variations in the concentration of OsO_4 solutions lead to erroneous interpretation of electron micrographs, particularly in quantitative electron microscopy. It is obvious that in order to know the concentration of OsO_4 within the tissue specimen, the exact concentration of the OsO_4 solution must be known. The concentration of this reagent in a solution can be determined through spectrophotometric methods by measuring at 649 μm the blue OsI_6^{2-} formed as a result of reduction of OsO_4 with I^-.

It should be noted that when OsO_4 solutions are stored in a refrigerator, all of the internal surfaces will be discolored by the leaking fumes, which may also affect other materials stored in the refrigerator. Osmium tetroxide fumes can penetrate plastics. If OsO_4 solutions need to be disposed of down a sink, large amounts of running water should be used. Burial in a guarded area designated for toxic waste is preferable, however.

REGENERATION OF USED OSMIUM TETROXIDE

Osmium is a rare and expensive metal and only a very minute fraction of the osmium present in a solution is incorporated into the specimen during fixation and staining. Furthermore, proper disposal of the used solutions is tedious. Therefore, it may be desirable to reclaim OsO_4 from used solutions, especially when this fixative is used routinely in large quantities in a laboratory. The procedures for the recovery are not especially hazardous if appropriate safety precautious (e.g., the use of a fume hood), as suggested in the references that follow, are taken.

Although procedures for recovering OsO_4 from used solutions in chemical laboratories (Grube, 1965) and in industry (Symmes, 1936) have been available for a long time, they require the use of equipment that is not readily available in most laboratories. Relatively simple procedures have been developed in recent years for recovering more than 80% of the osmium from used solutions (Jacobs and Liggett, 1971). Reduced OsO_4 can be regenerated by oxidizing it with hydrogen peroxide followed by distilling the newly formed OsO_4 as an aqueous solution. This method is fully explained by Schlatter and Schlatter-Lanz (1971).

A simple procedure was introduced by Kiernan (1978), which is presented below. This procedure involves, essentially, the oxidation of osmium residues in the used solution to OsO_4, which is then extracted into carbon tetrachloride (CCl_4). The solution of OsO_4 in CCl_4 is reduced with ethanol, precipitated, and recovered by filtration on the filter paper. Kiernan (1978) has discussed the chemical reactions involved in the procedure.

Procedure for Preparing OsO_2 (To Be Performed in a Fume Hood)

1. Five hundred ml of used OsO_4 solution is poured into a 1 liter flask to which are added stepwise with stirring 5 ml of concentrated H_2SO_4 and 200 ml of aqueous 6% solution of $KMnO_4$. The ingredients are mixed thoroughly, and the flask is covered with an inverted small beaker and allowed to remain at room temperature for $\sim$30 min. Brown color indicates the precipitation of MnO_2. Additional $KMnO_4$ solution may have to be added until the color becomes purple.

2. The mixture is extracted by shaking vigorously for 2–3 min in a separatory funnel with three changes (each 150 ml) of CCl_4. The heavier fraction, containing CCl_4 and osmium, is carefully separated from the aqueous phase (containing MnO_2) into a 500 ml conical flask.

3. On the addition of 10 ml of absolute ethanol to the flask, the color becomes darker, and black OsO_2 is precipitated. The flask is covered and kept at room temperature for 48 hr.

4. The colorless supernatant is decanted and the precipitate is retained.

5. Approximately 20 ml of absolute acetone is added to the flask containing the precipitate of OsO_2. With a glass rod, the precipitate adhering to the sides of the flask is scraped loose. The flask is shaken and the suspension is poured into a filter funnel containing a dry Whatman No. 1 filter paper. The flask is rinsed out several times with additional acetone, which is poured through the filter until almost all of the OsO_2 has been recovered.

6. The OsO_2 precipitate on the filter paper is rinsed with three aliquots of 50 ml of acetone.

7. On drying, the filter paper is placed on a porcelain evaporating dish, transferred to a vacuum desiccator, and allowed to remain overnight. It is preferable to have a desiccant (anhydrous calcium sulfate) in the desiccator, although drying is due to evaporation of the acetone.

8. Dry OsO_2 powder is collected in a small glass bottle that is tightly stoppered, and can be stored. The black powder is almost indefinitely stable, although it will be transformed to OsO_4 on contact with oxidizing agents, including prolonged exposure to atmospheric oxygen. Approximately 5 g of dry OsO_2 is obtained from each liter of used OsO_4 (2%) solution.

Preparation of 50 ml of 2% OsO_4 Solution

1. One g of dry, black OsO_2 powder is transferred into a clean glass bottle (having a glass stopper) with a 50 ml mark on it. 45 ml of distilled water is added to the bottle, which is shaken to suspend the powder.

2. One ml of 30% hydrogen peroxide is added and the ingredients are mixed rapidly. The bottle is covered with a piece of aluminum foil and placed in a refrigerator for 30 min.

3. One ml of 30% hydrogen peroxide is added drop by drop with continuous swirling. Almost all of the OsO_2 should be dissolved; however, the bottle is covered with aluminum foil and returned to the refrigerator for 10 to 30 min to ensure complete dissolution.

4. Add enough distilled water to reach the 50 ml mark and insert the glass stopper.

This procedure applies only to OsO_4 solutions containing buffers, electrolytes or nonelectrolytes, and tissue lipids. Osmium tetroxide solutions containing inorganic iodides and potassium and chromium compounds have not been subjected to the procedure.

Permanganates

Potassium permanganate was introduced as an alternate fixative to OsO_4 by Luft (1956) prior to the introduction of glutaraldehyde fixation. The primary reasons for the popularity of the permanganate fixatives were the clarity of membranes and the ease of fixation. It soon became known that the membranes stand out primarily because of extensive extraction of background cellular substances. This extraction takes place during fixation and/or subsequent dehydration. This point is elaborated later. Major alterations observed in $KMnO_4$-fixed tissues include swelling of mitochondria and plastids; loss of ribosomes, lipid droplets, microtubules (cytoplasmic and neurotubules), neurofilaments, myofilaments, nuclear annuli, inter- and perichromatic granules, and soluble cytoplasmic proteins; and enlargement of nuclear pores. Although internal membrane systems of mitochondria and plastids are sharply defined, the matrix of these organelles, which shows considerable density when fixed with glutaraldehyde and OsO_4, appears to be wholly removed by $KMnO_4$.

Permanganates penetrate into the tissue faster than do more commonly used fixatives, although their penetration into deeper cells of the tissue block is a problem. The rate of penetration by $KMnO_4$ into a gelatin–albumin gel model of protoplasm is 1 mm in 1 hr (Bradbury and Meek, 1960). In tissues, membrane systems are the first to be "fixed" (within 15–30 min) by $KMnO_4$; longer durations of fixation (1–4 hr) lead to an increase in the electron opacity of some other structures such as the nucleolus and chromatin. In fact, the appearance of the tissue fixed with $KMnO_4$ depends primarily on the duration of fixation.

POTASSIUM PERMANGANATE REACTIONS

Reaction with Membranes

Electron micrographs of tissue fixed by $KMnO_4$ exhibit high contrast of cytoplasmic membranes. The metal is in fact deposited on all the membranous surfaces in the form of a dense fine precipitate. Eddy and Johns (1965) have reported that after fixation for 1 hr in $KMnO_4$, 1 g of protein of erythrocyte membranes showed a deposition of ~0.5 mg MnO_2. They suggested that the ability of $KMnO_4$ to "fix" membranes may be due to formation of a complex network of hydrated MnO_2 coordinated by hydrophilic groups. Manganese dioxide possesses amphoteric properties and thus is able to interact with negative and positive groups of organic molecules (Mellor, 1932). Manganese dioxide may thus be deposited on the polar ends of lipids in the membrane. Such a deposition is compatible with the observed influence of polar groups in permanganate reduction. As an alternate explanation, it is conceivable that $KMnO_4$ reveals the lipid and protein components of the phospholipoprotein membranes by acting as an "unmasking" agent. These exposed proteins reduce the $KMnO_4$ and thus enhance the electron opacity of the membrane.

For the preceding reasons $KMnO_4$ is claimed to have special value in the study of cell membranes. Consequently, this reagent was widely used to study the myelin sheath, endoplasmic reticulum, Golgi apparatus, plasma membranes, mitochondria, and plastids. The continuity of endoplasmic reticulum throughout the cytoplasm and, in the case of plant tissues, from cell to cell can be demonstrated clearly with $KMnO_4$ fixation; in contrast, OsO_4 fixation usually fails to show this continuity.

In this connection it must be pointed out that the clarity of membranes obtained by $KMnO_4$ fixation may be due, in part, to the loss of the protein component of the membrane. Sjöstrand and Barajas (1968) demonstrated that fixation methods causing minimum conformational changes in proteins did not show the unit membrane structure in the tissue. On the other hand, fixation methods that caused denaturation of membranous proteins brought out the familiar unit membrane structure. It has also been demonstrated with Sephadex G-50 that $KMnO_4$ caused considerable oxidative cleavage of proteins and destruction of tyrosine and tryptophan residues (Hopwood, 1969a). Furthermore, complete loss of the α-helix in proteins by the reaction with $KMnO_4$ has been demonstrated by means of circular dichroism measurements (Lenard and Singer, 1968). Since $KMnO_4$ produces extensive conformational changes in protein molecules and oxidizes double bonds in the fatty acyl chains of phospholipids, it is entirely possible that the appearance of $KMnO_4$-fixed membranes is different from that of an unmodified membrane. The clarity of membranes may, in fact, be due to the presence of dense aggregates of unsaturated acyl chains. In view of the aforementioned remarks, data on the structure of cell membranes obtained with the aid of $KMnO_4$ fixation need reevaluation.

Reaction with Monoamines

Potassium permanganate reacts with different types of biogenic monoamines and related compounds and is reduced by hydroxyl groups in these compounds. Richardson (1966) introduced the use of concentrated solutions of permanganate salts for the demonstration of small granular vesicles in the monoamine terminals of the peripheral nervous system. This method has been applied successfully to the central nervous system (e.g., Bloom, 1973) and sympathetic ganglia (e.g., Eränkö, 1972).

It has been shown that monoamines *in vitro* react with $KMnO_4$ (Hökfelt and Jonsson, 1968). Adrenalin, because of the N-methyl group, yields the weakest reaction, whereas noradrenalin reacts rapidly and intensely. Indeed, $KMnO_4$ is able to show an increase or decrease of the noradrenalin content of sympathetic nerve terminals (Itakura *et al.*, 1975). Although the mechanism of this reaction is not certain, since $KMnO_4$ is a strong oxidizing agent, it probably reacts with the reducing amino groups to produce an electron-opaque precipitate. Metallic MnO_2 might be responsible for the electron opacity. There is little likelihood of oxidized amine being bound chemically to the precipitate; however, it may be retained to a certain extent within the precipitate. It is not clear to what extent monoamines are extracted during and after fixation. Furthermore, what proportion of the electron-opaque precipitate is oxidized monoamines is not known. Potassium permanganate may not even show a positive reaction in certain tissues containing monoamines. For example, $KMnO_4$ does not produce an electron-opaque precipitate in the rat adrenal medulla (Kanerva *et al.*, 1977). As a result, it is difficult to distinguish cells storing adrenalin from those containing noradrenalin in this tissue. All cells in this tissue show clear vesicles after fixation with $KMnO_4$. Considering the paucity of information on the chemical nature of the electron-opaque precipitates, caution is warranted while interpreting the significance of the presence or absence of vesicle densities.

As stated earlier, noradrenalin in amine granules does not react with $KMnO_4$ at pH 7.0. However, when the pH is lowered to 5.0, the amine granules of noradrenalin cells are well preserved (Rechardt *et al.*, 1977). Thus it is possible to differentiate noradrenalin cells from adrenalin cells at least in the rat adrenal medulla (Fig. 5.1). Three possible explanations have been forwarded for such a reaction (Rechardt *et al.*, 1977): (1) $KMnO_4$ is a more active oxidant at acid pH, (2) the penetration by the acid fixative is more rapid and uniform, and (3) the acid fixative might release the amine from its hypothetical tight binding into a carrier protein complex in the amine granule, which is then able to react rapidly with permanganate. A favorable effect of acid aldehyde fixation on peptide granules

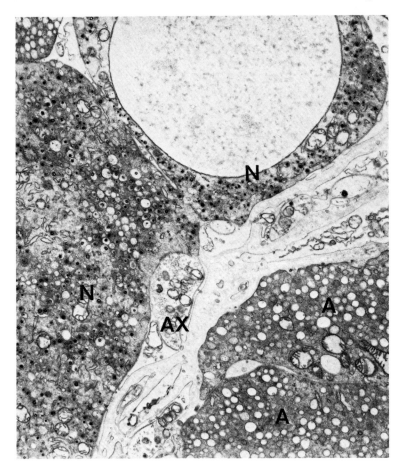

Fig. 5.1. Rat adrenal medulla fixed with $KMnO_4$ at pH 5.0. Two noradrenal cells (N) containing electron-dense cores in their amine granules are seen, whereas adrenal cells (A) show no precipitation in their granules. An axon terminal (AX) contacts the noradrenal cell. ×7225. (Rechardt *et al.*, 1977.)

(Mortensen and Morris, 1977), neurosecretory granules (Morris and Cannata, 1973), insulin granules (Howell *et al.*, 1969), and some granule types of the pars intermedia of the pituitary (Thornton and Howe, 1974) has been indicated.

Reaction with Other Cellular Substances

Nuclear magnetic resonance studies show that Mn^{2+} binds to the phosphate group of nucleotides. A possible model for a complex between the nucleotides and Mn^{2+} may involve the binding of the metal ion to the phosphate group of one nucleotide and simultaneously to the base portion of a second nucleotide (Kotowycz and Suzuki, 1973). Several studies have shown that Mn^{2+} interacts at N-7 of the nucleotide base (Berger and Eichhorn, 1971; Glassman et al., 1971, 1973). When the nitrogen at position 7 is replaced with a carbon atom, there appears to be little interaction with the purine ring. Clement et al. (1973) have also suggested chelation of Mn^{2+} between the vicinal phosphate group and N-7 of the guanine base. Manganese cations are known to show a marked preference for guanine. According to Yamada et al. (1976), some Mn^{2+} is bound to two phosphate groups with a Mn^{2+}–phosphorus distance of 0.33 nm, while other Mn^{2+} ions are simultaneously bound to the adenine ring. Biochemical studies of the interaction between Mn^{2+} and DNA suggest that the metallic cations interact electrostatically. However, the possibility of a nonelectrostatic interaction between Mn^{2+} and DNA cannot be eliminated.

On the basis of the rather limited data available, it appears that $KMnO_4$ is able to preserve a major portion of the DNA and prevents its clumping during dehydration. However, some of the basic protein content of the nuclei, especially amino acid arginine, is reduced after fixation with $KMnO_4$. Comparative studies indicate that OsO_4-fixed tissues retain more DNA than those treated with $KMnO_4$ (Bradbury and Meek, 1960). However, a longer fixation time with $KMnO_4$ usually results in more pronounced chromatin. Since ribosomes are absent in $KMnO_4$-fixed tissues, it is obvious that RNA is not preserved with this reagent. It has been shown that the coordination of Mn^{2+} in the tRNA sites causes conformational changes in this bipolymer (Vocel et al., 1975). In fact, $KMnO_4$ treatment results in an almost total absence of basiophilia in the cytoplasm.

As stated earlier, $KMnO_4$ causes most drastic conformational changes in the primary structure of protein chains. According to Hake (1965), the solubility of wool proteins after treatment with $KMnO_4$ increases, which may indicate that the fixative lacks the ability to produce cross-links. It is known that $KMnO_4$ acts as a denaturant and degrades the tertiary structure. Potassium permanganate probably reacts as a protein stain, but does not introduce intra- or intermolecular cross-links. It has been suggested that Mn^{2+} ions coordinated on tRNA are capable of coupling with the carboxyl groups of amino acids and dipeptides (Weiner et al., 1975; Vocel et al., 1975). It has been reported that 2 moles of $KMnO_4$ bind by coordination to imidazole groups of histidine in heavy meromyosin (Yazawa and Morita, 1974).

Potassium permanganate is known to react with the double bonds of fatty acid residues and is quite effective in fixing glycolipids, phosphatidyl glycerol, and phosphatidyl choline, provided the duration of fixation is sufficiently long (Ongun et al., 1968). Studies of the interaction of $KMnO_4$ with phospholipid and cholesterol monolayers clearly indicate that at pHs near neutrality, the double bonds in the fatty acyl chains are oxidized by this reagent, whereas the double bond of cholesterol is oxidized only at pH 2–3 (Shah, 1970). The presence of double bonds is necessary for the interaction of $KMnO_4$ with lecithin monolayers. Studies of the reaction between $KMnO_4$ and lipids indicate that 1 mole of $KMnO_4$ binds to one double bond in oleic acid, while 2.5 moles of $KMnO_4$ bind to one double bond in lecithin (Riemersma, 1970). Because of interaction with $KMnO_4$, unsaturated lipids probably yield ketohydroxy and diketo compounds (Coleman *et al.*, 1956). At a higher pH (11.0–12.0), dihydroxy compounds are formed.

Since $KMnO_4$ is easily reduced by many organic substances including fatty acids, it is pertinent to present the sequence of steps involved in this reaction. It has been proposed that cyclic manganic esters are formed as intermediates, although such compounds are unstable and have been neither isolated nor identified. The following scheme shows a hypothetical primary step in the reduction process:

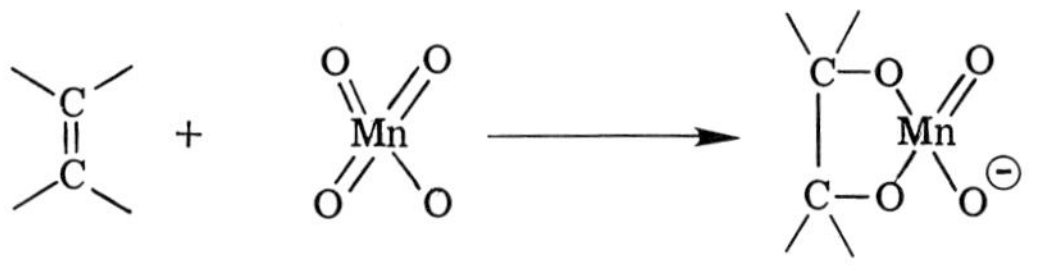

It has been suggested that $KMnO_4$ is unable to stabilize mucopolysaccharide substances of the intercellular connective tissue (Bradbury and Meek, 1960). However, the structures containing polysaccharides (e.g., the cell wall) are stained heavily with $KMnO_4$; glycogen is also fairly well preserved and stained by this reagent. It is known that $KMnO_4$ produces a precipitate when mixed with sucrose, a common constituent of plant cells. Other compounds, such as lipid droplets and nucleoplasm, are poorly preserved and appear as "washed-out" areas. Also, zymogen granules are not rendered electron opaque by $KMnO_4$ and appear as light areas. From the foregoing discussion it should be evident that $KMnO_4$ cannot be regarded as a general-purpose fixative.

CONTRAST AND LOSS OF CELLULAR SUBSTANCES

Potassium permanganate is a very strong oxidizing agent, and during the oxidation-reduction reaction it is reduced mainly to MnO_2, a solid brownish-black precipitate. Deposition of manganese or its oxides accounts for most of the observed contrast, and the rest is due to the removal of background proteins by dehydration agents. It has been demonstrated, for instance, that $KMnO_4$-fixed plastids show greater contrast caused by the leaching of the stroma than those fixed by glutaraldehyde (Diers and Schötz, 1966). This leaching is also responsible for the conspicuous discontinuous pattern of the plastid lamellae. Studies of animal tissues confirm the results obtained from plant tissues in that considerable amounts of cellular materials are lost during and after $KMnO_4$ fixation. Korn and Weisman (1966) have reported that after $KMnO_4$ fixation of amoeba all the neutral lipids and $\sim 25\%$ of the phospholipids were lost. The possibility that $KMnO_4$ does not act as a fixative and that the actual fixation of the tissue occurs in the alcohol during dehydration, cannot be discounted.

Several alternative interpretations of the presence of coarse granularity in the background of $KMnO_4$-fixed tissues have been proposed. One of the more plausible explanations is that due to dehydration, the deposition of rather large clusters (5-10 nm) of reaction products of permanganate occurs in the tissue and is responsible for this coarse granularity. It is also possible that this coarseness represents MnO_2 granules resulting from the decomposition of permanganate ions in aqueous or organic solvent solutions. This coarseness can be reduced to some degree by prefixing the tissue with a mixture of glutaraldehyde and acrolein and shortening the fixation time with $KMnO_4$. Since permanganates are relatively stable when dissolved in acetone, less granularity of the background may result from using permanganate dissolved in acetone instead of water. The use of acetone instead of ethanol as a dehydrating agent is also found to be helpful in minimizing the coarseness.

The overall quality of tissue preservation can be somewhat improved by prefixation in an aldehyde prior to fixation by $KMnO_4$ (Hayat, 1968b). This preserves not only membranous structures but also most other cell components. Satisfactory preservation of delicate tissues such as nerves (deF. Webster, 1971) and hard and dense specimens such as seeds (Mollenhauer and Totten, 1971) has been obtained by this method.

COMPARISON BETWEEN POTASSIUM PERMANGANATE AND OSMIUM TETROXIDE

Potassium permanganate is insoluble in oils and fats, as contrasted with OsO_4, which has a high solubility in and reactivity with these materials. Studies with gelatin–phospholipid and gelatin–albumin gels indicated that in both cases $KMnO_4$-fixed gels showed swelling, whereas those treated with OsO_4 exhibited shrinkage (Bradbury and Meek, 1960). Similar results were obtained with gelatin alone. The swelling caused by $KMnO_4$ in tissue specimens is most pronounced in mitochondria and plastids. Gels of gelatin produced by $KMnO_4$, in contrast to those produced by OsO_4, are soluble in hot water.

The distance between the pairs of membranes of endoplasmic reticulum is greater in $KMnO_4$-fixed than in OsO_4-fixed tissues. The difference in the effect of the two fixatives on these membranes were shown by Jard et $al.$ (1966). They demonstrated that frog bladder after treatment with $KMnO_4$ became completely permeable to sodium ions, whereas OsO_4-treated bladder maintained partial permeability to these ions. The most obvious difference between the results of fixation by $KMnO_4$ and OsO_4 is the presence of black reaction products of unsaturated lipids and ribosomes with the latter fixation, whereas the absence of such black deposits and ribosomes is characteristic of the former fixation. This difference in the lipid reaction of the two fixatives can be utilized for the identification of lipid droplets (Fig. 1.1).

The rate of penetration by $KMnO_4$ is slightly greater than that by OsO_4, and the former is considered to be a stronger oxidant. However, as with OsO_4 the rate of penetration falls off with time. This decline is probably due to the formation of insoluble electron-opaque reduction products that act as physical barriers to penetration.

FIXATION PROCEDURES

It has been claimed that the type of cation of the permanganate used influences the quality of tissue preservation. That sodium permanganate gives better preservation of the fine structure has been indicated. According to Wetzel (1961), sodium permanganate yields better preservation of plasma membranes. In this connection it is interesting to note that the concentration of potassium ions even in 1% $KMnO_4$ exceeds that found in the living systems of most organisms; whereas sodium permanganate shows greater similarity to a "physiological" extracelluar fluid. On the other hand, the sodium salt is less stable than the potassium salt, and sodium permanganate solution decomposes rapidly. This decomposition can be avoided by using only freshly made solutions.

Since divalent calcium ions tend to stabilize lipids, some workers prefer to use calcium permanganate over other permanganates. Afzelius (1962) compared the quality of preservation of the fine structure of rat liver fixed by calcium permanganate with $KMnO_4$. He found that the membranes in the specimen fixed with the former showed a width of ~ 10 nm, whereas similar membranes after fixation with the latter were ~ 7.5 nm wide. The trilaminar structure of the membrane was revealed much better when the tissue was fixed with calcium permanganate.

Lanthanum permanganate was used with the hope that its trivalency would provide an even greater stabilization of membranes than that produced by divalent calcium ions (Doggenweiler and Frenk, 1965). Lanthanum permanganate stains intercellular materials in vertebrate and invertebrate nervous tissue, and a surface layer of plasma membranes. The staining by this permanganate is much stronger in the presence of anions (e.g., phosphate) that can form insoluble salts with lanthanum (Dimmock, 1970). Therefore, specimens should be washed with a phosphate-free buffer prior to fixation with this permanganate. It should be noted that lanthanum permanganate does not penetrate into cells. A simple method for the preparation of lanthanum permanganate is described by Lesseps (1967).

Lithium permanganate has been used for preserving mammalian cells and virus (McDuffie, 1974). A short fixation time (10-15 min) is necessary for satisfactory preservation of the specimen. Similarly, a short fixation time is recommended for preserving tobacco mosaic virus and plant cells by $KMnO_4$ (Warmke and Edwardson, 1966). According to McDuffie (1974), lithium permanganate gives more fine detail of nuclear and viral structures than that yielded by $KMnO_4$. The reasons for the superiority of lithium permanganate over $KMnO_4$ are (1) highly charged membranes are more permeable to Li^+ than to K^+ (Eisenman, 1962; Meves, 1970) and (2) Li^+ is more strongly attracted to nucleic acids than is K^+ (Diamond and Wright, 1969). Thus Li^+ ions may penetrate into the cell faster than K^+ ions. Lithium permanganate is recommended for studying membrane-virus interactions.

Barium permanganate has been investigated as a fixative, and it was reported that it gave the same appearance of retina and sciatic nerve as that produced by $KMnO_4$ except that more contrast was achieved with the former (Doggenweiler and Frenk, 1965). Zinc permanganate was also tested but it produced substandard preservation of cytoplasmic details (Afzelius, 1962).

Until more information on the chemical effect of different permanganates on tissue is available, it is difficult to judge any one of the permanganates superior, in every aspect, to another. According to Millonig and Marinozzi (1968), the cation is of little importance in comparison with the anion MnO_4^- which is involved in all of the basic reactions of permanganate with the tissue. Presently, however, $KMnO_4$ and $LiMnO_4$ are used more often than other types of permanganates and appear to produce better overall preservation than that by other permanganates.

Although Luft (1956) originally prepared $KMnO_4$ in a buffer, unbuffered solutions yield equally satisfactory preservation of both plant and animal tissues. No obvious difference, as to general morphology, is observed when the tissue is fixed with $KMnO_4$ at a pH ranging from 6.0 to 7.5. Tissue appearance seems to change very little with various concentrations of permanganates. Also, the duration of fixation is not a critical factor in obtaining satisfactory preservation. One plausible reason for the minor importance of the buffer temperature, pH, and concentration is that permanganates act on the tissue so fast that these factors do not have time to influence the tissue before fixation.

For general fixation with $KMnO_4$, aqueous solutions having a concentration ranging from 0.6 to 3% are most suitable. Aqueous lithium permanganate is used in concentrations ranging from 5 to 10%. Adequate fixation can be obtained within 15-60 min depending on the type and size of the specimen and the objective of the study. The author prefers to use as short a duration as possible commensurate with adequate penetration of the tissue. Although temperature is not a critical factor during fixation, general morphology of the tissue appears slightly better preserved after fixation at low temperatures. Only fresh solutions should be used, for permanganates deteriorate rapidly on coming in contact with water. Specimens should be washed thoroughly prior to dehydration.

Permanganate Fixatives

Potassium Permanganate

Acid Permanganate. Small pieces of adrenal medulla (monoamine storing cells) are fixed in 3% $KMnO_4$ in 0.1 to 0.2 mol/l acetate buffer (pH 5.0) for 30 min at 4°C. After a brief rinse in the buffer, specimens are stained *en bloc* with uranyl acetate. Ultrathin sections are viewed without poststaining. The quality of ultrastructure preservation is adequate (Fig. 5.1).

Neutral Permanganate. Specimens are fixed in 3% $KMnO_4$ buffered with Krebs–Ringer–glucose (pH 7.0) for 2 hr at 4°C. After fixation, specimens are rinsed in the buffer several times and then allowed to remain in the buffer overnight prior to dehydration.

Lanthanum Permanganate

$La(NO_3)_3 \cdot 6H_2O$	1 g
$KMnO_4$	1 g
Veronal acetate stock solution	20 ml
Ringer's solution	6 ml
HCl (0.1 mol/l)	enough to obtain pH 7.6
Distilled water to make	100 ml

The fixative is used for preserving isolated cells and cell fragments including membranes, which are fixed for 1 hr at 4°C. In vertebrate and invertebrate nervous tissue, intercellular materials are stained distinctly.

Sodium Permanganate

NaCl	2.8 g
KCl	0.4 g
$CaCl_2$	0.2 g
$MgCl_2 \cdot 6H_2O$	0.2 g
NaH_2PO_4	0.16 g
$NaMnO_4$(0.83%) in veronal acetate (pH 7.5) to make	100 ml

The fixative was used for preserving plasma membranes of cells in free-living flatworms.

Lithium Permanganate

Specimens are washed with phosphate-buffered saline, and then fixed in 5-10% aqueous lithium permanganate for 10-15 min at room temperature. Alternatively, specimens are prefixed in 2.5% phosphate-buffered glutaraldehyde, followed by fixation in lithium permanganate. Sections are poststained with uranyl acetate followed by lead citrate. This method is recommended for studying mammalian membrane-virus interactions. Considerably longer fixation (3 hr at 4°C) was used for preserving autonomic nerves in order to study granular vesicles (Ishii, 1974).

Miscellaneous Fixatives

RUTHENIUM TETROXIDE

Because ruthenium tetroxide (RuO_4) is closely related to OsO_4, the possibility of its use as a stain in histology (Ranvier, 1887) and as a fixative for plant (Carpenter and Nebel, 1931) and animal tissue (Bahr, 1954) was explored. No further work on its application as a fixative was reported until much later. Membranes in rat kidney and liver (Gaylarde and Sarkany, 1968) and in the ventral lobe of rat prostate (Pelttari and Helminen, 1979) were successfully fixed with this reagent. These membranes appeared thicker than those preserved with other fixatives.

Ruthenium tetroxide is a strong oxidizing agent and decomposes readily. It dissolves slowly, and aqueous solutions decompose rapidly even when kept in the dark under cold temperatures. Fresh solutions show a golden yellow color, which after some time turns brownish as a result of the separation of black deposits of the lower oxides of ruthenium. After this change in color the solution is useless as a fixative. Like OsO_4, ruthenium tetroxide is superficial in its action, for it penetrates into the tissue very slowly.

The claimed advantage of fixation with ruthenium tetroxide is that like the plasma membrane, the nuclear and cytoplasmic membranes appear as triple-layered structures. These triple-layered structure can be clearly seen without additional staining. On the other hand, in many types of tissues fixed with OsO_4, although plasma membranes appear triple-layered, most other membranes commonly give the appearance of a single diffuse line. The width of various membranes show a remarkable uniformity in cells fixed with ruthenium tetroxide. Connections between the plasma membrane, membranes of the endoplasmic reticulum, and the inner and outer nuclear membranes have been demonstrated in rat kidney and liver tissues fixed with ruthenium tetroxide (Gaylarde and Sarkany, 1968).

In contrast to OsO_4, ruthenium tetroxide probably reacts strongly with some of the more polar lipids. The fixative also reacts strongly with proteins, glycogen, and monosaccharides (Gaylarde and Sarkany, 1968). The mechanism of interaction of ruthenium tetroxide with various cellular substances is not yet known. Further work is needed to elucidate, for example, the presence of an electron-opaque ''coating'' closely apposed to the surface of the plasma membrane in many types of cells fixed with ruthenium. According to a typical procedure, the tissue is prefixed with buffered 4% glutaraldehyde followed by postfixation with buffered 0.1–0.05% ruthenium tetroxide (pH 7.1) for 1 hr at 4°C. Penetration of ruthenium tetroxide into some tissues is poor.

DIMETHYLSUBERIMIDATE

Because glutaraldehyde is exceedingly effective in cross-linking the proteins, this dialdehyde is generally unsuitable for immunologic studies because it causes the loss of antigenicity. The dialdehyde also has the disadvantage of being an inhibitor of enzymes and introducing Schiff-positive aldehyde groups into tissue. To circumvent some of these limitations, dimethylsuberimidate (DMS) was introduced as a fixative for light and electron microscopy by Hassell and Hand (1974). McLean and Singer (1970) were the first to use diimidoesters as fixatives for immunoelectron microscopy. The chemical structure of DMS is as follows:

$$^-Cl^+H_2N \qquad NH_2^+Cl^-$$
$$\parallel \qquad\qquad \parallel$$
$$H_3CO-C-(CH_2)_6-C-OCH_3$$

Dimethylsuberimidate is a bifunctional reagent that cross-links proteins probably by reacting with α- and ϵ-amino groups. Probably a covalent bond is formed between the carbon adjacent to the amido group in DMS and an amino group in the amino acid, with the release of a mole of alcohol (Hunter and Ludwig, 1972; Wold, 1972). At about pH 7.5, the reaction with α-amino groups dominates, whereas at about pH 9.5, the reaction is primarily with ϵ-amino groups. More extensive cross-linking of proteins is expected at a higher pH because ϵ-amino groups exceed free α-amino groups in proteins. Since lysine is the major source of ϵ-amino groups, this amino acid may react preferentially with DMS (Hartman and Wold, 1967); thus proteins rich in lysine such as histones and collagen may be readily cross-linked.

Since the amid group (NH_2^+) is located close to the functional groups in the molecule, DMS probably does not alter the net charge of tissue proteins. Even after extensive reaction with certain imidoesters, significant amounts of enzymatic activity and immunologic properties seem to be retained (Hunter and Ludwig, 1972; Wold, 1972). Specimens fixed with DMS seem to show more accurately the aldehyde groups generated by the periodic acid–Schiff technique because DMS is not an aldehyde and does not introduce additional aldehyde groups.

It has been indicated that fixation with DMS results in a high retention of enzymatic activity (glucose-6-phosphatase, thiamine pyrophosphatase, and catalase) (Hand and Hassell, 1976). According to Yamamoto and Yasuda (1977), in specimens fixed with DMS, glutamate dehydrogenase retained 50% activity. It has also been demonstrated that DMS retains more glycogen than that observed after fixation with glutaraldehyde. Mitochondrial matrix shows increased electron density compared with that in the glutaraldehyde-fixed specimens (Fig. 6.1). In addition, the nuclei of DMS-fixed tissue specimens are strongly stained with the Feulgen method with little background reaction in the cytoplasm. Microtubules and neurofilaments are also known to preserve better in the presence of DMSO.

The DMS fixative solution is prepared by adding the various components in the following order (Hassell and Hand, 1974) (Pierce Chemical Co., Rockford, Illinois):

Distilled water	7.8 ml
NaOH (1 mol/l)	1.2 ml
Tris base	121–182 mg
DMS	160–200 mg

Adjust the pH to 9.5 with HCl or NaOH, and 1.0 ml of 0.2 mol/l $CaCl_2$ drop by drop. The buffer vehicle has an osmolality of 300 mosmols. Since DMS is unstable in aqueous solutions, the fixative solution should be prepared immediately prior to use. The duration of fixation should not exceed 2 to 3 hr at room temperature. Tris buffer appears to facilitate the penetration of this fixative into the tissue. The addition of Ca^{2+} reduces the extraction of proteins.

The size of the DMS molecule is larger than that of glutaraldehyde, and thus

Fig. 6.1. A, hepatocyte from a rat liver fixed with 20 mg of dimethylsuberimidate per ml in 0.15 mol/l Tris–HCl buffer (pH 9.5) and postfixed with 1% OsO_4. The nucleus (N) shows marginated chromatin. The rough endoplasmic reticulum (RER) is scattered throughout the cell. Lipid droplets (LD) and glycogen (GLY) are well preserved, but lysosomes (LY) appear pale. Mitochondria (M) appear typical except for a dense matrix. Bile canaliculus (B). ×7760. B, hepatocyte from a rat liver fixed with 2.5% glutaraldehyde in 0.1 mol/l cacodylate buffer (pH 7.4), and postfixed with 1% OsO_4. The ultrastructural organization is similar to that in part A, although the nucleoplasm is more granular, mitochondria have a lighter matrix, and glycogen is stained more intensely. ×6800. (Hassell and Hand, 1974.)

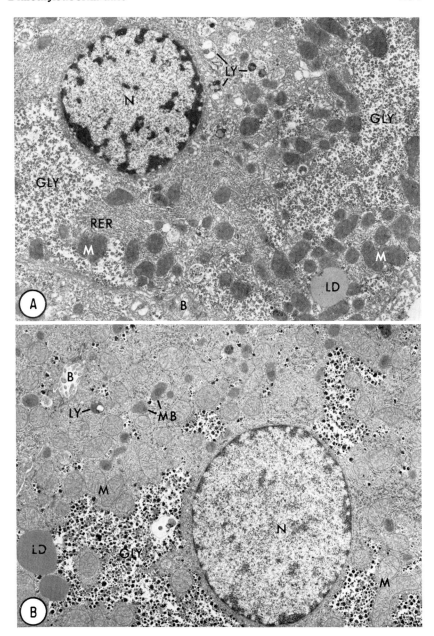

its penetration into the tissue is relatively slow. It should be noted that specimens fixed with DMS show swollen Golgi and smooth endoplasmic reticulum, and the quality of ultrastructure preservation is generally less satisfactory than that obtained with glutaraldehyde (Fig. 6.1).

CARBODIIMIDES

1-Ethyl-3(3-dimethylaminopropyl)carbodiimide-HCl (WSC) is a water-soluble bifunctional reagent. It has been suggested that it is preferable to glutaraldehyde as a fixative for cytochemical and immunocytological studies. It is thought to cross-link proteins with minimum alteration in their biological activity. It has been indicated that in the specimens fixed with WSC, the enzymatic activity retained is 15% for alcohol dehydrogenase, 70% for glucose-6-phosphatase, 73% for ATPase, and 61% for fructose-1,6-diphosphatase (Yamamoto and Yasuda, 1977). These percentages are much higher than those obtained after fixation with aldehydes.

Carbodiimide reacts with both carboxyl and amino groups in proteins at neutral pH, forming intermolecular cross-links. It carries a carboxyl group to an adjacent amino group in proteins. The proposed scheme for cross-linking the proteins (Yamamoto and Yasuda, 1977) is as follows:

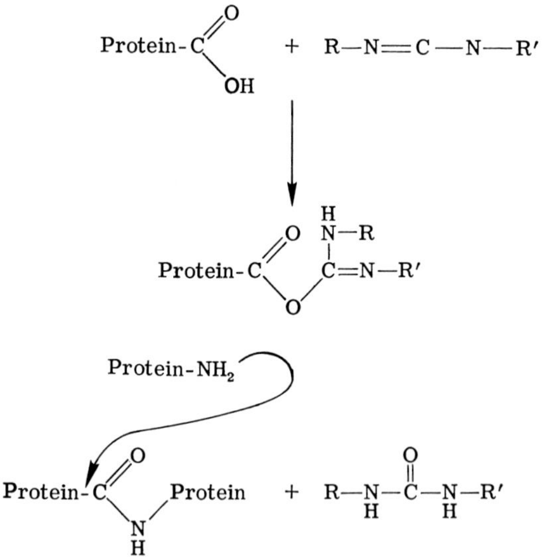

At acidic pH values, the protein groups that primarily react with WSC are carboxyl, sulfhydryl, and tyrosine (Carraway and Koshland, 1968; Carraway and

Triplett, 1970). The optimal pH for protein cross-linking seems to be 7.0 to 7.5, since amino groups in proteins are not very reactive at lower pH values. Carbodiimide was first used for the modification of carboxyl groups in proteins (Scheehan and Hlavka, 1956, 1957). It has been used in immunochemical studies (Goodfriend *et al.*, 1964; Johnson *et al.*, 1966; Linscott *et al.*, 1969) and for preserving tissue fine structure (Yamamoto and Yasuda, 1977). The preparation of the fixative solution (4%) is as follows:

WSC	400 mg
Phosphate buffer (0.1 mol/l, pH 7.4)	10 ml

The osmolality is 410 mosmols. Specimens are fixed in WSC for 2 hr at 4°C and postfixed with 1% OsO_4 for 1 hr at 4°C. The WSC solution should be prepared immediately prior to use.

TRIOXSALEN

Trioxsalen (4,5′,8-trimethylpsoralen) is a trimethyl derivative of psoralen which is a medically important furocoumarin known for its ability to photosensitize mammalian skin (tanning effect). Trioxsalen in conjunction with ultraviolet light of long wavelength (320–380 nm) can covalently cross-link pyrimidines in opposite strands of the DNA double helix (Cole, 1975). The major effect of trioxsalen on DNA migration is assumed to be DNA unwinding. The structure of trioxsalen is as follows:

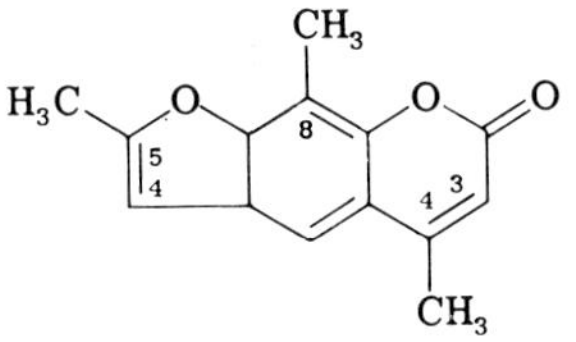

Because trioxsalen permeates plasma and nuclear membranes, it can cross-link DNA *in situ* in chromatin, in isolated nuclei, or in intact living cells (Pathak and Kramer, 1969). Since DNA within a nucleosome is protected, it is not crosslinked by trioxsalen. In other words, trioxsalen binds at the sites corresponding to the regular nuclease-sensitive regions of the chromatin in nuclei. Consequently, after treatment with trioxsalen, protected regions of DNA appear as a single strand, whereas unprotected regions appear as duplexes due to cross-linking (Lee, 1978). This method is useful in the study of chromatin structure with the TEM, for trioxsalen is able to preserve a linear record of its interaction sites with DNA (Hanson *et al.*, 1976; Wiesehahn *et al.*, 1977; Wiesehahn and Hearst, 1978) and RNA (Wollenzien *et al.*, 1978).

Methods of Fixation

DOUBLE FIXATION

Primary fixation with OsO_4 is undesirable because it has a damaging effect on the cytoplasmic matrix, microtubules, microfilaments, and the continuity of membrane systems. Primary fixation with OsO_4 definitely results in the breakdown of various cellular membrane systems into vesicles. Substantial evidence from several cell systems indicates that OsO_4 causes membranes of interdigitation of ciliary epithelium to become physically unstable. This results in the reorganization of these membrane sheets into sheets of tubules. On the other hand, in the glutaraldehyde-fixed tissues and presumably in the living state, these membranes are separated only by the intercellular space, but otherwise are continuous. The formation of rows of vesicles and tubular system in the ciliary epithelium are, therefore, considered artifacts of OsO_4 fixation (Tormey, 1964).

That primary fixation of amphibian oocytes with OsO_4 causes the outer nuclear membrane in the annulate lamellae to form vesicles has been demonstrated (Kessel, 1969). In oocytes fixed with glutaraldehyde, on the other hand, the outer nuclear membrane is continuous with lamellae. It is apparent that glutaraldehyde when used as a primary fixative is capable of preventing the breakdown of continuous membrane structures.

The X-ray diffraction pattern of actin-containing filaments is destroyed by osmication, whereas glutaraldehyde does not destroy this pattern (Page and Huxley, 1963a,b). Fixation with glutaraldehyde followed by OsO_4 is more effective than is OsO_4 alone in the demonstration of granular vesicles in the pineal bodies of the autonomic nervous system in rats (Machado, 1967). The vesicles in tissue fixed with the former exhibited an increase in the granularity of their membranes, in their size, and in the size and density of their cores. The advantage of double fixation has also been demonstrated convincingly in the study of mouse Paneth cell granules. When these cells are fixed with formaldehyde, glutaraldehyde, or OsO_4, the core of the granules is surrounded by a shell (halo) that is interpreted as a shrinkage artifact. On the other hand, double fixation with glutaraldehyde followed by OsO_4 reduces the shell to negligent proportions (Hampton, 1965). Many more examples are available substantiating the inadequacy of OsO_4 as a primary fixative.

Glutaraldehyde too has its limitations (discussed on p. 99). Certain cell components, especially lipids, are not fixed by glutaraldehyde, and thus are extracted during dehydration. As a result, membranes are not preserved in cells fixed with glutaraldehyde only. In the light of the aforementioned data, it is obvious that the best overall preservation of the fine structure is obtained by double fixation with glutaraldehyde or a mixture of formaldehyde and glutaraldehyde followed by OsO_4. Since aldehydes are potent reductants, excess glutaraldehyde or formaldehyde should be removed before osmication.

GLUTARALDEHYDE AND OSMIUM TETROXIDE MIXTURE

There are some indications that sequential double fixation, glutaraldehyde followed by OsO_4, has undesirable effects on certain specimens, especially on isolated cells and membranes. One possible explanation for these effects is that during prefixation the specimen is subjected to the detrimental effects of glutaraldehyde, such as shrinkage and lipid extraction, prior to the application of OsO_4. Moreover, OsO_4 induces marked conformational changes in proteins when the specimen has been prefixed with glutaraldehyde. It is known that cells fixed with glutaraldehyde followed by OsO_4 are less stable against mechanical stress (sonication) than those fixed with glutaraldehyde only.

Evidence is accumulating that indicates that a prolonged rinse in a buffer between the prefixation with glutaraldehyde and postfixation with OsO_4 produces undesirable effects such as uneven fixation, cell shrinkage with widened extracellular spaces, extraction or swelling of mitochondrial matrices, and clumping of chromatin. It has also been shown that the storage of glutaraldehyde-perfused tissue for as little as 30 min in cacodylate buffer, prior to hardening by OsO_4, has a profound flattening effect on agranular synaptic vesicles of cholinergic nerve endings (Bodian, 1970).

Certain plant cells (e.g., *Rubus fruticosus*), fixed with glutaraldehyde followed by OsO_4, show the presence of osmiophilic substance in the ground cytoplasm and in the stroma of certain plastids (Ljubešić, 1970). It is not clear whether such electron-opaque substance is a fixation artifact or represents lipids or some other substance abundant in these cells. Other reported disadvantages of sequential double fixation are the loss of arrangement of microtubules when fixed in the cold, myelinization of lipids, and a more dispersed pattern of ribosomes.

Lipids undergo profound changes during and after fixation with glutaraldehyde. Glutaraldehyde cross-links proteins without reducing the fluidity of the lipid bilayer (Jost *et al.*, 1973). Lipids in the cytoplasm as well as in the membranes, that have not been cross-linked by glutaraldehyde, are free to form not only single and multilayered vesicles, but also multivesicular mounds. Free blebs and vesicles or intramembrane particle-free membrane blebs (blisters) present in the aldehyde-fixed tissues are considered to be artifacts of aldehyde fixation (Bloom and Haegermark, 1965; Olah and Rohlich, 1966; Shelton and Mowczko, 1977; Hasty and Hay, 1978; Meyer, 1978); postfixation with OsO_4 does not prevent the formation of such artifacts. Such artifacts are virtually absent in freeze-fracture replicas of corneal fibroblasts that are frozen without aldehyde fixation and in sections of the tissue fixed with a mixture of glutaraldehyde and OsO_4 (Hasty and Hay, 1978). The mixture prevents postfixation movement of membrane lipids, especially negatively charged fluid lipids, which are capable of considerable mobility after aldehyde fixation (Fig. 7.1). However, according to Schook (1980), blebs are present in the lens placode of the chick embryonic eye even after fixation with a mixture of glutaraldehyde and OsO_4. Furthermore, the presence of blebs *in vivo* has been confirmed by observing them in unfixed specimens with differential interference contrast microscopy (Schook, 1980).

Fixation by the mixture results in sharp membrane definition and especially good preservation of nucleoprotein-, lipid-, and polysaccharide-containing structures. Granules and vesicles stain distinctly, but glycogen is defined poorly. This procedure seems to preserve polyribosomal structure particularly well in plant tissues, and cytoplasmic microtubules appear intact even after fixation at low temperatures. Fixation with the mixture yields distinctly better preservation of cellular materials in certain specimens such as single cells (Fig. 7.2) than that obtained by sequential double fixation with commercial glutaraldehyde and OsO_4.

The mixture has been used successfully for fixing white blood cells, which are not well fixed by sequential double fixation. The quality of preservation and contrast can be further improved by postfixing the cells with 2% uranyl acetate prior to dehydration. Several other types of cells in suspension or in monolayer cultures have been successfully fixed with this mixture (Hirsch and Fedorko, 1968; Hirsch *et al.*, 1968; Fedorko *et al.*, 1968). The mixture has been recommended for preserving ciliates (Shigenaka *et al.*, 1973), fragile structures in pathologically altered cells (Laiho *et al.*, 1971), *Tetrahymena* and chicken myoblasts (Kolb-Bachofen, 1977), and *Actinophrys* (Ockleford and Tucker,

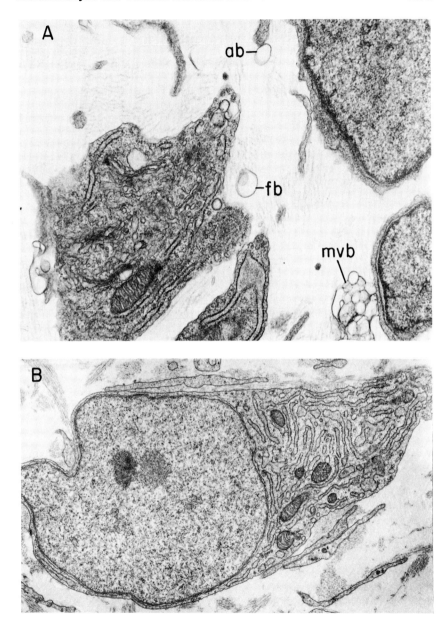

Fig. 7.1.   Cornea of 14-day-old avian embryo. A, fixed with glutaraldehyde followed by OsO $_4$. Free blisters (fb) and attached blisters (ab) are thought to be fixation artifacts. Multivesicular body is also present. These artifacts may result from the mobility of membrane lipids after aldehyde fixation alone. ×20,400. B, fixed with a mixture of glutaraldehyde and OsO $_4$; artifactual blisters are absent. ×8500. (Hasty and Hay, 1978.)

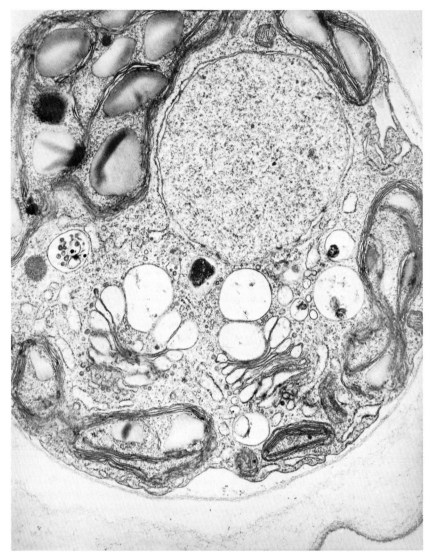

Fig. 7.2.   The mature spheroid (somatic cell) of *Volvox* (green alga) fixed with a mixture of 2% glutaraldehyde and 1% OsO_4 in the normal growth medium containing 1% sucrose (pH 7.7) for 45 min at 0°C. Cells were mordanted with 1.5% tannic acid in the growth medium (pH 7.6) for 30 min at room temperature, followed by a 5 min wash in 1% sodium sulfate in 0.05 mol/l cacodylate buffer (pH 7.0). Cells were agitated during processing and centrifuged between each step. Alternatively, cells can be processed in agar blocks. Note the sheath, plasma membrane, vacuoles, Golgi (hypertrophied), rough endoplasmic reticulum, pyrenoid, nucleus, and multivesicular body. ×22,800. (Dauwalder *et al.*, 1980.)

1973). Fixation with the mixture followed by osmication in the cold has been employed for processing algae, fungi, lichens, roots, stems, and leaves of higher plants, various animal tissues including insect midgut (Brunings and Priester, 1971), and isolated subcellular fractions (Franke *et al.*, 1969). A similar procedure has proved effective in the preservation of the interfacial zone that separates the intracellular structures of vesicular–arbuscular mycorrhizal fungi from the cytoplasm of the root cells of the host plant (Carling *et al.*, 1977). Contrary to previous reports, the interfacial zone is completely filled with a granular matrix. Excellent preservation of polyribosomes attached to the thylakoidal membranes in the chloroplast of bean leaf has been obtained by this method (Falk, 1969).

Neither conventional fixation with glutaraldehyde followed by OsO_4 nor primary fixation with OsO_4 is able to preserve satisfactorily tissues of the inner ear. It has been reported that glutaraldehyde damages the stria vascularis (Merck *et al.*, 1974) and that OsO_4 causes artifacts within the external sulcus (Santos-Sacchi, 1978). It has also been reported that fixation in this mixture followed by postosmication improves the retention of the structure of algal surface glycoproteins (Roberts, 1974; Dauwalder *et al.*, 1980) (Fig. 7.2).

A semisimultaneous protocol of fixation has proved more effective than the simultaneous method for studying mitosis in the diatom; mitotic spindle is especially well preserved (Tippit and Pickett-Heaps, 1977). The diatoms are fixed with 1% glutaraldehyde prepared in diatom medium (Reimann *et al.*, 1963); about 5 sec later, a small volume of 2% OsO_4 is added and the cells are fixed concurrently in this mixture for 20 min. The final concentration of OsO_4 in the mixture is 0.2%.

Although aldehydes reduce OsO_4 and it is usually desirable to remove excess aldehyde from the tissue prior to postfixation with OsO_4, these two fixatives can be used as a mixture because the reaction between glutaraldehyde and OsO_4 has a temperature-dependent lag period; at 18°C the lag is longer than 30 min (Hopwood, 1970). Thus, if fixation with the mixture is completed within ~30 min at 4°C, osmium precipitates may not develop. The mixture should be prepared immediately prior to use.

As stated earlier, OsO_4 and aldehydes are somewhat incompatible with each other. In the mixture, aldehydes are oxidized to an appropriate acid and OsO_4 is reduced to osmium blacks. When OsO_4 and glutaraldehyde are mixed in a 1:1 ratio, the mixture darkens because of the formation of osmium blacks. When OsO_4 is allowed to react to completion with a large excess of glutaraldehyde, the reaction product is a mixture of Os(VI), Os(IV), and Os(III); no Os(II) or Os(0) is formed (White *et al.*, 1976). Osmium tetroxide reacts more rapidly with glutaraldehyde than with formaldehyde (Hopwood, 1970). The rate of reaction increases with temperature and the concentration of the aldehyde.

As a result of the reaction between OsO_4 and aldehydes, the net concentration of fixative available to react with tissue is decreased. Furthermore, both OsO_4

and glutaraldehyde may compete for the same amino acid residues. Also, glutaraldehyde alone is more effective in cross-linking proteins than when used in combination with OsO_4. Other limitations in the use of the mixture include the presence of an electron-opaque coating on the plasma membrane, the availability of only a limited duration of fixation, and the necessity of lower temperatures of fixation, resulting in the loss of some microtubules. The aforementioned problems have prompted a reappraisal of the usefulness of the mixture. It has been indicated that by using vacuum-distilled glutaraldehyde followed by OsO_4, cells in suspension are well fixed and the preservation is reproducible (Gillett *et al.*, 1975).

The following method is used in preparing the glutaraldehyde–OsO_4 mixture:

    2% glutaraldehyde   1 part
    2% OsO_4 1 part

Both glutaraldehyde and OsO_4 solutions are prepared in a buffer, and the two solutions are mixed immediately prior to use. Specimens are fixed in the mixture at 4°C, washed thoroughly in the buffer, and postfixed in 2% OsO_4.

SIMULTANEOUS FIXATION FOR LIGHT AND ELECTRON MICROSCOPY

In histopathological studies it is sometime desirable to examine the tissue with the light microscope prior to correlative study with the TEM. Such combined studies may necessitate a prolonged storage of tissues of relatively large size. Thus, there is need for a fixative that is relatively stable and in which large tissue blocks can be fixed and stored for routine automated histologic processing as well as for electron microscopy.

The usefulness of electron microscopy in examining surgical and biopsy specimens for tumor diagnosis has been recognized. Electron microscopy is used extensively in the examination of renal biopsy specimens (Seigel *et al.*, 1973). Diagnostic electron microscopy is finding increased application in the study of liver, skin, muscle, nerve, and bone marrow, as well as other tissues used in biopsy and autopsy procedures. The importance and use of diagnostic electron microscopy are expected to increase rapidly in the near future. Methods for processing tissues for diagnostic electron microscopy have been discussed in at least two books (Trump and Jones, 1978; Johannessen, 1978).

Several fixatives have been developed for specimens to be examined with the light and electron microscopes (Lynn *et al.*, 1966; Chambers *et al.*, 1968; Zimmerman *et al.*, 1972; Carson *et al.*, 1973; Yanoff, 1973). The best fixative presently available seems to be a mixture of 4% commercial paraformaldehyde and 1% glutaraldehyde in a buffer having an osmolality of 176 mosmols

(McDowell and Trump, 1976). It is recommended that tissue blocks should not exceed 3 mm in width, and specimens for electron microscopy be taken from the outer parts of the tissue block. The gross penetration of this fixative into soft tissues (liver) is 2.5 mm in 4 hr, and the depth of good fixation is 1.75 mm in the same period of time. The appearance of tissue blocks stored in the fixative for as long as 12 months remained almost unchanged. The fixative is stable for at least 3 months when stored at 4°C.

Although the previously mentioned formulation is most satisfactory, an inexpensive and easily prepared fixative for light and electron microscopy is prepared by using a commercial solution of formaldehyde and Millonig's buffer (Carson *et al.*, 1973):

Technical grade formaldehyde (37–40%)	10 ml
Tap water	90 ml
Sodium phosphate (monobasic)	1.86 g
NaOH	0.42 g

Very large (2–3 cm³) tissue blocks require fixatives that have the maximum penetration power. Such fixatives can be developed by using acrolein or paraformaldehyde to which may be added trinitro compounds. One such formulation is as follows:

2% paraformaldehyde
2% glutaraldehyde
0.1 mol/l phosphate buffer (pH 7.3)
0.2% trinitrocresol
1.5 mM calcium chloride

Another such fixative consists of 10% paraformaldehyde in 0.2 mol/l collidine buffer, which is supposed to extract tissue components and thus facilitate the penetration (Winborn and Seelig, 1970). However, extraction of tissue components is undesirable.

ANHYDROUS FIXATION

The structure of dry tissues changes on coming in contact with aqueous fixatives and other solutions. It is difficult to be certain whether hydration effects precede the fixation or not. Therefore, attempts have been made to process dry specimens (e.g., seeds, spores, and lichens) by nonaqueous fixation methods instead of by conventional aqueous fixation (Perner, 1965; Hallam, 1976).

Specimens are fixed with a 6% solution of anhydrous glutaraldehyde in DMSO for 24 hr followed by treatment with chloroform for 4 hr to remove excess glutaraldehyde. Postfixation is carried out with 2% OsO$_4$ in chloroform for 2 hr.

Water is removed from 50% glutaraldehyde by treating it with molecular sieves. Chloroform should be warmed to facilitate the dissolution of OsO_4, and this procedure must be carried out in a fume hood. Infiltration is accomplished in a mixture (1:1) of acetone and Spurr's resin for 4 days prior to infiltration with pure resin for 3 to 4 weeks. Embedded specimens are very difficult to section; a diamond knife should be used and the trough should be filled with 70% solution of DMSO in water.

Modes of Fixation

There are four major modes of fixation: (1) vascular perfusion, (2) immersion, (3) dripping on the surface of the organ, and (4) injection into the organ. Although each of these modes has certain advantages and disadvantages, fixation by vascular perfusion is decidedly superior to other modes. This is true for both transmission and scanning electron microscopy. In fact, for certain types of tissues (e.g., lung, brain, and kidney), fixation by vascular perfusion is indispensable.

VASCULAR PERFUSION

Briefly, the advantages of vascular perfusion are as follows:

1. Fixation begins immediately after arrest of systemic circulation, resulting in minimum alterations in cell structure. It is always desirable to shorten the interval between death and fixation.

2. Under *in situ* conditions a rapid and uniform penetration by the fixative into all parts of the tissue is accomplished, resulting in increased rate and depth of actual fixation. The increased rate of penetration is related to a rapid and extensive flow of the fixative via vascular bed into the tissue.

3. Manipulation of tissues after arrest of systemic circulation but prior to fixation results in the introduction of artifacts. Traumatic effects of specimen handling prior to fixation on CNS is well documented. Such effects are avoided

because tissue subjected to vascular perfusion is sufficiently fixed and hardened prior to direct handling.

4. Rapid and even fixation of heterogeneous tissue components is achieved.

5. Fixed cells retain their *in vivo* shapes and relationships with one another.

6. Tissue is stabilized against excessive dissolution and translocation of cellular substances.

7. A brief but effective fixation by vascular perfusion results in increased preservation of enzyme activity.

8. An accurate estimation of enzyme activity is facilitated, for perfusion removes blood, enzymatic activity of the serum is eliminated, and natural inhibitors are suppressed. The presence of a natural inhibitor of hepatic glucose-6-phosphatase in fresh liver has been demonstrated (Beaufay *et al.*, 1954).

Vascular perfusion, however, has certain limitations. This method is obviously difficult to carry out with human beings, although an attempt has been made to fix human brain by vascular perfusion (Kalimo *et al.*, 1974). Organisms or tissues devoid of an efficient vascular bed cannot be perfused. Tissues located at a distance from blood vessels or surrounded by body fluids are best fixed by local immersion. A case in point is labyrinth hair cells encaged in fluid (Anniko and Lundquist, 1980). The fluid surrounding these cells dilutes the fixative. In fact, a vast majority of specimens cannot be perfused. Vascular perfusion requires appropriate anesthesia and analgesia. Sufficient amount of anesthetic should be injected intravenously to induce immediate deep anesthesia without a stage of excitation or heart arrest. However, too strong an anesthesia should be avoided. Animals must not be stressed any more than absolutely necessary during the operation.

Although available information on the effects of various anesthetics (when used for anesthetizing the animal) on the fine structure is quite meager, significant information of the effects of pentobarbitone on CNS and neuromuscular junctions is available. For example, it influences the electrophysiology (synaptic curvature): a marked increase in curvature negativity over the 0–80 mg/kg body weight dose range and a decrease in negativity at higher dose levels (Jones and Devon, 1978). The increase in the negativity is accompanied by an increase in synaptic length and dense projection numbers, with a consonant increase in the perimeter and area of the presynaptic terminal. Increasing doses of pentobarbitone seem to disrupt the presynaptic network. It has been suggested that pentobarbitone disrupts actomyosin structure (Ventilla and Brown, 1976). Levels of total acetylcholine increase in response to barbiturate anesthesia (Ksiezak *et al.*, 1974). Another effect of pentobarbitone is increased γ-aminobutyric acid concentration in the synaptic cleft (Cutler and Dudzinsky, 1974). The effects of various forms of euthanasia on the activity and distribution of different enzymes have been briefly discussed by Al-Azzawi and Stoward (1970). Overdoses of ether or chloroform may adversely affect the preservation of enzyme activity compared with that obtained in animals killed by decapitation.

Some of the treatments given to the organism immediately before and/or during vascular perfusion may alter the fine structure. Vasodilators (e.g., procaine) and anticoagulants (e.g., heparin) may affect both the enzyme activity and fine structure. The use of procaine, for example, in the rinsing solution may cause artifacts in the rat (Frenzel *et al.*, 1976). The perfusion method is rather elaborate and requires relatively sophisticated equipment. Furthermore, a considerable amount of skill and experience is needed to perform dissection and cannulation in the minimum possible time. Because vapors arising from the aqueous solutions of fixatives are damaging to the eyes, nose, and mouth, great care should be taken to avoid the fumes during the lengthy procedures of dissection, cannulation, and perfusion. Some workers prefer to carry out these procedures under a fume hood; others wear goggles to protect the eyes.

It is pointed out that even when vascular perfusion is carried out under optimal conditions, relatively large organs, such as brain and kidney, show uneven fixation. A case in point is the presence of solitary dark neurons observed with the light microscope, which are attributed to postmortem effects or pressure on the poorly fixed parts of the brain (Cammermeyer, 1978). The gray matter of the brain is fixed better than the white matter because the former is vascularized more extensively than is the latter. As a result, the gray matter is infiltrated rapidly and with larger amounts of the fixative. This artifact can be formed when the flow of the fixative is delayed or the brain is autopsied after termination of the perfusion. Similarly, various parts of the kidney exhibit different quality of fixation.

Considerations in Achieving Satisfactory Vascular Perfusion

In order to obtain consistently satisfactory fixation, it is essential that optimal conditions prevail during vascular perfusion. The important parameters include the complete exclusion of blood from the vascular system and prevention of vasoconstriction (blocking or narrowing of blood vessels), pH, temperature, durations of washing and fixation, composition and concentration of the perfusate, osmotic pressure of the perfusate, hydrostatic pressure employed during perfusion, actual rate of flow of the perfusate, and route of perfusate instillation.

An isotonic buffered saline or Ringer's solution can be used to flush out the blood from the vascular bed. For critical studies a complete Tyrode solution should be used. It is necessary that the optimal amount of saline be used. A delay in starting the flow of the fixative perfusate may be caused by the use of large amounts of saline or by the slow flow of the saline. The amount of saline and the duration needed to remove the blood depend on the extent of vasculature to be perfused. The amount can be calculated in milliliters equivalent to body weight in grams. In general, 15% body weight is desirable for perfusion of the whole body, 5% for perfusion of the head with or without forelimbs, and 1–2% for perfusion of the CNS exclusively (Cammermeyer, 1978). However, the amount of saline introduced is best controlled by timing the flow, rather than by introducing a certain volume of the saline. A duration of 10–30 sec is recommended.

Drugs such as papaverine, sodium nitrite, procaine hydrochloride, and lidocaine chloride can be employed as vasodilators with perfusates. Approximately 1% sodium nitrite in 0.9% sodium chloride can be added to the saline perfusate. An anticoagulant, such as heparin (150,000 U.S.P.) in a dose of 0.5–1.0 ml/kg body weight, can be used either in the perfusate or mixed with the injected anesthetic. Oxygenation of the prewash solution has been attempted in order to prevent artifacts that might be induced by oxygen deprivation during initial perfusion (Motta and Porter, 1974). In addition, artificial respiration has been carried out on narcotized animals (Cammermeyer, 1978). Tracheotomy is performed on the animal and air is insufflated with a respirator (Harvard Instrument Co., Dover, Mass.) from the moment the chest is opened until the heart is cannulated. However, possible effects of the aforementioned treatments on cell morphology need to be studied, and the necessity of employing all these procedures in all studies has not been established.

The best results are obtained when the pH is maintained within the physiological range (7.2–7.4). The temperature of the perfusate should not be below the body temperature of the animal; otherwise vasoconstriction may occur that would impair the effectiveness of the perfusion. Ideally, the temperature of the perfusate should be the same as that of the animal body.

The osmotic pressure of the perfusate should be identical to that of the blood of the animal under study. To accomplish this, however, is not easy because the osmotic pressure of blood varies among species. Optimal osmolarities for some tissue types are listed in Table 8.1. In order to raise the colloid osmotic pressure of the fixative, dextran or polyvinyl pyrrolidone (2%) or gum acasia (1%) can be added to the fixative. Such increased pressure minimizes the expansion of extravascular spaces. The addition of polyvinyl pyrrolidone to the fixative is especially helpful in improving the quality of preservation of brain tissue.

Agreement on the optimal concentration of the fixative during the initial fixation is lacking. According to some workers (Reese and Karnovsky, 1967; Peters et al., 1968; Brightman and Reese, 1969), aldehydes of low concentrations (1% formaldehyde and 1.25% glutaraldehyde) are desirable. Schultz and Case (1970), on the other hand, advocate the use of 19% glutaraldehyde preceding the main perfusion. The latter approach will result in rapid initial fixation.

The most effective route for vascular perfusion of a given organ is given later. However, if several organs must be perfused simultaneously in a single animal, the major part of the body can be perfused through the left ventricle and arch of

TABLE 8.1

Recommended Osmolality of Perfusate for Selected Tissues

Tissue type	Osmolality (mosmols)
Rat heart	300
Rat brain	330
Rat lung	330
Rat kidney	420
Rat renal medulla[a]	
Inner stripe of outer zone	700
Outer level of inner zone	1,000
Middle level of inner zone	1,300
Papillary tip	1,800
Rat developing renal cortex[b]	828
Rat (or cat and fowl) liver	450
Rat (or chicken) embryo liver	420
Rat skeletal muscle	475
Frog liver	300
Rabbit brain	820

[a] The addition of 3% dextran to the fixative helps to preserve interstitial structures and intercellular relationships (Bohman, 1974).
[b] With 6% glutaraldehyde (Larsson, 1975).

the aorta. When this route is used carefully, fixation of liver, heart, kidney, brain
and peripheral nervous system, gastrointestinal tract, spleen, and gonads can be
accomplished. This route is also ideal for studying proliferative liver lesions
because the method produces arterial perfusion of the liver, and the lesions have
almost exclusively an arterial blood supply (Jones *et al.*, 1977).

Perfusion Pressure

As stated earlier, in addition to the perfusate pressure, perfusion (hydrostatic)
pressure (which primarily controls the rate of flow of the perfusate) affects the
quality of tissue preservation. The ideal perfusion pressure apparently is equivalent to the pressure experienced by a given tissue *in vivo*. Severe endothelial
damage and artifacts at excessive (Fonkalsrud *et al.*, 1976) or insufficient
(Bylock *et al.*, 1977) perfusion pressure have been described. It is not easy to
determine and to control the ideal perfusion pressure for a given tissue. The exact
in vivo measurement of blood pressure in many tissues is difficult. Even when
perfusion pressure equal to the normal blood pressure of the animal is employed,
certain cell types of a complex tissue (e.g., kidney) may not be well fixed. A case
in point is the inner zone of medulla, which is fixed satisfactorily when perfusion
pressure (200 mmHg) higher than the normal blood pressure of the animal is
employed (Bohman, 1974).

Generally, fixation occurs more rapidly at higher perfusion pressures along with an ample flow of fixative solution. This practice will effectively force the blood out of the capillaries and improve the diffusion of the fixative into the tissue. The perfusion pressure should be at least equal to the systolic pressure. However, too high a perfusion pressure may increase the size of the organ and thus involve swelling of the tissue. Using different perfusion pressures (30–210 mmHg) it has been shown that rat liver is best fixed for scanning and transmission electron microscopy at a pressure less than 100 mmHg (Frenzel *et al.,* 1976). Satisfactory preservation of rabbit aorta for scanning electron microscopy is obtained when vascular perfusion is carried out at 80–125 mmHg, which are the systolic pressure and diastolic pressure, respectively, of this tissue (Swinehart *et al.,* 1976). Table 8.2 gives desirable perfusion pressure for some tissue types.

For obtaining optimal results, the maintenance of constant perfusion pressure, within the physiological range, during the entire procedure of perfusion is essential. A transition in the vascular perfusion pressure may also result in the introduction of artifacts. An example of such artifacts is the appearance of tendrils at the luminal surfaces of proximal tubules in the kidney (Andrews and Porter, 1974). It was indicated that a drop in the perfusion pressure prior to reexpansion of tubules produces numerous tendrils (Bulger *et al.,* 1976). This artifact seems to appear when two perfusates are used, which necessitates a switching from one perfusate to the other. Such a transition may be accompanied by a temporary decrease in the perfusion pressure. Kidney tissue is especially prone to this type of artifact because it is a functionally distended organ.

The problem of controlling the perfusion pressure during the entire procedure is very complicated. The resistance of the cannula–blood vessel–organ system differs from experiment to experiment. Thus, the perfusion pressure required

TABLE 8.2

Recommended Perfusion Pressure for Selected Tissues

Tissue type	Perfusion pressure (mmHg)
Rabbit spleen	110–120
Rabbit (or rat) brain	150
Monkey brain	110
Baboon lung	40
Rat or mouse heart	103–120
Rabbit arteries	100–120
Rat kidney	100–200
Rat (or cat and fowl) liver	100
Rat (or chicken) embryo liver	50
Rat skeletal muscle	100
Kitten liver	160–180

may not be the same for similar organs of the same animal species. The rate of flow of the perfusate is controlled not only by the perfusion pressure but also by several other factors, including the size of the needle used as a cannula and the diameter of the blood vessels. In addition, the pressure distal to the perfusion needle may not be equal to the proximal pressure of the supplying system. It has been shown that in the carotid arteries of rabbits the driving pressures proximal to the needle, in the needle, and in the contralateral carotid artery differ from one another (Hollweg and Buss, 1980). The pressure in the carotid artery appears to increase with increasing duration of perfusion. This is related to a decrease in the peripheral resistance. Therefore, to maintain a constant pressure during a prolonged perfusion, it may be necessary to reduce the driving pressure in the later stages. Reliable preservation is obtained only when pressure measurement and feedback control are carried out simultaneously, since the rate of inflow is as important as the rate of outflow.

It seems desirable to discuss the relationship between perfusion pressure and tissue volume changes, especially in the brain tissue. The volume of the fixed organ is affected by the perfusion pressure. In this respect an understanding of the factors that regulate the entrance of fixative to the tissue is essential. The phenomenon of volume changes can be explained on the basis of Starling's law, which applies to the regulation of interstitial fluid *in vivo* (Aukland, 1973). The net flow of the fixative solution across the capillary wall is controlled by the balance between the hydrostatic and oncotic (colloid osmotic) pressures within the vessel and the tissue. The intravascular hydrostatic pressure forces body fluid into the tissue, and this force is counteracted by the hydrostatic pressure that builds up in the tissue (Kalimo, 1976). Conversely, intravascular oncotic pressure tends to remove fluid from the tissue into the vessels. This movement is counteracted by the oncotic pressure in the tissue created by the osmotically active particles in the interstitial fluid.

At a high perfusion pressure (150 mmHg), more fluid is forced into the tissue by the intravascular hydrostatic pressure than at a low pressure (75 mmHg). Therefore, the former condition will cause an increase in the volume of the brain parenchyma and a decrease in the size of the ventricles. On the other hand, if the intravascular oncotic pressure of the perfusate is increased by using a more concentrated buffer or by adding osmotically active substances such as PVP, more fluid will return into the vessels. As a result, the brain will decrease in volume, while the ventricles will enlarge. The importance of choosing the perfusion pressure and buffer strength according to the specific requirement of the tissue or cell type under study is apparent.

Apparatus for Vascular Perfusion

Several types of units for vascular perfusion, both simple and complex, are used. These units are discussed below so that the reader may select the most appropriate one or modify an existing one to suit the objective of the study. The major improvement in this area has been to devise units that provide the desired perfusion pressure without any fluctuation. This improvement was necessary because perfusion pressure is an important factor in achieving satisfactory preservation of tissue morphology. Fluctuations in the pressure during perfusion can introduce artifacts including overdistension of the microvasculature in the tissue. Therefore, to prevent these artifacts, the pressure should be maintained at a constant level irrespective of alterations in the flow resistance within the body. Several devices have been introduced to maintain a constant pressure, and they are discussed next.

In the simpler systems the desired perfusion pressure is usually obtained by holding the perfusion containers above the animal (Fig. 8.1). The height of the container with respect to the animal determines the applied pressure. A height of approximately 102 cm produces a pressure of 78.5 mmHg. While using an

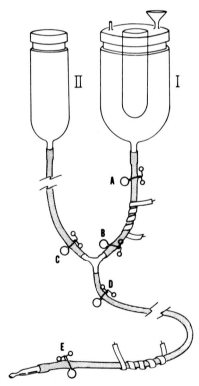

Fig. 8.1.   Diagram of a simple apparatus for general vascular perfusion. Containers I and II contain the fixative and Ringer's solution, respectively. (Palay *et al.*, 1962.)

intra-arterial gravity feed, the containers should be held 120–150 cm above the animal, whereas the intravenous route requires a level of only 20–30 cm above the animal. With this arrangement, however, it is difficult to control the pressure and temperature of the perfusate. Moreover, gravity flow becomes inconvenient when high perfusion pressures are needed. This difficulty can be alleviated to some extent by employing a peristaltic pump (e.g., Watson–Marlow MHRE 200) (Fig. 8.2). The pump is capable of delivering flow rates ranging from 1 to 100 ml/min. A peristaltic forced flow, however, may cause rupture of small blood vessels.

Another problem with the gravity flow method is detection of flow stoppage, which can be caused by the formation of a blood clot. If the flow stoppage is not detected for several minutes, the quality of fixation can be less than satisfactory. An immediate detection of flow stoppage can be achieved by a glass flow indicator (Fig. 8.3).

Since the perfusion pressure at a fixed flow rate depends on the vascular resistance encountered in the perfused animal, the peristaltic pump is not able to control the pressure, Also, the pressure produced by the pump is not constant and fluctuates rapidly. These limitations can be overcome by using a simple cushion-

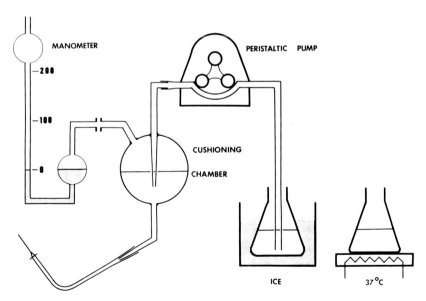

Fig. 8.2. Diagram of an improved apparatus for vascular perfusion. The peristaltic pump delivers a pulsating flow from one of the vessels at the right into the cushioning chamber. This moderates fluctuations in pressure and traps air bubbles. The manometer, indicating the pressure, is attached to the cushioning chamber. (Veerman *et al.,* 1974; © 1974, The Williams & Wilkins Co., Baltimore.)

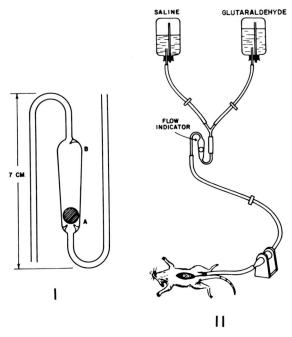

Fig. 8.3. I, diagram of vertical section showing construction of the flow indicator. II, the perfusion apparatus with the flow indicator connected. The middle section of the flow indicator is made of tapered glass tubing with 7 mm internal diameter at A and 9 mm internal diameter at B. Just below A there is a slight indentation in the glass tubing, to prevent a solid glass bead of 6 mm diameter from becoming wedged into the tubing. When there is no flow, the bead rests at A, and depending on the rate of flow, the bead will take a position between A and B. Glass tubing of 2 mm internal diameter is used in the connections from the flow indicator to the perfusion apparatus as indicated in Fig. 8.8. The flow indicator thus forms part of a closed system and is competely filled with fixative solution. (Goodman and Moore, 1971; © 1971, The Williams & Wilkins Co., Baltimore.)

ing chamber (Veerman *et al.*, 1974) (Fig. 8.2). The chamber has a volume of ~30 ml and moderates rapid fluctuations in pressure and traps air bubbles that may have entered through the suction tube. The pressure in the chamber is measured with a mercury manometer. A hydraulic pressure-control system (Gil and Weibel, 1969-1970) and later an electronic system (Gil and Weibel, 1971) were also introduced to achieve an accurate control of the perfusion pressure.

Since perfusion pressures provided by either peristaltic pumps or gravity feeds are not reproducible, Rossi (1975) devised an apparatus (Figs. 8.4 and 8.5) that avoids the use of peristaltic pumps or gravity to ensure perfusion (described on p. 223).

Another unit developed by Rossi (1975) and shown in Fig. 8.6 is more complex and contains an additional glass container and a third three-way valve for attaching a syringe (d) that can be used for a blood sample or for injecting

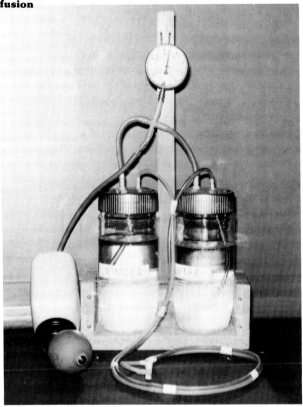

Fig. 8.4. A perfusion apparatus with two bottles for perfusates. (Rossi, 1975.)

solutions. The vascular system of the animal is rinsed with Ringer's solution through a three-way valve (c2) followed by perfusion fixation through turning the valve. Valves c1 and c2 are appropriately positioned to permit washing with the buffer.

The perfusion pressure can be adjusted by changing the speed of the peristaltic pump. The perfusion pressure can also be regulated by coupling the perfusate containers to a gas tank having compressed oxygen; the actual regulation is carried out by means of a reduction valve between the gas tank and the perfusate containers. Assimacopoulos and Kapanci (1974) described a simple glass tube that enables the pressure in the pressure circuit to be controlled accurately (Fig. 8.7). The glass tube is interposed between the peristaltic pump and the organ to be perfused. This tube acts as a safety valve.

The perfusion pressure can be kept constant by the aid of a water-filled cylinder (Fig. 8.8) (Pilström and Nordlund, 1975). The perfusate flow is checked with

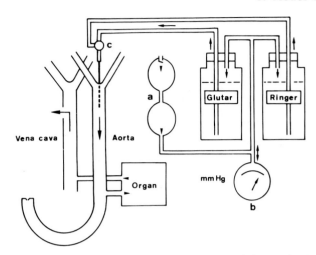

Fig. 8.5. Diagram of the perfusion apparatus shown in Fig. 8.4. a, rubber-ball syringe; b, manometer, c: three-way valve. (Rossi, 1975.)

a flow indicator as described by Goodman and Moore (1971). In this unit the temperature of the perfusate is kept constant by flowing through a glass tube spiraled 12 times and contained in a cylinder, through which water of constant temperature is pumped. The perfusate can be prechilled or prewarmed to 35°C, depending on the temperature desired.

Another important factor in achieving successful perfusion is that the perfusion pressure should not fall during the change in the perfusates. Usually a Y tube is employed to join the tubes from two containers, which permits a change from rinse to fixative solution without an intermediate fall in perfusion pressure. A three-way stopcock facilitates consecutive perfusion with rinse and fixative. The apparatus shown in Fig. 8.9 allows perfusion with three different perfusates in rapid succession. The unit shown in Fig. 8.2 permits the transfer of the suction tube from one perfusate to another while the pressure is maintained for several seconds by the cushioning chamber.

An apparatus for quantitative and reproducible control of the perfusion pressure was introduced by Lenn and Beebe (1977). This apparatus consists of a standard 600 ml blood transfer pack (Fenwal Lab., No. TA-1), a Y-tube solution administration set (Travenol Lab., Inc.), and a sphygmomanometer that is placed around the blood transfer pack (Fig. 8.10). Direct measurement of outflow pressure demonstrated that the sphygmomanometer used to supply external pressure to a plastic blood transfer pack achieved these features. Reproducibility of pressure is maintained by filling the bag with air between perfusions, or when smaller volumes of fixative solution are needed.

An inexpensive perfusion apparatus that allows sequential perfusion of several

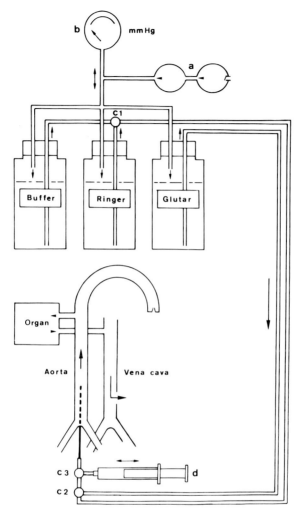

Fig. 8.6. Diagram of a more complicated perfusion apparatus. a, rubber-ball syringe; b, manometer; c1, c2, c3 = three-way valves; d, syringe. (Rossi, 1975.)

perfusates without an intervening drop in perfusion pressure has been devised by Bower (1981). This apparatus is useful for tracing of nervous pathways using horseradish peroxidase. The display of this enzyme requires perfusion of the brain with three different perfusates: (1) a saline prewash, (2) the fixative, and (3) cold sucrose-buffer (Rosene and Mesulam, 1978). These three solutions are perfused in succession, without an intervening drop in pressure.

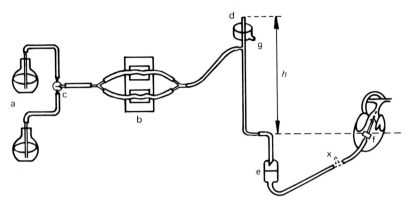

Fig. 8.7. Diagram of the apparatus used for vascular perfusion of lung. The glass tube (d) enables accurate control of the perfusion pressure irrespective of changes in the flow resistance and thus prevents the overdistension of the microvasculature. The open end of this tube acts as a safety valve. The density and viscosity of various perfusates should be similar, and the diameter of the tubes should be adapted to the required flow. Higher flow rates require tubes of wider internal diameter. (Assimacopoulos and Kapanci, 1974.)

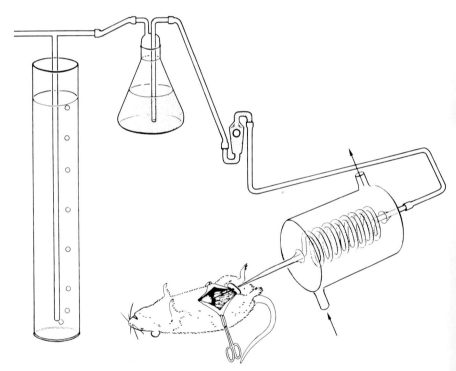

Fig. 8.8. Diagram of a perfusion apparatus showing a water-filled cylinder that can help to maintain constant perfusion pressure. The flow is checked by a flow indicator. Temperature of the perfusate is kept constant by flowing it through a glass tube, spiraled 12 times and contained in a cylinder, through which water of constant temperature is pumped. The perfusate should be either prechilled in a refrigerator at 4°C overnight or prewarmed to 35°C in a waterbath. (Pilström and Nordlund, 1975.)

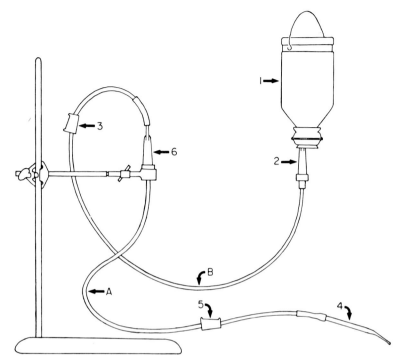

Fig. 8.9.   Diagram of the apparatus used for vascular perfusion of the CNS. The apparatus permits perfusion of three different perfusates in rapid succession. (Schultz and Case, 1970.)

Methods of Vascular Perfusion

General Method of Vascular Perfusion

A relatively simple and inexpensive apparatus (Fig. 8.5) was constructed by Rossi (1975) to carry out perfusion fixation in small animals such as mice, rats, hamsters, rabbits, and young dogs. All organs except lungs and those caudal to the catheter (needle) are perfused by using this apparatus. This apparatus and procedure suggested by Rossi (1975) are described here.

The apparatus consists of pressurized vessels for the solutions, devices for pressurization and for measuring pressure, vinyl tubing, and needles or catheters to deliver solutions into the vascular bed. Two glass containers (coffee jars) of 1 liter capacity with screw-on lids made airtight by suitable O rings are needed. Two holes (3–4 mm each) are pierced into each lid, through which vinyl tubings are inserted and cemented into place with an epoxy resin. One tube provides connection between the air spaces in the two jars through vinyl tubing in which is

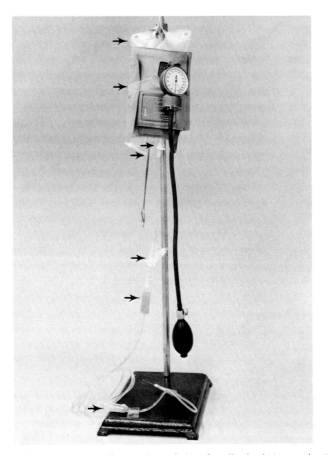

Fig. 8.10.   The apparatus set up for vascular perfusion of small animals (e.g., rodent brains from fetal to adult ages). The arrows, from top to bottom, indicate blood transfer pack, sphygmomanometer around blood transfer pack, outlet, hemostat clamping inlet, valve on outlet tubing, and drip chamber and distal valve. (Lenn and Beebe, 1977.)

inserted a four-way joint for connection with a manometer (Fig. 8.5b) and with a rubber-bulb syringe (Fig. 8.5a), fitted with a one-way valve for pressurization.

One end of the second tubing through each lid extends to the bottom of the jar and the other ends are joined by a three-way plastic valve (Fig. 8.5c). The third arm of this valve carries the catheter for entering the abdominal aorta. One jar is filled with Ringer's and the second with the fixative, and lids are screwed on tightly. The system is pressurized and freed of air bubbles by manipulating the three-way valve. The valve is turned to an intermediate position to prevent further loss of solutions and pressure is raised to 120–130 mmHg.

The abdomen of the anesthetized animal is opened, and the viscera are displaced until the aorta and vena cava are visible from the renal arteries distally. The aorta bifurcation is freed with rounded-tip forceps and the aorta clamped with rounded-edge surgical forceps just distal to the renal arteries. The aorta is opened with fine scissors at the bifurcation and a catheter or needle is inserted into it at the level of the surgical clamp, which is removed after the catheter has been locked in place with a grooved clamp.

Ringer's is admitted into the aorta through the three-way valve at ~120 mmHg to ensure flow against aortic pressure. Immediately the vena cava is cut and within ~2 min blood has been removed by Ringer's and the valve turned to admit the fixative solution. Fixation is completed within 10–15 min.

Aorta

The animal (e.g., rabbit) is killed by intravenous injection of 10% urethane anesthesia. The dorsal aorta from the arch to approximately the first intercostal artery is removed and placed in 0.9% saline. Each arterial section is tied to a hydrostatic column preset to a height calculated to produce a pressure head equivalent to either 80 or 125 mmHg. Containers for heparinized saline and for fixative are connected (Fig. 8.11). The tied arterial section is immersed in the saline bath and the T connection opened to allow heparinized 0.9% saline to flow through. Any leaks or opened side vessels are clamped.

After the leaks have been closed, the perfusion with the saline is terminated and 3% glutaraldehyde in 0.1 mol/l phosphate buffer (pH 7.4) is allowed to flow

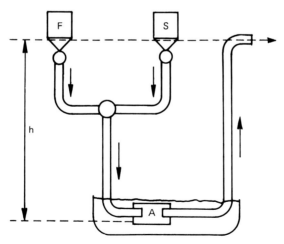

Fig. 8.11. Diagram of the apparatus used for vascular perfusion of aorta. Arrows indicate the direction of perfusate flow. (Swinehart *et al.*, 1976.)

through the vessel. The fixative can be dyed with methylene blue in order to monitor the proper flow of the fixative. The saline bath in which the section of the aorta has been immersed is replaced by 3% glutaraldehyde. The fixative can be recovered as it flows out of the system and poured back into the container for recirculation to maintain the desired height of the column (h in Fig. 8.11) of the fixative. The fixation by perfusion is maintained for ~45 min. The artery, still maintained under pressure, is immersed in fresh glutaraldehyde bath and left overnight at room temperature. Small pieces of the tissue are postfixed in OsO_4 for transmission electron microscopy. For further details, see Swinehart *et al.* (1976).

Note: The fine structure of thoracic aortic endothelium (e.g., rabbit) can be preserved by initiating fixation while the aortic circulation is intact and undisturbed (Garbarsch *et al.*, 1980). Continuous-pressure monitoring ensures that the luminal pressure is kept within 100 ± 5 mmHg. The pressure must be kept within this narrow limits from the moment fixative comes in contact until inelasticity of the vessel wall sets in. For fixation, glutaraldehyde is applied at decreasing concentrations (5-1.5%) in 0.11 mol/l phosphate buffer (pH 7.3) containing dextran (1.6-2.35%). The study is confined to the ventral wall of the vessel. This method is suitable for both the TEM and SEM.

Arteries

The artifacts caused by conventional fixation of arteries include wavy appearance of elastic laminae, narrowing of the lumen by a decrease in the inner circumference, partial detachment of endothelial cells, and disintegration of leucocytes and their adhering to subendothelial structures after removal of endothelium. The elastic laminae of aorta are considered to be stretched and straight *in vivo*. These fixation artifacts can be minimized by employing the following method of perfusion.

Animals are anesthetized and ~1 cm of the abdominal aorta near the renal arteries and 1 cm at the bifurcation is dissected, leaving ~7 cm of the vessel untouched. The vessel is perfused with 10 ml of Krebs–Ringer's solution through a silastic catheter that has been introduced proximally. During this washing, a second catheter is introduced distally. The outflow of the second catheter is placed 60 cm above the aorta (~82 mmHg pressure) (Haudenschild *et al.*, 1972).

The washing is followed by perfusion with 1.5% glutaraldehyde in 0.1 mol/l phosphate buffer (pH 7.4) having an osmolality of 500 mosmols for 20 min at room temperature. The containers with washing and fixation solutions are placed 100 cm above the aorta (~136 mmHg pressure). Small blocks from the perfused vessel are refixed by immersion with the same fixative for 1 hr and then postfixed with 2% OsO_4 in the buffer for 1 hr at room temperature.

Central Nervous System, Method I

Perfusates

Solution A. 10 ml of Tyrode solution containing 1% gum acacia prepared as follows:

Tyrode solution:

NaCl	8 g
KCl	0.2 g
$CaCl_2 \cdot 2H_2O$	0.27 g
$MgCl_2 \cdot 6H_2O$	0.21 g
$NaHCO_3$	1 g
Dextrose	1 g
$Na_2HPO_4 \cdot H_2O$	0.06 g
Distilled water to make	1000 ml

Add sodium bicarbonate after dissolving all other salts to avoid precipitation.

Solution B. 15–18 ml of 19% glutaraldehyde prepared as follows:

Sörensen phosphate buffer	62.5 ml
Gum acacia (20%)	12.5 ml
Polyphosphate solution	0.2 ml
Glutaraldehyde (25%) to make	250 ml

Solution C. 250 ml of 2% glutaraldehyde prepared as follows:

Sörensen phosphate buffer	62.5 ml
Gum acacia (20%)	12.5 ml
Glutaraldehyde (25%)	20 ml
Acrolein	2.5 ml
Polyphosphate solution	5 ml
Distilled water to make	250 ml

Solution D. Neutralized polyphosphate solution prepared as follows:

Sodium polyphosphate	179.2 g
NaOH (0.1 mol/l)	1.2 ml
Distilled water to make	1000 ml

Procedure. In Fig. 8.9, the intravenous bottle (1) is filled with solution C. The intravenous tube (B) attached to it is filled with solution B so that its drip chamber (2) is partly filled, and the clamp (3) is tightened. Tube A is filled with solution A by a syringe at the end where the cannula (4) is later attached after closing the clamp (5). The drip chamber (6) is filled with enough solution A to allow the clamp (5) to be manipulated so as to fill the cannula (4), and yet leave the chamber (6) about half full. It is important that tube A and the cannula (4) be completely filled without air bubbles. The perfusate bottle (1) is hung ~5 ft above the animal; alternatively, the bottle is pressurized with a regulated air supply to 4 psi (this generates a close to physiological pressure in the aorta by actual measurement). At perfusion the cannula (4) is inserted into the left ventricle and the root of the aorta, tied in place with a previously placed ligature, the right side of the heart opened widely, and the clamps (3) and (5) are released. Following a perfusion of ~15 min at room temperature, the brain is removed and placed in solution C, and small pieces of the tissue are carefully sliced. There is some evidence indicating that when brain tissue is postfixed with OsO_4 for 90 min or less, it is insufficiently stabilized, but when the duration is increased to ~3 hr, the specimen is stabilized so that ethanol or acetone dehydration followed by embedding in Epon, Araldite, or Vestopal produce similar results (Schultz and Karlsson, 1972). After standard duration of postfixation, both extracellular space and plasma membrane structure are affected by the embedding medium used. For further details of the perfusion method, see Schultz and Case (1970).

Note: Since cerebral cortex has a far more extensive capillary network than that in the white matter of the brain, the relatively less efficient intravascular perfusion method is effective for the former. However, parts of the brain (e.g., white matter) with poor capillary network can be reached more easily by the fixative when applied by intraventricular perfusion rather than by intravascular perfusion. Therefore, perfusion via cerebral ventricles suggested by Rodriguez (1969) is worth exploring, since this approach circumvents the blood–brain barrier.

Since thoractomy and consequently pneumothorax and cessation of respiration occur prior to actual vascular perfusion, some workers prefer artificial respiration until the tip of the cannula is inserted. This practice is unnecessary provided the operative procedures are performed as rapidly as possible. Artificial respiration may produce constriction of cerebral vessels (Kalimo, 1974).

Central Nervous System, Method II

The following method is recommended for simultaneously localizing the fluorescence with the light microscope and fixing the tissue for electron microscopy (Furness *et al.*, 1977, 1978). A mixture of glutaraldehyde and formaldehyde produce a fluorophore with catecholamines in the peripheral and central nervous systems. The fluorescence reaction is produced at room temperature and is stable in aqueous solutions.

Perfusate

Formaldehyde   4%
Glutaraldehyde   1%
0.1 mol/l phosphate or cacodylate buffer (pH 7.0)

Procedure. The animal (e.g., rat) is anesthetized with pentobarbital (40 mg/kg body weight) and injected with heparin (4000 IU/kg body weight). The chest is opened and the heart is exposed. The tip of an 18 gauge cannula connected to a perfusion apparatus (Fig. 8.12) is introduced into the aorta via the left ventricle and held in position with a clamp across the ventricles. The blood is flushed out with a 1% solution of sodium nitrite in 0.01 mol/l phosphate buffer (pH 7.0). This is accomplished in 10–30 sec of perfusion at 120 mmHg. This step is followed immediately by perfusion with the fixative solution for 10 min at the same pressure. Approximately 200 ml of the fixative solution is perfused. The fixed brain is placed in the fresh fixative at room temperature.

Specimens of suitable size for cryostat sections are frozen in isopentane cooled by liquid nitrogen. The cryostat sections are placed in the fixative at room temperature for ~15 min and then transferred to glass slides for examination with the light microscope. For electron microscopy small pieces of the fixed brain are further fixed for ~30 min and then postfixed with 1% OsO_4.

Note: Almost exclusive fixation of the brain and spinal cord can be obtained by blocking the flow of the perfusate into peripheral parts of the body. Several methods have been attempted to accomplish this. Sodium nitrite ($NaNO_2$) has

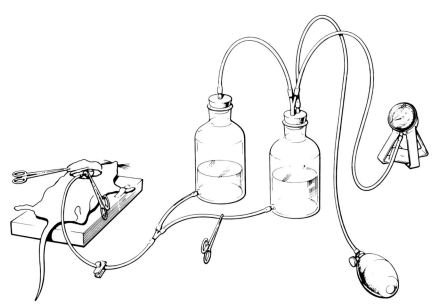

Fig. 8.12. The perfusion apparatus consists of two aspirator bottles, each of one liter, which are connected to a sphygmomanometer bulb and gauge to maintain and monitor the perfusion pressure. One bottle contains buffered sodium nitrite solution and the other carries the fixative solution. Clamps are used to direct the perfusate selectively from one or the other of the bottles. All tubing used to connect the apparatus is kept as short as practicable and is of wide bore (10 mm inside diameter) so that any pressure difference between the gauge and the cannula is kept to a minimum. (Furness *et al.*, 1978.)

been added to the saline or injected intracardially to enhance the degree of fixation of the CNS (Peters and Palay, 1966). Such a procedure has been combined with clamping of the aorta and subclavian arteries in order to restrict the supply of perfusate to the peripheral vascular bed (Morest and Morest, 1966). These procedures are not recommended because $NaNO_2$ causes a diversion of perfusates into the extracranial vasculature, resulting in a reduction in the amount of fixative reaching the brain.

The most effective procedure to obtain a preferential flow of perfusate to the CNS is to inject intracardially a high dose of epinephrine 1 min prior to perfusion with saline (Cammermeyer, 1968). Epinephrine (adrenalin chloride solution, 1:1,000, Parke, Davis), 1.5 mg/kg body weight, is injected into the left ventricle while the heart is still beating. With this application peripheral vasoconstriction is obtained; thereby circulation through the large muscular masses and several blood-rich visceral organs is blocked and the flow of perfusate is largely diverted into the blood vessels of the CNS. The CNS can be adequately fixed with a relatively small amount of fixative. Use of small amounts of perfusates may be helpful when isotope-treated animals are being perfused. Preferential flow of the perfusate may also be helpful in monitoring the action of a drug on the CNS.

Central Nervous System, Method III

Solution I

Sucrose	18.5 g
$NaH_2PO_4 \cdot 2H_2O$	4.1 g
Distilled water	400 ml
Calcium chloride (5%)	0.5 ml

Adjust the pH to 7.3 by adding a saturated solution of NaOH.

Solution II. Solution I containing 1% glutaraldehyde.

Procedure.   The following method seems to preserve the extracellular space (ECS) of the brain tissue. The perfusion apparatus (Fig. 8.13) used allows gradual increase (from zero to the desired value) of aldehydes in the perfusate (Cragg, 1980). The animal (e.g., rat) is rapidly anesthetized in a strong chloroform vapor. The perfusion must begin within 1 min of opening the chest. The cannula [No. 14 hypodermic needle squared off at the end and provided with a ground neck (C)] is inserted in a cut in the left cardiac ventricle and passed up into the ascending aorta. A circular clip (made by drilling a pair of artery forceps and covering the jaws with thin plastic tubing; see Fig. 8.13I) is applied to the aorta to grip the ground neck of the cannula. The right auricle is opened and the decending aorta is clipped (this will prevent the study of the spinal cord).

A roller pump (P) operating on 3 mm internal diameter tubing with two rollers at 200 rev/min is used to pump solution I from a reservoir (R) of 400 ml capacity

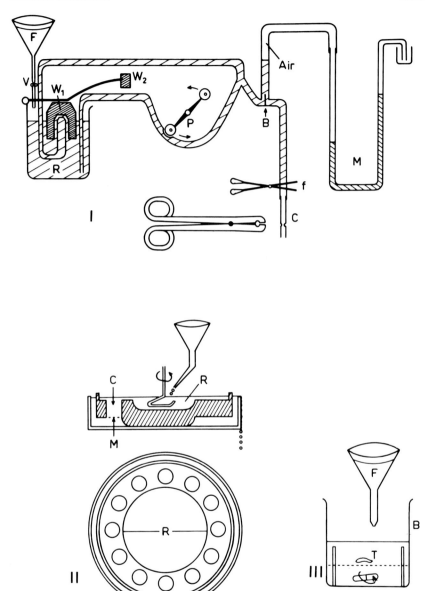

Fig. 8.13. I, A pumped circuit for perfusion at 100 or 300 mmHg with a continuous rise in concentration of fixative. II, A stirred reservoir and chambers for dehydrating tissues in a steadily rising concentration of alcohol. III, A stirred beaker for immersing tissue to be fixed in a steadily increasing concentration of aldehydes. (Cragg, 1980.)

back to a pressure release valve W_1 in the same reservoir. The aperture of this valve is 8 mm and the fitted weight W_1 of 77 g allows a pressure of 100 mmHg to develop in the flowing perfusate. A further weight W_2 84 g is adjustable on a hinged arm so that the pressure could be raised to 300 mmHg. The flow in the pumped circuit is 600 ml/min and part of this is bled off to the cannula (C) and also connected to a mercury manometer (M). This ensures that air bubbles do not enter the output, and the baffle (B) at the T junction together with the air in the connection to the manometer smooth out the pulsatile pressure from the pump.

With the cannula clamped in place, the clip (f) is removed and the pump is switched on and brought up to 200 rev/min over a period of 5 sec to avoid a rapid pressure transient. When 100 ml of perfusate has been passed and 100 ml is left in the reservoir (R), solution II containing fixative is poured into the funnel (F) and the flow rate is adjusted at V so that the level of fluid in the reservoir remains constant. The fluid escaping from the release valve W_1 provides rapid stirring of the reservoir, and the concentration of fixative in the perfusate increases smoothly. When 100 ml of fixative has been passed, the weight W_2 is slowly lifted off, and a second 100 ml of fixative is perfused at 100 mmHg pressure.

Tissue blocks (1 mm) are immersed in the same fixative for 48 hr. Tissue slices (less than 0.5 mm) are then cut and immersed in 1 ml of fixative in a vial that is cooled to 4°C. Then 0.2 ml of a 6% solution of OsO_4 in water (0.24 mol/l) is slowly added to give a final OsO_4 concentration of 1% in the fixative solution.

Embryo

The pregnant animal (e.g., mouse) is anesthetized by intraperitoneal injection of sodium pentobarbital (30 mg/kg body weight). The animal is laparotomized and the uterus is exposed. An opening showing the yolk sac is dissected in the uterine wall under a stereomicroscope. Care should be taken not to disturb the circulations of the conceptus. The beating heart of the embryo is located and a micropipette with a tip diameter of 25–50 μm is inserted through enveloping membranes and precardiac wall into the atrium (Fig. 8.14). The micropipette is immediately retracted, thus leaving an opening in the wall of the atrium through which the blood could flow out. The tip of the pipette is placed in the lumen of the ventricle and the microperfusion by the fixative is started. The start of the perfusion is accompanied by the escape of blood through the opening in the atrium. The perfusate consists of 2% glutaraldehyde in 0.1 mol/l cacodylate buffer (pH 7.4) containing 2% PVP.

The embryo is bleached and becomes yellowish immediately. The speed of fixation can be observed by adding alcian blue or astra blue 6GLL to the fixative; all the capillaries show fixative penetration within the first few seconds. The perfusion is continued for $\frac{1}{2}$ to 2 min. Small pieces of the embryo are fixed further for 2 hr in the same fixative. For further details, see Abrunhosa (1972). The

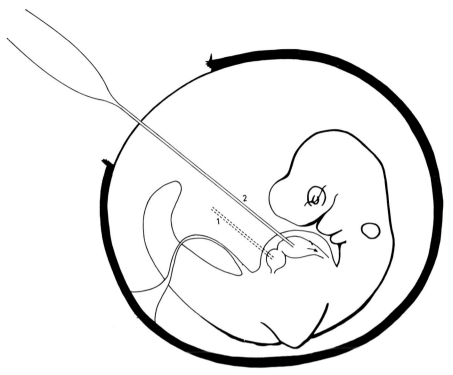

Fig. 8.14. Diagram of microperfusion fixation through the embryonic heart. The pipette is first introduced through enveloping membranes and precardiac wall into the atrium (1). It is then immediately retracted, thus leaving an outflow opening. In the second step, the tip of the pipette is forwarded into the lumen of the ventricle (2), and the perfusion is started. (Abrunhosa, 1972.)

quality of preservation of embryonic tissue obtained by the preceding method is seen in Fig. 8.15.

Fish (Whole)

Fish (e.g., catfish) is anesthetized by exposing it to a freshly prepared solution of tricaine methane sulfonate (M.S. 222) in water to a final concentration of 1:4000 (McFarland and Klontz, 1969). In ~5-10 min, the fish ceases all but opercular movement, loses equilibrium, and is insensitive to touch. Gauze sponges (5 × 5 cm) are soaked in the anesthetic solution and are positioned under the operculum and over the gill arches. An applicator bottle of 500 ml capacity is filled with the anesthetic solution.

The fish is removed from the solution and placed in the supine position if it has a depressed body form. A triangle is inscribed, making the sides by percutaneous

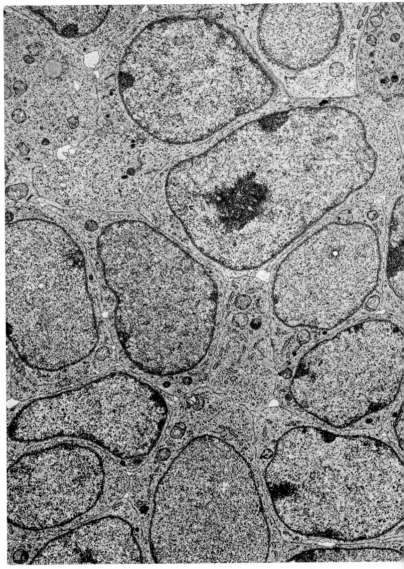

Fig. 8.15. Pancreatic anlage of day 11 embryo fixed by perfusion. The plasma membranes of different cells are separated by a space of rather constant width. The nuclei are rounded and have smooth contours, and cytoplasmic organelles are well preserved. ×9000. (Abrunhosa, 1972.)

incisions along the inferomedial body wall near both opercula and the transverse cut by a similar incision anterior to the bases of the pectoral fins. A scalpel and blunt forceps can be used to reflect the skin from the surgical area and reflect the skeletal muscle to expose the parietal pericardium. This layer is removed to uncover the bulbus arteriosus of the aorta. The triangle is enlarged and the heart and aorta are exposed by placing heavy surgical scissors transverse and superficial to the aorta and cutting away the cranial apex of the pectoral girdle. The aorta is exposed adequately without causing extensive trauma to gills and vessels supplying the body wall and liver. The applicator bottle is positioned and squeezed so that a continuous flow is introduced over the gills.

To start the perfusion, a moist cotton thread is passed dorsal to the aorta and is tied in a loose knot between the bulbus arteriosus and the first branchial arch artery. The lateral margin of the bulbus arteriosus is grasped with blunt forceps and, using scissors, a small opening is made in the wall of the vessel. Immediately, a cannula of polyethylene is inserted into the aorta to a point distal to the thread. Perfusion is immediately begun with a mixture of glutaraldehyde (2%) and formaldehyde (2%) and the thread is tied securely around the vessel and cannula to prevent loss of perfusate from the vessel aperture. A transverse incision is made through the ventricular wall to permit escape of blood from the vascular system upon displacement of the fixative. Sufficient pressure (60–100 mmHg) is maintained to give a flow rate of 10 ml/min. A total of 20–100 ml of the fixative is perfused per fish, depending on the size of the fish.

The gravity feed device comprises a ring stand, a 50 ml syringe (perfusate reservoir), and an 18 gauge needle over which polyethylene tubing (internal diameter, 0.1 cm; external diameter, 0.2 cm) is fitted. A height of 1–2 cm allows sufficient pressure.

After the perfusion, the fish is covered with moist towels and 30–60 min are allowed for *in situ* fixation. Yellowish coloration in the tissue indicates the completion of fixation. For further details, see Hinton (1975).

Heart

Perfusates
1. Ringer's solution containing 0.1% procaine. The pH is adjusted to 7.2–7.4 with 1 mol/l HCl, and the osmolality to 300 mosmols with NaCl. Perfusion takes ~3–5 min.
2. 2% glutaraldehyde in 0.045 mol/l cacodylate buffer. The pH is adjusted to 7.2–7.4 and osmolality to 300 mosmols. Perfusion takes 5 min.

Procedure. In Fig. 8.16 containers (B) filled with various perfusates are placed in a thermoregulated bath (A). The height of the containers from the working table should be adjustable to regulate the rate of perfusion. The perfu-

Sol. I   Sol. 2   Sol. 3

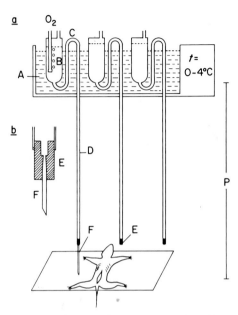

Fig. 8.16.   Diagram of the apparatus used for vascular perfusion of heart. a: A, thermoregulated bath; B, perfusate containers; C, siphon; D, plastic tubing; E, middle piece; F, cannula. b: Details of the middle piece. (Forssmann *et al.*, 1967.)

sates maintained at 0–4°C flow through a siphon (C) made up of soft plastic tubing. The cannula (F) made up of Teflon is attached to the tubing (D) by a middle piece that has a conical distal opening in which slides the cannula. This system allows rapid change from one solution to another, and the same cannula can be used for all perfusates. The distal end of the cannula is obliquely pointed.

Since the tubing (D) is not heat insulated, the temperature of the perfusates will rise to ~6°C. With adequate flow and pressure the temperature of the perfusate coming out of the animal does not exceed 10°C. The containers are placed ~140 cm above the working table considering the average arterial pressure of 103 mmHg in rats. During perfusion, a flow rate of ~40 ml/min is necessary.

The rats are anesthetized by intraperitoneal injection of 125 mg urethane/100 g of body weight. The animal on its back is tied to the working table and the ventral abdominal wall is opened widely with an incision along the white line. The intestine is pushed aside gently and the abdominal aorta is exposed. Two silk threads are slipped under the aorta, the first under the renal arteries and the second above the bifurcation of iliac arteries. The two threads separated by a distance of ~5 mm delineate the segment of the aorta where a slit is made for the insertion of the cannula.

The cannula filled with Ringer's solution containing procaine is inserted into the lumen of the aorta. The cannula is then tied in place to the aortic wall with the first thread. The second thread is used to tie the aorta; this prevents hemorrhage. After the cannula is in place, perfusion is started. The proximal end of the cannula is slipped into the conical opening of the middle piece (D), and the solution flows into the circulatory system of the animal. Simultaneously, the inferior vena cava is cut to allow the evacuation of blood and perfusate. For further details, see Forssmann *et al.* (1967).

Note: Although excellent fixation throughout the heart of small animals can be obtained by vascular perfusion, in large animals the fixation of different regions of the heart, especially the atrium, varies considerably. Depending on the type of animal, papillary muscle should be used because the myofibrils in it are mostly well aligned in one direction.

Kidney, Method I

Perfusate

Glutaraldehyde (25%)	40 ml
0.135 mol/l Sörensen phosphate buffer to make	1000 ml

The solution should have a pH of 7.2 and an osmolality of ~420 mosmols.

Procedure. The animal is anesthetized intraperitoneally with 35 mg sodium pentobarbital/kg of body weight. The anesthetized animal is tied onto the operating table with its back down. The abdominal cavity is opened wide by midline incision, and the intestine is moved to the left side. The aorta below the origin of the renal arteries is carefully exposed for retrograde perfusion. A simple intravenous infusion set (a bottle, a drip chamber, and intravenous tubing with a squeeze-type flow regulator) can be employed.

A hypodermic needle (gauge 10–16), bent at a right angle, is inserted into the abdominal aorta proximal to its distal bifurcation. The needle is connected to a tubing that, in turn, is connected to a bottle containing the fixative. A hydrostatic pressure equal to 150 cm water is adequate. Immediately after the perfusion has started, the aorta is clamped just below the diaphragm. After perfusing for ~10 min, small pieces of the tissue are fixed by immersion in glutaraldehyde for several hours to a week. For further details, see Maunsbach (1966).

Kidney, Method II

The animal (e.g., cat) is anesthetized with Nembutal (40 mg/kg body weight), and a tracheal tube is inserted and one of the femoral arteries is catheterized with PE 100 polyethylene tubing. The catheter is advanced until the tip lies at the junction of the left renal artery and the aorta. A loose ligature is placed around the aorta above the left renal artery, and another one below the right renal artery. The placing of these ligatures minimizes an accidental stimulation of nerves at the hilus. The renal arterial system can thus be isolated with the tying of these two ligatures. Another PE 100 polyethylene catheter is inserted into a femoral vein and pushed forward until the tip lays in the vena cava between the two renal veins.

A loose ligature is placed below the junction of the vena cava and the right renal vein. Another ligature is placed around the vena cava above the left renal vein. When these two ligatures are tied during perfusion, they can substantially isolate the renal venous drainage from the remainder of the venous circulation. Approximately 45 min are needed to complete the surgical procedures.

Heparinized saline is perfused by means of a Sigma peristaltic pump into the animal via the catheter inserted into the femoral artery. During saline perfusion, the four ligatures described above are tied as rapidly as possible; this can be accomplished in less than 1 min. This is followed by perfusion with the fixative solution. The fixative is 1% glutaraldehyde in 0.1 mol/l phosphate buffer (pH 7.2) with an osmolality of 320 mosmols. The rate of perfusion is set at 40 ml/min, which is approximately equal to the renal blood flow in the cats of this size. The effluent is usually free of blood within 3–4 min. After perfusing for 10 min, the kidneys are removed and cut into small pieces which are fixed for a further 2 hr in the fixative at room temperature. This is followed by washing in the buffer and postfixation with 2% OsO_4 in the buffer for 1–2 hr for transmission electron microscopy. For further details, see Yun and Kenney (1976). Figure 8.17 shows the quality of tissue preservation using the preceding method.

Kidney (Large Animals), Method III

Perfusate

> 1% glutaraldehyde
> 2.5% PVP

Both solutions are prepared in Tyrode's solution at pH 7.3 (p. 411). PVP is added in order to adjust the colloid osmotic pressure of the fixative solution to a level equal to that of the blood.

Procedure.   The animal (e.g., adult pig) is fasted for 12 hr prior to the experiment. About 30 min before anesthesia, Azaperonum, NFN (Sedaperone), 2.8 mg/kg body weight, is injected intramuscularly. Sodium pentobarbital, 10–15 mg/kg body weight, is injected into an ear vein until surgical anesthesia is achieved. Atropine, 3 mg/kg body weight, is injected intravenously.

The animal is placed on its back and the hind legs are tied down. Laparotomy is made by an incision starting 3–4 cm to the right and caudally to the xiphoid

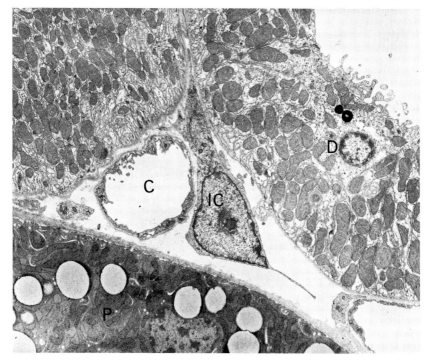

Fig. 8.17. Cat kidney fixed with vascular perfusion using isosmotic fixative (320 mosmols). Note interstitial cell (IC), capillary (C), distal tubule (D), and proximal tubule (P). ×3884. (J. Yun, unpublished.)

process, continuing to the left parallel with the costal curvature. The incision is continued caudally 5 cm from the transverse processes of the lumbar region until 5 cm cranial to the pubic pecten, ending 3–4 cm on the right side of the ventral midline. The right kidney is removed. The perfusion of the left kidney requires three assistants.

The descending aorta is exposed by an incision through the parietal peritoneum just cranially to the anterior mesenteric artery at position A (Fig. 8.18). The aorta is freed from its attachment to the diaphragmatic crura by blunt dissection with the fingers. Dissection dorsally to the aorta at this place should be avoided because of the danger of causing pneumothorax. The aorta is exposed in a way that enables assistant I to occlude the blood flow with his fingers. The decending aorta is then exposed by blunt dissection using forceps 10 cm caudally to the anterior renalis sinister at position B. Without occluding the blood flow, a loose ligature is tied at position B.

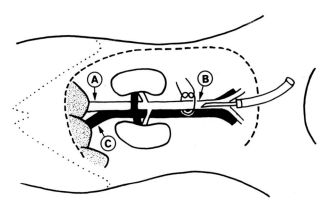

Fig. 8.18.   Schematic drawing of the dorsal abdominal cavity in the pig. Dashed line shows abdominal incision. For details see the text. (Elling, *et al.*, 1977.)

Heparin, 5000 IU, is given intravenously to prevent intravascular coagulation, since clotting cascade is very easily triggered in the pig (Rowsell, 1969). Assistant I is ready to occlude the blood flow through the aorta at position A. Assistant II moves the intestines to the left to expose the caudal vena cava just caudally to the liver at position C. Assistant III is ready at this stage to cannulate the aorta at position B with a 5 mm gauge cannula connected with a container with the fixative solution placed 1.4 m above the animal. This height provides a perfusion pressure at the blood pressure level, i.e., 120–140 mmHg. About ½ sec prior to cannulation, the flow is started to prevent air bubbles being injected. The cannula is placed in the aorta behind the ligature at position B.

Immediately after the perfusion flow has started, the caudal vena cava is incised transversely at position C by assistant II. Simultaneously, assistant I occludes the aorta at position A with his fingers. The loose ligature around the aorta at position B is tightened around the cannula to prevent flow of the fixative into the hind legs or leakage to the abdominal cavity. At this stage a free flow of blood from the abdomen and fixative solution from the caudal vena cava should be ensured by extending the right cranial abdominal incision further dorsally. Perfusion is continued for 5 min at the rate of 0.6 liter/min (10% of the cardiac output per kidney). Satisfactory tissue preservation obtained with this procedure is indicated by the presence of open lumina of tubules, evenly arranged brush borders in proximal tubular cells, and narrow and constant width of both lateral and intercellular spaces. For further details, see Elling *et al.* (1977).

Liver, Method I

Perfusate

Glutaraldehyde (25%)		40 mi
Phosphate buffer (0.15 mol/l, pH 7.2) to make		500 ml

The solution should have an osmolality of ~450 mosmols, which can be adjusted with sucrose.

Procedure.   The anesthetized animal (e.g., rat) is tied to the operating board with its back down. The abdominal cavity is opened by a midline incision with lateral extensions, and the intestine is gently moved to the left side. The portal vein is exposed and two ligatures are passed behind it. The distal ligature is tied to block the flow of venous blood from the portal vein and the flow of hepatic arterial blood to the liver. At the site where the portal vein is branched to different liver lobes, a 1 cm, 20 gauge syringe needle is inserted and secured by the second ligature. Before inserting the needle, the perfusion pressure is adjusted to 20 mmHg, which subsequently falls to ~10 mmHg during perfusion. The needle has been previously connected via an ordinary clinical intravenous infusion set to a flask containing the perfusate.

The inferior vena cava below the diaphragm is cut open to relieve the pressure in the right heart. The flask containing the fixative is hung 25–30 cm above the animal. The rate of flow should be 5–10 ml/min. Care should be taken to prevent air bubbles from entering the portal vein or the perfusion system.

Within 3 min after the start of the flow of fixative through the portal vein, the thoracic cavity of the animal is opened and the right thoracic wall is removed. A syringe needle is inserted into the main stem of the hepatic vein and secured with a ligature. The needle has been connected previously to a second flask of fixative, which is hung 15 cm above the animal. The retrograde perfusion should begin ~5 min after the start of flow of the fixative through the portal vein. The flow rate of the retrograde should be ~5 ml/min.

The perfusion is usually completed in 10–15 min, at which time only clear fixative solution comes out of the portal vein. Within 30–40 sec after the start of perfusion, the color of the liver changes from dark reddish brown to light brown. The completion of fixation can be detected grossly on the surface of the liver. The consistency changes from soft to rather stiff, resembling that of a boiled egg. Uniform overall fixation can be checked by immersing the liver slices in distilled water; poorly fixed areas will show white discoloration. The uniformity of fixation can also be checked by examining 0.2–0.5 μm thick sections stained with 1% toluidine blue in 1% borax with a light microscope. Immediately after the completion of perfusion, the liver is removed and cut into small segments (1 mm^3) that are immersed in glutaraldehyde for ~2 hr.

In the well-fixed liver, Disse and sinusoidal spaces are clearly dilineated by extensions of endothelial and Kupffer cells, and the sinusoids are patent and practically free of blood cells and floccular material. In addition to hepatocytes, endothelial, Kupffer, and fat-storing cells are well preserved and easily identified. Endothelial cells are smoothly apposed to the sinusoidal wall. Kupffer cells are attached to broad gaps in the endothelial lining. Fat-storing cells are intercalated between hepatocytes and cells lining the sinusoid. Glycogen is stained dark and is present in clumps (Fig. 8.19).

Liver, Method II

Perfusate

Glutaraldehyde	2.7%
Formaldehyde	0.8%
Sodium bicarbonate buffer (pH 7.4)	0.2 mol/l

The osmolality is 770 mosmols.

Procedure.   The animal (e.g., rat) is anesthetized by intraperitoneal injection of 0.06 ml/100 g body weight of 60% solution of Nembutal. After exposing the liver by laparotomy, two ligatures are loosely placed around the portal vein and hepatic artery, immediately caudal to the branches of the portal vein going to different lobes of the liver. A syringe needle (0.6 × 25 mm) is introduced into the portal vein ∼1 cm caudal to the ligatures. The ligatures are tightened, which action simultaneously constricts the wall of the portal vein around the needle and ligates the hepatic artery.

The fixative is introduced into the portal vein via a plastic tube connected to the syringe needle. The perfusion flow is regulated by a Watson Marlow pump (type MHRE) at the rate of 5 ml/100 g body weight per minute. Immediately after the start of perfusion the inferior caval vein is dissected to allow the perfusate to escape from the vascular system. The temperature of the perfusate is equal to room temperature, and the duration of perfusion is ∼2 min.

At the end of perfusion no traces of blood are found in the effluent solution. Small pieces of the liver (1 mm³) are fixed by immersion in the same fixative for an additional 2 hr at room temperature. After a rinse in the buffer, the specimens are postfixed either in 1% OsO_4 or in a mixture of 1% OsO_4 and 1.5% potassium ferrocyanide for 1 hr.

Liver, Method III

As an alternative to methods I and II, the perfusion can be performed via the inferior caval vein instead of via the portal vein. The advantages of the former are that there is no decrease in blood supply through the liver during surgical manipulation; both the arterial and venous sides are perfused, thereby increasing the surface area that is fixed; and uniform perfusion of the liver (Glaumann, 1975). According to this method, a small tube connected to a peristaltic pump and a container for fixative is inserted into the inferior caval vein and anchored with 2 to 3 ligatures. The portal vein and the hepatic artery are cut open, the perfusion is started, the thorax is opened, and the superior caval vein is clamped. It should be noted that liver can be perfused through the caval vein (or the portal vein) or through the aorta with equally good results.

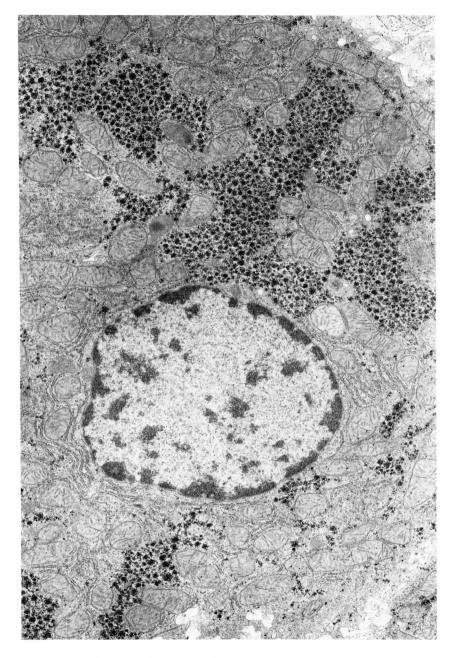

Fig. 8.19. Rat liver fixed and stained according to conventional procedures except that dioxane was used for dehydration. Note nucleus, mitochondria, ribosomes, rough endoplasmic reticulum, microbody, and glycogen. (T. P. Shearer, unpublished.)

Liver (Embryos and Very Small Animals), Method IV

Perfusate

Glutaraldehyde, 2.5%, buffered with calcium acetate (pH 7.2), with a final osmolality of 420–450 mosmols.

In very small animals the blood vessels are likewise very small, and thus consecutive perfusion through the portal vein and the hepatic vein is difficult to perform in these animals. In animals such as adult fowls, the liver and its blood vessels are hidden behind the thick muscles of the chest wall. Because of the lengthy time required for the dissection, it is necessary to apply artificial respiration until the perfusion can be begun. In embryos having patent ductus venosus, perfusion through the umbilical, portal, or hepatic veins may not yield satisfactory results because of the long bypass through the ductus venosus. Vascular perfusion of the liver in the aforementioned specimens can be performed directly through the hepatic substance.

Procedure. The apparatus consists of two containers (carrying saline and fixative solution separately) coupled to a gas tank containing compressed oxygen. The perfusion pressure can be regulated by means of a valve between the gas tank and the fluid containers. The perfusate is led through a piece of plastic tubing to a hypodermic needle. Connection with the liver circulation is achieved by inserting the needle superficially into the hepatic parenchyma. A branch of the superior mesentric vein is cut open with scissors and the perfusion is immediately started with saline. The perfusion pressure is raised from 0 to 50 mmHg. The rinse with saline should not last longer than 40 sec. This is followed by perfusion with glutaraldehyde and the pressure is raised to 150–175 mmHg. The perfusion with glutaraldehyde is continued for 10–15 min. Small tissue blocks are cut from the liver and fixed further in glutaraldehyde for several hours to a week for scanning electron microscopy and 1–2 hr for transmission electron microscopy. The latter specimens are postfixed with 1% OsO_4. The tissue parts adjacent to the perfusion needle are discarded. For further details, see Sandström (1970).

Liver (Biopsy), Method V

Liver biopsy and surgical samples can be preserved by transparenchymal fixation. Vonnahme (1980) introduced this method for the fixation of human biopsies weighing 2–8 g. Immediately after collecting the wedge biopsy, it is immersed in Ringer's solution containing heparin. This is done to prevent the accumulation of neutrophils in the sinusoids. A needle (0.9 mm dia.), connected to a perfusion apparatus by means of a plastic tube, is inserted from the capsule side into the sample. To avoid precipitation of serum proteins the heparinized Ringer's solution is perfused for 4 min at a rate of 1.5–10 ml/min, depending on the sample size. This is followed by perfusion with 2% glutaraldehyde in 0.1

mol/l cacodylate buffer (pH 7.4, 340 mosmols). During perfusion, the cut surface of the sample should face upward. In this way the perfusate (restricted on either side by the capsule) builds up and flows out over the cut surface. After removal of the needle, a thin slice containing the puncture part is discarded. The remaining tissue is further fixed for 24 hr for scanning electron microscopy.

Lung, Method I

Perfusates

1. Ringer's solution containing papaverine sulfate 25 mg/liter, heparin 5,000 U-USP/liter, and 1.5% dextran (mol. wt. 70,000).

2. 2.5% glutaraldehyde in phosphate buffer (pH 7.4) (340 mosmols) containing 1.5% dextran.

3. Ringer's solution without papaverine and heparin.

Procedure.   In Fig. 8.20 the solution in glass flasks (a) is aspirated by a peristaltic pump (b) (Harvard App. Model 1210). The appropriate solution is selected by the double inlet stopcock (c). The outlet of the pump is connected by means of a rubber tube to the perfusion pressure control tube (d). The glass tube (400 mm in length and 3 mm in internal diameter) is provided with a moat near the upper end and a 100 mm long T connection beneath the moat. The upper end of the glass tube is open, and its lower end is connected via a drip-set (e) to a

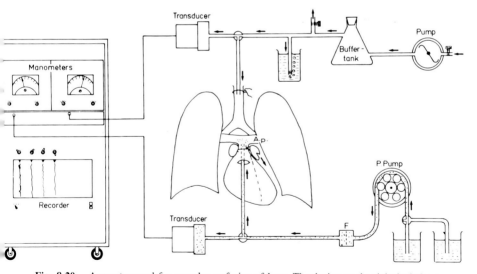

Fig. 8.20.   Apparatus used for vascular perfusion of lung. The device on the right includes a peristaltic pump, another pump for adjustment of transpulmonary pressure, and pressure transducers; the one on the left includes a recorder and manometers. Perfusion can be completed without the aid of transducers or recorders, which are rather expensive. (Gil and Weible, 1969–1970.)

needle (Luerlock 19), which is introduced into the pulmonary artery. The level at which this needle is situated is noted. The height (h) that determines the hydraulic driving pressure of solutions to be perfused is measured between this level (f) and the upper open end of the perfusion pressure control tube (d).

A hydraulic pressure of 35 mmHg is desirable; any desired pressure can be obtained by lowering or raising the tube (d). The pressure developed by the peristaltic pump (b) must be higher than the required pressure. Its output is regulated by altering the speed of the pump in such a manner as to achieve a slight overflow from the open end of the tube. The overflow solution is collected in the moat (g), which can be recirculated by connecting the outlet of the moat to the corresponding glass flask by means of a tube.

The animal is deeply anesthetized and artifically ventilated (Gil and Weibel, 1969-1970). Immediately after the needle is inserted into the pulmonary artery, the left atrium is cut. The lung microvasculature is rinsed with the perfusate 1 for 1-2 min. This is followed by fixation with perfusate 2 for $\sim$ 20 min and then rinsing with perfusate 3. These perfusates are changed by means of the stopcock (c). Small pieces of the tissue are postfixed in 1% OsO_4 for 1 hr at 4°C for transmission electron microscopy.

Note: If the objective is to visualize the secretion on the surface of bronchioles in the lung with the TEM or SEM, the fixation should be accomplished by intravascular perfusion or by retrograde filling of the bronchioles after injection of the fixative into the periphery of the lung. In contrast, if the fixative is introduced into the trachea, the surface secretion is washed away. For scanning electron microscopy of lung, the airway fixation procedure provides better preservation of surface details of cellular and epithelial lining layers (Warheit *et al.*, 1981).

Lung, Method II

Perfusates
1. Normal saline   300 ml
   Procaine hydrochloride   3 g
   Heparin   3 ml
2. 3% glutaraldehyde in 0.1 mol/l phosphate buffer (pH 7.4) having an osmolality of 500–550 mosmols
3. 2% agar in perfusate 2.

Procedure. After the animal (e.g., rat weighting 250–300 g) is anesthetized, a tracheostomy is performed. The animal is ventilated with a respirator and pure oxygen is administered for 30 min before a midline abdominal incision is made exposing the diaphragm. The diaphragm is incised and the respirator discontinued, allowing the fully degassed lungs to collapse. Heart is exposed by a thoractomy and using a standard hospital IV infusion set the lungs are washed with perfusate 1 through the right ventricle. Concurrently, the left auricular appendage is removed to allow escape of the pulmonary perfusate. The wash is carried out to flush and vasodilate the circulatory system and is continued until a clear affluent from the left atrium appears.

Immediately after the wash is completed, perfusate 3 is instilled through the tracheostomy tube at a pressure of 23 mmHg to reinflate the collapsed lung and thus prevent fluid leaving through the respiratory tree. The lungs are removed, and 1 mm^3 pieces are cut from each right middle lobe and fixed in perfusate 2 for an additional period of 2 hr at 4°C. Postfixation is accomplished in 1% OsO_4 in the buffer for 2 hr at 4°C.

In conventional vascular perfusion fixation, seepage of the fixative into the alveoli may result in the disruption of the surfactant from the alveolar surface. The loss of surfactant can be diminished by replacing the air with agar into the degassed lung as described earlier. This procedure prevents the occurrence of an air/liquid interface at the alveolar level, and the mixture acts both as a barrier preventing the loss of surfactant and as a fixative (Callas, 1974).

Lung (Small Animals), Method III

The mercury manometer is preset at 15–20 mmHg (Fig. 8.21). This is accomplished by filling the flask with 200 ml of fluid and then calibrating the mercury manometer. The calibration is done by measuring up from the mercury level with

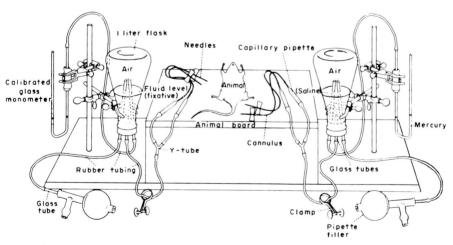

Fig. 8.21. Diagram of the apparatus used for vascular perfusion of lungs of small animals. The apparatus can supply perfusates at pressures that are at approximately physiologically normal levels. The components of the unit are 1 liter Erlenmeyer flasks, glass manometers, propipette pipette fillers, Butterfly needles, cannuli, glass and rubber tubings (5/32 in bore diameter) of various lengths, and ring stands or other types of supportive frameworks. (Brody and Craighead, 1973.)

a millimeter ruler (2 mm divisions for 1 mmHg pressure) and marking the tube before applying the pressure. The pressure is created (both on the fluid and on the mercury) by forcing air into the flask with the pipette-filler. The tubing from the fluid remains clamped until flow is needed. The mercury level is adjusted with the pipet-filler to any previously described line on the manometer. When the fluid line is unclamped, air pressure drives the fluid through the cannuli. The dissection board is raised to approximately the same height as the perfusate level so that gravity is not a factor in the perfusion pressure.

The thoracic cavity of a Nembutal-anesthetized animal is opened to expose the heart and lungs. The needle cannula is inserted into the right ventricle of the beating heart, and phosphate-buffered saline is perfused from the right-hand unit in Fig. 8.21. After ~30 sec, most blood cells are washed out from the pulmonary vasculature. The cannula is rapidly replaced by the cannula from the flask filled with the fixative (4% glutaraldehyde). The heart usually ceases to beat after 10–15 sec, although the fixative continues to flow through the patent pulmonary artery into the lungs. This apparatus does not require sophisticated pumps or gauges and allows intratracheal or intra-arterial perfusion either simultaneously or successively. For further details, see Brody and Craighead (1973).

Lung (Postmortem Fixation of Human Lung), Method IV

Within 30 min after death an isoosmolar mixture of 5% glutaraldehyde and 0.025 g of indocyanine green (dissolved in 19 ml of distilled water and 1 ml of pasteurized plasma protein solution for stabilization, mixed to equal parts) are injected into peripheral air spaces of the intact lung without opening the thorax. The dye is introduced to facilitate recovery of the fixed samples. Injection is done with a long, thin needle (gauge 20, 10 cm) introduced into the lung through an intercostal space. After passing the rib, which serves as a landmark for the proximity of the pleura, the needle is introduced into the lung to a depth of 2–3 cm. While gently injecting 5–10 ml of the solution, the needle is slowly withdrawn to achieve adequate distribution of the fixative. Axillary, lateral, or dorsal injection sites are chosen, the latter not far below the scapular angle. An injection into the pericardium should be avoided; otherwise the fixative would be drained through the bronchial tree. The tissue specimens fixed *in situ* maintain their geometric dimension after opening of the chest and are removed at autopsy and further fixed. For further details, see Bachofen *et al.* (1975). The results of this procedure are shown in Fig. 8.22.

Lung (Human) Biopsy Specimens, Method V

In Fig. 8.23 the liter flasks (F) contain ~100 ml of 4% glutaraldehyde and can be moved up and down on the ring stands so that the fixative level approximates the height of the tissue. When squeezed, the bulb (B) forces air through the bulb line

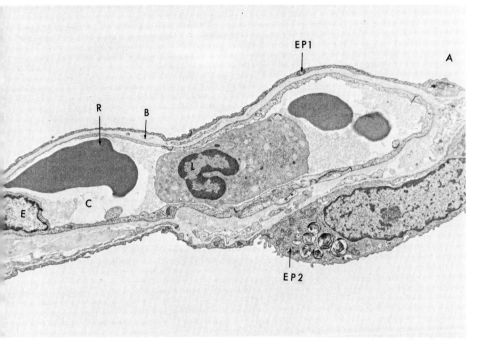

Fig. 8.22. Interalveolar septum of human lung (55-year-old male) fixed by the transcutaneous injection method. Alveolus (A), epithelial cell type I (EP1), epithelial cell type II (EP2), capillary (C), basement membrane (B), endothelial cells (E), leucocyte (L), and red blood cell (R). ×5005. (Bachofen *et al.*, 1975.)

(BL) and exerts pressure on the fixative. This pressure is transferred through the line to the manometer (LM) and is measured by the mercury gauge (M) (5–10 mmHg is recommended). When the fixative line (FL) is unclamped, fixative passes into the cannulated tissue at a predesignated pressure.

The specimen is placed in a Petri dish and the orifices of small blood vessels on the cut surfaces are identified with a dissecting microscope. One to several of the vessels are cannulated and perfused with 4% glutaraldehyde for 15–20 min at the pressure mentioned earlier. The tissue block is then cut into smaller pieces and further fixed for 2–8 hr. For further details, see Brody and Craighead (1975).

Muscle (Skeletal Muscle of Rat Hind Limb)

Perfusates
 Solution A: Locke's solution containing 0.1% $NaNO_2$
  Locke's solution:
  NaCl   9.2 g
  KCl   0.42 g

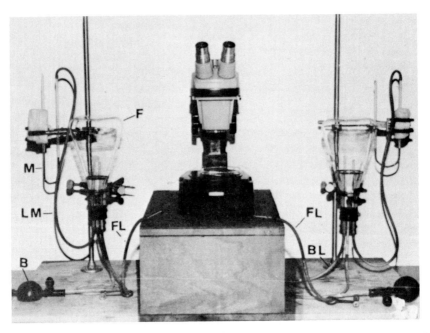

Fig. 8.23. The perfusion apparatus for lung biopsy specimens. (Brody and Craighead, 1975.)

$CaCl_2 \cdot 2H_2O$ 0.24 g
$NaHCO_3$ 0.15 g
Sucrose   1.0 g
$NaNO_2$ 10 g
Distilled water to make   1000 ml
The final solution contains 2 mM $CaCl_2$.
Solution B: 12% glutaraldehyde
  Glutaraldehyde (25%)   60 ml
  Cacodylate buffer (0.1 mol/l pH 7.2) to make   500 ml
The final solution contains 2 mM $CaCl_2$ and has an osmolality of 474 mosmols.

Procedure.   The animal (e.g., rat) is anesthetized intraperitoneally with 40 mg of Nembutal/kg of body weight. The abdominal aorta is cannulated just above the bifurcation to the hind limbs, and the vena cava is cut to allow free flow of solutions. The temperature of the perfusate is maintained at 40°C in a water bath. The perfusate is pumped with the aid of a peristaltic pump (Hughes Hi-Lo) at a constant pressure of 100 mmHg into the hind limb. Perfusion with solution A is carried out for 2–3 min to dilate the vascular bed and remove most of the blood. This is followed by perfusion with solution B for ~10 min. The tissue is cut into small pieces and fixed in solution B for an additional period of 1 hr, rinsed in the buffer, and postfixed in 1% OsO_4 for 1–2 hr for transmission electron microscopy. For further details, see Bowes *et al.* (1971).

Ovary

Perfusate

Glutaraldehyde	1%
Paraformaldehyde	1%
2,4,6-Trinitrocresol	0.01%
Cacodylate buffer	0.1 mol/l

The animal (e.g., guinea pig) is anesthetized by an intraperitoneal injection of sodium pentobarbital (30–35 mg/kg of body weight). Each of the two ovaries is perfused independently of the other. A Holter peristaltic pump (Extracorporeal Medical Specialties, Inc., King of Prussia, Pennsylvania; model 911, fitted with size D pump chamber) can be used to deliver the perfusates. The perfusion can be started at a pressure of 35–50 mmHg. The abdomen is opened with a U-shaped incision extending from pubis to the ribs. A ligature (4-0 surgical silk) is placed around both the ovarian artery and vein and tied loosely in an overhead knot. A length of the uterine artery is carefully separated from its companion vein, and a second ligature is tied loosely around the artery. Ligatures are placed in similar locations around the blood vessels of the other ovary.

The uterine artery is grasped somewhat caudal to the ligature with fine forceps, lifted, and pulled slightly taut. The cannula (26 gauge needle), with Krebs–Ringer–bicarbonate buffer (containing 2 mg glucose/ml, pH 7.4, 300 mosmols) flowing from it at a rate of 2 ml/min, is inserted into the uterine artery, guided past the ligature, and secured into place by tightening the ligature. Immediately a nick is made in the uterine vein to provide an outflow for the perfusate, and the ligature around the ovarian artery and vein is tightened, isolating the ovary from the circulatory system.

After a 2–3 min buffer wash, fixation is begun and continued for 10–30 min. Shortly after the fixative has reached it, the ovary begins to harden, indicating a favorable perfusion. After the perfusion is under way on one side, perfusion of the other ovary is started by using the same sequence. The ovary is cut into small pieces that are placed directly in 0.1 mol/l cacodylate buffer containing 5% sucrose for a wash, and postfixed with 1% OsO_4 in the same buffer for 1 hr in the cold. For further details see Paavola (1977).

Spinal Cord

Perfusate: 4% glutaraldehyde

Glutaraldehyde (25%)	32 ml
Distilled water	168 ml
$NaH_2PO_4 \cdot 2H_2O$	3.75 g

The pH is adjusted to 7.4 by adding ~0.9 g of NaOH.

Procedure.   A 6–9 cm long goldfish is anesthetized by placing it in a 1:1000 solution of MS-222 (Sandoz, Hanover, New Jersey) in tap water. Anesthesia is maintained by perfusing a 1:2000 solution of MS-222 through the mouth and out of the gills. The fish is then placed, ventral side up, in a special stand and an incision is made in the midline. With a syringe, 5 ml of the perfusate is perfused through the heart, and the tail is cut off to allow the perfusate to escape. The duration of perfusion is ∼3–5 min, depending on the resistance offered by the vascular system of the fish.

The spinal cord is removed and cut into 1 mm segments while immersed in the perfusate. The specimens are rinsed briefly in phosphate buffer and postfixed in 1% OsO_4 for 1 hr at 4°C for transmission electron microscopy.

Spleen

Perfusates
Method I. Modified Ringer's solution

KCl	0.3 g
NaCl	8.78 g
Distilled water to make	1000 ml

Procaine hydrochloride is added to a final concentration of 0.1%. The pH is adjusted to 7.4. Procaine is added to prevent arteriolar spasm and thus allow free flow of the perfusates.

Method II. Fixation fluid consists of glutaraldehyde (1.7%) in Sørensen's buffer (0.08 mol/l) with an osmolality of 310 mosmols. Procaine hydrochloride (5%) in water is added to a final concentration of 0.1%.

Procedure.   The animal (e.g., rabbit) is anesthetized by intravenous pentobarbitone sodium (∼30 mg/kg body weight). A nylon catheter with a three-way stopcock is inserted in the right femoral artery with the tip between the left renal and coelic arteries. The aorta is dissected free in order to control the position of the tip of the catheter and to prepare the clamping of the aorta above the origin of the coelic artery, just below the diaphragm.

Modified Ringer's solution containing 0.1% procaine is perfused at a perfusion pressure of 110–120 mmHg. This pressure is equal to the systolic blood pressure. The aorta is clamped above the coelic artery and a small cannula is inserted into the inferior vena cava, or one or two mesenteric veins are severed to allow free blood flow. After ∼20 min of perfusion (300–400 ml of Ringer's solution), the splenic vessels appear pale, and the perfusate is changed to the fixation fluid at the same perfusion pressure. The perfusion fixation is maintained for 15–20 min. After fixation, the spleen is removed and small tissue blocks are immersed in 2.5% glutaraldehyde for 1–2 hr, followed by postfixation in 1–2% OsO_4 for 1–2 hr for transmission electron microscopy. For further details see Elgjo (1976).

Testes (e.g., Guinea Pig and Rat), Method I

Rinsing solution

NaCl	9 g
Polyvinylpyrrolidone	25 g
Heparin	0.25 g
Procaine–HCl	5.0 g
Distilled water to make	1 liter

Adjust the pH to 7.35 with 1 mol/l NaOH, and filter twice through Millipore filters (3/μm or smaller pore size).

Fixative solution I

0.2 mol/l monosodium phosphate (NaH_2PO_4)	45 ml
0.2 mol/l disodium phosphate (Na_2HPO_4)	405 ml
25% formaldehyde	60 ml
25% glutaraldehyde	60 ml
Polyvinylpyrrolidone	25 g
Distilled water to make	1 liter

Adjust the pH to 7.35 and filter as earlier.

Fixative solution II. As earlier except that 120 ml each of formaldehyde and glutaraldehyde are used and 0.5 g picric acid is added.

Procedure. The perfusion apparatus and instruments required for the method to be described are shown in Fig. 8.24. The details of the abdominal aorta and its main branches and the location and function of ligatures are shown in Fig. 8.25.

The animal is anesthetized and its abdominal cavity is exposed. The viscera is displaced to the animals's right and the aorta is cleared from connective tissue and fat. Two ligatures of cotton surgical threads are placed loosely around the lower part of the aorta ~6 mm apart. A ligature is placed around the aorta above the left renal artery. The first ligature of the aorta is made by tightening the lowermost knot and then the middle ligature is lifted to stop the blood flow.

The aorta is incised obliquely with fine scissors below the middle ligature and the cannula is quickly inserted in a retrograde manner. The blood flow is established by releasing tension on the loose middle ligature. The cannula is made from polyethylene tubing (Intermedic P.E. 160) and one end is cut obliquely and the other end is closed with a clamp. One end of the cannula is filled with a solution of 0.9% heparin and the closed end is left with a trapped air space. At the same time, the cannula is pushed further into the aorta beyond the point of the ligature and secured by tightening the middle ligature.

The vena cava is first exposed between the right renal vein and the liver and then opened with a large incision. The aortic cannula is cut 3 cm from its insertion and the blood flow is reversed by connecting it to the adaptor of the rinsing solution. The ligature of the aorta is closed above the left renal artery.

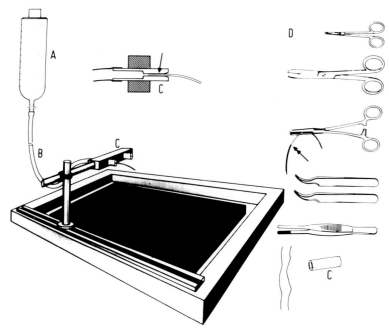

Fig. 8.24. The perfusion apparatus and instruments needed for the fixation of testis. A graduated reservoir (A) to contain the perfusates is connected to a tubing 200 cm in length and 0.4 cm in diameter (B). The adaptor system (C) is used to join the cannula to the tubing. A cutaway view of the adaptor is seen above (C). Instruments and material used in setting up the perfusion apparatus are shown at (D). At the double arrow note the partially filled (dark) portion of the cannula containing the heparin solution. (Forssmann *et al.*, 1977.)

The rinsing solution is perfused for 1–1½ min at a pressure of 100 mmHg, followed by fixative solution I for 2–2½ min, and finally by fixative solution II at a pressure of 140 mmHg for 2–5 min. For further details see Forssman *et al.* (1977).

Testes, Method II

The perfusion is carried out through the testicular artery. The animal (e.g., rat) is put under light ether anesthesia and a 30 gauge needle is inserted into the abdominal portion of the artery, a short distance above the proximal end of the pampiniform plexus. The needle is attached to standard venipuncture tubing which runs to a three-way stopcock connecting the main tubing with two bottles, one containing a mammalian Ringer's solution with 0.4% procaine added and the other containing 2% glutaraldehyde in 0.133 mol/l phosphate buffer (pH 7.4).

The Ringer's solution is allowed to flow by pressurizing the bottle with a rubber atomizer bulb. The rate of flow is ~2 ml/min. After a few minutes, the testis will blanch. The Ringer's solution is replaced by the fixative, and soon after the hypogastric vein is cut allowing the fixative to drain freely. This vein

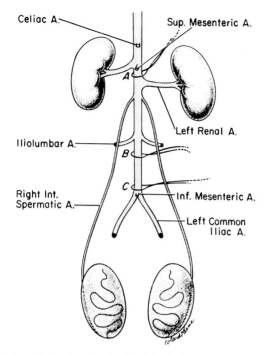

Fig. 8.25. A schematic drawing showing the abdominal aorta and its main branches. Note that the right and left internal spermatic arteries originate immediately below the renal arteries. The three ligatures A, B, and C are indicated. The cannula is inserted retrogradely into the aorta between the two lowermost ligatures and the tip is pushed beyond the point of the middle ligature (B). During the perfusion procedure the three ligatures are tied (see text). Ligature A prevents fixative from reaching the GI tract and thorax; ligature B secures the cannula in the aorta and prevents fixative from reaching the iliac vessels and legs; ligature C secures the cannula in the aorta and prevents displacement. The only portion of the aorta that receives fixative is the 2–3 cm region between ligature A and B—the left renal artery, the two internal spermatics (testicular artery), and the lumbar vessels originating from this part of the aorta. (Forssmann *et al.*, 1980.)

receives the venous drainage of the pampiniform plexus and flows into the external iliac vein. After fixation by perfusion for ~15 min, the testis is excised and cut into small pieces that are fixed in glutaraldehyde for an additional period of at least 2 hr. After a thorough rinse in the buffer, the specimens are postfixed with 1% OsO_4 in the buffer for 2 hr for transmission electron microscopy. For further details see R. V. Clark (1976).

Testes, Method III

This is a retrograde abdominal aorta vascular perfusion. The abdominal cavity of the animal (e.g., rat) is opened and the abdominal aorta is carefully exposed and cleared of connective tissue and fat. The inferior vena cava and testicular vessels visible on the posterior abdominal wall should not be touched. The left kidney is reflected anteromedially and the retroperitoneal space immediately above the kidney is opened with a curved blunt iris forceps. After identifying the left renal artery and tracing it to its origin, a silk suture (000) is placed around the aorta immediately above the origin of the left renal artery and left untied.

The lowest portion of the aorta is grasped with a curved Halsted forceps and slightly elevated. A needle (18–25 gauge, depending on the size of the aorta) is inserted in a retrograde direction into the abdominal aorta near its bifurcation. Immediately the flow of saline is established at a pressure of 130 cmH$_2$O, the suture around the aorta is tied, and the inferior vena cava or the left renal vein is incised to permit an egress of fluid. As soon as the testicular vessels have been cleared of blood, saline is followed by the fixative. The volume of the fixative ranges from 100 to 150 cm^3. The fixative used is 5% glutaraldehyde buffered in collidine.

Following perfusion, the hardened testes are removed from the animal and ∼1 mm blocks are cut and fixed in the same fixative solution for an additional period of 30 min at 4°C. After a brief wash in the buffer (15 min), the specimens are postfixed in buffered 2% OsO$_4$ for 2 hr at 4°C. For further details the reader is referred to Vitale *et al.* (1973).

IMMERSION FIXATION

The disadvantages of fixation by immersion are many, some of which are discussed here. Varying periods of time necessarily elapse between the removal of tissue from the body and the actual fixation of that tissue. On dissection, the immediate alterations in blood pressure induce volume changes in the tissue. Moreover, anoxia and spontaneous postmortem changes may cause alterations in the tissue morphology. The fixed outer surface of the tissue block may offer resistance to further penetration by the fixative. The diffusion rates of various components of the buffered fixative may differ. Thus, the rate and depth of penetration by fixatives into the tissue block are less than satisfactory. In addition, actual fixation is rather slow and uneven. Diffusion barriers are more important than are chemical reactions in the determination of rate of fixation.

It is known that fixatives such as glutaraldehyde and OsO$_4$ penetrate the tissue slowly, and thus the core of the tissue block is rarely well fixed (Fig. 2.1). Both internal and external surfaces of the tissue may show artifacts. Tissue specimens fixed by vascular perfusion show more mitotic figures than those fixed by immersion. The preservation of certain intracellular macromolecules is adversely affected by immersion fixation. For example, glycogen may be completely depleted in the cerebral tissue of mice 1 min after decapitation (Lowry *et al.*, 1964). On the other hand, glycogen can be preserved in the brain by vascular perfusion.

Immersion fixation may cause artifactual dark neurons in CNS. The appearance of clear cells and mixed cells in adrenal medulla fixed by immersion is well documented. Such a population of cells does not appear in the tissue fixed by perfusion. However, normal clear cells (e.g., the undifferentiated cells in the gastrointestinal epithelium) that show a completely undifferentiated cytoplasm have been described in specimens fixed by perfusion.

Nonhepatocytes in liver parenchyma are associated with the sinusoid and the Disse space and are difficult to identify in exsanguinated tissue specimens fixed by immersion. The extracellular spaces appear larger in liver fixed by perfusion compared with those found in the immersion-fixed liver tissue. It has been shown that the dimensions of extracellular spaces in the vascular-perfused liver are as follows: the lumina of sinusoids (10.6%), the space of Disse (4.9%), and the bile canaliculi (0.4%); the percentages are based on the total liver volume (Blouin *et al.*, 1977; Reichen and Paumgartener, 1976).

The mode of fixation affects the endothelial lining in the kidney. Both a completely closed endothelium and a discontinuous endothelium having gaps of variable diameter are artifacts caused by immersion fixation. On the other hand, a fenestrated endothelium differentiated into sieve plates and endothelial cell processes obtained with vascular perfusion is considered to be closer to the condition *in vivo*. The latter morphology of the endothelial lining is also obtained after freeze-etching. The collapse of the proximal tubule in kidney tissue specimens fixed by immersion is well documented. This results in considerable change in the size and shape of cells.

In the rat myocardium, fixed in diastole state by vascular perfusion, sarcomeres appear relaxed, whereas fixation of a similar tissue by immersion may cause the sarcomeres to contract (Vodovar and Desnoyers, 1975). Bandlike particle-free areas of the sarcolemma of muscle fibers fixed by immersion probably indicate a change of the membrane structure related to commencing necrosis (Schmalbruch, 1980).

Mechanical pressure exerted during dissection and cutting of the unfixed tissue prior to fixation by immersion may introduce artifacts. One of the results of mechanical trauma arising from slicing the tissue with a razor blade is disrupted membranes and the possibility of the liberation of mitochondria and secretory granules into the extracellular space and capillary lumina. The mobile cytoplasmic components of the local homogenate resulting from razor blade traumatization are free to move from the region traumatized to any morphologically normal region by Brownian motion before fixation has been completed (Wacker and Forssmann, 1972). The occurrence of such artifacts is further favored by the slow penetration by the fixative during fixation by immersion. It has been shown that the rate of penetration by the fixative is 10 μm/sec, whereas the average displacement of an organelle (e.g., mitochondrion) under similar conditions is of the order of 0.3 μm/sec (Forssmann, 1969).

As stated earlier, in spite of the drawbacks of fixation by immersion many tissue types must be fixed by this method. It is absolutely imperative that immersion fixation be carried out under optimal conditions in order not to compound the problem. Certain tissue types, such as skin and bone, can tolerate a brief interruption to their blood supply and still retain their structure; these can be fixed satisfactorily by immersion. In certain other cases the quality of preservation by immersion can be improved by immersing an intact part of the body in the fixative to accomplish *in situ* fixation instead of immersing its small slices. For example, sartorius muscle shows better preservation when an intact skinned frog's leg is totally immersed in ~4% glutaraldehyde compared with the preservation of muscles fixed even after the most careful dissection. Such a procedure preserves the order in the muscle fiber. Fixation by immersion is intrinsically easy, but its drawbacks may become painfully evident after such specimens are viewed with the electron microscope.

DRIPPING METHOD

The advantage of the dripping method is that the tissue remains attached to the body of the animal at least during prefixation. Thus, normal blood supply and metabolism of the cells are maintained until the instant of fixation. Because fixation is initiated *in situ,* autolytic changes are minimized. Furthermore, maintenance of *in situ* conditions during fixation appears to maximize the depth of satisfactory fixation. Also, there is evidence indicating that the adverse effects of hypotonic or hypertonic fixing solutions are minimized by an intact blood circulation during fixation. *In situ* fixation is especially important in the fixation of muscles that remain attached and extended during fixation.

The disadvantage of the dripping method is that it allows the study of only the surface layers of an organ, since the fixative does not penetrate deeper layers. In the case of kidney, for instance, only the outer 200–400 μm of the cortex is well fixed, which comprises two to four layers of tubules and the outer row of glomeruli. Obviously, the method is more useful for scanning electron microscopy than for transmission electron microscopy.

For fixation by dripping, the animal is anesthetized and the tissue exposed by dissection. In some cases it is necessary to expose the selected tissue completely by removing the overlying tissue and membranous covering, in order to facilitate direct flooding of the tissue. The enclosing connective tissue membranes can be removed by peeling them off with a fine pair of forceps. However, utmost care is necessary during the removal of the tissue covering, otherwise this operation may damage the structure of the organ. Also, injury to blood vessels and nerves should be kept at a minimum during dissection and subsequent procedures.

Immediately after the tissue is exposed, it is flooded with an ample supply of a suitable chilled (~2°C) fixative in a fume hood. In general, buffered (pH 7.2–

7.4) glutaraldehyde (3%) or a mixture (1:1) of glutaraldehyde and formaldehyde is dripped immediately on the exposed tissue. To maintain a constant supply of fresh fixative to the tissue, drops of fixative solution should be added every few seconds. After 2-3 min, a thin pad of cotton is placed over the tissue surface and the fixative is added less frequently (a few drops after every 2-3 min). After fixation and hardening have continued for ~20 min, the organ is removed and immersed in a fresh cold fixative. While immersed in the fixative, the organ is cut into small pieces that are fixed for an additional period of 1-3 hr to complete the primary fixation.

To minimize the contamination and dilution of the fixative, it is necessary that oozing blood and tissue fluids be flushed away with excess fixative. The animal should be kept slightly tilted during the whole procedure so that the used fixative can drain away slowly from the tissue being fixed. In view of the harmful nature of the fixative fumes, one must avoid inhaling them.

INJECTION METHOD

Since the injection method is also carried out *in situ*, it has the same advantages as those of the dripping method. This method is used for preserving the internal structure of organs. A suitable, chilled fixative is injected slowly and directly into the selected living organ for ~2-20 min after exposing it from the overlying tissues. The fixative is injected by using a micropipette or hypodermic needle. The organ is removed and cut into small pieces that are further fixed by immersion for 1-3 hr. Manual injection has the disadvantages of subjecting the tissue to variations of pressure and/or of flow of the fixative solution. The injection method has been employed for fixing sinus venosus in frog heart (Baldwin, 1970), pleural and abdominal cavities in mouse diaphragm (Karnovsky, 1967), lumen of rat kidney tubule (Maunsbach, 1966), rat pituitary cranial cavity (Smith and Farquhar, 1966), knee joint in rabbit cartilage (Palfrey and Davies, 1966), ventricle of goldfish heart (Robertson, 1963), bat stomach (Ito and Winchester, 1963), posterior chamber in rabbit eye (Tormey, 1963), cat pancreas (Sjöstrand and Elfvin, 1962), and cat spinal cord (Bunge *et al.*, 1961).

It has been reported that blood and tissue fluid can be washed out from small tissue pieces ($\sim 3 \times 5$ mm) by puncture perfusion (Murakami, 1976). After the tissue block is rolled in a wet cellophane sheet, it is punctured by a syringe needle (diameter, ~0.3 mm), and a Ringer's and glutaraldehyde solution is gently perfused. The specimen is further fixed for ~24 hr by immersion with glutaraldehyde.

Fixation of cat autonomic ganglia was obtained by a modified injection method (Koelle *et al.*, 1974). The abdominal aorta of the anesthetized animal is tied just below the diaphragm, the heart is exposed rapidly, the right superior vena cava is nicked, and ~100 ml of cold fixative solution are injected via the left ventricle. Ganglia are removed, cut into small pieces, and fixed for an additional period of 2 hr.

Human lung tissue can be preserved immediately postmortem by transthoracic injection of stained glutaraldehyde. The following method was used by Bachofen *et al.* (1975). The fixative solution consists of equal parts of 5% glutaraldehyde and 0.025 g of indocyanine green dissolved in 19 ml of distilled water and 1 ml of pasteurized plasma protein solution for stabilization. The dye acts as a label to facilitate collection of the fixed parts of the lung.

The fixative solution is injected (within 30 min after death) into peripheral air spaces of the intact lung without opening the thorax. Injection is accomplished with a long, thin needle (gauge 20, 10 cm) introduced into the lung through an intercostal space. After passing the rib, which serves as a landmark for the proximity of the pleura, the needle is introduced into the lung to a depth of 2–3 cm. While gently injecting 5–10 ml of the fixative solution, the needle is slowly withdrawn to achieve adequate distribution of the fixative. The success of the procedure depends partly on the exact location of the injection. Axillary, lateral, or dorsal injection sites are preferred; the latter should not be far below the scapular angle. Ventral injection sites have the disadvantage of the possibility of injecting the solution into the pericardium, which may result in a rapid draining of the solution through the bronchial tree and inadequate spreading of the fixative.

Tissue samples are removed at autopsy and fixed with 2.5% glutaraldehyde in potassium phosphate buffer (pH 7.4) for an additional period of 1 hr. The specimens are postfixed with 1% OsO_4 buffered at pH 7.4 and adjusted to an osmolality of 345 mosmols with sodium cacodylate for 2 hr.

Since the tissue is fixed *in situ,* the geometric arrangement of air spaces and capillaries is preserved in the well-inflated lung. However, caution is required while injecting the fixative solution; excess force should be avoided to ensure that the fixative spreads in the air spaces with as little pressure as possible.

ANESTHESIA

Sublethal anesthetization is preferable to decapitation or cervical dislocation because the latter two initiate cell death prior to arrival of the fixative. Several equally effective anesthetics are available for animals of different body weights. Small animals (e.g., rat and guinea pig) can be anesthetized with ether or halothane. However, barbital anesthesia is preferred to ether, since incomplete perfusion of the medulla has been shown to occur in animals anesthetized with ether (Bohman, 1974). For large animals, an intravenous injection with Inactin, urethane, pentobarbital (Nembutal), Pentazocine, or chloral hydrate is recommended. Pentobarbital anesthesia usually results in better fixation than that obtained after ether anesthesia. Animals can be anesthetized by injecting 50–60 mg of pentobarbital per kg of body weight or by injecting 129 mg of Inactin per kg of body weight. Anesthesia and analgesia can also be accomplished by injecting 30 mg of Pentazocine followed by 30 mg of pentobarbital per kg of body weight.

The type and dose of anesthetic required is species-dependent. The correct dose for commonly used laboratory animals can be found in books by Lumb (1963), Soma (1971), and Barnes and Etherington (1973).

If detrimental effects of chemical anesthetics are suspected, low temperatures as anesthetic may be used. Above-freezing temperatures may also be desirable for very small or newborn animals. Advantages of such temperatures include the production of analgesia, reduction of metabolic rate and motor activity, and elimination of potential harmful effects of chemical agents. This approach may also be useful for puncture perfusion.

Effects of Fixation

EFFECT OF FIXATION ON ACTIN

Fixation with OsO_4 causes disorganization of actin filaments both *in vitro* and *in situ* in muscle and nonmuscle cells; primary fixation with glutaraldehyde does not prevent this damage, which includes both shrinkage and formation of filament networks. Both longitudinal and lateral shrinkage of actin in the muscle is caused by OsO_4. The interfilament distance between OsO_4-fixed muscle actin seen in an electron micrograph is ~70% of that *in vivo* (as deduced from X-ray diffraction data) (Huxley, 1971). Fixation with OsO_4 causes a 10% reduction in the length of actin-containing filaments in muscle (Page, 1964). Since damaged filaments do not lie straight, the effects of OsO_4 are particularly deleterious with respect to the longitudinal organization of the filament.

The longitudinal shrinkage of actin occurs during initial fixation with OsO_4 even when the muscle ends are held at a fixed length during fixation. If the ends of the muscle fiber are not held during dehydration, an additional longitudinal shrinkage of ~11% of the initial length takes place (Page and Huxley, 1963b). It should be noted that dehydration with ethanol causes shrinkage if the ends of the ⟨muscle are not held, but dehydration with acetone does not.

On the other hand, glutaraldehyde fixation causes only ~4% reduction in the length of actin-containing filaments in muscle (Page, 1964) in the absence of postfixation with OsO_4. Glutaraldehyde-fixed specimens do not shrink during dehydration in ethanol even when the ends are not held. This demonstrates that

OsO_4-fixed filaments are susceptible to dehydration in ethanol. Relevant data on actin filaments in muscle fixed with glutaraldehyde followed by OsO_4 are not available. However, it is known that glutaraldehyde fixation does not alter the X-ray diffraction pattern of actin-containing filaments, but subsequent postfixation with OsO_4 virtually destroys its X-ray diffraction pattern (Reedy, 1971).

The cytoplasmic matrix of most nonmuscle cells contains actin that constitutes 10–15% of the total cell protein, and yet only a few cell types fixed with standard procedures show actin filaments. Since actin is a major structural protein in the cytoplasm, its nonpreservation indicates substantial damage and alteration in the appearance of ground cytoplasm. When actin filaments are seen in fixed nonmuscle cells, they are found in two different configurations: straight and unbranched (e.g., in the microvilli of intestinal brush border) and short, bent entities that form a disordered microfilament network. The latter configuration is thought to be an artifact caused by the action of OsO_4 on actin filaments. That OsO_4 causes progressive fragmentation of actin *in vitro* is well established. Among all the steps in specimen processing, OsO_4 causes the most damage. In 1% (40 mM) OsO_4 in 100 mM cacodylate buffer (pH 7.2) at room temperature, the destruction of actin filaments *in vitro* begins immediately, whereas under mild conditions (1 mM OsO_4 at pH 6.0 at 4°C) the filaments retain their normal appearance and viscosity for ~30 min (Pollard and Maupin, 1978).

The presence of straight actin filaments in nonmuscle cells treated with OsO_4 can be explained by assuming that other proteins (e.g., tropomyosin) associated with actin are responsible for the stability of these filaments during fixation. Tropomyosin can inhibit damage to actin by OsO_4. However, the foregoing explanation is insufficient to provide an understanding of the biochemical variables that control the preservation of actin filaments in different cell types; the following elaboration may be helpful.

The reaction between OsO_4 and actin requires the presence of free OsO_4 and appears to be related to the oxidation of sulfur-containing amino acids (cystine, cysteine, and methionine). This presumably results in a slow cleavage of the actin molecule into discrete peptides (Maupin-Szamier and Pollard, 1977). It seems that a large amount of OsO_4 is reduced by actin, but that very little OsO_4 binds to actin.

At least three factors play a dominant role in the preservation of actin for electron microscopy: (1) successful cross-linking during primary fixation with glutaraldehyde, (2) optimal conditions of fixation by OsO_4, and (3) the presence or absence of proteins associated with actin. The conditions of fixation include OsO_4 concentration, temperature, buffer type and osmolarity, and pH. Proteins (e.g., tropomyosin) associated with actin inhibit its destruction. Primary fixation with glutaraldehyde introduces cross-links between actin molecules within a filament as well as between actin and tropomyosin. Thus, associated proteins probably contribute to the preservation of actin in the muscle (Huxley, 1957), stress fibers (Goldman *et al.*, 1975), sperm acrosomal processes (Tilney *et al.*, 1973a; Tilney, 1975), and the brush border (Mooseker and Tilney, 1975). Actins lacking associated proteins tend to be stabilized by the addition, during fixation, of exogenous tropomyosin (Maupin-Szamier *et al.*, 1975) or heavy meromyosin (Ishikawa *et al.*, 1969). However, intact cells resist the permeation of these materials. Under favorable conditions, actin can be preserved by OsO_4.

A prerequisite to the preservation of actin is its effective cross-linking by glutaraldehyde prior to osmication, followed by osmication with a low concentration of OsO_4. Actin filaments *in vitro* can be preserved with 1% glutaraldehyde in 50 mM phosphate buffer (pH 7.0) followed by osmication with 0.1% OsO_4 in the same buffer (pH 6.0) containing 50 mM KCl and 5 mM $MgCl_2$ for 10 min at 1°C (Pollard and Maupin, 1978; Maupin-Szamier and Pollard, 1978). *En bloc* treatment with uranyl acetate is not recommended.

The structure of actin filaments (6–8 nm wide) *in vitro* is well preserved by negative staining with uranyl acetate after a brief treatment with OsO_4. According to this procedure, actin filaments are treated with 0.2% OsO_4 in 50 mM phosphate buffer containing 50 mM KCl and 5 mM $MgCl_2$ for 5 min at 20°C, followed by negative staining with 1% uranyl acetate. Methods for preparing actin from skeletal muscle have been presented by Fujime and Ishiwata (1971) and Forer (1978).

Actin filaments in amoeba can be preserved by fixation with 3% glutaraldehyde in 100 mM phosphate buffer (pH 7.0) containing 1 mM $CaCl_2$ for 1 hr at room temperature followed by osmication with 1% OsO_4 in the same buffer (pH 6.0) for 1 hr at room temperature (Pollard and Maupin, 1978).

For adequate preservation of actin configuration in intact nonmuscle cells, the use of glutaraldehyde containing tannic acid followed by a mild treatment with OsO_4 is recommended. This method has been used to visualize actin filaments (LaFountain *et al.*, 1977). Such a fixative clearly preserves the arrowhead configuration of S-1-decorated actin filaments (Begg and Rebhun, 1978). This fixative, compared to glutaraldehyde alone, preserves the filamentous structure of actin in pure pellets and in crude plasma membrane preparations even when followed by a "harsh" treatment with OsO_4. An alternative approach may be to expose the cells to heavy meromyosin prior to fixation to protect the actin filaments. Such an approach has proved successful for preserving filaments in isolated membranes that are fixed and embedded (Pollard and Korn, 1973; Pollard and Maupin, 1978). Intact cells can be made permeable to actin-binding heavy meromyosin or other proteins by treatment with a detergent or glycerol. However, these treatments are likely to damage the fine structure. The aforementioned approaches are worth testing, since present fixation procedures fail to preserve actin filaments satisfactorily in most intact cells.

EFFECT OF FIXATION ON BRAIN TISSUE

Approximately 20% of adult mammalian brain volume is taken up by the extracellular space (ECS) found between axons, dendrites, glial processes, and their cell bodies (Van Harreveld, 1972). The volume of ECS in fetal brain tissue is greater than 20%, whereas in aged rats the volume is reduced to ~10%. Conventional methods of fixation result in the preservation of only ~5% of the ECS because fixatives cause a movement of sodium and chloride ions from the extracellular fluid into the neuronal and glial cell processes (Van Harreveld and Khattab, 1968). This ionic diffusion is accompanied by the influx of water from the ECS into cellular processes, resulting in the distortion of their profiles to polygonal shapes. This expansion of cellular processes occupies the ECS (Fig. 9.1). As a result, fixed tissue shows swollen cellular processes and drastically reduced ECS. Of all the cellular elements, astrocytic processes and neuronal dendrites show the maximum swelling (Cragg, 1979). Global ischemia also causes reduction in the ECS. It should be noted that the ECS is unevenly distributed among various cellular processes. For example, synaptic glomeruli contain little ECS, whereas they are widely separated from the surrounding cell bodies (Cragg, 1980).

The reduction of ECS can be minimized (Fig. 9.2B) by replacing the sodium chloride in the extracellular fluid by a substance, such as sucrose, that is unable to enter cells during fixation. In other words, this substance is osmotically active. This substance is expected to replace the extracellular fluid without affecting the size of the cellular processes. The impermeant solution should be used at a concentration that is isotonic with cerebrospinal fluid (0.28 mol/l). However, since sucrose does not enter the cells, the cytoplasm becomes hypotonic and the cells shrink (e.g., condensed axon terminals). Such shrinking can be reduced by replacing part of the sucrose with sodium phosphate. The desirable ratio of permeant to impermeant solute is 1:1.

The replacement of extracellular fluid by an isotonic sucrose solution should be carried out within a minute or so of interrupting the blood supply (the ECS is considerably narrowed after 2 min of circulatory arrest), and before any fixative reaches the tissue. Since sucrose does not cross the blood–brain barrier, it must first be opened or bypassed by immersing the tissue blocks in the impermeant solution. This barrier can be opened by perfusing the brain from the aorta with isotonic sucrose solution at a pressure of 300 mmHg. After applying the high pressure for a few minutes to open the barrier, perfusion is continued for several minutes at physiological pressure to complete the replacement of the extracellular fluid. This is followed by perfusion with the fixative solution (1% glutaraldehyde and 1% formaldehyde in phosphate buffer) containing 0.28 mol/l sucrose and 2-5 mM CaCl$_2$. According to Hossman et al. (1977), hypertonicity of the perfusate

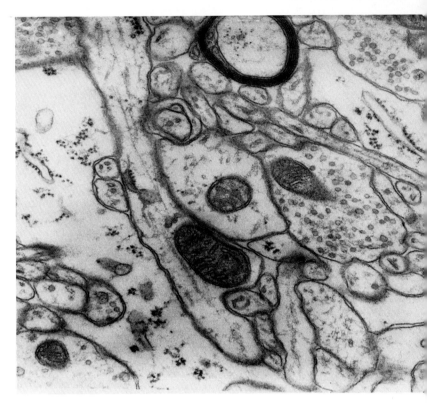

Fig. 9.1. Normal human cerebral cortex fixed by immersion in a mixture of 1% glutaraldehyde and 4% formaldehyde in 0.1 mol/l sodium phosphate buffer. The extracellular space is lost. ×33,600. (B. Cragg, unpublished.)

Fig. 9.2. A, the median eminence from a rat perfused with 3% glutaraldehyde in 0.15 mol/l cacodylate buffer (pH 7.4) containing 8 mmol/l CaCl₂ and 2% PVP (average mol. wt. 30,000–40,000) having an osmolality of 650 mosmols. The perfusion pressure was 150 mmHg (200 cmH₂O). Note large extracellular spaces between the axonal processes. The cytoplasm of a glial cell (short arrow) and a neurosecretory axon (long arrow) appear condensed. ×17,160. (Kalimo, 1976.) B, normal human cerebral cortex (tissue block size 0.5 mm) was immersed in a mixture of 0.135 mol/l sucrose, 0.135 mol/l sodium phosphate buffer (pH 7.3), and 0.002 mol/l CaCl₂ for 2 min in a magnetically stirred beaker. This is followed by fixation with 2% glutaraldehyde in the preceding solution for 16–24 hr. After washing in the buffer, the tissue is postfixed with 1% OsO₄ and stained *en bloc* with uranyl acetate. The extracellular space is preserved. ×31,500. (B. Cragg, unpublished.)

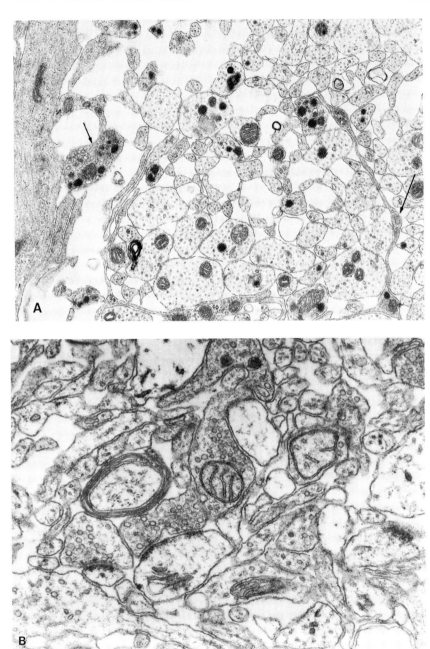

(1000 mosmols) does not affect the volume of the ECS. The brains of unanesthetized and pentobarbital-anesthetized rats have osmolalities of 307 and 316 mosmols, respectively; the plasma value from the same animal is 298 mosmols.

The blood-brain barrier can also be circumvented by perfusion via the cerebral ventricles (Rodriguez, 1969). Since the ions of the cerebrospinal and cerebral interstitial fluids are rapidly equilibrated, intraventricular perfusion facilitates the penetration of fixative into areas of the brain (e.g., white matter) which cannot easily be reached by intravascular perfusion. This approach is worth studying in greater detail.

The numerous cellular processes in the brain tissue act as an effective barrier to the penetration of fixatives by diffusion. Blocks of brain tissue fixed by immersion, therefore, show delayed fixation in the core region, resulting in the production of a large number of dark neurons. Dark neurons can also be produced by premortem trauma incurred during removal of the calvarium and extraction of the unfixed brain. Such artifacts can be almost totally eliminated by perfusion fixation with large amounts of perfusates and by delaying the autopsy several hours after the perfusion. However, even perfusion fixation can cause the production of solitary dark neurons in the transitional zones between gray and white matter when the saline perfusion is delayed or slow, or the autopsy is performed shortly after the perfusion (Cammermeyer, 1978). The remote action of postmortem trauma seems to be the main cause of the formation of solitary dark neurons.

Ligation of the aorta or intracardial injection of epinephrine may cause restriction of the perfusate flow to the head (Cammermeyer, 1968). The use of excessively large amounts of saline may also delay the flow of the fixative. Other factors that may cause problems include slow flow of the saline following either obstruction of the perfusion system with air bubbles and preparatory difficulties such as the preparation of very large or very small animals (Cammermeyer, 1978). The amount of saline introduced is best controlled by timing the flow; this duration should not be longer than 10-60 sec.

When the autopsy is performed immediately after the perfusion, incompletely fixed neurons are very sensitive to mechanical trauma; a 2 hr delay is suggested for various fixatives (Stensaas et al., 1972). It is worth remembering that artifacts introduced during perfusion fixation cannot be reversed either by continued fixation by immersion or by postfixation.

Different regions of the brain prefer fixatives of different osmolarities. It has been demonstrated that the regions of the median eminence and anterior and posterior pituitary glands, which are vascularized by capillaries having a thin, fenestrated endothelium and a distinct perivascular space, require a more dilute buffer in the fixative than that required by hypothalamus or frontal cortex, which are vascularized by capillaries of the blood-brain barrier type having a thick, nonfenestrated endothelium with no perivascular space (Kalimo, 1976). Fixatives with isotonic buffer are ideal for hypothalamus, whereas such fixatives cause marked shrinkage of cells and widening of the extracellular space in the endocrine regions. Fixatives made up in 0.15 mol/l cacodylate buffer yield excellent preservation of hypothalamus and frontal cortex, whereas the endocrine regions, median eminence, and anterior and posterior pituitary are severely damaged. A possible explanation for this difference is that capillaries of the blood–brain barrier type do not allow leakage of osmotically active particles (e.g., cacodylate molecules) to such an extent that the extracellular fluid becomes excessively hypertonic. The degree of permeability of different types of capillaries affects the behavior of various parts of the CNS during vascular perfusion. There is a difference in the rate of penetration of the fixative into the gray matter and white matter. The richly vascularized gray matter is fixed more rapidly than the less-vascularized white matter.

Even when the best perfusion method is used, the myelin sheath shows poor preservation. The damage to the sheath includes wrinkling, separation, and breakage of lamellae. Improved preservation of the myelin sheath can be obtained by adding 1.5% potassium ferricyanide to OsO_4 during postfixation (Langford and Coggeshall, 1980) (Fig. 3.5). However, this method may not be useful for general fixation of neural tissue, since the use of potassium ferricyanide may result in the extraction of cytoplasmic ground substance as well as nuclear materials. The fixation of myelin can also be improved by adding 2 mM $CaCl_2$ to the fixative solution.

EFFECT OF FIXATION ON KIDNEY TISSUE

Fixation of renal tissue by immersion results in gross distortion or collapse of both tubular and vascular elements. Fixation by the dripping method produces satisfactory preservation, but only of a limited number of outer cell layers. Fixation by vascular perfusion is the method of choice for the preservation of nephrons over their whole length. Satisfactory fixation can be obtained by the micropuncture method (Murakami, 1976) for scanning electron microscopy.

However, kidney tissue is too complex and varied both structurally and functionally to be satisfactorily and uniformly fixed throughout the organ even by vascular perfusion. A fixative solution of any one osmolarity cannot yield good fixation of the entire organ. At least a fivefold range of osmolarities may be present between different tissues of this organ. Glomeruli from poorly perfused and poorly fixed areas of the kidney show endothelial, epithelial, and mesangial swelling similar to that seen in pathological (glomerulonephritis) or immersion-fixed specimens, or during early autolysis. Mesangial cell swelling is an important index of poor fixation of renal glomeruli and corresponds to the rapid swelling of proximal tubular epithelial cells (Johnston et al., 1973). Swelling of mesangial cells is absent in well-fixed specimens. Endothelial and epithelial cells may show swelling, even with good fixation of other parts of the organ. A very distinct layering of the basement membrane of capillaries and loops of Henle seems to be a fixation artifact. Multilayered appearance of this membrane is usually absent in well-fixed tissues.

Glomeruli are thought to be relatively resistant to the effects of poor fixation, in contrast to the more labile proximal tubular epithelial cells. The former are less prone to the effects of differences in the fixative osmolarity. The most striking effect of osmolarity differences is observed on the size of intercellular space of tubular elements. The intercellular space is generally narrower after using isosmotic fixatives (320 mosmols for cat) than that obtained after using fixatives hypertonic to plasma (650–2000 mosmols) (Yun and Kenney, 1976) (Fig. 8.17).

Developing renal cortex of rat is best fixed with 6% glutaraldehyde in 0.1 mol/l cacodylate buffer (pH 7.2) having an osmolality of 828 mosmols (Larsson, 1975). In contrast, good fixation of adult proximal tubules is obtained with 1% glutaraldehyde. Less-developed vascular supply in the outer cortex than in the inner cortex of the embryonic renal tissue requires higher concentrations of the fixative. On the other hand, the presence of extensive vascular supply in the adult renal cortex does not necessitate higher concentrations of an aldehyde.

The standard perfusion protocol used may not satisfactorily fix vessels in the kidney, i.e., they may not be preserved in their natural expanded or contracted state. In order to achieve satisfactory fixation of vascular bundles in the outer medulla, the fixative has been recommended to be applied directly to the bundles by means of a thin polyethylene catheter (Moffat and Creasey, 1971). The fixative used is 4% glutaraldehyde in 0.1 mol/l cacodylate buffer containing sucrose having a total osmolality of 720 mosmols. The fixative is delivered at body temperature and the pressure used is that produced by the fixative container suspended 2 m above the animal (rat). The fixed area of the tissue is recognized by its brownish coloration.

Renal Medulla

Fixative osmolarity that yields good preservation of other tissues in the kidney does not produce satisfactory fixation of renal medulla, because of the specific osmotic conditions in the latter. The tubular elements and the collecting system of the medulla contain fluid markedly hypertonic to plasma; the interstitial fluid is also hypertonic to plasma. Interstitial cells (rich in lipids) tend to shrink less and swell more than other types of cells in the medulla. Even various levels of the medulla require different fixative composition. Optimal osmolalities of the 3% glutaraldehyde fixative solution for various levels of rat medulla are as follows (Bohman, 1974):

Inner stripe of outer zone	700 mosmols
Outer level of inner zone	1000 mosmols
Middle level of inner zone	1300 mosmols
Papillary tip	1800 mosmols

For each experiment the fixative osmolality should be adapted to the diuretic state of the animal, since the osmolality of medullary tissue fluid fluctuates between diuresis and antidiuresis. A perfusion pressure (200 mmHg) higher than the normal blood pressure of the animal is recommended for the fixation of medulla. Fixative osmolalities (700–1800 mosmols) higher than those used for fixing other parts of the kidney are desirable for medulla. Osmolality can be increased by adding NaCl. The presence of 3% dextran in the fixative solution is considered beneficial for the preservation of interstitial structures and intercellular spaces (Bohman, 1974). However, available information on the real size of these spaces in the living tissue remains incomplete.

Biopsy Specimens

Immersion fixation of biopsy specimens from human kidney results in a poor preservation of tubules and interstitial tissue. Improved fixation of surgically removed human kidney can be obtained by perfusing it through the renal artery within 3–5 min after removal (Møller et al., 1980). The perfusion under pressure control is preceded by a brief rinse with Tyrode solution. The fixative is 2% glutaraldehyde in 0.1 mol/l cacodylate buffer containing 2% dextran (mol. wt. 40,000) and Lissamine green to indicate uniformity of perfusion. The preservation of brush border region of proximal tubule cells is less regular because of the interruption of blood flow prior to perfusion.

EFFECT OF FIXATION ON MEMBRANE FRACTURE

The freeze-fracture technique is an important morphological method for analyzing membrane ultrastructure. Information obtained by this technique has helped us to understand cell structure and function at the molecular level. This is exemplified by splitting and high-resolution replication of lipid bilayers. Freeze-fracture is a sophisticated complex technique, and thus the investigator must have at least a limited understanding of various steps such as ultrarapid freezing, chemical fixation, freezing, cryoprotection, cleaving, etching, replication, cleaning, and mounting. The discussion of all these steps is outside the scope of this volume; the role of chemical fixation is discussed below. For a detailed discussion on the preparation of specimens for freeze-fracture the reader is referred to Rash and Hudson (1979).

Since OsO_4 and $KMnO_4$ alter the hydrophobic properties of the lipids, they cannot be used for freeze-fracture. When OsO_4 is used as a prefixative, a significant reduction occurs in the number of membrane particles (Parish, 1975). Moreover, after OsO_4 fixation, the loss of particles is greater from the E-face than from the P-face. After fixation with OsO_4, most cells cross-fracture, and only a few fracture within the membrane. The normal fracture plane within the membrane depends on the ordered packing of lipids. Osmium tetroxide appears to disrupt the orientation of lipids perpendicular to the plane of the membrane and to cause a loss of membrane fluidity. Another possibility is that OsO_4 strengthens the bonds holding both leaflets together. Yet another possibility is that OsO_4 brings about cross-linking of transmembrane proteins and their associated lipids as well as lateral fixation within the lipid bilayer. It has also been suggested that certain components of the membrane are lost, others are cross-linked, and the hydrophobic region is disrupted by OsO_4 fixation (Parish, 1975). As a consequence of any one or several of these effects, either the cell cross-fractures, or, when it fractures within the membrane, there is a deviation of the usual fracture plane.

Aldehydes also have certain disadvantages, but much less serious than those of OsO_4. Since aldehydes do not fix lipids, they are unable to immobilize rapidly physical movement of membranes. Thus, membrane lipids could be left sufficiently fluid to undergo varied changes during subsequent exposure to cryoprotectants. These changes include membrane fusion, blistering, and phase separation.

Many membranes undergo major changes in partition coefficients (Satir and Satir, 1979) and other membranes develop characteristic artifactual alterations after glutaraldehyde fixation. An example, as stated earlier, is the production of membrane blisters lacking intramembranous particles after fixation with glutaraldehyde (Hasty and Hay, 1978). The majority of plasma membrane fracture faces in glutaraldehyde-fixed chick embryo specimens showed artifactual blisters that ranged in size from 100 to 250 nm and were devoid of particles. Membrane particles were present in the surrounding area of the fracture face (Stolinski et al., 1978).

Similar blisters, observable with the SEM, are produced when specimens have been fixed with glutaraldehyde followed by OsO_4 (Shelton and Mowczko, 1978). Such artifacts were absent when cells were fixed with glutaraldehyde or OsO_4 alone. Simultaneous fixation in a mixture of glutaraldehyde and OsO_4 also prevented blister formation. It is important that such blisters should be distinguished from real blebs, which are rounded cytoplasmic processes containing cytoplasm and intramembranous particles. However, whether the occurrence of clusters of large clear vesicles underlying the plasma membrane in other cell systems is physiological or artifactual has not been determined (Pfenninger, 1979). According to Kalderon and Gilula (1979), the particle-free regions (at least in the myoblast plasma membrane) are not artifacts, but rather important elements in the normal fusion process of membranes.

In spite of the aforementioned inherent limitations of aldehydes, these reagents exert certain beneficial effects on membranes. Fixation with glutaraldehyde minimizes cryoprotectant-induced artifacts as well as temperature-dependent changes in membrane particle distribution. The cryoprotection of fixed specimens can be carried out for relatively long periods of time without excessive autolytic changes. In addition, if necessary, subsequent cryoprotection and freeze-fracturing of the fixed tissue can be delayed. Fixation with glutaraldehyde facilitates cell fracture within the membrane (between the leaflets of the membrane). Unfixed red blood cells may show a ratio of 60:40 between fracture within the membrane and cross-fracture.

After fixation with glutaraldehyde, the loss of particles seems to be the same from both fracture faces. Semiquantitative studies of glutaraldehyde-fixed intact human erythrocytes and ghost membranes indicated a mean reduction of 17% in total particle number on the E- and P-faces of the membrane (Pricam et al., 1977). However, glutaraldehyde-fixed and glycerol-treated cells show more intramembranous particles on both the P- and E-faces than those shown by unfixed and glycerol-treated cells. Intramembranous particles are multimeric associations of proteins, and the cross-linking property of glutaraldehyde probably causes the particles to anchor in the membranes. The use of glycerol alone tends to result in aggregation and reduction in the number of particles.

Glutaraldehyde fixation increases the density of particles; this increase appears to be roughly the same on both fracture faces. The particle density could also be altered by changes in cell volume and surface area. Fixation reduces the diameter of particles; E-face particles are more influenced by fixation than are P-face particles. Generally, E-face particles are larger than those of the P-face. This reduction in size also seems to be related to the cross-linking property of glutaraldehyde. According to Satir and Satir (1979), glutaraldehyde fixation alters specific associations of rosette particles at the E-face in a manner that is different from the alterations seen at the P-face.

From the foregoing discussion it is apparent that fixation with glutaraldehyde alone has the least effect on particle frequency on the membrane fracture faces. For routine studies, a brief fixation with glutaraldehyde is recommended. However, it is recognized that even fixation with glutaraldehyde can cause substantial alterations in the appearance of membrane fracture faces. The position of the fracture plane and the distribution of particles on membrane fracture faces both seem to be altered, depending on the type of fixative and the conditions of fixation used. The possibility of artifacts introduced not only by chemical fixation but also by cryoprotection and/or freezing must be borne in mind when interpreting freeze-fractured membrane faces.

The alternative approach to chemical fixation is ultrarapid freezing to halt cellular activity virtually instantaneously. In essence, this technique involves bringing a specimen rapidly into firm contact with a pure copper surface cooled to $-269°C$ (4 K), thus avoiding the harmful effects of nonvolatile chemical cryoprotectants and chemical fixatives (Heuser et al., 1976, 1979). The rate of freezing depends primarily on the size of the specimen and specimen support and the cooling medium used. Generally the smaller the size of specimen, the greater the rate of freezing. The same is true for the size of the metal specimen support. However, general use of this approach is restricted by the complexity of the equipment and constraints of working with liquid helium.

EFFECT OF FIXATION ON MESOSOMES

The controversy over the presence, absence, or variation of mesosome structure caused by chemical fixation is still unresolved. Mesosomes not only are few in number but also show considerable variations in their morphology in unfixed freeze-etched cells, whereas chemically fixed and freeze-etched cells show a significant increase in the frequency of occurrence of mesosomes (Ghosh and Nanninga, 1976). It has been proposed that the large, complex mesosomes observed more frequently in bacteria fixed with OsO_4 may be artifacts produced by the fixative (Fooke-Achterrath et al., 1974). Silva et al. (1976) proposed further that both the complex and simple mesosomes are artifacts produced by OsO_4 or glutaraldehyde fixation. These authors showed that bacteria prefixed with uranyl acetate did not show mesosomes, a continuous plasma membrane being the only membranous structure present.

Chemical fixation and ionic manipulation cause condensation of the very extensive DNA strand within the bacterial cell. Such condensation is expected to pull on the cytoplasmic membrane, resulting in its inward deformation and invagination. This deformation may be enhanced by direct effects of the fixative on the membrane. These mechanisms may be responsible for mesosome formation in some cases.

On the other hand, because unfixed, negatively stained *Staphylococcus aureus* cells reveal mesosomes, the presence of mesosomes at least in these cells is not a fixation artifact (Kawata et al., 1978). According to Higgins and Daneo-Moore (1974), there is a decrease in the frequency of mesosomes when cells are prefixed with OsO_4 instead of with glutaraldehyde. This reduction is supposedly attributable to autolysis that occurs during OsO_4 fixation. According to these workers, most mesosomes remain invisible in unfixed freeze-etched cells because they cross-fracture; glutaraldehyde and/or OsO_4 fixation promotes mesosome surface fracturing, which results in their visualization.

Mesosomes located near the periphery of the cell are thought to be less adversely affected by chemical fixation than those having intracytoplasmic locations. Therefore, it has been suggested that mesosomes located adjacent to the plasma membrane are more closely representative of the living state of this organelle.

In spite of the controversy over the presence or absence of mesosomes *in vivo,* it is accepted that mesosomes' configuration, number, and location in chemically fixed, thin-sectioned bacteria are dependent on the fixation conditions. Important parameters of OsO_4 fixation that influence mesosome characteristics include concentration of the fixative (Silva, 1971, 1975), temperature at which fixation is carried out, and presence or absence of Ca^{2+} in the fixative solution (Burdett and Rogers, 1970). According to Ghosh (1978, personal communication), the influence of OsO_4 concentration is not so dramatic as that of other parameters.

In OsO_4-fixed cells, mesosomes are usually larger and more frequent than in unfixed cells. The undesirable effect of OsO_4 fixation can be minimized by prefixation with glutaraldehyde, indicating that proteins somehow play an important role in the stability of mesosomes. It has been shown that bacterial cells fixed with glutaraldehyde, acrolein, or formaldehyde prior to OsO_4 fixation exhibit smaller and simpler mesosomes compared to those seen in cells fixed only with OsO_4 (Silva et al., 1976). Low concentrations of $KMnO_4$ act to reveal large, complex mesosomes in Bacillus, whereas high concentrations (0.6–1.0%) show a continuous plasma membrane but no mesosomes (Silva, 1975). Chilling the cells prior to fixation has been suggested to reduce the number of artifactual mesosomes.

EFFECT OF FIXATION ON MICROTUBULES AND MICROFILAMENTS

Microtubules are unstable structures in vivo, and several factors are known to affect their preservation. The factors that may produce disassembly of microtubules include autolysis, low temperatures, tissue preparation trauma, certain types of fixatives and buffers, higher or lower than the optimal osmolarity of fixative solutions, excessive divalent cations (especially Ca^{2+}), improper use of gaseous or local anesthetics or barbiturates, inadequate oxygen, increased pressures, presence of antitubulin agents or mitotic spindle inhibitors, and certain neuronal degenerative conditions.

Various classes of microtubules react differently to fixation and temperature treatments. In general, axonemal microtubules are considered to be stable, whereas cytoplasmic microtubules tend to be labile. Even microtubules of the same type may differ in their stability. This is examplified by the differential stability that is a characteristic feature of spindle microtubules. It has been shown that microtubules attached to the kinetochores of chromosomes seem to be more stable at cold temperatures than are microtubules not attached to chromosomes (Brinkley and Cartwright, 1975).

It is generally believed that microtubules are not recognizable in most tissues when they are fixed with OsO_4 without glutaraldehyde prefixation. However, OsO_4 is not inherently incapable of preserving labile microtubules. Microtubules were preserved, for example, in freshwater green algae (*Eudornia illinoienses*) fixed with 31% OsO_4 in carbon tetrachloride (a nonpolar solvent) either at room temperature or in the cold (Hobbs, 1969), although chloroplast and mitochondrial membranes were not well preserved. Microtubules in the brain were well preserved with unbuffered 4% OsO_4 followed by unbuffered 12% glutaraldehyde (Westrum and Gray, 1977). It is conceivable that the 1–2% aqueous solutions of $Os\dot{O}_4$ commonly in use act too slowly at low temperatures to preserve microtubules. Thus, the rate of penetration by the fixative as well as the actual fixation are both important for the preservation of microtubules. For routine preservation of cytoplasmic microtubules, prefixation with glutaraldehyde at room temperature is recommended. Furthermore, the temperature should be decreased gradually when ice cold OsO_4 follows glutaraldehyde fixation at room temperature; this is accomplished by gradually lowering the temperature to 4°C before postfixation. Also, gradual acclimatization of specimens to OsO_4 is desirable.

Even when microtubules are preserved, their appearance and size are affected by changes in the duration of fixation. This phenomenon has been demonstrated, for instance, in the spindle microtubules of *Allomyces neomoniliformis* (Olson and Heath, 1971). After 25–30 min fixation at 4°C in a mixture of glutaraldehyde (3%) and acrolein (3%), microtubule walls appeared to be ''solid'' and composed of a single layer of osmiophilic material having a thickness of 6.5–7.5 nm. On the other hand, after less than 15 min, or more than 40 min, fixation in the same mixture, the microtubule wall appeared to be composed of two layers of globular subunits with a total diameter of 8 nm. Similar shape and dimension of the microtubules were obtained after fixation in glutaraldehyde followed by OsO_4. According to Olson and Heath (1971), glutaraldehyde and OsO_4 fixation tends to condense the subunits, thus enhancing the appearance of ''hollow wall,'' whereas the intermediate duration in glutaraldehyde–acrolein tends to loosen their structure, thus obscuring the dimer configuration.

In many types of cells a large percentage of microtubule population is of short length (i.e., < 1 μm long). Such observations have been made in root cells (Hardham and Gunning, 1978), a fungus (Heath and Heath, 1978), and heart muscle (Goldstein and Entman, 1979). A twofold increase in the percentage of long microtubules (> 2 μm) in developing radish root hairs has been reported following fixation with 1% glutaraldehyde in the PM buffer (instead of in phosphate buffer) (see p. 48) (Seagull and Heath, 1980). It is not clear whether or not the longer profiles of microtubules reflect *in vivo* condition, because this lengthening could result either from enhanced preservation or from artifactually increased polymerization, or both. It has been suggested that the presence of guanosine triphosphate (GT) in the PM buffer maintains a pool of GT-charged, polymerization-competent tubulin subunits, thereby facilitating the synthesis and/or maintenance of microtubules during fixation (Seagull and Heath, 1980).

However, it should be noted that glutaraldehyde-treated tubulin may polymerize into microtubules. During early stages of fixation, glutaraldehyde could yield a high level of GT-charged, assembly-competent subunits, which could artifactually increase the length of preexisting short microtubules (Seagull and Heath, 1980). It is known that the rate of microtubule polymerization is quite rapid, and thus fixation time is sufficiently long to permit such polymerization.

In vitro and *in vivo* studies of the effect of fixatives on actin filaments (Szamier *et al.*, 1975; Kuczmarski and Rosenbaum, 1976; Pollard, 1976) have raised doubts whether these structures also occur in living cells. Studies of cultured cells indicate that the organizations of microtubular and microfilamentous systems are identical whether or not they are postfixed with OsO_4 after glutaraldehyde fixation (Temmink and Spiele, 1978). Microfilaments consisting of actin alone, or of other proteins, are preserved by most fixation procedures. These structures are, however, most easily studied in unembedded critical-point-dried cells. It should be noted that unembedded (unsupported) whole mounts of cells are vulnerable to deformation under the electron beam.

EFFECT OF FIXATION ON MITOCHONDRIA

Mitochondria undergo configurational or conformational changes concomitant with the energy cycle. It is well documented that mitochondria change their configuration in different energy states and that a particular configurational state of the mitochondrion corresponds to its functional state. For instance, the cristae undergo a transition from the orthodox (scalloped) to the aggregated configuration (flattened sheets) in the mitochondria of adrenal cortex when its function changes from steroidogenesis to ATP synthesis (Allman *et al.*, 1970). The cristae of mitochondria in radiant-heat receptors of *Agkistrodon piscivorus* undergo configurational changes that parallel the sensory activity of the receptor organ (Meszler and Gennaro, 1970). Sarcoplasmic reticulum preparations are also known to undergo configurational changes under various conditions (Deamer and Baskin, 1969). It is known that fixation procedures modify morphological appearance of cell organelles. Therefore, to study the correlation of structure and function or biochemical activity of a cell organelle, it is necessary to determine specific fixation conditions that preserve a specific unstable, dynamic functional state of an organelle. The following discussion deals only with mitochondria.

It is appropriate to present a brief description of various types of cristal configurations prior to discussing methods that stabilize and preserve these types. There are two nonenergized configurations of the cristae: orthodox (nonphosphorylated state) and aggregated (condensed) (Fig. 9.3). By definition, the matrix space is maximally expanded in the orthodox configuration and maximally contracted

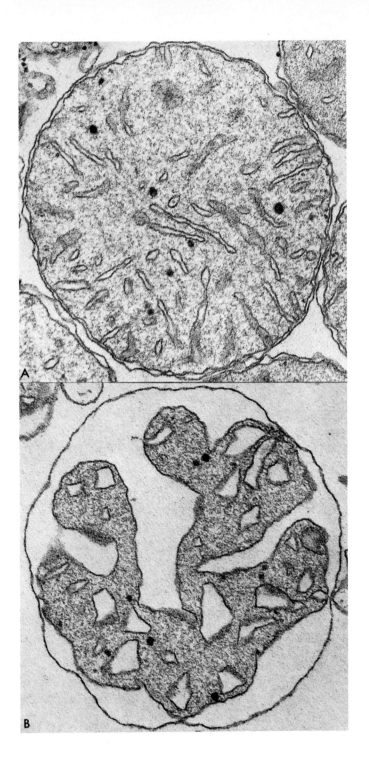

(~50%) in the aggregated configuration. As the intracristal space expands, the neighboring cristae come closer to one another, hence the term "aggregation." Thus, in the aggregated configuration, the intracristal space is maximal and the matrix space is minimal, and just the opposite is true in the orthodox configuration. Since the latter configuration almost always appears when mitochondria *in situ* are processed according to standard fixation procedures, the term "orthodox" is applied for its description. At least two variations of the energized-twisted configurations (Fig. 9.4) have been identified: tubular and zigzag in isolated mitochondria and in *in situ* mitochondria, respectively. As for the tubular mode, heart mitochondria show highly regular tubules that are not much wider than orthodox cristae, whereas liver mitochondria exhibit irregular tubules that are many times wider than orthodox cristae (Williams *et al.*, 1970).

Mitochondria with relatively little matrix protein (e.g., in heart muscle) exhibit the most regular configurational changes, whereas those with abundant matrix protein (e.g., in liver) show a more complex pattern of configurational changes. The overall volume of mitochondria remains more or less constant in spite of the changes in the interior of mitochondria in the orthodox and in the aggregated configurations. However, according to Muscatello *et al.* (1972), the overall volume of *in situ* phosphorylating mitochondria was reduced.

As stated earlier, when *in situ* mitochondria under anaerobic conditions are processed according to standard fixation procedures, the cristae almost always assume the orthodox configuration. The reason for this uniformity of appearance of mitochondria is that excised tissue specimens are processed in such a way that anaerobiosis is inevitable by the time the fixative reaches the specimen. The exhaustion of the available oxygen supply occurs within 5-20 sec. Therefore, it is almost impossible to preserve the energized configuration of *in situ* mitochondria as well as in those isolated prior to fixation.

When the tissue is fixed with glutaraldehyde only, almost no intracristal space appears. On the other hand, if the tissue is postfixed with OsO_4, the intracristal space becomes substantial. The intracristal space is therefore related to some extent to fixation with OsO_4. Intracristal spaces may also expand in connection with the isolation of mitochondria. These conditions seem to favor accumulation of water in the interior of the cristae.

The osmotic pressure of the suspending medium is important in the determination of cristal form in the nonenergized configuration as well as in the energized-twisted configuration. More specifically, the orthodox to aggregated transition in

Fig. 9.3. A, the orthodox conformation of a rat liver mitochondrion. The matrix shows a decrease in electron density and the volume of the matrix increases by 100% during the condensed-to-orthodox mechanochemical ultrastructural transformation. High energy state. B, the aggregated (condensed) conformation. The volume of the matrix is ~50% of the total mitochondrial volume and shows maximal electron density. The inner membrane is irregularly oriented and contacts between inner and outer membranes are present. Low energy state. (Hackenbrock, 1968.)

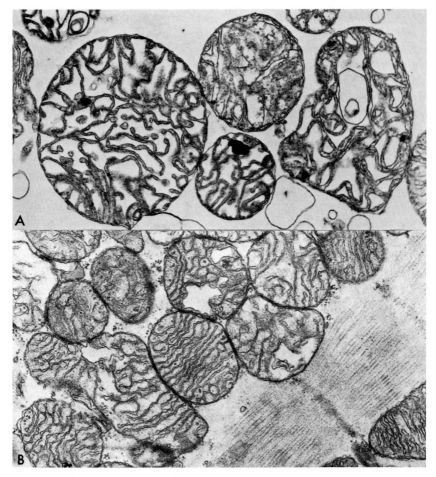

Fig. 9.4. A, beef heart tissue exposed to Krebs–Ringer phosphate solution prior to fixation with 0.05% glutaraldehyde. The energized-twisted (zigzag) configuration of cristal membrane is clearly seen. ×26,100. B, Beef heart mitochondria *in situ* exposed to Krebs–Ringer phosphate solution for 10 min and then fixed with 0.01% glutaraldehyde followed by postfixation with 2% OsO₄. The energized-twisted (tubular) configuration of cristae is apparent. ×27,000. (T. Wakabayashi, unpublished.)

isolated mitochondria under anaerobic conditions is independent of the energizing cycle and is related to the osmotic pressure exerted by sucrose in the suspending medium.

When sucrose is added to the suspension medium, the osmotic gradient between the intracristal and matrix spaces is raised to the point at which the former space is forced to expand. Phosphorylation also changes the permeability properties of the inner surface membrane, which is reflected by extrusion of water from the matrix space. Under such conditions the orthodox configuration changes to the aggregated mode. On the other hand, under standard conditions the osmotic pressure of the cytoplasm bathing the mitochondrion is insufficient to compel the expansion of the cristal membranes, resulting in the appearance of the orthodox mode.

Special methods of fixation are available for the study of configurational changes in *in situ* as well as in *in vitro* mitochondria (Williams *et al.*, 1970). Not only energized transitions (nonenergized to energized) but also nonenergized transitions (orthodox to aggregated) can be captured by proper fixation conditions. In fact, some of the cristal configurations can be captured within a single type of mitochondrion by using specific fixation conditions. These conditions are (1) small size of the tissue specimen (less than 1 mm in any dimension), (2) use of a modified Krebs–Ringer phosphate solution as the suspending medium, (3) aerobic conditions to encourage the energized state, (4) anaerobic conditions to facilitate the nonenergized state, and (5) rapid fixation.

The orthodox configuration can be achieved either by imposing anaerobic conditions or by adding an uncoupler (carbonyl cyanide *m*-chlorophenyl hydrazone) under aerobic conditions, whereas the aggregated configuration can be obtained in the sucrose medium. Energizing conditions include endogenous plus added substrate and inorganic phosphate in the presence of oxygen. These conditions lead to a mixture of energized and energized-twisted configurations. Inorganic phosphate induces the transition from the energized to the energized-twisted configuration. When the osmotic pressure of the suspending medium is increased by the addition of sucrose, the energized-twisted configuration of mitochondria *in situ* becomes tubular rather than zigzag.

The energized configuration of mitochondria *in vivo* can be preserved when fixation is accomplished before the oxygen supply is exhausted. This is carried out by direct addition of glutaraldehyde to the mitochondrial suspension prior to sedimentation. Fixation with glutaraldehyde appears to be completed within 1–2 sec even at 4°C (Wakabayashi, 1972). In the fixation of mitochondria in suspension, the rate-limiting factor seems to be the speed of mixing the fixative with suspension rather than the speed of interaction between the fixative and mitochondria.

A mitochondrial suspension containing 4 mg of mitochondria in 4 ml of reaction mixture is mixed with an equal volume of 1% glutaraldehyde in 0.05 mol/l K-cacodylate buffer (pH 7.4) containing 0.25 mol/l sucrose, and allowed to incubate for ~1 hr at room temperature. Glutaraldehyde concentration as low as 0.01% can be used with satisfactory results. The mitochondria are sedimented for ~30 min and the pellet is washed in the buffer containing 0.25 mol/l sucrose and postfixed with 0.5% OsO_4 (final concentration) for ~1 hr. Before dehydration, the pellet is exposed to 1% uranyl acetate in 25% ethanol.

Tissue blocks (less than 1 mm^3) of heart are first exposed to 0.25 mol/l sucrose solution and then fixed with 2% glutaraldehyde followed by 1% OsO$_4$. After such a treatment mitochondria show the aggregated configuration, but the quality of preservation of myofibrils is poor.

If OsO$_4$ is used as a primary fixative, concentrations higher or lower than the optimal concentration range of this fixative tend to transform *in vitro* mitochondria into the orthodox configuration. However, after *in vitro* or *in situ* mitochondria have been prefixed with glutaraldehyde, OsO$_4$ concentration does not affect mitochondrial configuration. *In vitro* mitochondria are also changed to the orthodox configuration with glutaraldehyde only when its concentration is lower than the optimal range. For example, beef heart mitochondria fixed *in vitro* with 0.01% glutaraldehyde show the aggregated configuration. If glutaraldehyde concentration is lower than 0.01%, the aggregated state is changed to the orthodox configuration (Wakabayashi, 1972). Conventional fixation of *in situ* mitochondria with 2% glutaraldehyde results in the orthodox configuration, and the lowest concentration of glutaraldehyde that results in the orthodox configuration is ~0.05%. When tissue blocks are fixed with glutaraldehyde concentrations lower than 0.05% in the presence of 0.25 mol/l sucrose, mitochondria show the aggregated configuration. Similar configuration is obtained when tissue blocks are first exposed to 0.25 mol/l sucrose solution and then fixed with 2% glutaraldehyde. When tissue blocks are exposed to Krebs–Ringer phosphate solution prior to fixation with glutaraldehyde, the twisted configuration is induced.

It is apparent that caution is warranted in the interpretation of structural modifications that can be influenced by the fixation procedure, and reproducibility of the results with a given fixative does not invalidate this cautionary statement. Basic criteria for assessment of the results must be established prior to correlating the structure with biochemical activity.

EFFECT OF FIXATION ON MYELIN

The repeat period of myelin membranes in tissue sections is smaller than that found in the living state. Second major change occurring during preparatory procedures is the appearance of a single electron dense line representing the cytoplasmic boundary. X-ray diffraction studies indicate that these changes are induced by preparative treatments. Both glutaraldehyde and OsO$_4$ seem to induce the formation of small, localized regions of disordered stacks of membrane bilayers, that probably are extracted during dehydration (Kirchner and Hollingshead, 1981). These regions are devoid of intramembrane particles, and are thought to consist mainly of lipid. Such changes can occur even at very low concentrations (0.025%) of glutaraldehyde.

The center-to-center separation of myelin bilayers is decreased by about 0.9

nm after fixation with either glutaraldehyde or OsO_4. However, external and cytoplasmic surfaces of membranes react differently to a fixative. Membranes fixed with either glutaraldehyde or OsO_4 retain a capacity for swelling at external membrane boundries. Thus, glutaraldehyde or OsO_4 alone does not stabilize external membrane surfaces. Stable cross-linking between these surfaces is accomplished when fixation with glutaraldehyde is followed by OsO_4. Stabilization of membrane arrays can be achieved by individual fixatives in the presence of DMSO. Generally myelin configuration is altered much less by glutaraldehyde than that by OsO_4. Dehydration causes further alterations within the myelin arrays. Embedding medium also affects the final structural image of the myelin. The preceding effects of fixatives also apply to other types of membranes (also see p. 284).

EFFECT OF FIXATION ON PLANT VIRUS

Plant viruses *in situ* are difficult to preserve when tissue specimens are fixed according to standard procedures. The primary reason for this difficulty is that viruses are rich in nucleic acids that do not interact with glutaraldehyde except at elevated temperatures, and even then are not rendered completely insoluble. Fixation of nucleoproteins occurs through the cross-linking of protein components in which are trapped nucleic acids. The same reason can explain the difficulty in visualizing the DNA in prokaryotes; very little, if any, protein is directly associated with DNA in most bacterial cells.

Only a few viruses have been found in intracellular crystalline arrays, although purified viruses can often be crystallized. Small isometric virus particles are difficult to identify in the tissue section unless they occur in crystalline aggregates or in regions of the cell where they are not mistaken for ribosomes. Viral crystallization *in situ* has been induced by wilting the tissue prior to fixation (Milne, 1967). Plasmolysis of the tissue has also been employed to allow more control over conditions (Hatta and Mathews, 1976). Ultracentrifugation of virus-infected tissue is still another method used to induce viral crystallization (Favali and Conti, 1971).

Pretreatment of leaf tissue with sucrose induces crystallization of tobacco yellow mosaic virus (Hatta, 1976). This is accomplished by floating 3 mm diameter disks of leaf tissue on 40% sucrose solution for 1 hr at room temperature prior to fixation. Thin sections should be cut from the tissue near the edges of the disk. Leaf age, duration of treatment, and sucrose concentration affect crystallization.

A very brief fixation in glutaraldehyde brought about the retention of some of the crystalline inclusions of cowpea mosaic virus (Langenberg and Schroeder, 1973). Brome mosaic virus (Paliwal, 1970) and other plant viruses (Edwardson *et al.*, 1966) were shown to occur in orderly arrays in the tissue fixed with chromic acid–OsO_4. Extremely labile tobacco mosaic virus inclusions were preserved by using a mixture of chromic acid–formaldehyde (Langenberg and Schroeder, 1973) (Fig. 3.7). However, chromic acid–OsO_4 is a less satisfactory general-purpose fixative than glutaraldehyde followed by OsO_4. Aggregates of several types of viruses can be preserved by chilling the tissue (at 5°C for ~2–5 hr) prior to fixation in glutaraldehyde at 4°C (Langenberg, 1979). This treatment is thought to gel tissue proteins, including those associated with the virus particle. In the noncooled tissue, intracellular redistribution of viruses probably occurs immediately after cell death.

EFFECT OF FIXATION ON PLASMA MEMBRANE

Although available knowledge of the extent of alterations introduced in the morphology of cells by fixation is insufficient, membrane reorganization is regarded as one of the major known changes that is induced even under the best conditions of fixation. Fixation causes rapid changes in the membrane permeability and in the osmotic behavior of the cell. Cell membranes of various cell types may differ in the manner in which they behave after fixation, but all plasma membranes undergo changes in osmotic equilibria after fixation.

Osmium tetroxide destroys the differential permeability of the plasma membrane, making it permeable to low-molecular-weight substances, including vital dyes. This fixative causes loss of fluidity of membranes, and disturbs lipid chains and fixed relative positions of amphipathic proteins (Jost *et al.*, 1973). The electrical resistance of cell membranes is lowered after fixation especially when OsO_4 is used; membrane-bound OsO_4 is in itself conductive. Any injury to the plasma membrane will usually lower its electrical resistance.

Fixation with glutaraldehyde, on the other hand, does not result in the loss of relative impermeability of plasma membranes to the majority of the ions. One of the exceptions is K^+, which shows considerable leakage during fixation of human red blood cells in glutaraldehyde or formaldehyde (Vassar *et al.*, 1972). It is likely that even after the tissue has been "fixed" with glutaraldehyde, reactive groups persist in membranes. Considerable evidence indicates that glutaraldehyde-fixed cells remain osmotically active (Fahimi and Drochmans, 1965b; Elbers, 1966; Millonig and Marinozzi, 1968; Bohman and Maunsbach, 1970; Bone and Denton, 1971; Bone and Ryan, 1972).

According to Jard *et al.* (1966) and Grantham *et al.* (1971), glutaraldehyde- or formaldehyde-fixed cells remain partly impermeable to ions such as Na^+. Heller *et al.* (1971) indicate that glutaraldehyde-fixed specimens are able to undergo volume changes.

However, some evidence indicates that fixation with glutaraldehyde may render some cells osmotically inactive (Carstensen *et al.*, 1971; Penttilä *et al.*, 1974). According to Pentillä *et al.* (1974), within a few minutes after fixation cellular ATP disappears, normal intracellular and extracellular values for ions are disrupted, and active transport mechanism for ions and the passive permeability properties of the cell membrane are disrupted. Nevertheless, although fixatives in the concentrations commonly used markedly increase the permeability of plasma membranes (this is especially true in the case of single cells), cell membranes of glutaraldehyde-fixed cells maintain partial selective permeability. This is the reason that the osmolarity of solutions used after fixation with glutaraldehyde may affect the cell structure.

In certain types of tissues fixed with OsO_4 only, plasma membranes that are apposed along their external surfaces break down into chains of vesicles or tubules; examples of such tissues are toad spinal ganglia (Rosenbluth, 1963), ciliary epithelium (Tormey, 1964), proximal tubule cells (Maunsbach, 1966), and stria vascularis (Santos-Sacchi, 1978). Other parallel membrane systems in certain tissues also respond in the same manner to OsO_4. Examples are smooth endoplasmic reticulum in testicular interstitial cells (Christensen and Fawcett, 1961), gastric parietal cells (Ito, 1961), T-system in fish muscle fiber (Franzini-Armstrong and Porter, 1964), and nerve sheaths (Doggenweiler and Heuser, 1967). Primary fixation with glutaraldehyde does not produce such artifacts. However, there are exceptions; for instance, prefixation with glutaraldehyde did not prevent the vesiculation in prawn nerve sheaths produced by postfixation in OsO_4. These differences apparently reflect the variations in the reaction of membranes of different organisms and tissues to similar fixation methods.

Several factors are assumed to be responsible for the structural reorganization of membranes and the artifactual production of vesicles. Although the exact mechanism responsible for the breakdown of plasma membranes by OsO_4 is not known, osmium probably damages certain sensitive regions of the membrane. It should be noted that the appearance of rows of vesicles does not seem to represent accurately the three-dimensional nature of the artifact. Primary osmication possibly causes selective regions of the membranes to coalesce, forming a sheet of tubules, which appear as a row of vesicles when viewed in cross section. The difference in the rates of penetration by glutaraldehyde and OsO_4 does not seem to be responsible for the aforementioned superiority of glutaraldehyde over OsO_4. It is possible that OsO_4 creates physical instability in these membranes by denaturing their surface proteins. There is another possibility, as suggested by Robertson (1959), that it is only the innermost of the three layers of a plasma membrane that is stabilized by OsO_4, whereas the two external layers break down into chains of vesicles. Other factors include the direct chemical action of the fixative and the osmotic forces that expand or contract the cytoplasm in some manner that in turn causes disruption and recombination of membranes

(Doggenweiler and Heuser, 1967). Finally, the reactive groups of adjacent plasma membranes presumably interact, and this results in the reorganization of membranes during subsequent processing of the tissue. For example, organic solvents, unpolymerized embedding media, and temperature may serve to catalyze these interactions.

However, not all paired plasma membranes (e.g., mesaxon of unmyelinated fibers or neuron–satellite cell interface) of normal tissues exhibit this type of breakdown under any standard conditions of fixation (Rosenbluth, 1963). Furthermore, not only does fixation with different reagents result in varied dimensions of the cell membrane but also fixation with a specific reagent can result in varied dimensions of different membranes in a cell. It is obvious that not all the factors that determine the phenomenon of structural reorganization of membranes are known as yet.

The structural validity of the typical trilaminar appearance of membranes obtained after double fixation, i.e., glutaraldehyde followed by OsO_4, has been questioned (Korn, 1969; Sjöstrand and Barajas, 1968, 1970; McMillan and Luftig, 1975). The reasons for this skepticism are these: (1) proteins that are basic components of the membrane are extensively solubilized or denatured by OsO_4 (Trump and Ericsson, 1965; McMillan and Luftig, 1973) and (2) a massive reorganization of the lipid bilayer is caused by OsO_4 (Jost and Griffith, 1973). These and other studies suggest that a more accurate view of membrane ultrastructure may be obtained by using glutaraldehyde alone, provided lipid extraction is either prevented or reduced significantly during dehydration and embedding. This can be partly accomplished by employing special dehydration and embedding chemicals such as ethylene glycol and Vestopal, respectively (Sjöstrand and Barajas, 1968; Sjöstrand, 1976). Glutaraldehyde (5%) alone has been used to reveal the location of membrane proteins that are removed or obscured by OsO_4 fixation (McMillan and Luftig, 1973, 1975). However, in these studies no attempt was made to use special procedures to prevent or minimize lipid extraction during dehydration and embedding. It may be worthwhile to note that conventional dehydration solvents cause damage by removing the lipid components of the membrane, whereas detergents exert their effect by changing the physical properties of lipids, which, in turn, causes a detachment of the integral proteins from the membrane.

Ethylene glycol is used because it is the mildest denaturing organic solvent; globular protein molecules have been shown to tolerate up to 90% ethylene glycol without exhibiting conformational changes. Vestopal is employed because it is the most polar embedding medium that allows the obtaining of ultrathin sections. Nevertheless, even these procedures do not prevent altogether the extraction of lipids.

Alternatively, membrane damage can be minimized by using exceedingly low concentrations of OsO_4 during the postfixation. Such low concentrations fix certain lipids but do not significantly denature membrane proteins. Osmium tetroxide solution having a concentration of 0.025% has been used as a postfixative for human erythrocyte ghosts (McMillan and Luftig, 1975). This concentration reportedly preserved almost as much of the phospholipids as did 1.0% OsO_4. With this procedure the average diameter of membranes is 9–20 nm, depending on the source of membrane, instead of the conventional 7–9 nm. According to Hasty and Hay (1978), a mixture of glutaraldehyde and OsO_4 improved the preservation of plasma membrane and eliminated many of the artifacts that might result with primary aldehyde fixation. However, Kalderon and Gilula (1979) suggested that this mixture disrupted membrane structure and produced "lesions" in the cell.

Another approach is to use urea or urealike reagents (3.3–5 mol/l), which form polymerization compounds with glutaraldehyde (Pease and Peterson, 1972; Heckman and Barrnett, 1973). The use of these compounds avoids conventional dehydration. However, the use of these compounds at high concentrations should be avoided, for they could cause extensive denaturation of membrane proteins. A procedure that prevents extraction and alteration of membrane proteins and lipids has yet to be developed (also see p. 282).

EFFECT OF FIXATION ON STAINING

Various fixatives affect differently the staining characteristics of the tissue. Lowering of the isoelectric point of proteins as a consequence of fixation is certain to alter the physical nature of proteins, with a resulting increase or decrease in the number of sites available for stain uptake. Furthermore, since bound or free fixative in the tissue may react with the stain, the staining results may vary, depending on the type and concentration of the fixative present and on whether the staining precedes or follows the embedding. The possible reaction between the fixative and the stain may affect the quality of fixation as well as that of the staining. For this reason every effort should be made to wash out the unbound fixative prior to staining.

A fixative may modify the physicochemical state of the tissue and thereby change the reactivity of tissue components with the stain. It is conceivable, for instance, that binding sites may become blocked through reaction with active groups in the fixative, thus leading to decreased staining. Conversely, a fixative may introduce reactive groups (e.g., aldehyde), resulting in increased staining. A few examples follow.

Some cell components may show either an increased or decreased affinity for lead stain, depending on the type of fixative used. This is true for glycogen (Minio *et al.*, 1966) and cytoplasmic granules (polysaccharide) (Robertson *et al.*, 1975). After lead staining, the appearance of glycogen in the tissue fixed with formaldehyde or OsO_4 is similar; after similar staining, however, ribosomes and nucleoli are more dense in formaldehyde-fixed than OsO_4-fixed tissue (Daems and Persijn, 1963). Increased staining of glycogen by lead is obtained in the tissue fixed with glutaraldehyde followed by OsO_4 compared with that in the tissue fixed by either of the fixatives alone.

Bismuth is another stain whose reaction is influenced by the fixative. The type of aldehyde used affects the binding of bismuth to specific reactive groups in the cell. Formaldehyde fixation permits bismuth binding to some amine and amidine groups and probably to primary phosphates (Brown and Locke, 1978). Thus, fixation with the monoaldehyde facilitates staining of nucleolus with bismuth by exposing amino groups. On the other hand, glutaraldehyde cross-links amines and thus blocks the staining of nucleolus with bismuth. After glutaraldehyde or formaldehyde fixation, perichromatin and interchromatin staining with bismuth is not affected.

The use of stains is limited by the type of fixation employed prior to staining. It is known, for example, that lipids will not be visible when the tissue is fixed with aldehydes, since they are extracted during dehydration. Double fixation with an aldehyde followed by OsO_4 before dehydration, on the other hand, does not prevent the staining of at least some of the lipids. Osmium tetroxide, of course, acts as not only a stain for unsaturated lipids but also a general fixative. However, fixation with aldehydes alone generally improves the specificity of most stains because the aldehyde molecules themselves do not stain, and thus the specificity of the subsequently used stains is not obscured.

Although the best overall staining is achieved by using specimens fixed with glutaraldehyde followed by OsO_4, this staining lacks specificity. Fixation or postfixation with OsO_4 encourages nonspecific staining because this reagent itself is a general stain. This general staining is partly the result of the presence of bound osmium itself, which is deposited in the tissue during treatment with OsO_4, and occurs partly because of the enhancement of staining by other heavy metals in the presence of bound osmium. For example, uranyl, lead, and indium salts stain somewhat specifically nucleic acid-containing structures, provided that bound osmium is absent. Similarly, reduction of silver yields specific staining of mucosubstances in the absence of bound osmium; bound osmium can cause nonspecific deposition of silver (e.g., silver methenamine reaction).

Particular caution is warranted in the use of OsO_4 in cytochemical studies of carbohydrate-containing macromolecules, for carbohydrate-containing moieties of these macromolecules are vulnerable to the destructive influence of this metal. Osmication may destroy or solubilize some of these macromolecules; the inherent osmiophilia of certain other cellular structures may confuse the identification of these macromolecules, and bound osmium may alter the effect of stains on them. A case in point is acinar secretory granules in rat submandibular glands. In aldehyde-fixed tissue the granules stain readily with methods for carbohydrates, whereas the granules of postosmicated specimens do not (Simson *et al.*, 1978). In other words, the majority of the cytochemically demonstrable carbohydrate of the acinar granule is destroyed by OsO_4.

If the specimen must be treated with OsO_4, the staining specificity can be somewhat enhanced by employing dilute solutions of strong oxidizing agents such as hydrogen peroxide or periodic acid, which remove the lower oxides of osmium from the fixed tissue. However, this treatment may damage the cell structure, introducing its own untraceable artifacts.

Although carbohydrate-containing structures are studied in the tissue fixed by aldehydes, even these reagents may cause false staining. Both glutaraldehyde and acrolein (but not formaldehyde) can introduce free aldehyde groups into the specimen. These groups can destroy the specificity of the periodic acid oxidation methods for detecting muco- and glucoproteins and certain other mucosubstances. Specimens prepared for periodic acid oxidation, Feulgen hydrolysis, and the silver–methenamine treatment should not be fixed with acrolein, for this extremely reactive monoaldehyde rapidly introduces free aldehyde groups into the tissue. The first two methods detect aldehyde groups and the third method detects sulfur-rich proteins as well as aldehyde groups.

As pointed out earlier, even fixation with glutaraldehyde can make the staining results confusing, for it may introduce Schiff-positive groups into the tissue. Glutaraldehyde is known to cause false-positive Feulgen staining of DNA in spite of vigorous washing after fixation. Free aldehyde groups introduced by this dialdehyde are responsible for this false staining. These free aldehydes can be blocked by reducing them to alcohol groups with freshly prepared 0.5% sodium borohydride ($NaBH_4$) in 1% NaH_2PO_4 for 1 hr at room temperature (Lillie and Pizzolato, 1972; Kasten and Lala, 1975).

En bloc staining and fixation of tissue specimens with uranyl acetate prior to or during dehydration may cause extraction and/or clumping of glycogen to various degrees. Homogeneous distribution of glycogen is preserved by conventional fixation in the absence of *en bloc* treatment with uranyl salt. Although the exact mechanism responsible for this damaging effect by uranyl acetate is not known, it has been suggested that a cellular component responsible for preventing the clumping and extraction of glycogen is separated or destroyed by *en bloc* treatment with uranyl acetate (Rybicka, 1977). Since glycogen is present usually in combination with protein, this protective cell component may be proteinaceous. The presence of such a protein(s) may also explain the binding of lead to glycogen complex, since glycogen alone is not known to form coordination complexes with metal ions. Thus, in the absence or separation of this protein from glycogen by uranyl acetate, clumps of glycogen remain unstained by lead (Fig. 9.5).

Primary fixation or postfixation with OsO_4 should be avoided when exploring the cytochemistry of cellular proteins and glycoproteins. Some of these sub-

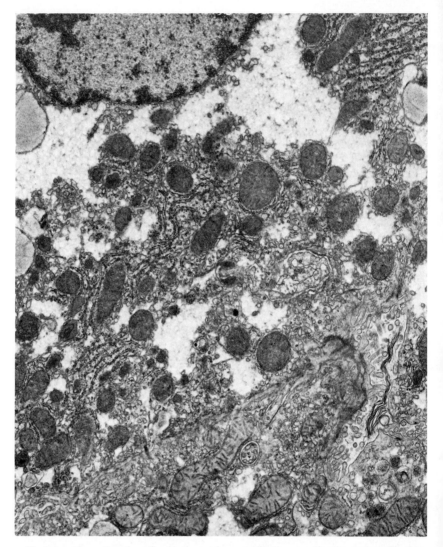

Fig. 9.5. Mouse liver tissue fixed by immersion according to conventional procedure. The tissue block was stained *en bloc* with uranyl acetate. Note unstained clumps of glycogen; other cell organelles show satisfactory preservation and staining.

stances in the aldehyde-fixed tissue can be stained readily and somewhat selectively with metal-containing stains. A few examples follow. Iron hematoxylin has been employed for staining granules in the aldehyde-fixed enteroendocrine cells at the subcellular level (Nichols *et al.*, 1974). Iron is deposited on spherical and nonspherical granules measuring 200–500 nm in diameter. This staining is probably due to the presence of basic protein residues in the granule, which is capable of binding ferric ions in the absence of bound osmium. Iron hematoxylin has also proved effective in the staining of elastic fibers fixed with glutaraldehyde only (Brissie *et al.*, 1974). Although the nature of the group in elastic fibers interacting with the stain is not known, desmosine, isodesmosine, and tetracarboxylic or tetraamino acid residues in elastic fibers (West *et al.*, 1966) may account for this staining. The staining of aldehyde-fixed parotide secretory granules with iron hematoxylin has been attributed to the presence of a protein similar to elastin, or perhaps to a metalloprotein (Simson *et al.*, 1974).

In conclusion, although all fixation methods cause false positive staining to various degrees and almost all stains are essentially nonspecific, by a careful choice of fixative and stain, the specificity of most stains can be improved. A mixture of glutaraldehyde and formaldehyde is the best fixative for achieving "specificity" in the staining of most cellular structures. The best fixation method for each of the stains has been presented by Hayat (1975) and Lewis and Knight (1977).

Buffer or vehicle solutions used prior to, during, or after fixation influence the staining of many cellular substances. The composition of the vehicle used either during primary or during secondary fixation may influence the morphology and cytochemistry of the specimen. The type of buffer used plays an important part, for instance, in the preservation of substructure of secretory granules. Filaments in acinar secretory granules aggregate into fibrils in aldehyde fixatives containing phosphate buffer or calcium; vesicles and filaments are prominent with fixatives containing collidine buffer; and with cacodylate buffer filaments are the major granule constituent in females, whereas condensed fibrils are common in the male acinar granules (Simson *et al.*, 1978).

Another example is the effect of different buffers on the staining of nodes of Ranvier in mammalian peripheral nerves. Ferric ions bind to the extracellular nodal gap substance when nerves have been pretreated with phosphate buffer, whereas these ions bind to the cytoplasmic side of the nodal axolemma after pretreatment with cacodylate or veronal acetate buffer (Quick and Waxman, 1977). The former binding occurs presumably because the phosphate, which is loosely bound to the gap substance during fixation, complexes with ferric ions, resulting in the formation of iron phosphate precipitate. These and numerous other examples testify to the dependence of cytochemical phenomena on the chemical nature of buffers employed prior to the staining reactions.

EFFECT OF FIXATION ON SYNAPTIC VESICLES

When aldehydes are used as primary fixatives, each axon shows a mixed synaptic vesicle population. Two types of spherical vesicles, large and small, can be recognized. In addition, depending on the axon type and the conditions of fixation, flat or elongated synaptic vesicles can be distinguished. It should be noted that each axon type, under given fixation conditions, exhibits a constant and characteristic vesicle ratio. These different types of vesicles have been found in both vertebrates and invertebrates.

It is postulated that round synaptic vesicles are typical within excitatory nerve terminals and elongated vesicles within inhibitory nerves, in crayfish leg muscle nerves (Atwood *et al.*, 1972). This difference in shape after fixation in the same fixative solution indicates that the transmitter content of the round and elongated vesicles causes them to react differently to the same fixative. Synaptic vesicles in the inhibitory nerve terminals are definitely smaller than those in the excitatory nerve terminals. Thus, although the difference in size and in shape has been used to classify synaptic vesicles, the difference in size seems to be more important in characterizing them. Sufficient evidence is available indicating that the size and shape of synaptic vesicles change in a complicated manner according to the preparatory method used. Therefore, knowledge of the fixation procedure is vital when one is characterizing the vesicle size, shape, and content of particular synapses.

When aldehydes are used as primary fixatives, the ratio between different types of vesicles is dependent primarily on the osmotic effect of the buffer. However, the vesicle ratio is not directly proportional to the osmotic pressure. The concentration of the fixative (within certain limits) has little effect on the shape or size of the synaptic vesicles. This observation suggests that aldehydes pass fairly freely through the membrane of the cell and the vesicles so as to produce little osmotic effect on them. According to one study, if the phosphate buffer has an osmolality of 200–800 mosmols, it does not affect the shape/size ratio of vesicles (Valdivia, 1971). However, if the osmolality is lower than 200 mosmols, the ratio of round vesicles is increased and the ratio of flat vesicles is decreased, whereas the reverse is true if the buffer has an osmolality higher than 800 mosmols.

The osmolarity of both the vehicle in the primary fixative solution and the washing solution affects the terminals as well as the vesicles. An increase in the vehicle osmolarity from isotonicity to hypertonicity produced severe shrinkage of the nerve terminals in the crayfish stretch receptor organs, but had far less effect on the vesicles (Tisdale and Nakajima, 1976). The osmolarity of the washing solution, on the other hand, had little effect on the terminals, but altered the vesicle structure. This evidence indicates that after glutaraldehyde fixation the differential permeability of synaptic vesicle membranes is not destroyed, and that they remain sensitive therefore to osmotic stress, whereas nerve terminals become less sensitive after glutaraldehyde fixation. It is deemed likely that both the vesicles and the nerve terminals become shrunken and flattened after hyperosmotic primary fixation, and that the vesicles are reversed to more spherical shapes during the isosmotic wash, while the nerve terminals remain shrunken (Tisdale and Nakajima, 1976). A few studies illustrating the effects of fixation on synaptic vesicles *in situ* and *in vitro* are reviewed next.

It has been shown that large, dense core vesicles purified from bovine splenic nerve (sympathetic) readily swell or shrink in response to changes in osmolarity of sucrose, potassium phosphate buffer, or both (Thureson-Klein *et al.*, 1975). The control vesicles in a medium at 330 mosmols had a mean diameter of 70–75 nm; they shrank to 60 nm and swelled to 105 nm after a 50% increase and decrease, respectively, in osmolarity. The core of the vesicle responded to changes in osmolarity, but did so less markedly than the vesicle membrane. Vesicles *in vitro* are also affected by incubation conditions and postmortem delay prior to chilling the nerves.

In the motor endplates (cholinergic) from rat diaphragm muscle fixed with aldehydes in an isotonic buffer, fewer than 10% of the vesicles were elongated (Korneliussen, 1972). The number of elongated vesicles increased in the presence of a hypertonic buffer, whereas the number decreased in a hypotonic buffer. Such effects can be produced by either phosphate buffer or cacodylate buffer.

Since primary fixation with OsO_4 results in a population containing fewer than 10% elongated vesicles, irrespective of the vehicle or buffer osmolarity, it is concluded that such conditions cause vesicles to lose their osmotic sensitivity, as well as their ability to change shape (Korneliussen, 1972). According to Korneliussen, varying the pH from 6.0–8.0, or varying the temperature from 4 to 37°C, did not significantly change the percentage of elongated profiles; however, more elongated profiles were seen when fixation was carried out at 47°C. The slight increase in the number of elongated profiles at pH 6.0 may be due to increased osmolarity caused by the addition of HCl. According to A. W. Clark (1976), the classic OsO_4–veronal fixative yields the most accurate representation of osmotically stressed synaptic terminals.

It should be noted that amine-storing vesicles in various cell and tissue types respond differently to a fixation method, probably because of intrinsic chemical differences among vesicles. In addition to catecholamine, other substances may be present in the vesicles, and these might precipitate in contact with various fixatives. Additional studies are needed to clarify the effects of fixation on vesicles in various cell types and at different developmental stages.

It was stated earlier that anesthetics influence synaptic morphology, possibly because of membrane recycling. There may be variations in the ultrastructural characteristics of central synapses even when specimens are obtained from unanesthetized animals. The variations seem to depend on whether the animal is killed by stunning across the back of the neck or by the cannulation technique (Devon and Jones, 1979). The area and perimeter of the presynaptic terminal are greater, and the percentage of terminals displaying vesicles attachment sites increase, following cannulation as opposed to stunning. The rapid fixation with cannulation more rapidly interrupts metabolic pathways and vesicular mechanisms than does fixation following stunning. This is reflected in the greater percentage of terminals having vesicles attachment sites in the cannulated specimens.

EFFECT OF FIXATION ON TIGHT JUNCTIONS

Most of our knowledge of tight junctions has been obtained by freeze-fracturing tissue specimens that were fixed with glutaraldehyde and cryoprotected with glycerol. The structure of the tight junction varies depending on whether fresh or aldehyde-fixed tissues are used. The type of aldehyde used also affects the appearance of the tight junction. Conventional fixation of tight junction membranes of liver and small intestine with glutaraldehyde (2–3%) produces a structure containing fibrils on the inner face and a few particles on the outer faces (van Deurs and Luft, 1979). With decreasing concentrations of glutaraldehyde (0.1%) and durations of fixation, the percentage of fibrils on the inner face decreases and the number of particles simultaneously increases on the outer face of the membrane.

After fixation with 1.5% formaldehyde, the appearance of the junctions is similar to that seen after a mild fixation with gutaraldehyde. Unfixed (but glycerol-cryoprotected) specimens also show some fibrils. In fresh frozen specimens (without cryoprotection), fibrils are absent and a large population of particles is present on the outer face. It is possible that cross-linking of proteins with glutaraldehyde fixation may result in the formation of fibrils and binding of the proteins to the inner face of the membrane. In order to minimize these artifacts, only dilute solutions of glutaraldehyde or formaldehyde should be used for fixation for brief periods of time. It should be noted that even conventional fixation with glutaraldehyde shows tight junction strands of particulate form in certain tissues (Simionescu *et al.*, 1976). Other possible artifacts include loss of particles and plastic deformation of proteins during the process of freeze-fracturing.

EFFECT OF FIXATION ON SPECIMENS FOR X-RAY MICROANALYSIS

The total or partial loss of one or more elements from the tissue during chemical fixation is inevitable. Although the loss of elements during processing has seldom been accurately measured, the loss rates can be exceptionally high. The degree of loss of most elements is controlled by the type of specimen and fixative used (Vassar *et al.*, 1972; Penttilä *et al.*, 1974), the type of buffer, the concentration of the fixative (Vassar *et al.*, 1972; Penttilä *et al.*, 1974; Howell and Tyhurst, 1976), the processing temperature (Penttilä *et al.*, 1974), and the specimen size (Hall *et al.*, 1974; Harvey, 1980).

Significant differences in the effects of glutaraldehyde, formaldehyde, and acetaldehyde on the leakage of potassium from human erythrocytes have been indicated (Vassar *et al.*, 1972). More calcium was lost during fixation with OsO_4 than with glutaraldehyde (Howell and Tyhurst, 1976). According to Penttilä *et al.* (1974), glutaraldehyde and OsO_4 caused qualitatively similar but kinetically different changes in the magnesium and potassium permeability of Ehrlich ascites tumor cells. Fixation generally results in the loss of K^+ and a net uptake of Na^+.

Although elements are lost at each step in the preparative regime, the major loss occurs during the initial fixation stage. For example, the rate of loss of potassium from Ehrlich ascites tumor cells during the first 1 min of fixation in 3% glutaraldehyde was on the order of 15% and 40% of the original potassium content at 0°C and 40°C, respectively (Morgan, 1980). Apparently major losses of elements occur within the period of time necessary for satisfactory preservation of the fine structure. Even a drastic reduction in the duration of fixation is not a practical solution for the retention of labile cell components.

The substantial initial loss of elements is due to leaching of unbound ionic phases followed by the relatively slow loss of less readily exchangeable compartments. This phenomenon is exemplified by the studies carried out by De-Filippis and Pallaghy (1975). They showed that the loss of [65]zinc and [203]mercury labels from leaf tissues during processing was significantly reduced when the radioisotopes present in the free spaces of the tissue were previously exchanged with unlabeled ZnCl and $HgCl_2$ solutions. Morgan (1980) aptly points out that care must be exercised in the interpretation of sequential loss data, because a negligible recorded loss during a late stage of processing does not unequivocally exonerate that stage of processing. It may simply mean that compartments soluble in that particular medium have already been extracted.

The chemical state of binding of an element with cell molecules and macromolecules influences its solubility. Biologically important electrolytes are more prone to leaching than are structural elements such as phosphorus and sulfur. Available microprobe data suggest that nuclear phosphorus (located primarily in nucleoproteins) is not so easily extracted as cytoplasmic phosphorus (located primarily in membrane lipids) (Andersen, 1967; Morgan *et al.*, 1975).

Conventional fixation is satisfactory for preparing specimens for the localization of ions in water-insoluble compounds. It was shown that over 90% of the total lead present in the shoots of *Potamogeton* was retained during fixation and dehydration (Sharpe and Denny, 1976). However, the majority of low molecular-weight ionic compounds are water soluble. Such ions can be studied by using precipitation techniques. By using these techniques, water-soluble and usually electron transparent ions are precipitated during fixation. For example, phosphates and Cl^- can be precipitated with $Pb(NO_3)_2$ and $AgNO_3$, respectively (Läuchli, 1972; Stelzer *et al.*, 1978). Sodium, Ca^{2+}, and Mg^{2+} can be precipitated with $KSb(OH)_6$ (Simson and Spicer, 1975). These techniques have been discussed in detail by Hayat (1975).

In addition to loss of elements, their redistribution (displacement) is a major problem encountered during chemical fixation. If the movement of elements occurs across distances too small to be resolved by the instrument used, this movement to new loci can be tolerated. Also, if the movement is confined within the volume of even relatively major structures (whole organelles or cells) that are being analyzed, it can be tolerated. Since the effects of fluid phases of fixation on cellular materials is extremely complex, it is almost impossible to pinpoint and measure elemental redistribution. Moreover, little is known about the accurate distribution pattern *in vivo* for comparison purposes. Loss of elements during flotation in the knife trough and staining of specimen sections cannot be disregarded. Coleman and Terepka (1974) proposed at least four criteria that may prove useful in identifying redistribution artifacts.

The most effective method for minimizing alterations in the tissue elements is rapid freezing. The rate of freezing should exceed the self-diffusion rate for water, so that there is minimal ionic movement and ice crystal formation. For a detailed discussion of the freezing methods for X-ray microanalysis, the reader is referred to Marshall (1980).

EFFECT OF FIXATION ON VASCULAR ENDOTHELIUM

It is thought that damage to vascular endothelium is involved in atherosclerosis. The earliest changes appear to be in cell shape, orientation, and integrity. Since preparatory procedures also cause changes in cell structures, it is necessary to distinguish between the two types of damage. Correct evaluation of early stages of pathological development depends on the knowledge of artifacts introduced during specimen processing. The following discussion pertains to fixation and prefixation procedures used for scanning electron microscopy.

The first step is to anesthetize the animal, preferably with sodium pentobarbital (Nembutal). It is administered intravenously without smooth muscle relaxant and has been used in animals as diverse as pigeons and calves. Alternatively, Xylocaine can be used. It is thought that vessels treated with this reagent are found in a dilated state without spasm, although whether such a state exists *in vivo* or not is uncertain. Following this, the blood must be removed from the vessel prior to actual fixation; otherwise artifactually deposited plasma proteins and blood cells will confuse the endothelial surface and create the appearance of thrombosis. Furthermore, if the blood is not removed, it may partly or completely block the vessels, thereby changing the perfusion pressure. This is a critical step because unfixed endothelial cells are extremely sensitive to the washing process. These cells are affected by variations in chemical composition, pH, osmolarity, and temperature of washing and holding solutions, as well as by the duration of washing. For example, fewer hairlike projections were found after rinsing at 20°C than at 40°C (Edanaga, 1974), and larger craters were found with a brief prewash with saline solution than without (Gregorius and Rand, 1975).

Two general approaches have been used to remove blood: (1) perfusion of vessel with a washing solution prior to perfusion with glutaraldehyde solution; and (2) *very* rapid perfusion of an isolated segment of vessel with the fixative. The perfusion must be sufficiently rapid to remove the blood before fixation is appreciable. Also, this perfusion must be carried out at physiological pressure (reproducing the normal physiological pressure of the vessels).

Washing solutions that have been used include physiological NaCl (Edanaga, 1974), 0.9% NaCl containing heparin (Swinehart *et al.*, 1976), Ringer's solution containing heparin (Buss *et al.*, 1976), Hank's balanced salt solution (Fishman *et al.*, 1975), phosphate-buffered saline (PBS) (Katora and Hollis, 1976), phosphate-buffered Tyrode solution (Svendsen and Jorgensen, 1978), and cacodylate buffer. The surface of the endothelium can be rapidly and drastically altered by contact with washing solutions. A consensus on the ideal washing solution and whether or not it should be at all does not exist.

Tyrode's solution adjusted to pH 7.4 with bicarbonate is preferred as the washing solution, as the vehicle for glutaraldehyde, and in all other steps in processing. The osmolality is 290 mosmols without glutaraldehyde. For perfusion fixation, 1% gluraldehyde is satisfactory. The time required to complete the fixation depends on the vessel type. Vessels with thin walls are fixed in a shorter time than those with a thick wall. An average duration is 1 hr to overnight fixation under pressure. The vessels can be tied off under pressure and immersed in a container of fixative after 5–15 min of perfusion. After this initial fixation, the vessels may be opened and tied (or pinned) out flat for further fixation in 2.5% glutaraldehyde (few hours to overnight). The length of the vessel can be maintained by pinning the isolated segment to tong depressors to their *in situ* length; pinning is done beyond the tie so as not to lose pressure. Small segments of the specimen are rinsed, postfixed in 1% OsO_4, and dehydrated for critical point drying from liquid CO_2 (Hayat, 1978).

Artifacts

Collapse and circular contraction of the vessel wall are unavoidable when immersion fixation is used. Also, wrinkling of normal longitudinal folds, crenation of red blood cells, aggregation of platelets, and the presence of artifactual projections (ridges) and undulations in the vascular endothelium are artifacts. The so-called cross-bridges linking two individual cells are doubtless artifacts. A comparison of Figs. 13.4A and B demonstrates that bridgelike structures on the specimen surface are artifacts produced by immersion fixation.

The production of such artifacts is expected because fixation by immersion is carried out at or near zero hydrostatic pressure, whereas under *in vivo* conditions the aorta (rabbit) experiences systolic as well as diastolic pressures of 125 and 80 mmHg, respectively. Most of these artifacts are produced by retraction of the wall before or during fixation. Also, the media of the artery contracts more than the intima and the internal elastic lamina.

However, it is possible that longitudinal ridges do form if medium and large arteries undergo significant contraction *in vivo*. This area remains to be clarified. The endothelial cell membrane is capable of considerable response to environmental influences. The response often takes the form of pseudopods, blebs, and craters. It is necessary to obtain control vessels that are free of such features before one can attribute these features to either experimental conditions or stimulation of the endothelium during fixation.

Most of the artifacts mentioned can be eliminated by fixing the vessels in the physiological (i.e., distended) state by vascular perfusion at an appropriate perfusion pressure. Both excessive and low perfusion pressure cause severe endothelial damage. Also, if the pressure is not maintained for a sufficiently long time, cellular damage is likely.

Various anesthetics may produce different side effects. In certain cases the physiological conditions may vary in different parts of the same vessel system. Important differences between species cannot be disregarded. The method of sacrificing the animal is also important. Hollweg and Buss (1980) have discussed in detail the problems encountered during the preparation of blood vessels for scanning electron microscopy.

Changes
in Specimen Volume

Tissue specimens undergo transient changes in volume in each of the preparatory steps (Fig. 10.1). An increase or decrease in specimen weight also occurs during processing. A knowledge of these changes is important because any increase or decrease in specimen size is an indication of the quality of preservation of cellular structures. Information on volume changes is particularly important in quantitative morphometric studies. Prior to discussing changes in specimen volume caused by swelling or shrinkage during and after fixation, a general description of the mechanism of cellular swelling is in order.

Plasma membrane is permeable to water and to all the solutes that contribute significantly to the osmolarity of the extracellular fluids. It is known that intracellular proteins exert an oncotic pressure that attracts extracellular water and causes tissue swelling. But swelling does not occur *in vivo*. The reason may be that although water is capable of entering the cells rapidly, it is actively extruded with equal rapidity from the cells with the help of the Na–K pump in the presence of energy. Thus, cells *in vivo* maintain a higher intracellular osmotic pressure (compared with extracellular) as well as their normal volume.

It is known that inhibition of respiration without a change in the osmolarity will result in tissue swelling. At the time of fixation, respiration is blocked immediately, and thus the tissue might swell unless the fixation solution is hypertonic. This is one of the reasons that slightly hypertonic solutions are desirable during fixation. Although tissue swelling is solely the result of net

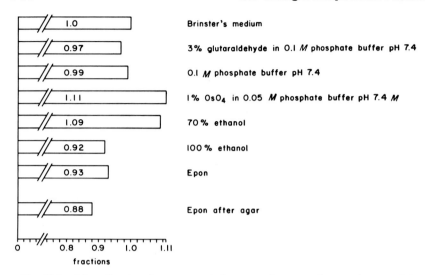

Fig. 10.1. Mean diameter of mouse ova during successive steps of preparation regarded as optimal (expressed as fraction of their initial mean diameter in Brinster's medium). The specimen showed the lowest volume in 100% ethanol and remained unchanged after embedding in Epon. (Konwiński *et al.*, 1974.)

movements of water, swelling is associated with an uptake of ions followed by water. Cell interiors are electrically negative relative to the cell exterior. As a result, Cl⁻ and other negatively charged small ions are maintained in relatively low concentrations in the cell interior. Intracellular negativity repels the negative charge of Cl⁻ from outside.

Energy and specific ion pumping systems are required to maintain low concentrations of Na⁺ in the cell interior. When the tissue is separated from the body for immersion fixation, the respiration is inhibited and the energy required for maintaining specific ion concentrations is diminished. Sodium diffuses into the cells and cannot be extruded, and simultaneously, K⁺ leaks out of the cells. As a result of Na⁺ accumulation inside the cells and loss of K⁺, the negative potential of the plasma membrane is reduced, so that Cl⁻ then can enter the cells. There may be a greater entry of Na⁺ than loss of K⁺ by the cells. The net gain of solutes (especially of osmotically active solutes) by the cells results in a gain of water and swelling of the tissue. Autolysis also results in increased intracellular osmotic pressure, with resultant tissue swelling. The extent of its contribution to the problem of swelling is not quite clear.

The degree of shrinkage or swelling is influenced by the osmolarity and the species of ions present in the fixation vehicle, the concentration of the fixative (too high or too low), and the type of specimen under study. The final size of the specimen is also influenced by the conditions of dehydration and embedding. A change in cell size is primarily a result of changes in the amount of water. A change in the size of subcellular components may also contribute to the volume of the specimen, although it cannot be assumed that subcellular components necessarily undergo the same transient changes in volume as the whole cell. It is likely that the membranes of subcellular components have values of permeability different from those of the plasma membrane. Consequently, the plasma membrane and membranes of cell organelles would be expected to respond differently to the same osmotic gradient. Furthermore, the osmotic gradient faced by the subcellular component is different from that encountered by the plasma membrane (Eisenberg and Mobley, 1975). Therefore, the correct volume of subcellular components cannot be deduced from measurement of the whole cell.

As mentioned earlier, the initial changes in cell volume are primarily the function of vehicle osmolarity during fixation with aldehydes. Cell volume may decrease only slightly in this step, provided a vehicle with optimal osmolarity is used. The following discussion explains why moderate hypertonicity is considered to be optimal osmolarity for most tissue specimens. Hypotonic fixative solutions (regardless of aldehyde concentration) tend to cause swelling of the tissue. The swelling is easily detectable in mitochondria, for example. On the other hand, hypertonic vehicles bring about shrinkage. Although the method to avoid swelling or shrinkage of the tissue would seem to be the use of isotonic solutions, in practice "isotonic" fixative solutions have proved less than satisfactory for intact tissues. Indeed, the total osmotic pressure exerted by the solute particles in a fixative solution is not always related to the swelling or shrinkage of cells. Some of the reasons for this phenomenon are considered in the following discussion.

Since the available information on extracellular osmolarity is insufficient (and particularly so on intracellular osmolarity), it is not easy to prepare isotonic fixative solutions. (Isotonic fixatives for cells grown in culture are easy to prepare.) The problem is further compounded because, in some cases, the intracellular osmolarity is considerably different from that of the plasma. For instance, kidney cortex cells are isotonic to 0.23 mol/l NaCl, whereas plasma osmolar values range from 0.31 to 0.34 mol/l. Furthermore, various cell types in a tissue (e.g., kidney) possess different intracellular osmotic pressures.

Fixative solutions made isotonic to blood cells are not necessarily isotonic for other tissues; this is particularly true for tissues that have been excised. The osmolarity of an excised tissue that has been removed from the body and deprived of its blood supply, perhaps brought to a lower temperature than its normal physiological range, and probably physically damaged, may not be the same as that *in situ*. Moreover, during fixation new ionized groups are formed in the tissue.

Although membrane permeability undergoes a rapid change during fixation, intracellular colloid–osmotic equilibrium is maintained. Thus, fixative osmolarity needs to be adjusted to this equilibrium, which may differ from the isosmotic surrounding of the cell. In addition, when tissue blocks are fixed with an isotonic fixative solution, cell membranes are penetrated relatively slowly by the fixative. This is undesirable, since the rate and quality of fixation (including swelling) are partly dependent on the speed of membrane penetration by the fixative. Slower penetration of isotonic fixatives may cause anoxia, resulting from the continuation of metabolism, and this in turn, may cause swelling.

The use of isotonic vehicle with OsO_4 is also questionable, since immediately after exposure to such a fixative, cell membranes become freely permeable to ions and small molecules. The complete destruction of osmotic activity of the reflecting cells of teleost scale after fixation with OsO_4 has been documented (Bone and Denton, 1971). It is known that membranes lose their relative impermeability, for example, to Na^+ when exposed to OsO_4. Chemical analyses have shown that fixation with OsO_4 leads to a considerable uptake of Ca^{2+} by the tissue; this Ca^{2+} is localized in part at the plasma membrane. Even ferritin molecules are reported to pass through the membrane after fixation in OsO_4 (Tormey, 1965). Since fixation with OsO_4 causes an inhibition of active transport and a change in passive permeability, even isotonic fixatives incorporating osmium could cause distortions.

Several studies indicate that so-called isotonic fixative solutions fail to prevent swelling (e.g., Millonig and Marinozzi, 1968). Marked swelling of the cellular processes in all levels of the cortex of rat brain fixed with isosmotic glutaraldehyde solution (280 mosmols) has been demonstrated (Sumi, 1969). Furthermore, certain tissues show swelling even in isotonic solutions if their respiration is inhibited, either by the addition of metabolic inhibitors or by lowering of the temperature to 0–4°C (Trump and Ericsson, 1965). This swelling is probably related to the inactivation of active Na–K transport at the cell membrane, causing a gain in Na^+, Cl^-, and water and loss of K^+ from the cell.

Slightly hypertonic fixatives yield superior results in morphometric studies. The effectiveness of such solutions is exemplified by a constant surface-to-volume ratio maintained by cell components under hypertonic conditions of fixation. Mitochondria show a constant surface-to-volume ratio in specimens fixed with hypertonic solutions (550 mosmols for liver fixed by perfusion) at low temperatures. Hypotonic solutions yield a lower ratio irrespective of the temperature used during fixation. A decrease of this ratio indicates a change in the form of mitochondria to a more spherical shape, since spheres have the smallest surface-to-volume ratio among the geometric solids.

Both the theoretical and practical reasons presented earlier indicate that "isosmotic" fixative solutions often fail to prevent swelling of cells in a tissue block. On the other hand, slightly hypertonic solutions effectively prevent swelling. In this connection, not only the effects of osmolarity but also the effects of specific ions employed should be considered. It is known that osmotic pressure

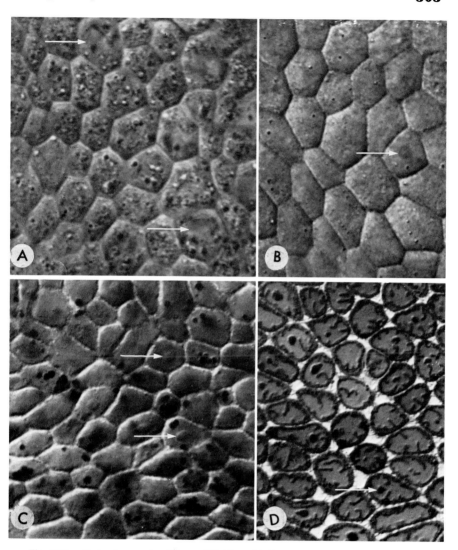

Fig. 10.2. Photomicrographs of frog gallbladder epithelial cells obtained by Nomarski differen-tial interference transmitted-light microscopy. Morphological changes, especially progressive shrink-age and production of intercellular spaces, occurring in various steps of processing can be seen. Maximum shrinkage seems to occur during infiltration and embedding. A, no lateral intercellular spaces are seen in Ringer's solution; B, the geometry of cells is not significantly altered after OsO_4 fixation, although darkening and slight swelling of cells are visible; C, some shrinkage and intercellu-lar spaces are visible after dehydration and embedding in Epon (whole mount); D, intercellular spaces and shrinkage are clearly seen in a semithick (1 μm) section. Focusing was done at the nuclear level; arrows indicate nucleoli. ×1250. (Frederiksen and Rostgaard, 1974.)

exerted by a given concentration of protein decreases with increasing concentrations of neutral salts in the medium. Thus, at least theoretically, it is possible to prevent swelling by placing the tissue in hypertonic salt solutions.

However, isotonic fixative solutions will diminish the injury to the surface layers of a specimen caused by osmotic imbalance. Furthermore, in the case of delicate and sensitive specimens such as lung, embryo, and tissue culture cells (into which penetration is not a problem), fixation under isotonic conditions is desirable, provided that appropriate ions are added to the fixative mixture. A relatively small departure from isotonicity can cause severe damage to the fine structure of lung tissue (Gil and Weibel, 1968). Red blood cells are known to be best preserved under isotonic conditions. The presence of an electrolyte (NaCl) at isotonic concentration is considered imperative to preserve the shape of these cells during fixation.

A brief comment on how to determine whether tissue has been fixed in a solution that is excessively hypertonic is in order. The adverse effects of excessive hypertonicity include artifactual widening of extracellular spaces and separation of cytoplasm from the nuclear membrane. Generally, cell organelles not open to the interstitial fluid (e.g., mitochondria and sarcoplasmic reticulum) shrink in volume almost in proportion to the decrease in cell volume in hypertonic solutions, whereas organelles open to the interstitial fluid (e.g., transverse tubules) either swell or remain unchanged. In myocardial cells of *Limulus*, for example, transverse tubules swelled, whereas sarcoplasmic reticulum shrank in hypertonic fixatives; the sarcoplasmic reticulum shrank from an average luminal diameter of 46 nm to 32 nm (Sperelakis, 1971; Forbes *et al.*, 1972).

In conclusion, generally tissue specimens shrink during prefixation with aldehydes, followed by swelling in the buffer wash and during postfixation with OsO_4. The increase in cell volume is influenced by the duration, temperature, osmolarity, and concentration of OsO_4. This increase is more pronounced in 2% OsO_4, and the addition of sucrose enhances the rate of increase. During the initial stages of dehydration, the cell volume increases slightly, followed by a drastic reduction in higher concentrations of ethanol-propylene oxide or acetone. A procedure of slow and *continuous* change of concentration of dehydration agents is superior to the conventional step procedure of dehydration. The cell volume

---

Fig. 10.3. Frog gallbladder epithelium. A, dehydration and infiltration by the conventional step procedure. The cross-sectional area of the epithelial cell is small, the microplicae are bent against the lateral cell surface, and only a rim of cytoplasm is found at nuclear level. B, Dehydration and infiltration by the *continuous* procedure. Note radiating pattern of microplicae (arrows), which interdigitate with microplicae of neighboring cells. The cells exhibit a light ectoplasmic zone and a relatively wide zone of cytoplasm rich in mitochondria at the level of the nucleus. This procedure causes less shrinkage and yields superior ultrastructural preservation compared with that obtained by the step procedure. ×11,750. (Rostgaard and Tranum-Jensen, 1980.)

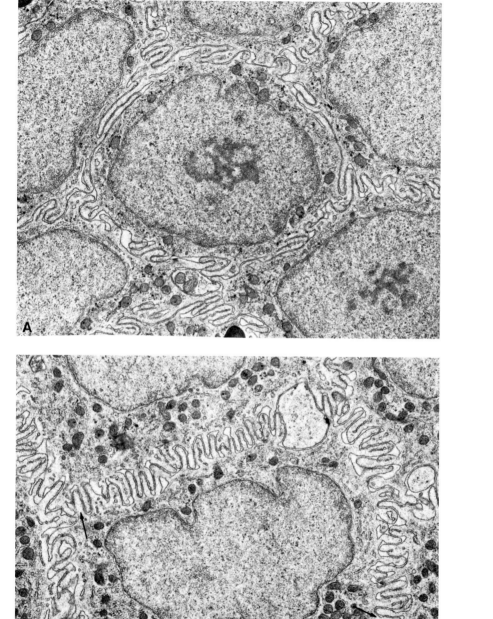

also decreases during *en bloc* treatment with uranyl acetate. Cultured smooth muscle cells shrank 13–20% after treatment with uranyl acetate (Lee *et al.,* 1979, 1980); tissue blocks tend to shrink less.

A slight swelling occurs (at least in cultured cells) during infiltration, and some shrinkage takes place during polymerization of the resin (see Fig. 10.2). Figure 10.2 clearly shows that intercellular spaces between mucosal cells are not present in the living frog gallbladder. The presence of widened intercellular spaces as a consequence of cell shrinkage is due to preparatory procedures. This shrinkage and other undesirable effects of a conventional step procedure of infiltration are considerably minimized by a slow and *continuous* infiltration (Fig. 10.3) (Rostgaard and Tranum-Jensen, 1980). The previously mentioned sequence of volume changes is true in most tissues; however, there are exceptions. Various apparatuses for continuous fixation, dehydration, and infiltration have been designed (Hanzon, 1959; Evans, 1966; Turler and Frei, 1969; Noel, 1972; Brown, 1977; Cragg, 1980; Rostgaard and Tranum-Jensen, 1980).

It is apparent that changes in the tissue volume caused by the primary fixative solution bear little relationship to the final size after subsequent osmication, dehydration, and embedding. However, the net amount of final shrinkage or swelling depends in part on the ability of the primary fixative solution to maintain the original volume.

The ultimate goal in monitoring the dimensional alterations of specimens in various steps of preparatory procedures is to determine the correlation between the size of the cells embedded in the resin and their size *in vivo*. Apparently, the ability to maintain the original *in vivo* size throughout the preparatory procedure is the criterion of choice for testing the reliability of the preparatory procedure and the quality of specimen preservation.

Postmortem Changes

After death or removal of tissue from the body, anoxic and spontaneous postmortem effects cause alterations in ultrastructure and enzyme activity. Ion displacement occurs as soon as tissue is separated from the body. Autolytic effects involve alterations in the shape, size, electron density, and location of cell components. An increase (initial) (e.g., ornithine transcarbamylase and acid ribonuclease) or decrease in enzyme activities also occurs. Certain cellular materials may appear or disappear because of autolysis; examples of such materials are lipids, glycogen, and lysosomes. Lipid droplets, for instance, accumulate mainly in the proximal convoluted cells in canine kidneys as early as 30 min after perfusion (Magnusson *et al.*, 1975). Since cellular death and ultrastructural changes frequently accompany pathological processes, an understanding of the normal postmortem changes is necessary in order to distinguish between cellular alterations caused by disease and those resulting from normal postmortem. For example, in assessing lesions in muscle at autopsy, postmortem alterations should be taken into account.

When organs are initially fixed *in situ* by vascular perfusion, it is necessary to complete the fixation by immersion of small pieces of tissue in the same fixative for an additional period of time. Even with this procedure, the core of the tissue block may show autolytic (postmortem) alterations, because fixatives, in comparison with vehicles, are slow to penetrate the core. Thus, the vehicle may cause damage before the fixation has effectively begun. It is apparent that the nature and extent of autolytic changes need to be appreciated.

Factors implicated in postmortem changes include ischemia, postmortem glycolysis, and proteolysis. Autophagocytosis may be accompanied by activation or increased amounts of digestive enzymes. The number and volume of lysosomes increased in rat hepatocytes up to 1 hr during autolysis (Riede *et al.*, 1976). Similarly, increased autophagocytosis was observed in rat pancreatic acinar cells during autolysis *in vitro* (Jones and Trump, 1975). On the other hand, recent studies by Nevalainen and Anttinen (1977) indicated no increased autophagic (lysosomal) activity in similar cells during autolysis. It has been reported that autophagocytosis requires some energy and production of ATP (Arstila *et al.*, 1974). An autophagic phenomenon was not increased in the latter study, probably because there was a rapid decline of cellular energy in the tissue specimens kept at room temperature. Trypsin activation was also not observed in this study during *in vitro* autolysis. The previously mentioned discrepancy indicates that additional studies are needed to understand the role of various internal and external factors in the onset of autolytic changes.

It is likely that structural alterations in cell components following death may render organelles physiologically ineffective long before any significant changes in certain biochemical constituents are detected. An absence of the appearance of any structural damage after death does not therefore necessarily mean the absence of physiological or biochemical damage. In other words, morphological integrity of cell organelles may not have a direct bearing on their functional activity. For example, loss of enzyme activity is not directly related to morphological disruption of cell organelles. It should be noted, however, that metabolic efficiency of an organelle may be impaired immediately after death, not because of a loss of any particular cellular material but because of disorientation of enzymes. This is exemplified by mitochondria, which lose their efficiency as respiratory units rapidly after death, although the amount of their enzymes remains almost constant.

Although autolysis usually causes a gradual loss of the highly ordered structural organization, various tissue types differ in their response to autolysis. Striated muscle and liver tissues are examples of tissue types that resist postmortem changes. Muscles kept in moist atmosphere at 6°C for as long as 24 hr exhibit good preservation of myofibrils and ribosomes and fairly good preservation of nuclear membrane, sarcoplasmic reticulum, and sarcolemma. Heart muscle is more susceptible than is skeletal muscle to postmortem effects.

The morphology of hair cells in crista ampullaris is unusually well preserved, even when they are kept up to 30 min at room temperature; after 90 min, severe autolytic changes occur (Anniko and Bagger-Sjöbäck, 1977). However, most inner ear structures show high sensitivity to postmortem effects. It has been shown that 15 min after the animal is sacrificed, significant ultrastructural alterations occur within the organ of Corti (Wersäll *et al.*, 1965). Relatively rapid postmortem cell breakdown occurs in the brain, lung, and pancreatic tissues.

Scanning electron microscopy studies showed that in the small intestine of a neonatal calf, swelling of villus tips and denudation of a few villi occurred 10 min after severing the carotid arteries (Pearson and Logan, 1978). Such alterations in villous surface morphology were more pronounced in the proximal and middle small intestine than in the distal small intestine. However, similar artifacts were not seen in specimens removed under general anesthesia.

Postmortem changes in mouse lung tissue, left in the cadaver, begin to show up after 1 hr (Pattle *et al.*, 1974). According to Bachofen *et al.* (1975), a satisfactory ultrastructural evaluation of human lung tissue can be carried out when the autopsy is performed within ~8 hr, except that the pericycle seems to be edematous. Apparently, this cell type is especially sensitive to hypoxia. Postmortem changes in the lung tissue have been described by Finlay-Jones and Papadimitriou (1973), Pattle *et al.* (1974), and Bachofen *et al.* (1975). Effects of prolonged autolysis at 4°C on the ultrastructure and enzyme activity in human CNS have been studied by Mann *et al.* (1978).

Not only different tissue types but different cell types in a tissue or an organ may show different degrees of susceptibility to autolysis. A well-known example is kidney tissue. Another example is crista ampullaris, where secretory epithelium is less affected by autolysis than is sensory epithelium after the same time interval (Anniko and Bagger-Sjöbäck, 1977). Various cell organelles also differ in their response to autolysis, and this response seems to be related to their chemical composition. Proteins are most resistant, whereas lipoproteins (membranes) are least resistant. It has been suggested that the presence of greater enzyme activities on the membrane surfaces may be one of the reasons for this vulnerability (Lim and Solomon, 1976). Zymogen granules in pancreatic acinar cells remained morphologically intact for at least 3 hr at room temperature in a moist atmosphere after death, but cell membranes showed evidence of damage (Nevalainen and Anttinen, 1977). Peripheral zones of zymogen granules may become less electron dense than their cores, thus indicating onset of autolysis.

In striated muscle stored at 40°C, the Z and I bands degenerate in less than 4 hr, whereas the A band is resistant to change up to 24 hr. Generally, actin shows greater susceptibility to proteolytic digestion during postmortem than does myosin. Probably, bonds between the Z band and I band are weakened during postmortem. Degradation of Z bands in the muscle during postmortem is thought to be caused by endogenous proteolytic enzyme. Catheptic enzyme has also been implicated in the degradation of myofibrils. A proteolytic enzyme secreted by a bacterium (*Pseudomonas fragi*) may also be responsible for the degradation of Z bands (Tarrant *et al.*, 1973).

Transverse tubules are relatively resistant to postmortem effects. In striated muscle, glycogen either disappears or decreases rapidly, depending on the speed of fixation. During initial ischemia, glycogen may be decreased to 60%, but after the arrival of the aldehyde fixative, glycogen remains more or less constant.

Effects of prolonged autolysis on the fine structure and enzyme activity in rat myocardium at room temperature have been studied (Penttilä and Ahonen, 1976).

It is known that autolysis is accompanied by degenerative nuclear changes. The earliest and most prominent nuclear change is clumping and condensation of chromatin along the nuclear membrane and around the nucleolus followed by its complete disappearance. A uniform distribution of chromatin within the nucleus is closer to *in vivo* conditions. Autolytic degradation involves the loss of fibrillar appearance of perichromatin fibrils and the disappearance of perichromatin granules. Nuclear fibrils are less resistant to autolysis than are nucleolar granules. In the initial stage, the nuclear membrane may show crenation, and during the advance stage of autolysis it may be disrupted.

In renal cortex, measurable changes occur in size and number of membrane intercalated particles of the freeze-fractured outer nuclear membrane after 20 min of ischemia (Coleman *et al.*, 1975, 1978). Such changes are not observed in the inner nuclear membrane, nor in the plasma membrane, indicating that the latter membranes do not respond in the same manner to ischemic changes as does the outer nuclear membrane.

Using human central nervous tissue at 4°C, it has been determined that although the nucleus undergoes shrinkage following cell death, this may not be accompanied by any significant loss of DNA or nuclear RNA up to 96 hr after death (Mann *et al.*, 1978). The decrease in cytoplasmic RNA after cell death probably involves mRNA and tRNA, which are distributed throughout the cytoplasm. This loss may result from the action of lysosomal enzymes, for these species of RNA are not protected by a membrane.

Mitochondrial changes include rounded shape, condensation, swelling, rupturing of membranes, formation of myelin figures, and appearance of flocculent densities or loss of matrix density. The swollen mitochondria, while retaining their outer double membrane, exhibit a progressive disruption of the cristae. For example, in the liver, initial postmortem changes consist of the loss of cristae and swelling in mitochondria of the parenchymal cells. Mitochondria in different cell types exhibit different degrees of alterations during the same time interval after death. This organelle in cardiac muscle is relatively resistant to the effects of autolysis both at room temperature and at 4°C. Morphology of mitochondria in the muscle kept in the cold for 24 hr remained intact, but after 48 hr delay the cristae showed degeneration (Lim and Solomon, 1976). Well-preserved mitochondria were seen when rat pancreatic acinar cells were kept at room temperature in a moist atmosphere for 24 hr (Nevalainen and Anttinen, 1977). Similar findings have been reported in rat heart muscle (Penttilä and Ahonen, 1976). On the other hand, mitochondria of the hair cell in crista ampullaris showed autolytic changes as early as 10–15 min after death of the animal (Anniko and Bagger-Sjöback, 1977). According to deF Webster and Ames (1965), reproducible, generalized, and progressive alterations occurred in the mitochondria of nervous tissue only 3 min after withdrawal of both oxygen and glucose. The appearance of prominent flocculation of mitochondrial matrix usually indicates irreversible cell damage. These flocculent densities appear to contain denatured proteins of the mitochondrial matrix.

Smooth endoplasmic reticulum is less resistant to postmortem changes than is rough endoplasmic reticulum; the former begins to form chains of vesicles within 30 min after the liver sample has been collected (Trump *et al.*, 1962). Autolytic changes in rough endoplasmic reticulum cisternae comprise degranulation, swelling, vesiculation, and dilation. Changes in the Golgi complex are not very conspicuous, although its cisternae may become dilated.

Plasma membrane may become indistinct and ruptured, which may allow cell organelles to escape to the extracellular space. Microvilli appear swollen and myelin figures may form at their tips. Microvilli of the bile canaliculi may disappear within 1 hr after death, but before their disappearance, dense cytoplasmic bodies containing amorphous material may become apparent in them. Microtubules and microfilaments may break down completely. Sinusoidal spaces may dilate as early as 1 hr after death.

Postmortem tissues that have been allowed to recover for several hours in a warm, oxygenated, physiological solution may show better preservation of ultrastructure compared with that shown by control specimens from the same cases. It has been indicated, for example, that tissues can be revived after 5 hr of anoxia and that these specimens can then be utilized in either physiological or morphological studies (Ferguson and Richardson, 1978). Tissues taken from autopsy cases are placed in a modified Krebs–Henseleit solution for at least 3 hr at 37°C. During this period a mixture of oxygen (95%) and CO_2 (5%) is bubbled through the solution at a rate of 4-6 liters/min. This physiological solution contains the following components in millimoles per liter:

Ca	2.6
Na	143.3
K	5.9
Mg	1.2
Cl	128.3
H_2PO_4	2.2
HCO_3	24.9
SO_4	1.2
Dextrose	10.0
EDTA	0.21

Small tissue pieces are fixed in 3% glutaraldehyde in 0.1 M phosphate buffer (pH 7.4) containing 0.1 mM $CaCl_2$ for 3 hr.

Both qualitative and quantitative ion shifts occur during postmortem. Gener-

TABLE 11.1

Conditions of Subcellular Structures of Different Tissue Types Kept at 6°C for a Period of Time and Fixed in Glutaraldehyde[a,b]

Tissue type	Neuron	Astrocyte	Oligodendrocyte
A. Brain			
Time delay (hr)	24	24	24
Chromatin material	+	−	+
Golgi apparatus	−	−	−
Microtubule	+ +	+ +	+ +
Neurofilament	+ +	+ +	+ +
Nuclear envelope	+	+	+
Nuclear pore	+	−	+
Plasma membrane	+	+	+
Rough endoplasmic reticulum	+	+	+
Ribosomes	+ +	+ +	+ +
Smooth endoplasmic reticulum	+	+	+
B. Neuropil			
Myelin sheath	+		
Synaptosome	+ +		
Spine apparatus	−		
Synaptic vesicles	+		

	Cardiac muscle	Liver	Glomerulus	Proximal tubule	Distal tubule	Skeletal muscle	
C. Other tissues							
Time delay (hr)	24	24	24	24	24	24	48
Basement membrane			+	+	+	+	+
Chromatin material	+	+	+	+ +	+		
Endoplasmic reticulum	−	+		+	+ +	+	−
Foot process			−				
Golgi apparatus	−						
Microvilli in lumen				+ +			
Mitochondria	+ +	+		+ +	+ +	+	−
Myofibril	+ +					+ +	+ +
Nuclear membrane	+	+	+	+	+	+	+
Nuclear pore	−	−	−				
Ribosomes	+ +	+ +		+ +	+ +	+ +	+ +
Sarcolemma/plasma membrane	+	+	+	+	+ +	+	+
Sarcotubule	+					+	−

[a] From Lim and Solomon (1975).

[b] Blank space indicates subcellular structure not observed; very well preserved, + +; fairly well preserved, +; poorly preserved, −.

ally, K and Mg intracellular concentrations decrease, whereas those of Na, Cl, and Ca increase. On the other hand, P and S show less change. These changes resemble those found in many pathological conditions. It is now well established that a number of diseases are associated with changes in elemental distribution at the cellular and subcellular levels (see Kuypers and Roomans, 1980). It is clear that caution should be exercised in interpreting the results of X-ray microanalysis of autopsy specimens, so as not to confuse ion shifts caused by a disease with those due to postmortem effect.

Although the information available on the length of time cells of different tissue types remain unchanged morphologically after death of the animal is insufficient, and in certain cases is contradictory, this duration for most cells is probably in the range of less than 1 min to 30 min in a moist atmosphere. The degree of resistance of various tissues and cell organelles to autolysis is given in Table 11.1. All organelles of the same type in the same cell may not show damage in the early stages of autolysis. In a cell, for instance, some mitochondria may show damage and others may appear normal. When the tissue remains in the cadaver, it is exposed to fewer physicochemical stresses compared with those imposed on excised tissues. Autolytic changes set in sooner in diseased organs compared with normal organs. Low temperatures retard the onset of autolysis. In spite of the previously mentioned relative resistance of certain cells and organelles to autolysis, it is stressed that the selected tissue should be placed in the fixative as rapidly as possible (see p. 318 for information on plant cells).

12

Plant Specimens

Fixative formulations used for the preservation of plant specimens in general have been those developed for fixing animal tissues. Although such formulations give adequate preservation of a variety of plant tissues (Fig. 12.1), fixatives specifically suited for the best preservation of plant specimens have yet to be developed. Differences in the protein contents of plant and animal tissues suggest that ideal fixatives for the latter may not be best suited for the former. Generally, animal tissues are richer in protein in comparison with plant tissues and viruses. Proteins occur in plant tissues in concentrations that may vary from 0.3 to 2.0% of the fresh weight (Loomis, 1974), although local protein concentrations may exceed 2%. For instance, more than half the protein content of a cell can reside in chloroplasts (Salisbury and Ross, 1969).

The relatively low protein content of plant cells must be taken into account when devising formulations, since studies of model protein systems indicate that glutaraldehyde cannot cross-link low concentrations of protein. For example, it has been shown that glutaraldehyde does not fix albumin at concentrations of less than 3–5% (Millonig and Marinozzi, 1968), nor will it fix gelatin concentrations of 2% or less (Langenberg, 1979). It therefore can be inferred from these studies that a minimum protein concentration in a cell is necessary in order for it to be fixed by glutaraldehyde. Meristematic cells show better fixation than that exhibited by specialized cells. One reason for this difference may be that the former usually do not contain a central vacuole, and thus the protein concentration per unit volume should be higher in these cells.

Other major differences between typical plant and animal tissues that affect the

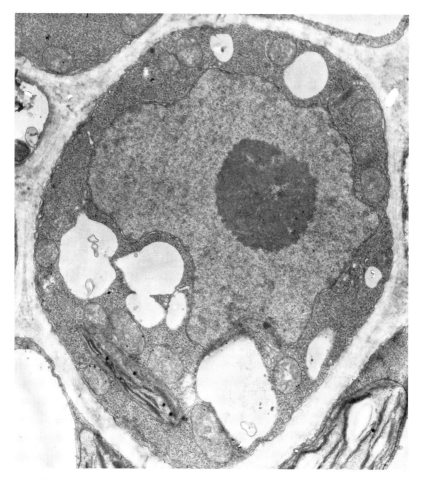

Fig. 12.1. A plant cell showing satisfactory preservation of various cell organelles. Note dense ground cytoplasm rich in ribosomes.

fixation process include the presence in plants of a nonliving outermost boundary (cell wall) and a large central vacuole(s). These two characteristics create a problem in obtaining a reliable and satisfactory fixation of plant cells, since they confer an osmotic pressure to the cell that is considerably higher than that of the surrounding media. If the plant cell is killed before fixation, compartmentalization of acids is lost because of damaged vacuolar membrane. It is known that vacuolar sap has a pH of ~5-6, whereas the average pH of plant cytoplasm is 6.8-7.1 (Kramer, 1955; Drawert, 1955). Thus, in order to minimize the adverse effect of the released sap, fixative solutions having a relatively high pH (~8.0) give superior preservation. The best approach seems to be to use a higher pH during prefixation with glutaraldehyde, followed by a lower pH ($\sim$7.0) during postfixation with OsO_4. Protozoa also are better fixed at a higher pH. In fact, certain animal tissues that are highly hydrated require a higher pH during fixation. Helander (1962) indicated that gastric mucosa was preserved much better at pH 8.5.

A second adverse effect on fixation of the plant cell's central vacuole is a considerable dilution of the fixative by the large volume of the vacuolar sap. This problem is essentially unique to cells of higher plants and, apparently, to highly hydrated specimens. In order to counter this dilution, it is desirable to use fixative and postfixative solutions of relatively high concentrations (e.g., 5% glutaraldehyde and 2% OsO_4). NaOH–PIPES buffer is the most desirable vehicle; the next most desirable is cacodylate; and the least desirable is phosphate buffer.

Many plant tissues contain large amounts of air (including gases) especially in the intercellular spaces (e.g., spongy layers in the leaf). The presence of air in the tissue is a hindrance in the penetration of the fixative solution. Therefore, it is desirable to carry out initial prefixation under a gentle vacuum (1 atm for roots). Within a minute or so after the vacuum is applied, one can see a stream of tiny bubbles being released from the tissue. Higher vacuums will cause damage, such as separation of plasma membrane from the cell wall (Fig. 12.2).

Air present in plant tissues may be absorbed by using boiled buffers; alternatively, buffers can be degassed using a vacuum aspirator. Such buffers may be used for both rinsing and preparing glutaraldehyde solution.

Surfaces of most plant organs (leaves, stems, and flowers) are usually covered by waxes that resist the penetration of aqueous solutions. In certain cases these substances are hydrophobic. The presence of tiny appendages (e.g., hairs) on the surfaces of certain plant organs may trap air and thus impede the penetration of the fixative solution. These problems can be alleviated somewhat by dissolving the impervious substances with a very brief treatment in dilute household detergent. Sometimes gentle rubbing with a brush while in the fixative solution helps to dislodge the air bubbles and even some surface substances. Even vigorous shaking of the specimens in the fixative solution in a vial may help. Total immersion of the tissue blocks in the fixative solution within a few minutes should be considered a prerequisite to satisfactory fixation.

What role, if any, plasmodesmata (cytoplasmic connections between plant cells) play during fixation is not known. However, they could transmit harmful effects of acidification by the released sap to cells farther away before the arrival of the fixative solution within these cells.

Although plant tissues do not undergo postmortem changes per se, desiccation occurs rapidly, and this causes changes at the ultrastructural level. Therefore, like animal tissues, plant specimens should be fixed by immersion immediately after their removal from the plant. Attempts have been made to initiate fixation *in vivo,* i.e., to start the fixation prior to removing the specimen completely from

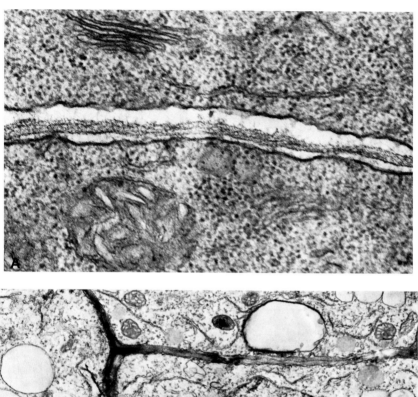

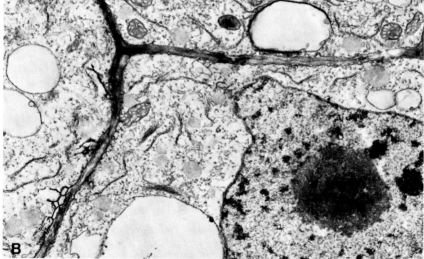

Fig. 12.2. Bean root tip cells. A, excessive vacuum has caused separation of plasma membrane from the cell wall. B, plasma membrane remains apposed to the cell wall under optimum vacuum.

the plant. During excision of differentiating xylem in willow, the stem was irrigated with cold fixative solution (Robards, 1968). Another example is fixation of a relatively thin leaf by sealing a metal ring with lanolin onto its upper surface and then filling the ring with the fixative solution (see p. 401 for details). Other procedures can be devised to begin the fixation before separating the tissue from the plant organ.

Plant plasma and vacuolar membranes are particularly sensitive to fixation conditions; the appearance of the latter membrane can be used as a criterion for evaluating the quality of fixation. The smooth contours of vacuolar membranes are thought to be more representative of the *in vivo* condition than are irregular contours. In a well-fixed cell, the tonoplast is clearly visible as an intact single membrane that forms a distinct boundary with the cytoplasm. Fixation quality is considered substandard when the tonoplast has pulled away from the cytoplasm or when the plasma membrane has shrunk away from the cell wall (Fig. 12.2). Several studies indicate that the vacuolar membrane is easily damaged during fixation; the tonoplast of cells with large central vacuoles seems especially sensitive (Fowke and Setterfield, 1968; Pickett-Heaps and Fowke, 1969; Catesson, 1974; Lawton and Harris, 1978).

Since vacuoles contain hydrolytic enzymes, such as nucleases and proteases (Berjak, 1973; Matile, 1975), their release because of the rupturing of the tonoplast during fixation may cause cytoplasmic damage. This damage, which is of course an artifact, should be distinguished from the changes occurring due to the natural senescence that precedes the differentiation of cells such as vessel elements. It should be noted that tonoplast breakdown is an almost universal phenomenon during cell differentiation, disease, nutrient deficiency, and chemical and physical damage.

Living plant cells are ideally suited for the evaluation of fixation procedures with the aid of phase contrast and Nomarski interference contrast microscopy. Events such as cessation of streaming, loss of control over cytoplasmic organization, and change in the structure of the cytoplasm caused by the fixation process can be monitored. Such an approach has been used to study the sequence of events that occur during the conversion of protoplasm to a stable gel during fixation (O'Brien *et al.*, 1973; Mersey and McCully, 1978). These studies indicate that standard glutaraldehyde fixation inevitably introduces structural alterations in the cytoplasm but that such changes can be detected only by comparison of the fixed tissue with the living specimens.

As in the case of animal cells, plant cells undergo degenerative changes due to natural senescence. Examples of cells that undergo natural degenerative changes during normal development are vessel elements, fibers, sieve cells, and storage cells in cotyledons. Structural changes caused by natural senescence and those induced by infection (e.g., by virus and fungus), mineral deficiency, chemical or physical treatment, etc. (reviewed by Butler and Simon, 1971) may be similar.

Since similar degenerative changes can also be caused by preparatory procedures, it is necessary to elaborate briefly on the structural changes associated with plant cell death or senescence. It has been shown, for instance, that an incomplete tonoplast in young fiber cells is an artifact of fixation and is not due to the result of natural senescence (Lawton and Harris, 1978). A massive amount of data is available on fixation artifact in animal cells, whereas very little information is available on plant cells. Although an adequate amount of information covering the physiological and biochemical aspects of plant senescence is available, the data at the ultrastructural level are meager.

Regardless of the cause of senescence, the sequence of senescent events in plant cells follows a fairly standard pattern, and the symptoms seen in each organelle are very much the same. The first detectable cellular change is a reduction in the population of ribosomes. Free ribosomes disappear first, followed by those attached to rough endoplasmic reticulum. In detached wheat leaves floating on water in weak light, the first degenerative change occurs in the mesophyll cells after 2 days (Shaw and Manocha, 1965). This change comprises the swelling of the tips of rough endoplasmic reticulum, which results in the production of vesicles. The breakdown of chloroplasts begins either at this time or immediately afterward. The chloroplast stroma disappears, the thylakoids swell and disintegrate, and osmiophilic droplets increase in size and number. Mitochondria are more resistant than other cell components and are still present at later stages of senescence.

The vacuolar membrane breaks down after 4 days, long before the degeneration of other organelles. This results in the mixing of vacuolar sap and the cytoplasm. The Golgi complex undergoes hypertrophy and then disappears. Although the nucleus is relatively stable, at a later stage its membrane becomes irregular and breaks up into vesicles (Butler and Simon, 1971), and the nuclear chromatin shrinks and disappears. The plasma membrane is usually one of the last components to degenerate; this leads to the release of vacuolar sap into the intercellular spaces. A similar pattern of breakdown is observed when the leaf is allowed to senesce under natural conditions while still attached to the plant (Barton, 1966). Hydrolytic enzymes present in lysosomes and spherosomes probably play a significant role in the process of senescence. Certain types of senescence in the early stages can be reversed.

Fixation for Scanning Electron Microscopy

Biological specimens, with few exceptions, are fixed prior to examination with the scanning electron microscope (SEM). The reasons and criteria for satisfactory fixation are the same as those for transmission electron microscopy. Only a few types of living organisms can be examined in the SEM, by dint of their ability to withstand the severe environmental conditions of vacuum and irradiation encountered in the SEM (Crowe and Cooper, 1971). These specimens apparently retain sufficient water while in the specimen chamber of the SEM to remain alive during and after examination (Hartman and Hayes, 1971). Such specimens are, of course, examined without fixation, dehydration, or metal coating. However, no study has been published claiming the examination of the living vertebrate tissue.

If the information needed is contained in the surface of a biological system, then the specimen does not usually need to be sectioned or dissected. Natural surfaces are relatively easy to prepare. Leaf, insect, and bone surfaces, blood cells, and protozoa are examples of the specimen types that do not require elaborate preparatory treatments, although the natural surfaces of these specimens should be thoroughly cleaned prior to viewing with the SEM.

Most biological specimens, however, are extremely complex three-dimensional systems and usually require dissection and/or sectioning prior to fixation. Stripping the epidermis from leaves with the aid of a fine tweezers to expose the interior is an example of such dissection (Shih, 1974). These steps are necessary both to reveal the surfaces of interest and to obtain specimens of manageable size.

The dissection and sectioning processes can inflict severe mechanical stress on the unfixed tissue, causing deformation of surface topography of the final specimen. Both skill and care are needed to minimize this type of deformation. The use of a tissue sectioner (chopper) is helpful in cutting tissue blocks of large size (see p. 12). Sophisticated ultradissection techniques for neuronal tissue (Lewis, 1971) and microdissection in the SEM (Pawley *et al.*, 1975; Irino *et al.*, 1978) have been attempted. Low-temperature ashing techniques have been employed for clearing the specimen of unwanted tissue components. Cytological techniques are now available for isolating cell organelles for scanning electron microscopy (e.g., Lima-de-Faria, 1978; Kirschner, 1978; Marchant, 1978).

After an appropriate specimen is selected, the preparatory procedures generally include fixation, dehydration, critical point drying or freeze drying, and conductive coating. Only the topic of fixation is discussed in this volume; the remaining three steps have been considered elsewhere (Hayat, 1978). Fixation must be carried out with utmost care to maintain the surface integrity. It must always be remembered that the internal and external surface morphology of tissue blocks as well as the external morphology of tissue sections or individual cells can be significantly affected by fixation procedures used (Figs. 13.1–13.5).

CLEANING OF SPECIMENS

The surface of biological specimens, both hard and soft, is usually covered with extracellular materials, which may include mucus, blood, lymph, cell sap, cell debris, gum, wax, and gelatinous capsules of bacteria. Certain cells are surrounded by protein or polysaccharide-containing fluids. These materials are considered contaminants, since they form an opaque layer on the surface of specimens in the SEM. During specimen preparation, fluids necessary for fixation, dehydration, and intermediate procedural steps may stabilize and toughen some of these materials, although some surface materials may be dissolved by these fluids. The surface materials must, therefore, be removed prior to fixation.

The majority of surface materials can be removed from the specimen surface by rinsing with an appropriate isotonic solution (e.g., buffered saline solution) or distilled water. The reason for using isotonic solution is that it prevents the swelling or shrinkage of the cell that accompanies osmotic stress, since washing is carried out prior to fixation. The washing solution should ideally be delivered at the same temperature as that of the living cells in their normal environment, should contain all essential ions present in the extracellular fluid, should be buffered to the appropriate physiological pH (7.2–7.4), and should possess the same osmolarity as that of the fluid normally in contact with a given cell. In

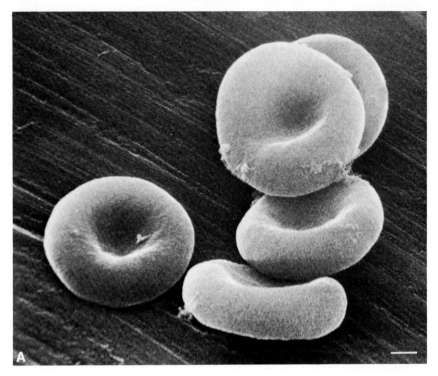

Fig. 13.1.   Human red blood cells showing the effect of different concentrations of glutaraldehyde. A, fixed with 0.1% glutaraldehyde for 15 min. The cells are close to *in vivo* shape. B, fixed
with 2.5% glutaraldehyde for 1 hr. The cells have deformed shape. The bar represents 1 μm. (M.
Horisberger, J. Rosset, and M. Weber, unpublished.)

short, an attempt should be made to design a washing solution that reproduces the
natural environment of a given cell or tissue. Distilled water is adequate for
removing most of the contaminants from the surface of hardy specimens such as
bacterial spores. Gentle water rinsing was used in preparing attached rumen
bacteria (Costerton *et al.*, 1978). For studying plaque, teeth should be washed
with distilled water prior to fixation; the surfaces of unwashed specimens are
obscured by an amorphous material.

The surfaces of dry specimens (e.g., bones, leaves, stems, petals, and wood)
can be cleaned with puffs of air or gas. The lumina of cavities of certain animal
specimens (e.g., intestine, blood vessels, and heart) should be flushed out. Blood
plasma can be at least diluted, if not washed away completely, by perfusion of
the blood vessels of an organ with a buffered physiological saline solution.
Puncture perfusion method (Murakami, 1976) can be used for removing blood
and tissue fluid from small tissue samples including biopsy samples. The tissue

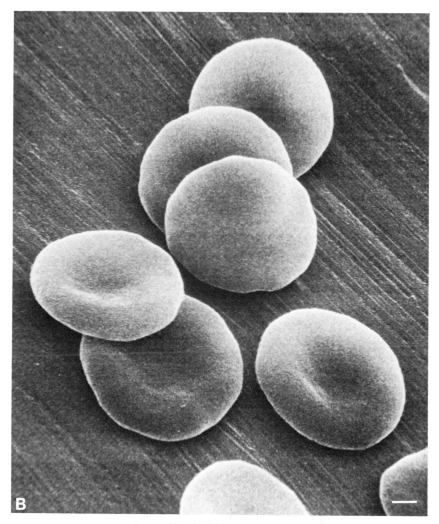

Fig. 13.1 (*Continued*)

block is placed on cotton gauze, punctured by a syringe needle (26 G), and gently perfused with Ringer solution until it becomes pale. The tissue is then perfused with 2% glutaraldehyde.

Generally, materials such as mucus are partly dispersed by the washing solution or the fixative, and thus do not form a coherent layer to obscure the surface details. Tenaciously adhering mucus can be removed by squirting the washing solution onto the cells. This type of mucus in some cases may have to be removed by treating the tissue with 1–4% HCl and/or ultrasonication. For exam-

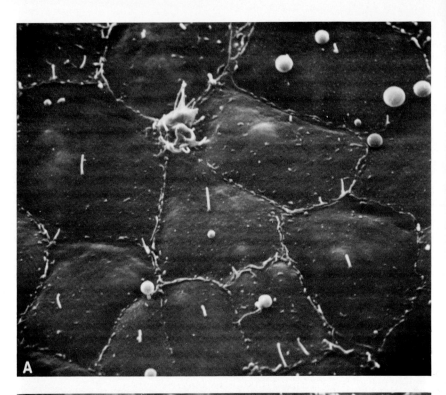

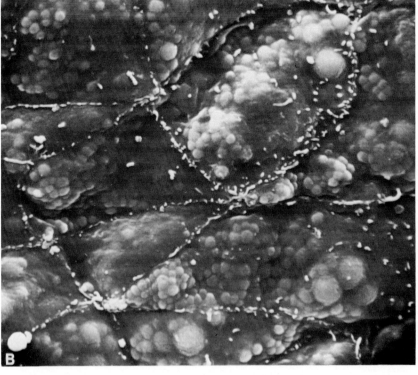

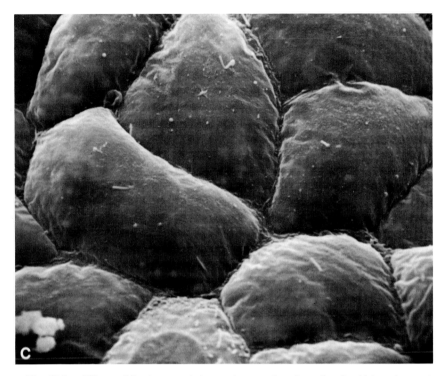

Fig. 13.2. Effects of fixative osmolarity on the ventral surface of early chick embryos. A, fixation with modified Hank's solution (final osmolality 394 mosmols) containing 1% each of glutaraldehyde and formaldehyde for 2 hr, and postfixation with 2% OsO_4. Relatively flat surfaces of cells are preserved by the isotonic fixative. A single cilium on each cell and microvilli marking cell boundaries can be clearly seen. Surfaces do not show blebbing, whereas strongly hypertonic Karnovsk's fixative (2010 mosmols) causes surface blebbing (see B). C, immersion in 3% glucose for 5 min followed by fixation with hypotonic Tyrode's solution (250 mosmols) containing 1% glutaraldehyde. Cell surfaces are protruded and bordering microvilli are not seen. Somewhat shortened cilia are present. The cells are clearly swollen. ×2470. (Litke and Low, 1977.)

ple, a jet of compressed air was employed to remove mucus from sensory papillae of tongue (Shimamura and Tokunaga, 1970), and sonication was used for cornea and other tissues following a brief fixation with OsO_4 (Kuwabara, 1969, 1970). Cetylpyridinium chloride can be used to remove mucus from amphibian skin. Mucus can also be removed during fixation with aldehydes containing sucrose. Fixed coelenterate structures were treated with 16% glycerol in water to remove the mucus and then with 20% ethanol for 24 hr to remove the glycerol (Mariscal *et al.*, 1978). Heavily mucus-coated sperms of marine invertebrates were cleaned with hyaluronidase in seawater prior to fixation (Atwood *et al.*, 1975). Mammalian sperms can be cleaned by gentle repeated centrifugation

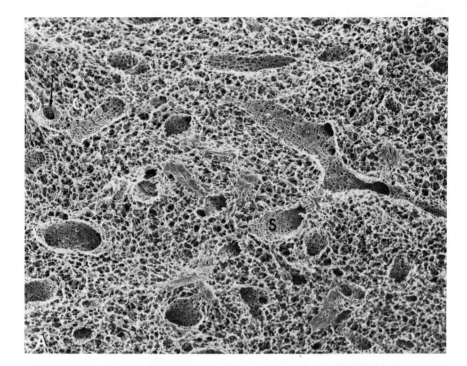

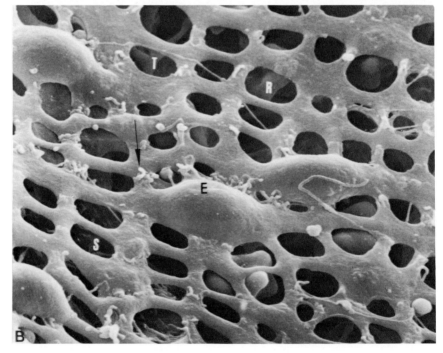

in balanced saline solution (Tyrode's) (Cohen, 1979). Since cleaning treatments (saline rinsing, mechanical scrubbing) prior to fixation of canine stomach mucosa could produce alterations, Fallah *et al.* (1976) teased the mucus from the tissue surface.

The presence of adherent mucus on the surface of certain tissues, such as epithelium, is a serious problem, for it obscures the examination of the underlying tissue. Eisenstat *et al.* (1976) presented a procedure for removal of such mucus layer which results in little damage to the underlying epithelium. First, gross debris from a hollow organ is removed by gentle irrigation with 0.2 mol/l saline. Small tissue pieces are immersed in 0.2 mol/l buffer (pH 7.3) for 1 min and then subjected to several bursts of the buffer delivered from a jet bottle. Subsequently, at the beginning of dehydration (in 10% ethanol or acetone) mucus fragments can be carefully removed with fine forceps. After critical point drying, small residual mucus fragments can be removed by using compressed Freon, and, if necessary, by a fine needle under a dissecting microscope. If the mucus still adheres to the epithelium, it can be lifted off in its entirety by inserting dissecting needle into the mucus layer (while monitoring the procedure with the dissecting microscope), followed by a few bursts with compressed Freon to remove remaining mucus fragments.

In certain cases, even vigorous washing and other mechanical treatments may not remove the contaminant; digestion with appropriate enzyme may solve the problem. Stomach mucus can be safely removed from gastric surface epithelial cells by enzymatic digestion followed by a mechanical action (Wood and Dubois, 1981). Organs or tissues are exposed to 1% glycosidic solution for 3–5 min, followed by vortexing for 3 min in the same solution. Specimens are vortexed three times for 1 min each in PBS, that removes the residual network of mucus and washes away the enzyme.

Papain and other enzymes have been used to remove contaminants from various types of tissues. Cell surfaces in the intact tissue covered by collagen and basement membrane can be unmasked by exposing the specimen to HCl and collagenase. According to one method (Evan *et al.,* 1976), fixed tissue speci-

Fig. 13.3. Dog spleen fixed by vascular perfusion, followed by freeze-fracturing. The fractured specimens were exposed to amyl acetate, critical point dried with CO_2, and coated with carbon and gold. A, red pulp showing a synoptic view of various profiles of the splenic sinuses (S) and the spongy appearance of the splenic cord in which profiles of the cordal vasculature of relatively large caliber are recognized. Numerous fenestrations and openings of the branching sinuses (arrow) are present in the sinus wall. ×170. B, inner aspect of a splenic sinus. Long spindle-shaped endothelial cells (rod cells) with focal nuclear elevations (E) arranged parallel to the long axis of the sinus traversing this field. Fenestrations (stomata) (S) are present between the rod cells with a relatively regularly spaced latticelike pattern. Irregularly curled slender cytoplasmic projections (arrow) are often recognized near the junctions of the rod cells. Note that RBCs (R) are parts of cordal reticular tissue (T) and can be observed through these stomata, suggesting that these are pathways of blood between the cordal space and the lumen of the sinus. ×4600. (Suzuki *et al.,* 1977.)

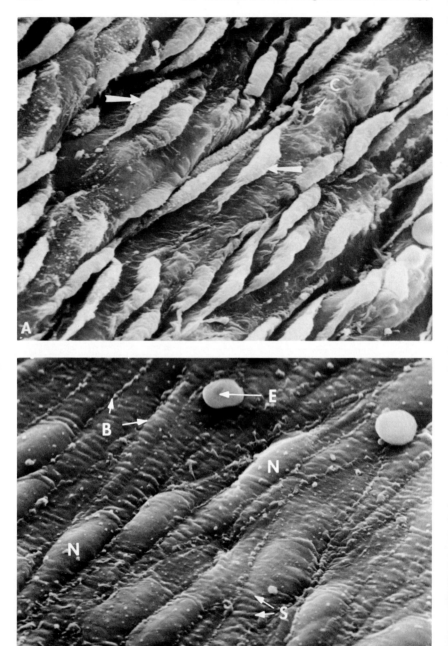

mens (e.g., kidney and skin) are rinsed in the buffer to remove the fixative and then exposed to 8 N HCl for 50–70 min at 60°C. After washing the specimens in the buffer to remove the acid, they are placed in 0.1 mol/l phosphate buffer (pH 6.8) containing 10 mg collagenase/10 ml of buffer for 3–8 hr at 37°C. The specimens are washed thoroughly in the buffer, placed on a coverslip previously coated with gelatin (Vial and Porter, 1974), and secured to the gelatin by immersion in 2.5% glutaraldehyde in 0.075 mol/l cacodylate buffer for 15 min at room temperature. The duration of the treatment will vary, depending on the tissue under study. Mucilaginous materials can also be removed by treating the specimens with glusulase (Endo Lab., Garden City, New York) diluted (1:50) for 1–2 hr at room temperature prior to fixation.

Some of the other enzymes that have been used are chymotrypsin, for intestinal epithelium (Kanazawa *et al.*, 1972); trypsin, for isolation and cleaning of parietal cells of stomach mucosa (Lee, 1972); diastase, for cleaning the organelles isolated by mechanical means (Sandborn and Makita, 1972); trypsin, for chromosomes (Wolfe and Martin, 1968); pronase and elastase, for removing neuronal sheaths (Lewis, 1971); and crude bacterial α-amylase (Steven *et al.*, 1971), for removing glycoproteins and other noncollagenous components that adhere to collagen and elastin fibers of the aorta (Finlay *et al.*, 1971; Minns and Steven, 1975).

Besides enzymes, acids and other reagents can be employed for removing certain contaminants and facilitating the examination of specimens. Palynomorphs have been treated with hot HCl after digestion with hydrofluoric acid to remove extraneous mineral and organic debris from the surface (Leffingwell, 1974). Siliceous diatoms can be cleaned of organic matter by using potassium persulfate ($K_2S_2O_8$) and oven heating (Ma and Jeffrey, 1978). About 0.2–0.4 μg of $K_2S_2O_8$ is added to 5 ml of well-rinsed sample in a test tube. To prevent excessive evaporation, a loose-fitting glass cap is placed over the test tube mouth. The tube is left on a sand bath in an oven for 6–8 hr at 333–363 K. The sample is removed from the oven and centrifuged or allowed to stand until particles settle out. The supernatant is decanted carefully, followed by thorough rinsing with distilled water. Other methods used for cleaning phytoplanktons include pancreatin (Reimann, 1960); heat (Zoto, 1973); H_2O_2 and ultraviolet irradiation (Swift, 1967); HNO_3 (50%) and heat (Crawford, 1972); HNO_3, $NaNO_3$, or H_2O_2 (Hargraves and Guillard, 1974); H_2SO_4, $KMnO_4$, and $(COOH)_2$ (Hasle and Fryxell, 1970); and H_2SO_4 (Patrick and Reimer, 1966). The use of $KMnO_4$ and HCl is one of the gentler methods. Hydrochloric acid (5%) has been

---

Fig. 13.4. Endothelial cell surface of a rabbit common carotid artery. A, fixed by immersion and showing marked distortion of surface fine structure. Bridgelike artifacts (arrows) are apparent. Crater (C). B, fixed by intravascular perfusion and showing structural organization characteristic of satisfactorily prepared tissue. Endothelial cell borders (B); nuclear protrusions (N); surface striations (S); erythrocytes (E). ×1100. (Gertz *et al.*, 1974.)

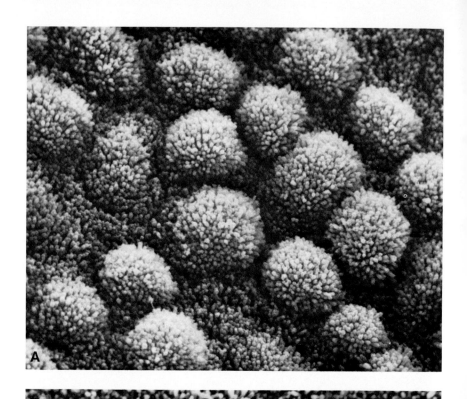

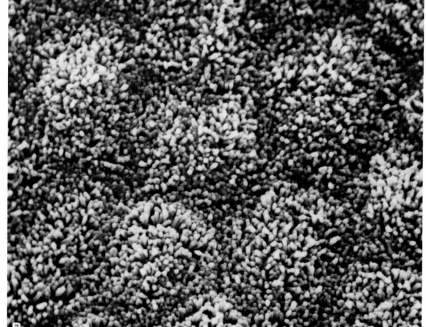

employed to dissolve the gelatinous layer around sensory cilia and cupulae of the semicircular canals (Lim and Lane, 1969). In order to expose the underlying sensory cells in the inner ear, saccular and utricular otoliths and cupulae of the crista ampullaris were removed during dehydration (overnight treatment with 70% ethanol to impart rigidity to the gelatinous membrane) by flushing with ethanol from a fine pipette; fine-tipped tweezers were used to initiate separation before flushing (Lim, 1976).

Calcareous structures can be solubilized by HCl (1 mol/l), ascorbic acid solution (1%), or EDTA (10% in distilled water). Deciliation can be accomplished by dibucaine hydrochloride or by aspirating the organisms through the bore of a hypodermic needle. Dilute EDTA has been used for separating nerve cell processes (Lewis, 1971). Some of the other treatments that have been used are prolonged maceration with glycerol and dilute ethanol to remove milk from lactating mammary glands (Nemanic and Pitelka, 1971), N-acetyl cysteine as a mucolytic agent (Williams *et al.*, 1973), and sucrose for gastrointestinal mucosa (Landboe-Christiansen and Parapat, 1972). Hydra cells have been isolated by omitting Ca and Mg from the medium while adding sucrose (Westfall and Enos, 1972). Oxidation in high-frequency gas discharge has been used to dismantle organic material, layer by layer, in order to make the cell wall structure accessible for scanning electron microscopy (von Barthlott *et al.*, 1976). This method has limitations, for etching effects cause structural disturbances.

Sonication has been used for aldehyde-fixed hen's oviduct (Bakst and Haworth, 1975), nematodes (Behbehani *et al.*, 1977), and sperms (Baccetti, 1975). Such an approach was also used for marine trematodes (Cohen, 1979). Sonication should be used carefully and with frequent observations.

Caution is required in the use of some of the treatments mentioned, because they can cause unwanted and unpredicted alterations in the surface topography. As mentioned, it should be remembered that specimens are especially vulnerable to the action of these reagents when exposed to these treatments prior to fixation. Vigorous mechanical treatments may damage the surface of fragile specimens, and, in fact, even a rinse in saline solution may change the appearance of cultured cells.

Preservation of Microbial Association to Intestinal Epithelium

One disadvantage of mucus removal to expose the surface of intestinal epithelium is the loss of microbial communities associated with the tissue. This association is facilitated by bacterial polysaccharide filaments. Bacteria are also

Fig. 13.5. Frog gallbladder epithelium. A, dehydration and infiltration by the conventional step procedure. The luminal surface presents a cobblestone appearance recognized as an artifact. B, dehydration and infiltration by the *continuous* procedure results in regularly aligned microvilli covering a smooth luminal surface in which individual cell contours are delicately outlined. ×5700. (Rostgaard and Tranum-Jensen, 1980.)

entrapped in the mucus secreted by the goblet cells. A technique is available that allows the removal of mucus, leaving the association between bacteria and the mucosal surface fairly intact (Schellenberg and Pangborn, 1980). The abdominal cavity of the animal (rat) is exposed and the intestine severed 2 cm above and below the ileo-cecal junction, without damaging the mesentry. After inserting the catheters (3 mm in diameter) into the severed ends and securing with surgical thread, 2.5% glutaraldehyde in 0.1 mol/l phosphate buffer (pH 7.2) is injected with a syringe until the intestine and cecum are slightly inflated. Simultaneously, the tissue is bathed with the fixative solution.

After 5 min the tissue is removed from the animal, leaving the catheters in place, and immersed in the fixative in a glass container. The same pressure is maintained in the intestinal sample throughout this procedure. The tissue is fixed for 2 hr at 4°C. Specimens should be handled gently to avoid dislodgement of microorganisms. Fragments of intestinal contents are discarded as they are dislodged. After attaching the specimen to the stub with silver paint, a light stream of compressed Freon gas is used to gently blow away mucus adhering to areas of interest.

RELAXATION PROCEDURES

Some organisms, especially of aquatic habitat, are highly motile and free swimming; others contract when exposed to fixative solution. It is desirable to quiet motile organisms or relax those that contract. In order to preserve these organisms in a relaxed state, they may need narcotization. Anesthesia should be applied gradually and should last only until the organism stops contracting on stimulation. A high dose of anesthetic that might cause death prior to fixation is undesirable. Fixation should follow immediately after the organism has reached the anesthetized state. Possible damage caused by anesthetic reagents can be confirmed by simultaneous processing of controls.

An ideal relaxant is $MgCl_2$; the recommended doses are 20% $MgSO_4 \cdot 7H_2O$ and 2.5% $MgSO_4 \cdot 6H_2O$ for marine and for freshwater organisms, respectively (Maugel et al., 1980). Magnesium salts can be added as crystals or in solution gradually to water containing the organisms until the appropriate anesthesia has been attained. Chloretone is another useful depressant used as a 0.5–1.0% aqueous solution. A mixture (1:1) of chloral hydrate and menthol has been used to relax ostracod crustacea (Sandberg, 1970). Local anesthetics (e.g., cocaine hydrochloride, procaine hydrochloride, lidocaine, and eucaine hydrochloride) can also be used as a 1.0% aqueous solution to produce narcosis. A more potent anesthetic for fish is tricaine methanesulfonate (0.02–0.25%). The use of low temperatures has the advantage of avoiding the potential harmful effects of reagents. The previously mentioned methods have been discussed by Galigher and Kosloff (1971) and Maugel et al. (1980).

FIXATION PROCESS

Generally, tissue specimens should be fixed prior to or immediately after removal from the organism or after their rapid cleaning or washing. This urgency becomes apparent when one considers that an appreciable amount of time can elapse between the moment the tissue specimen is separated from its normal environment and the time it is ready for fixation. During this interval the tissue specimen that has been separated from its normal environmental conditions, such as proper temperature and pH, and deprived of necessary nutrients and oxygen, undergoes rapid autolytic changes. Even when the cleaning is accomplished with an ideal solution, the tissue specimen remains unstable and its macromolecular systems remain in a mobile state.

Ideal fixatives for scanning electron microscopy do not differ significantly from those used for transmission electron microscopy except with respect to osmolarity. The ideal osmolarity of the fixatives for the former is lower than that recommended for the latter. Fixatives commonly used for transmission electron microscopy should have higher osmolarities because they become diluted with tissue fluids at the time of actual fixation of deeper layers of cells in the tissue block. The so-called hypertonic fixatives being used today for transmission electron microscopy are actually either isotonic or hypotonic because it is not the total osmolarity, but the vehicle osmolarity that is primarily responsible for the osmotic effects of the fixative solution. This point and the reasons for using slightly hypertonic fixatives for transmission electron microscopy have been discussed earlier.

Ideal fixative solutions for scanning electron microscopy of single cells and the outer layer of cells, on the other hand, should be isotonic, although information concerning the osmolarity of many cell and tissue types is not available. The reason for using isotonic fixatives is that usually only surface layers of the tissue block are viewed with the SEM, whereas ultrathin sections for transmission electron microscopy may contain cells from all layers of the tissue block. The cells of the surface layers require isotonic fixatives to avoid osmotic shock, whereas those of the deeper layers require hypertonic fixatives. The slow rate of penetration of isotonic fixatives is not a problem for surface cells, nor for unicellular organisms and particulate specimens.

Specimens should be protected from osmotic shock (Fig. 13.5). Like cells of the surface layers of a bulk tissue, cell monolayers grown in culture are prone to osmotic shock. Also, when fixation is carried out by vascular perfusion, osmotic shock becomes a problem for cells lining the blood vessels and cells close to blood capillaries. The importance of the use of isotonic fixatives in preparations for scanning electron microscopy is thus apparent.

The concentration of the fixative seems to be an important factor in obtaining satisfactory fixation of delicate cells in suspension, such as red blood cells (Fig. 13.1). It has been reported that mouse red blood cells were well preserved with

0.1% glutaraldehyde, whereas 2% glutaraldehyde distorted the cells (Arnold *et al.*, 1971). Even when the osmolality of these two fixative solutions was adjusted to the same value (320 mosmols, which is the value of mouse serum), the latter damaged the cells. This evidence suggests that not only the final osmolality of the fixative and its vehicle, but also the concentration of the fixative itself are important in obtaining satisfactory preservation of certain specimens.

It must be pointed out, however, that there is no general agreement on the importance of the contribution of glutaraldehyde to the effective osmotic pressure of the fixing solution. There is no doubt that vehicle osmolarity is an important factor in obtaining satisfactory fixation of a wide variety of specimens (Fig. 13.2). In fact, it is true that the total osmolarity of the fixing solution within reasonable limits) is less important than the vehicle osmolarity. In the case of cultured cells, it has been demonstrated that satisfactory fixation is obtained when the osmolarity of the vehicle is equal to the osmolarity of the growth medium for the specimen under study, regardless of the glutaraldehyde concentration used (Bell *et al.*, 1975). Umeda and Amako (1975) indicated that *E. coli* cells were best preserved in a 0.15 mol/l vehicle; at higher vehicle molarities, the surface of the cells was adversely affected. It is concluded that the osmotic pressure that results from the buffer, sucrose, and electrolyte components is a major factor in achieving satisfactory fixation, since osmotic pressure is exerted effectively only by those molecules or particles that do not penetrate the plasma membrane.

It should be noted that both the total and vehicle osmolarities are more critical for scanning than for transmission electron microscopy. In other words, a satisfactory range of vehicle osmolarity for transmission electron microscopy may yield unsatisfactory fixation for scanning electron microscopy. The type of ions of a buffer has a profound effect on specimens viewed with the SEM. The use of different buffers results in differences in the surface morphology (surface villi) and shape of lymphocytes.

PARAMETERS OF FIXATION

Mode of Fixation

Fixation by vascular perfusion, when possible, is superior to fixation by immersion. Figure 13.3 is an example of the excellent quality of tissue preservation obtained by vascular perfusion. The advantages of vascular perfusion and limitations of fixation by immersion have been discussed in Chapter 8. Some of the disadvantages of fixation by immersion for scanning electron microscopy are presented here.

The presence of extracellular materials may hinder the penetration of the fixative during immersion. A case in point is intestinal absorptive cells that have a fixation barrier because of the mucus surrounding the villi. This barrier prevents the fixative from penetrating the apical side of the absorptive cells. This problem can be minimized by using jet fixation instead of immersion fixation (Boom *et al.*, 1974). According to this procedure, the fixative is squirted from a Pasteur pipette (internal diameter 0.2 mm) onto a predetermined site. Some of the examples of artifacts caused by fixation by immersion are discussed later.

Platelet aggregates are known to be abundant in the immersion-fixed tissue sinuses, which probably are formed because of blood stagnation. When similar tissues are fixed by vascular perfusion, the platelet aggregates are smaller and fewer. Wrinkling of normal longitudinal folds, crenation of red blood cells, and the presence of bridges in the vascular endothelium fixed by immersion have been reported. These and other distortions of endothelial cell morphology have been observed with the SEM in the luminal surface of arteries fixed by immersion (Gertz *et al.*, 1975). A comparison of Fig. 13.4A and 13.4B demonstrates that bridgelike structures on the specimen surface are artifacts produced by immersion fixation. These artifacts can be eliminated by vascular perfusion at an appropriate perfusion pressure (Wolinsky, 1972; Edanaga, 1974; Davies and Bowyer, 1975; Clark and Glagov, 1976; Swinehart *et al.*, 1976; Hollweg and Buss, 1980; also see p. 296).

Specimen Size

The period of time elapsed between the moment the tissue specimen is collected and the time of its actual fixation is far more critical when preparing specimens for transmission electron microscopy because in this case fixation must take place throughout the depth of the tissue block. In contrast, for scanning electron microscopy, one is concerned usually with cells of the surface layers of the tissue block. For this reason, specimens of relatively large size can be fixed satisfactorily for scanning electron microscopy. However, though it is tempting to use large tissue specimens for scanning electron microscopy, generally this accomplishes little and creates problems. Fixation of large specimens requires prolonged periods of time and even then may be incomplete. For certain types of studies, such as examination of organelles *in situ* and internal surfaces of tissues, the specimen size should be as small as possible. Plant and animal tissue blocks larger than 1 mm^3 tend to show poor fixation in the core region. In general, the smaller the size of the specimen, the better and more uniform will be the quality of fixation. Very nearly uniform fixation can be obtained by vascular perfusion; detailed procedures for fixation of various types of tissue by vascular perfusion have been presented in an earlier chapter.

Large specimens also invite the danger of incomplete dehydration and incomplete removal of the intermediate fluid during critical point drying. Specimens of large size, moreover, are poor electrical conductors even after coating with a metal, and thus are prone to charging. It is apparent, therefore, that specimen size should not be larger than that absolutely necessary. Furthermore, tissue specimens should have small vertical dimensions.

Temperature

The ideal temperature at which fixation should be carried out is close to the *in vivo* temperature of the specimen, especially when using vascular perfusion. The situation is more complex when the specimen is fixed by immersion, since both specimen type and size, as well as the objective of the study (i.e., whether internal or external portions of the tissue block are needed), play important roles. The role played by temperature in the rate of penetration by the fixative into the tissue block and in the onset of autolytic changes has been discussed earlier.

Certain specimens can be preserved satisfactorily either in the cold or at room temperature, whereas large specimens or specimens having delicate projections, which must be preserved instantaneously, should be fixed at warm temperatures. It has been demonstrated that surface projections of cells are better preserved at 37°C than at lower temperatures (Lin *et al.*, 1973). Delicate cell retraction processes and filamentous projections are also better preserved at an initial temperature of 25–37°C (Barber and Burkholder, 1975). Similarly, human breast tumor cells are preserved satisfactorily at 38°C (Westbrook *et al.*, 1975). Swelling is slightly more pronounced when the tissue is fixed in the cold, because of the deleterious effects of suboptimal temperature on surface structure. It is apparent that, in general, a higher temperature of fixation is preferable for specimens prepared for scanning electron microscopy than that needed for routine transmission electron microscopy.

Duration of Fixation

The optimal duration of fixation for the majority of tissues is not known. For most purposes, an arbitrary duration of 2–8 hr is currently in use, provided the specimen size is not too large. For large specimens, the duration of fixation can be extended to ~24 hr. It is desirable that specimens not be allowed to remain stationary in the fixative; agitation during fixation may be confined merely to gentle swirling of the fixative at 15 min intervals during the first hour of fixation.

An advantage of long fixation periods is the eventual hardening of the specimen. In this state, the specimen is less vulnerable to mechanical damage during subsequent handling and better withstands the damaging effects of the electron beam and vacuum in the column of the SEM. Also, hardened specimens can be trimmed or dissected with a razor blade to expose desired structures with a minimum of crushing or distortion. However, very little is known about other effects of long fixation times, except that they result in the extraction of tissue constituents, particularly at higher temperatures. Since the extraction caused by chemical fixation is progressive in time, the need to determine the optimal duration is apparent.

SPECIMEN SHRINKAGE

Almost all specimens processed for scanning electron microscopy show shrinkage to various degrees. An agreement on the degree of shrinkage introduced during each of the three major steps (fixation, dehydration, and critical point drying) of specimen processing is lacking. The shrinkage has been interpreted as an effect of fixation (Ulmer and Honjo, 1973) or the result of dehydration prior to critical point drying (Bessis and Weed, 1972). In other studies, critical point drying has been implicated as the major factor contributing to specimen shrinkage (Waterman, 1972; Schneider, 1976; Gusnard and Kirschner, 1977; McGarvey *et al.*, 1980). Specimens undergo shrinkage to varying degrees during each of the three major steps mentioned earlier, although shrinkage introduced during fixation may be reversible in the low concentrations of alcohol during dehydration. Shrinkage during dehydration can be minimized by using a slow and *continuous* procedure instead of a conventional step procedure of dehydration. Figure 13.5 shows superior preservation of tissue morphology by using a slow and *continuous* procedure of dehydration with ethanol.

Various types of specimens differ in the degree of shrinkage incurred during each step of processing, even when processed under similar conditions. Dense or compact specimens undergo relatively less shrinkage in each of the three steps of processing, although the amount of shrinkage is not directly related to the "softness" of the tissue. Specimens containing relatively large amounts of water or lipids may show a higher shrinkage. Solubility of lipids in acetone and ethanol is well known. Steroids and terpenes are soluble in CO_2.

The amount of shrinkage also depends on whether specimens are intact tissue blocks or are cells in culture or isolated organelles. These two types of specimens react differently to osmotic changes in their external environment and allow different degrees of extraction of their constituents. Isolated cell organelles are resistant to osmotic changes, whereas cells in culture and tissue blocks are not. It should be noted that the extent of shrinkage may differ in different directions of a cell, i.e., the shrinkage in length may differ from that in width of a cell.

As stated earlier, various types of specimens show varied amounts of shrinkage. Some of the examples are as follows: ~12% for CNS, 12–13% for embryonic and fetal tissues of vertebrates (Waterman, 1974), 22% for canine endocardium (Wheeler *et al.*, 1975), 20% for soft tissues (lung, brain, spleen, liver, kidney, and skin) (Niimi *et al.*, 1975), 21% for mammalian membranous cochlea (Hunter-Duvar, 1977), 45% and 31% in the rat abdominal aorta and thoracic aorta, respectively (McGarvey *et al.*, 1980), 43% for isolated human lymphocytes (Schneider, 1976), and 25–30% for human erythrocytes and isolated mouse liver nuclei (Gusnard and Kirschner, 1977). Different cell types within a tissue, furthermore, may show different amounts of shrinkage. Erythrocytes, for instances, shrink less than lymphocytes; this is confirmed by the following data.

Overwhelming evidence indicates that both intact tissue blocks and cells in culture undergo the major part of the net shrinkage during critical point drying. Isolated cell organelles (e.g., nuclei) and cells in suspension (e.g., erythrocytes) shrink maximally during critical point drying, as opposed to fixation and dehydration (Gusnard and Kirschner, 1977). This is especially true when amyl acetate is used as an intermediate fluid. Shrinkage of these specimens during dehydration can be almost eliminated by replacing ethanol with ethylene glycol–cellosolve. However, it should be noted that the advantage of avoiding shrinkage during fixation and dehydration is eliminated during critical point drying. In other words, even when optimal fixation and dehydration procedures are used, a certain amount of shrinkage will occur during critical point drying or freeze-drying. The shrinkage occurring during critical point drying seems more to be the effect of substitution of the intermediate fluid than the result of high pressure during critical point drying. Poorly fixed cells undergo more shrinkage during subsequent processing than that shown by well-fixed cells (also see pp. 98 and 299). Shrinkage during dehydration and critical point drying can be substantially minimized by treating the cells with tannic acid and uranyl acetate, following sequential fixation with aldehydes and OsO_4.

FIXATIVES

It must be emphasized that fixatives should be prepared in solvents that correspond, when possible, to the natural environment of a given organism. Fixatives for mammalian tissues, for instance, should be prepared in a normal saline solution. As stated earlier, most commonly used fixatives for scanning electron microscopy have been adapted from those used for transmission electron microscopy. Aldehydes are used commonly for scanning electron microscopy. Glutaraldehyde alone or in combination with formaldehyde is the most effective fixative. This dialdehyde is most effective in preserving cellular proteins, for it introduces both inter- and intramolecular cross-links. The dialdehyde brings about fixation rapidly, whereas its rate of penetration into the tissue is slower than that by formaldehyde. This difference in the rate of penetration is expected, since the size of the formaldehyde molecule is smaller than that of the glutaraldehyde molecule. The ideal fixative is a mixture of glutaraldehyde and formaldehyde, since glutaraldehyde fixes rapidly and formaldehyde penetrates rapidly.

The optimal ratio of the two aldehydes in the mixture for a given specimen must be determined by trial and error, although mixtures for general use are given later.

Acrolein, an unsaturated monoaldehyde, can also be used in combination with other aldehydes, but not alone. Because it penetrates and acts extremely rapidly, acrolein has a special advantage over some other aldehydes that have a lower penetration capability. Consequently, this aldehyde is especially effective in the fixation of specimens of large size. Acrolein may also prove to be a better fixative for certain specimens for which other aldehydes are unsatisfactory. A case in point is provided by plant cells and embryos growing in culture; these are preserved satisfactorily with a mixture of acrolein (1.5%), glutaraldehyde (2%), and formaldehyde (1.5%) (Homes, 1974). It should be noted that acrolein is a hazardous chemical because of its flammability and extreme reactivity and that it must be used in a fume hood (also see p. 141).

Glutaraldehyde followed by OsO_4 is also used. Postfixation with OsO_4 has an additional advantage in that it imparts modest conductive property and chemical hardening to the fixed tissue. A mixture of glutaraldehyde and OsO_4 has proved useful for fixing delicate structures such as cilia and glandular and simple hairs. Aldehydes followed by OsO_4 and thiocarbohydrazide are also used (Kelley *et al.*, 1975; Malick and Wilson, 1975; see Hayat, 1978, for a review). The use of OsO_4 as a primary fixative is not recommended, although Horridge and Tamm (1969) used OsO_4 for preserving cilia of *Paramecium.* Osmium tetroxide–mercuric chloride (Parducz, 1967) is useful for some protozoa. It is a satisfactory fixative for ciliates, flagellates, and other motile specimens. The fixation is virtually instantaneous and thus facilitates the study of protozoan motile systems. This fixative, however, has been reported to be destructive to some protozoa (e.g., marine foraminifera) that possess only delicate, unitary surface limiting membranes (Marszalek and Small, 1969). These protozoa can be fixed satisfactorily with aldehydes. Osmium tetroxide-mecuric chloride is also useful for hardening the free surfaces of soft tissues, isolated cells, and cell cultures. Fixation with this mixture followed by freeze-drying preserves infected plant tissue quite well (Murphy *et al.*, 1974).

Cultured cells are best fixed with 2% glutaraldehyde in 0.1 mol/l cacodylate buffer (pH 7.2) containing 0.1 mol/l sucrose (Fig. 13.6). The total osmolality of this fixative is 500 mosmols, and the vehicle osmolality is 300 mosmols. Murine erythrocytes are best fixed with 0.1% glutaraldehyde in Kreb's buffer followed by postfixation with 2% glutaraldehyde (Arnold *et al.*, 1971). Lymphocytes are best fixed with 1% glutaraldehyde in 0.1 mol/l cacodylate buffer at 25–30°C followed by fixation in 4% glutaraldehyde for 6–12 hr at 4°C (Barber and Burkholder, 1975). Human leukemia cells are best preserved with 1% glutaraldehyde in 0.1 mol/l phosphate buffer (pH 7.2), which has an osmolality of 320 mosmols (de Harven *et al.*, 1975). It is recommended that tissue specimens be

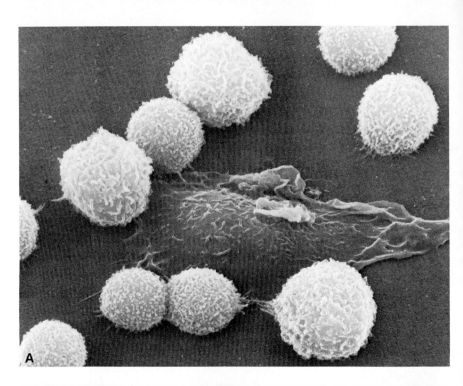

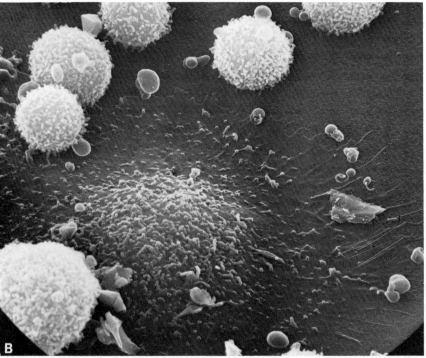

fixed with aldehydes of low concentrations for 30–60 min followed by further fixation with aldehydes of higher concentrations.

Soft marine organisms are best fixed with low concentrations (0.1%) of glutaraldehyde or OsO_4 in a balanced, buffered salt solution that is isosmotic with respect to the natural environment of cells (Ulmer and Honjo, 1973). For soft animal and plant tissues, glutaraldehyde can be used in concentrations ranging from 1 to 3%. The most commonly used concentration of glutaraldehyde and formaldehyde mixture (1:1) is 2%.

Aldehyde-fixed biological specimens remain osmotically active unless the duration of fixation is considerably prolonged, or until they are postfixed with OsO_4. Thus, aldehyde-fixed specimens remain vulnerable to distortion during the immediate postfixation treatment. It should also be remembered that aldehydes do not preserve most of the lipids, and substantial amounts of phospholipids are extracted during dehydration. This loss, however, can be minimized by adding calcium chloride to the dehydration solution.

The cross-linking properties of various fixatives used for transmission electron microscopy can be utilized to produce the desired degree of hardness in a tissue prior to dissection for scanning electron microscopy. Fixation with glutaraldehyde–formaldehyde makes the tissue tough, whereas OsO_4–mercuric chloride makes the tissue very hard and brittle. The chemistry of interactions between various fixative and cell components has been presented. Development of ideal fixatives for various types of cells and tissues to be studied with the SEM is urgently needed. The most reliable test for the versatility of a fixative for scanning electron microscopy is examination of the fixed specimen with the TEM.

Fig. 13.6. Mouse peritoneal cells. A, fixed with 1% glutaraldehyde in cacodylate buffer (pH 7.2) for 10 min at 22°C. A flattening macrophage and several lymphocytes show satisfactory surface morphology after fixation with glutaraldehyde alone. B, fixed with 5% glutaraldehyde in cacodylate buffer (pH 7.2) for 72 hr at 4°C, and postfixed with 1% OsO_4 in the buffer for 30 min at 22°C. A flattening macrophage and several lymphocytes are present. Note that artifactual blisters (0.2–0.6 μm in diameter) decorate the lymphocytes. These blisters appear to result from expansion of the plasma membrane caused by glutaraldehyde fixation. Postfixation with OsO_4 stabilizes this artifact; whereas in the absence of postfixation, dehydration and critical point drying treatments obliterate these blisters. The appearance of the blisters is not dependent on the osmolarity of the vehicle. Horizontal field width = 15.2 μm. (E. Shelton, unpublished.)

14

Fixation for Enzyme Cytochemistry

The objectives of an ideal fixation procedure for enzyme cytochemistry are (1) adequate preservation of the steric conformation at active site(s) of the enzyme molecule, (2) satisfactory morphological preservation, (3) immediate termination of metabolism so that autolysis is avoided, (4) stabilization of cell structure to withstand incubation and subsequent treatments, (5) breakdown of membrane permeability barriers to facilitate transmembranic diffusion of substrates and trapping agents, and (6) anchoring the enzyme to the site of its location (without appreciably decreasing its activity) so that its diffusion is prevented. A fixation procedure that could satisfy all the preceding objectives is not currently available.

The effect of fixation on enzyme activity constitutes a critical limiting condition for cytochemical detection of enzymes. The fixation must not destroy, either by chemical reaction or by physical extraction, the reactive groups in the enzyme molecule. Loss of activity of the enzyme molecule results from destruction of its active site(s).

For enzyme studies the preservation of chemical activity must take precedence over ultrastructural integrity, but there must be (at the least) adequate preservation of cellular structure. The exact location of an enzyme must be determined in reference to the preserved cellular organelles. When several enzymes possess overlapping substrate specificities, the activity of each enzyme can be identified by relating it to a specific identifiable cell organelle. This is the primary reason for expending so much effort to develop methods that preserve enzyme activity and cellular fine structure. New reactive groups should not be introduced by the fixative, since they may yield a false positive reaction. Finally, in certain studies it may be necessary to avoid introducing heavy metals through fixation; these may confuse the localization of reaction products resulting from the use of metallic capturing agents such as copper, bismuth, and manganese. The use of *en bloc* staining with uranyl acetate may create problems. This problem may become more serious, since the use of metals with low atomic numbers as capturing agents in enzyme cytochemistry is expected to increase in the near future.

At present, aldehydes and OsO_4 are the reagents of choice for preserving cellular structure for conventional electron microscopy. Because OsO_4 is a heavy metal oxide whose oxidizing can readily destroy enzymatic activity, its use in enzyme cytochemistry is limited, whereas many cytochemical reactions can be performed on tissues after aldehyde fixation. Also, relatively large tissue blocks can be prefixed especially with formaldehyde. Since aldehyde fixation imparts some degree of firmness to these blocks, the prefixed blocks can be cut into smaller pieces with minimal damage prior to further treatment. Prefixation with an aldehyde thus lessens the distortions introduced by mincing fresh tissues, prior to their incubation and postfixation with OsO_4. Fixation with aldehydes can preserve from 10 to 75% of the enzyme activity, depending on the enzyme, the tissue type, and the fixation conditions. It is advisable to use purified glutaraldehyde, at a low temperature, for a fairly short duration. For these and other reasons discussed earlier and later, aldehydes are used widely in the study of enzyme activity at the subcellular level.

Although formaldehyde and other aldehydes have been employed since 1951 (Seligman *et al.*, 1970) for the preservation of enzymatic activity at the light microscope level, their use in electron microscope cytochemistry is of relatively recent development. A serious attempt to explore the usefulness of several mono- and dialdehydes as fixatives for the preservation of both enzymes and fine structure was made in 1963 and 1964 by Sabatini, Barrnett, and associates. Since then rapid progress has been made in developing better methods for the preservation of enzyme activity at the subcellular level.

Presently, fixation with aldehydes is a prerequisite for many of the methods that are available for satisfactory fine structural demonstration of enzymes. The difference in the reactivity of various aldehydes with proteins should be considered in the selection of an appropriate aldehyde for the preservation of a given enzyme. Also, the duration of fixation and concentration of the fixative can be varied. Thus, it is possible to demonstrate a wide spectrum of enzymes, ranging from highly resistant alkaline phosphatases to labile succinic dehydrogenase, by choosing conditions appropriate for the adequate preservation of both enzyme activity and fine structure.

Cross-linking does not always inhibit enzyme activity, and even if only a small percentage of the activity is preserved, this usually is sufficient for precise localization of the enzyme. It is thought that losses up to 75% of total enzyme activity can still be compatible with a meaningful cytochemical localization. A recovery of 12% of acid phosphatase activity, for example, is sufficient to be visualized by both light and electron microscopy. It is possible that only a few of the amino groups present in the active site of enzyme molecule can maintain the activity of the enzyme. However, false positive results could be obtained when enzyme activity is too low. Under optimal conditions of fixation, the recovery of enzyme activity ranges from 75 to 10%, depending on the type of enzyme.

Different components of a cell react differently to the fixative with respect to the preservation of enzymatic activity. The degree of inhibition of enzyme activity may vary from one site to another in a given tissue. Such variation may be the result of quantitative differences in the initial levels of activity, differences in the sensitivity to fixation (which might reflect the differential inactivation of isozymes), or differences in the degree of fixation because of unequal penetration of the fixative. A case in point is 5-nucleotide phosphodiesterase, which shows greater inhibition in the cytoplasm than in the nucleus after glutaraldehyde fixation. The effect of glutaraldehyde fixation on this enzyme has been studied by Tsou et al. (1974) and Tsou (1975). Systematic studies on the effects of glutaraldehyde on phosphatase activities have been carried out (Janigan, 1965; Anderson, 1967; Ericsson and Biberfeld, 1967; Hopwood, 1967a, 1969b; Essner, 1973). Various degrees of inhibition of the activity of different enzymes by glutaraldehyde are given in Table 14.1.

Various enzymes differ in their susceptibility to fixation. In general, hydrolytic enzymes (e.g., phosphatases) are more resistant to aldehyde fixation than are oxidative reductases (e.g., succinic dehydrogenase). This is the reason that in many studies of the former, fixed tissues have been used, whereas the latter have been best studied in unfixed tissues. Since succinic dehydrogenase is one of the more insoluble or tightly bound dehydrogenases, fixation is not required to prevent its displacement or extraction from the tissue. Hajós and Kerpel-Fronius (1970) preserved succinic dehydrogenase in mitochondria of unfixed rat myocardium by histochemical incubation of 45 min, prior to fixation in 4% depolymerized paraformaldehyde followed by OsO_4, and Altman and Barrnett (1975) preserved the same enzyme in unfixed cryostat sections of rat liver and heart. It has been claimed that as long as the duration of incubation does not exceed 20–50 min, the intracellular components retain their acceptable morphological appearance. However, some other dehydrogenases, which are less susceptible to aldehyde fixation, should be studied in fixed specimens. Generally, labile enzymes (e.g., cytochrome oxidase, glucose-6-phosphatase, and succinic dehydrogenase) are not well preserved by glutaraldehyde. It should be noted that although cardiac muscle withstands incubation prior to fixation, the

TABLE 14.1

Comparative Inhibition of Enzymes in Various States by Glutaraldehyde[a]

Enzyme	State	Reaction time	% Activity remaining	Temp. (°C)	% Glutaraldehyde (g/100 ml)	pH	References
Acid phosphatase	Tissue blocks	1 hr	40	4	4	7.3	Anderson, 1967
		2 hr	28	4	4	7.2	Hopwood, 1967a
		6 hr	19–14	2	4	7.2	Janigan, 1965
		18 hr	15	4	4	7.2	Hopwood, 1967a
		24 hr	12	2	4	7.2	Janigan, 1965
	Perfused tissue	5 min	12	4–20	1.5	7.4	Arborgh et al., 1971
N-acetyl-β-glucosaminidase	Tissue blocks	18 hr	54[b]	2	4	7.2	Janigan, 1964
Alanine aminotransferase	Tissue blocks	1 hr	50	4	4	7.3	Anderson, 1967
Aryl sulfatase	Perfused tissue	5 min	30	4–20	1.5	7.4	Arborgh et al., 1971
Aspartate aminotransferase	Homogenate film	5 min	30–32	4	1	7.2	Papádimitrio and van Duijn, 1970
ATPase, Na- and K-activated	Homogenate	40–60 min	0	4	0.5	7.3	Ernst and Philpott, 1970
ATPase, Mg-activated	Homogenate	40–60 min	15	4	0.5	7.3	Ernst and Philpott, 1970
Carboxypeptidase	Crystalline	1 hr	40	23	1	7.5	Quiocho and Richards, 1966
Catalase	Crystalline	1 hr	12	23	4	7.2	Schejter and Bar-Eli, 1970
	Tissue blocks	2 hr	12.5	4	4	7.2	Hopwood, 1967a
		18 hr	6.6	4	4	7.2	Hopwood, 1967a
Cholinesterase	Tissue blocks	1 hr	75	4	4	7.2	Anderson, 1967
α-Chymotrypsin	Crystalline	1 hr	0.4–1.2	0	2.3	6.2	Jansen et al., 1971
iso-Citric dehydrogenase	Tissue blocks	1 hr	20	4	4	7.3	Anderson, 1967
Creatinine phosphokinase	Tissue blocks	1 hr	12	4	4	7.3	Anderson, 1967
β-Galactosidase	Tissue blocks	7 hr	43	2	4	7.2	Janigan, 1964
		24 hr	24	2	4	7.2	Janigan, 1964
β-Glucuronidase	Tissue blocks	2 hr	38	4	4	7.2	Hopwood, 1967a
		7 hr	24	4	4	7.2	Janigan, 1964
		18 hr	15	2	4	7.2	Hopwood, 1967a
		24 hr	12	4	4	7.2	Janigan, 1964
Glycogen phosphorylase b	Crystalline	10 min	40	23	0.05–0.01	7.5	Wang and Tu, 1969
α-Hydroxybutyrate dehydrogenase	Tissue blocks	1 hr	30	4	4	7.3	Anderson, 1967
Lactate dehydrogenase	Tissue block	1 hr	13	4	4	7.3	Anderson, 1967
Ribonuclease	Crystalline	—	32	4	4	7.2	Sachs and Winn, 1970
Subtilisin	Crystalline	30–60 min	10–15	23	2	6–8	Ogata et al., 1968

[a] From Hopwood (1972b).
[b] Tissue washed in running water before assay.

majority of the other tissue types (e.g., nervous tissue) disintegrate even with a short incubation prior to fixation.

Relatively slow penetration of the reaction medium into unfixed tissues has been established. The penetration of the reaction medium into the unfixed tissues is certain to be impeded by the natural permeability barriers of the intact cells. In particulate specimens and cultured cells, however, penetration does not constitute a significant problem. Diffusion artifacts in unfixed tissues make interpretation difficult. It is known that liver esterase tends to diffuse from unfixed sections; the thinner the section, the more diffusion will occur (Barrow and Holt, 1971). It is known that, in unfixed ultrathin sections, when the limiting membrane of certain organelles (e.g., lysosomes) is cut open during sectioning, the contained enzymes are released. A brief fixation prevents such leakage. Certain enzymes (e.g., mitochondrial adenosine triphosphatase) remain inactive in unfixed sections and thus are inaccessible to substrates; activation may be achieved by brief fixation with an aldehyde. Fixed tissues generally require shorter incubation durations than those required for unfixed tissues. It is concluded that incubation without prior fixation may result in (1) inadequate preservation of the fine structure, (2) diffusion and displacement of enzymes within the cell, as well as into the surrounding medium, (3) maintenance of the permeability barrier of membranes, and (4) difficulty in handling the specimens.

COMPARISON OF DIFFERENT ALDEHYDES

Various aldehydes differ in their ability to preserve enzyme activity and morphology of cell components. The activity of a given enzyme may be best preserved by a certain aldehyde that may not be suitable for the preservation of the activity of most other enzymes. The selective preservation of a certain enzyme by an aldehyde can be useful in the study of overlapping enzymes. A few examples will suffice. Succinic dehydrogenase is most susceptible to glutaraldehyde but can withstand a brief fixation in glyoxal, hydroxyadipaldehyde, crotonaldehyde (Sabatini *et al.,* 1963; Ogawa and Barrnett, 1965), or formalin (Walker and Seligman, 1963). Although mitochondrial LDH in the muscle tissue is well preserved after paraformaldehyde fixation, extramitochondrial LDH in skeletal muscle and liver is considerably inhibited (Fahimi and Amarasingham, 1964). In peroxisomes, catalase and some oxidases withstand glutaraldehyde fixation, but most mitochondrial oxidative enzymes are completely inactivated by such treatment. Hanker (1975) has presented a detailed discussion on the preservation of oxidoreductases.

Generally, glutaraldehyde is a stronger inhibitor of enzyme activity than is formaldehyde. Inhibition of enzymes in various states by glutaraldehyde is given in Table 14.1. Quantitative studies of the effect of glutaraldehyde and formaldehyde on several enzymes (β-galactosidase, N-acetyl-β-glucosaminidase, acid phosphatase, β-glucuronidase, catalase, various dehydrogenases, and amino transferases) indicated that the loss of enzymatic activity in the former was approximately twice that following the treatment with the latter (Janigan, 1964, 1965; Hopwood, 1967a; Anderson, 1967). This and other studies indicate that glutaraldehyde yields only moderate preservation of enzyme activity in comparison with the far better preservation yielded by fixation in formaldehyde. Generally, however, enzymes seem to be more firmly stabilized within the cell after fixation with glutaraldehyde rather than being somewhat diffusable after formaldehyde fixation.

The difference in the preservation of enzyme activity by glutaraldehyde and by formaldehyde is related to the difference in the molecular structure and size of these aldehydes. Being a monoaldehyde, formaldehyde is a smaller molecule than dialdehyde glutaraldehyde. Glutaraldehyde penetrates slowly but cross-links rapidly; in contrast, formaldehyde diffuses rapidly but cross-links slowly. In this connection, the difference in the rate and/or nature of denaturation of enzymes by these two aldehydes is also important. The presence of aldehyde groups on both sides of glutaraldehyde molecule makes it an extremely efficient cross-linking agent for enzymes. Inter- and intramolecular cross-links introduced by glutaraldehyde are very stable. The existence of α, β-unsaturated aldehydes in glutaraldehyde may account for its property of strongly cross-linking enzymes, although this proposition has been questioned (see p. 78).

It is generally thought that the quality of preservation of the fine structure is proportional to the number of protein cross-links introduced by the aldehyde, whereas the preservation of enzyme activity is inversely proportional to the number and speed of cross-links introduced. Therefore, enzyme activity is inhibited more by glutaraldehyde than by formaldehyde. However, increasing concentrations of formaldehyde decrease enzyme activity more than do comparable higher concentrations of glutaraldehyde. Because purified glutaraldehyde cross-links relatively slowly, inhibition of enzyme activity by this dialdehyde seems to increase progressively with increasing degrees of impurity.

Although it is a strong cross-linking agent, glutaraldehyde is a very useful fixative for enzyme localization. In fact, by a suitable choice of conditions, even the most labile enzymes can be detected with glutaraldehyde. Controlled application of glutaraldehyde should, indeed, be very useful in preserving enzymatic and antigenic activities. Since the best preservation of fine structure is obtained with glutaraldehyde, it has become the most widely employed aldehyde fixative for enzyme studies.

The ideal approach apparently would be to develop a formulation that will give satisfactory preservation of both enzyme activity and fine structure. The development of such a formulation should not be very difficult, since sufficient biochemical information regarding the differences in the reactivities of various aldehydes with proteins is available (presented earlier). By employing appropriate mixtures of aldehydes and their concentrations, it should be possible to demonstrate cytochemically a wide spectrum of enzymes as well as satisfactory preservation of the fine structure (Fig. 14.1). For instance, a mixture of formaldehyde and glutaraldehyde (Karnovsky, 1965) has been used for the localization of peroxidase (Cotran and Karnovsky, 1968), AcPase, and ATPase in the Golgi and endoplasmic reticulum in amoeba (Wise and Flickinger, 1971), and glutamic oxalacetic transaminase in the heart (Lee *et al.*, 1971).

Besides aldehydes, fixatives used for preserving immunoreactivity could prove useful for preserving enzyme activity. Examples of such fixatives are carbodiimide (Kendall *et al.*, 1971), cyanuric chloride (Goland *et al.*, 1969), dimethylsuberimidate (McLean and Singer, 1970; Davies and Stark, 1970), and parabenzoquinone (Pearse and Polak, 1975). It seems worthwhile to test the usefulness of these fixatives either alone or in combination with aldehydes in preserving enzyme activity. Preliminary reports indeed indicate that carbodiimide and dimethylsuberimidate are useful in preserving enzyme activity (see pp. 196). Recently, it was reported that parabenzoquinone, when used in combination with aldehydes, preserved both the immunoreactivity and fine structure. Another approach is to use reagents that selectively inactivate specific isoenzymes that otherwise are difficult to distinguish from each other (Koelle *et al.*, 1970).

EFFECT OF ALDEHYDES ON ENZYMES

For the validity of cytochemical reactions, it is important to know whether or not fixed and unfixed enzymes possess the same characteristics. Available evidence indicates that enzymes are modified by aldehyde fixation. A case in point is glutaraldehyde-fixed hepatic glucose-6-phosphatase, the optimum pH of which is displaced by 0.2 units toward the neutrality and its sensitivity to acid pH is increased (Berteloot and Hugon, 1975). However, this modification does not seem to have any effect on the cytochemical procedure. Aldehydes may change the adsorptive properties of enzymes. How and to what extent such a change in turn affects the capturing agent and the final reaction product is not known.

The reduced enzyme activity seen in the tissue fixed with aldehydes could be because these reagents depress the rate of hydrolysis of certain substrates by the enzyme and may also lower the affinity of an enzyme for a substrate. The extent of denaturation (loss of quaternary, tertiary, or secondary structure) of an enzyme molecule occurring during fixation is not known. As long as the conformation of an adequate number of active sites on the enzyme molecule is maintained, even though other regions may be denatured, sufficient enzymatic activity is retained

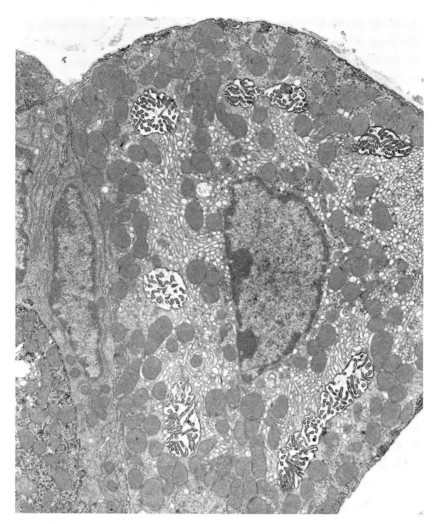

Fig. 14.1. A parietal cell from mouse gastric mucosa showing dense carbonic anhydrase reaction product in the microvilli of the intracellular canaliculi, basal folds, and cytoplasmic matrix. Specimens were fixed with a mixture of 4% formaldehyde, 3% glutaraldehyde, and 0.05% picric acid in 0.1 mol/l cacodylate buffer (pH 7.3) for 3 hr at room temperature, washed overnight with 8% sucrose in 0.2 mol/l cacodylate buffer at 4°C, frozen-sectioned 15 μm, incubated for 8 min in Hansson's (1967) medium, postfixed with 1% OsO₄ for 7 min at 4°C, and *en bloc* stained with 1% uranyl acetate for 10 min. × 8800. (Sugai and Ito, 1980.)

to yield positive reaction. Available information on the effects of aldehydes on the biochemical behavior of enzymes is quite meager.

FACTORS AFFECTING THE PRESERVATION OF ENZYME ACTIVITY

Satisfactory preservation of enzyme activity is controlled by many factors, which include (1) the type, ionic strength, and pH of the buffer, (2) the presence of neutral salts or hydrophilic compounds (e.g., sucrose and DMSO); (3) the presence of reducing or oxidative compounds or of heavy metal ions; (4) the type, concentration, and rate of penetration of the fixative solution; (5) the duration and mode of fixation and the prefixation wash; (6) the rate of penetration of the incubation medium; (7) the mode of incubation; (8) temperature; and (9) specimen size. In addition, the method of obtaining the specimen and the period of time elapsed between the death of the organism and processing of the tissue (i.e., postmortem interval) cannot be disregarded. Certain enzymes may undergo quantitative changes, depending on the anesthesia, decapitation, or surgical manipulation used. Activation of rat liver lysosomal enzymes has been demonstrated as a result of ether anesthesia. Failure to achieve rapid fixation falsifies the estimation and localization of certain enzymes.

Specimen Size

In enzyme cytochemistry, the size of tissue specimens is more critical than it is for routine study of the fine structure only. For an accurate localization of enzyme activity, tissue specimens should be sufficiently small ($\sim$50 μm) to be rapidly penetrated by both the fixative and the substrate. In this respect, unicellular organisms present little difficulty. Rapid and uniform fixative penetration of tissues can be achieved by vascular perfusion; this, however, is not always possible or practical to carry out. Therefore, tissues should be cut into the smallest blocks possible without disrupting the fine structure. Mechanical tissue sectioners are available, which yield more uniform and thinner and smaller specimens than those obtained by manual dicing of either fixed or unfixed tissues. Furthermore, the latter approach is likely to inflict mechanical damage on the tissue. Mechanical tissue sectioners can provide sections of even unfrozen and unfixed tissues.

Tissue sectioners in use are (1) TC-2 tissue sectioner (Smith and Farquar, 1965; Smith, 1970), which is a modification of the McIlwain Mechanical Chopper (McIlwain and Buddle, 1953); (2) Oxford Vibratome (Smith, 1970); (3) Om U Reichert microtome fitted with the Frigistor System; and (4) a simple and inexpensive tissue chopper that can be built in any laboratory (Shnitka *et al.*,

1968). The Vibratome has the advantages of cutting serial sections, providing uniformly thin sections ranging from 5 to 20 μm, and cutting unfixed and unfrozen tissue blocks. However, this instrument is slower than the TC-2 and may not cut certain fixed tissues (e.g., pancreas, muscle, and intestine). The TC-2 yields sections ranging from 25 to 250 μm and cannot cut sections of unfixed tissue thinner than 100 μm (also see p. 12).

Concentration of Aldehyde

The concentration of aldehydes is extremely important in achieving accurate and maximum preservation of enzyme activity. A higher concentration than the optimal provokes an excess of cross-links, inhibiting enzyme activity without improving morphological preservation. Low enzymatic activity could cast doubt on the validity of the cytochemical results. For example, a relatively low concentration of glutaraldehyde (1.3%) perfused for 6 min yields the best results for the preservation of ultrastructure and recovery of hepatic glucose-6-phosphatase activity. Fixation with low concentrations of glutaraldehyde (0.2–0.5%) strengthens the binding of certain enzymes to membranes. Studies by Ellar *et al.* (1971) indicate that after fixation with 0.5% glutaraldehyde, the release of enzymes such as ATPase, NADH dehydrogenase, and polynucleotide phosphorylase from the plasma membrane of *Micrococcus lysodeikticus* by washing with buffers is prevented and that these enzymes become more strongly attached to the membranes. The maximum enzyme activity is maintained after treatment with low concentrations of aldehydes for moderate periods of time, since these conditions permit sufficient diffusion of the aldehyde into the tissue specimen but limit the extent of cross-linking.

Although the optimal concentration of an aldehyde for maximum preservation of enzymatic activity in a given tissue is determined by trial and error, a general range of concentration from 0.5 to 2.0% is satisfactory for the majority of enzymes. According to Nagata and Murata (1972), the preservation of lipase activity in the pancreatic acinar cells of mice and rats was not affected when the specimens were fixed with 2.5–5.0% glutaraldehyde. Nevertheless, it should be noted that although a fairly wide range of glutaraldehyde concentrations is safe for morphological preservation, most enzymes cannot afford this luxury.

Duration of Fixation

Duration of fixation is critical in the preservation of maximum activity of enzymes. This is expected because aldehydes are very efficient cross-linking agents, and thus are potential inhibitors of enzymatic activity. Generally, as stated earlier, the longer the duration of fixation, the less the preservation of enzyme activity. Cholinesterase in crustacean muscle, for example, is completely inhibited when the specimen is fixed in 3% glutaraldehyde for 2 hr, whereas some cholinesterase activity is preserved after 20 min in the fixative (Spielholz and Van der Kloot, 1973). A marked reduction in the glucose-6-phosphatase activity in fresh-frozen sections of rat liver occurred as early as 1 min after immersion fixation with 3% glutaraldehyde; a corresponding reduction occurred after 5 min with formaldehyde and after 10 min with hydroxyadipaldehyde (Ericsson, 1966). The maximum preservation of the activity of this enzyme was obtained after the liver was fixed with glutaraldehyde by vascular perfusion of a short duration (1–2 min).

On the other hand, the increase in the inhibition of activity of certain enzymes is not substantial when fixation is prolonged within reasonable limits. Most of the inactivation of acid phosphatase is generally completed within a short period of time, and longer fixations exert a relatively minor inhibitory effect, but are essential for preserving cellular morphology. Most enzymes show loss of activity during the first 10–20 min exposure to the fixative, the loss being much more gradual thereafter. Apparently, the best approach is to use a duration of fixation that will yield satisfactory preservation of both the enzymatic activity and cellular morphology. However, in order to preserve excessively sensitive enzymes, morphological preservation may have to be sacrificed. The cross-linking of proteins (including enzymes) by glutaraldehyde can be terminated by adding hydrazine, which is not harmful to proteins, has no side reactions, and reacts rapidly with glutaraldehyde.

Temperature

In general, higher temperatures enhance the speed of chemical reactions between the fixative and the enzyme. Higher temperatures also increase the rate of penetration (diffusion) of the fixative into the tissue (and also speed up the rate of autolytic changes). Since aldehydes are potent cross-linking agents, in order to prevent excessive cross-linking of enzymes, fixation is generally carried out at 4°C.

pH

The pH of the fixative solution is a critical factor in the maximal preservation of enzyme activity. Not only do different cells and tissues tend to require a specific pH value, but various organelles in the same cell also react differently at a given pH with regard to the preservation of enzyme activity. In other words, at a given pH, the preservation of an enzyme in a particular type of organelle may be maximum, but at the same pH other organelles may not show maximal activity. It has been shown, for instance, that the optimal pH for demonstrating glucose-6-phosphatase in the endoplasmic reticulum, nuclear envelope, and

Golgi (in cultured chicken heart cells) ranges between 5.8 and 6.2, whereas in the nucleus the reaction product is produced only at pH 6.8 (Schäfer and Hündgen, 1971). Similar information on other enzymes and tissues is needed. Small changes in the pH of a buffer or a fixative solution may bring about large changes inside the cell, including the activity of its enzymes.

Buffers

The role played by buffers in the preservation of enzyme activity has not been fully appreciated. It is quite likely that a buffer wash prior to and/or following fixation with aldehydes alters enzyme activity. Phosphate buffer is known to inhibit completely the activity of glucose-6-phosphate dehydrogenase (Löhr and Walker, 1963), whereas cacodylate buffer tends to inhibit β-glucuronidase activity somewhat (Smith and Fishman, 1968). Lactate dehydrogenase shows ~25% lower activity in 0.4 mol/l Tris–HCl buffer (pH 7.4) than that in 0.1 mol/l phosphate buffer (pH 7.4) (Dahl and From, 1971). This reduced activity may be the result of conformational changes in the enzyme molecule.

It is appropriate to mention that, although information on enzyme activities as obtained through biochemical studies is significant, enzymes in an isolated state may possess different sensitivities to reagents, including buffers, from that seen *in situ*. It has been pointed out that in the case of succinate dehydrogenase, 35% higher preservation than that measured with biochemical methods is necessary to yield an observable difference in the tissue sections at the light microscope level (Riecken *et al.*, 1969). No such information is available at the subcellular level.

Little information is available regarding the effects of buffer washing of variable duration on the activity of various enzymes. Such information should be important for understanding the role played by buffers in the fixation for enzyme localization. Furthermore, not only the type of ions in the buffer, but also the electrolytes added to the fixation solutions exert a profound effect on enzyme activities. It cannot be assumed that various soluble enzyme systems in cells respond the same way to different buffer systems.

Arborgh *et al.* (1971) reported that immersion of the tissue following prolonged fixation with glutaraldehyde in a chilled buffer (0.1 mol/l Tris–maleate) resulted in a significant increase in the activity of certain enzymes (e.g., acid phosphatase and aryl sulfatase). A buffer wash, however, may inhibit the activity of other enzyme systems.

Mode of Fixation and Incubation

The available data indicate that the extent and quality of preservation of enzyme activity and ultrastructure are affected not only by the characteristics of the fixative but also by the method of applying the fixative to the tissue. Under *in vivo* conditions, the rate and depth of fixation are increased. Fixation by immersion, on the other hand, results in uneven preservation of enzyme activity due to slow penetration by the fixative. This limitation is serious in enzyme studies because longer durations needed to allow penetration of the tissue block are bound to inactivate the enzyme. Slicing the tissue into very small pieces to facilitate rapid penetration may cause excessive mechanical damage.

In the tissue fixed by immersion, the total activity of blood and tissue enzyme is measured, which introduces a nonsystematic fault by the variability of the amount of blood remaining in the organs. Vascular perfusion, in contrast, removes blood almost completely from the organs. This has been confirmed by von Matt *et al.* (1971), who quantitatively determined rat liver acid phosphatase activity in perfused and decapitated animals. They found that the perfusion method significantly increased the reproducibility of the results. Vascular perfusion eventually supresses natural inhibitors of enzymes. Hepatic glucose-6-phosphatase is known to have a natural inhibitor in the fresh liver. If it is necessary to use fixation by immersion, frequent changes of purified glutaraldehyde solution during fixation may help to minimize the destruction of enzyme activity.

Since enzyme inactivation is controlled, in part, by the duration of fixation, a mild but effective fixation can be obtained by a short vascular perfusion. Thus, more of the original enzyme activity can be preserved by vascular perfusion. Venous perfusion (Ericsson, 1966) and transparenchymal perfusion (Kanamura, 1971) with glutaraldehyde have been employed for cytochemical demonstration of glucose-6-phosphatase activity in rat liver. As much as 70% of the total activity of this enzyme has been preserved after vascular perfusion of the liver with 1% glutaraldehyde (Casanova *et al.*, 1972). After vascular perfusion with glutaraldehyde (2%) for 3 min followed by perfusion with the incubation medium, 55% of phosphatases' activity has been obtained in rat liver (Glauman, 1975). A mixture of glutaraldehyde (1%) and formaldehyde (3.7%) was employed as a perfusate in localizing glutamic oxalacetic transaminase activity in rat heart (Lee *et al.*, 1971).

Limited evidence indicates that the addition of hydrogen peroxide to glutaraldehyde results in better preservation of both fine structure and enzyme activity. Studies of nucleoside diphosphatase and thiamine pyrophosphatase activities in hepatocytes and other cells of rat indicate that glutaraldehyde–hydrogen peroxide preserves enzyme activity better than conventional glutaraldehyde (Goldfischer *et al.*, 1971). Improvements in the tissue preservation as well as in the speed and depth of fixative penetration have also been reported in the same study. In addition, frozen sections of tissues fixed in the presence of hydrogen peroxide have been reported to be less friable and easier to cut. Nevertheless, additional evidence is needed before the use of hydrogen peroxide can be recommended.

Another approach is to incubate the tissue during fixation (whether by immersion or vascular perfusion). In this procedure, the substrate protects the active site of the enzyme by competing with the aldehyde. Inactivation of aspartate aminotransferase, for example, was delayed by adding ketoglutarate to the aldehyde (Papadimitriou and van Duijn, 1970). Leskes *et al.* (1971) successfully incubated rat liver during fixation by perfusion via portal vein and succeeded in recovering glucose-6-phosphatase activity at a level that was ~80% of that present prior to fixation. The perfusion was carried out with 2% glutaraldehyde and lasted for 3–5 min. The exposure of the tissue to the substrate during fixation did not contribute any detectable background precipitate. Kanamura (1971) also indicated that, in the absence of the substrate from the fixation solution, the deposition of glucose-6-phosphatase reaction product in cells of proximal convoluted tubule of rat kidney was diminished. Still another approach is to carry out a short vascular perfusion with glutaraldehyde, followed by perfusion with a medium containing the enzyme substrate (Glaumann, 1975). This method, employed to preserve phosphatases in rat liver (Fig. 14.2), is as follows.

Under ether anesthesia, the abdomen is opened and a small tube connected with a peristaltic pump and a container for fixative is inserted into the inferior vena cava and anchored with two or three ligatures. The portal vein and the hepatic artery are cut open and perfusion with oxygenated 0.1 mol/l cacodylate buffer (pH 7.2) containing 0.1 mol/l sucrose is carried out for 1 min; the amount of this rinsing solution is 5 ml/100 g body weight. This is followed by perfusion with 2% glutaraldehyde in the same buffer–sucrose mixture for 3 min at 37°C.

After fixation, the buffer–sucrose mixture is perfused for 5 min, followed by perfusion with the incubation medium for 5 min. After waiting 10 min, the latter step is repeated two to four times (total incubation time is 30–60 min). Finally, the liver is perfused with the buffer–sucrose mixture for 3 min. Dimethyl sulfoxide (2%) is added to the incubation medium to increase the permeability of the membranes to the substrate and lead ion.

In order to minimize the loss of enzyme activity, exposure to excess, unbound fixative should be stopped uniformly and immediately after fixation is completed. As described, this can be accomplished by perfusing with sucrose–buffer mixture immediately after perfusion fixation. According to Rosene and Mesulam (1978), fixation should be carried out at 20°C to facilitate diffusion of the fixative, and sucrose–buffer mixture should be perfused at 4°C to cool the tissue, thus diminishing the disruption of either the tissue or the enzymatic activity. The procedure results in more homogeneous fixation and in addition offers a more precise control over the fixation process.

SINGLE-CELL SPECIMENS

When the cell envelope presents a barrier to the penetration by incubation media, prefixation before incubation becomes a necessity. Fixation with an aldehyde breaks this barrier by making the cell envelope more permeable. In

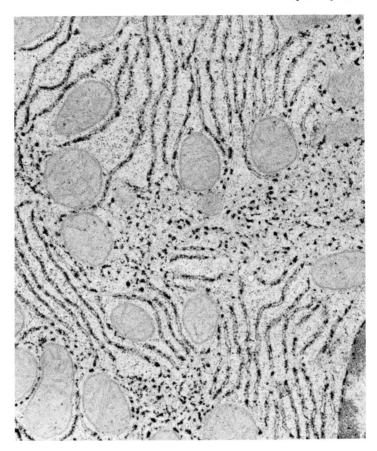

Fig. 14.2.   Portion of a hepatocyte from rat liver after perfusion incubation via the inferior caval vein for the demonstration of glucose-6-phosphatase. The reaction product is present throughout the endoplasmic reticulum (both rough and smooth) and the nuclear envelope. Note the relatively well-preserved ultrastructure in spite of the short fixation time and long incubation time. ×10,000. (H. Glauman, unpublished.)

addition, prefixation with glutaraldehyde minimizes enzyme displacement. Prefixation has been shown to bind enzymes (e.g., ATPase, NADH dehydrogenase, and polynucleotide phosphorylase in *Micrococcus lysodeikticus*) to membrane fragments (Ellar *et al.*, 1971). On the other hand, organisms that allow the substrate to reach the enzyme do not need prefixation, provided exposure of the living cells to the incubation medium is not excessively damaging. Also, for the localization of enzymes that are tightly bound to membranes, prefixation can be dispensed with. An example of such an enzyme is succinate reductase in *Micrococcus lysodeikticus*.

For accurate results it is recommended that the effects of fixation on enzyme activity be determined by assay methods and that the effect of fixation on enzyme location be determined by comparison with unfixed cells.

In single cells, enzyme activity has been localized by carrying out incubation either prior to or following fixation. For example, alkaline phosphatase in bacteria has been localized either without prefixation (McNicholas and Hulett, 1977) or with prefixation (Ghosh *et al.*, 1971; Wetzel *et al.*, 1970; Molnar and Szepessy, 1972). Gross inhibition of enzyme activity (especially soluble enzymes such as alkaline phosphatase) may occur when cells are prefixed with an aldehyde. Alkaline phosphatase in *B. licheniformis* showed ~68% inhibition within 6 min after treatment with 3% glutaraldehyde (McNicholas and Hulett, 1977). ADPase and ATPase in *Vitreoscilla* showed ~80% inhibition within 20 min after exposure to 1% glutaraldehyde (Burnham and Hageage, 1973). Cheng *et al.* (1970) reported that fixation in 5% glutaraldehyde for ~2 hr caused the periplasmic alkaline phosphatase to shift to the outer regions of the outer membranes in *Pseudomonas aeruginosa*.

Fixation parameters for localizing phosphatase activity in gram-negative organisms (Done *et al.*, 1965; Kushnarev and Smirnova, 1966; Munkres and Wachtel, 1967; Voelz and Ortigoza, 1968; Cheng and Costerton, 1973) and gram-positive organisms (Wood and Tristram, 1970; Okabayashi *et al.*, 1974) have been presented. Costerton and Marks (1977) have presented a detailed methodology for the preservation of enzyme activity in prokaryotes.

SUBCELLULAR FRACTIONS

Tissue fractions need not be fixed prior to incubation, since prefixation is not necessary to ensure satisfactory preservation of morphological details and enzyme activity. The criterion for the preservation of fine structure is obviously less stringent in this case than in intact tissue blocks. Some enzymes in particulate specimens are inactivated even by a brief fixation with low concentrations of aldehydes. Studies by Widnell (1972) showed that after a fixation of 15 min with 2% glutaraldehyde, 5'-nucleotidase was extensively inactivated (~95%) in plasma membranes and rough microsomes, whereas the nonspecific phosphatase was inactivated to a lesser extent (~60%). In these studies, the enzymes were investigated in unfixed cell fractions. Substantial destruction of phosphatase activity occurs in tissue fractions even in the presence of 1% glutaraldehyde (Marchesi and Palade, 1967; El-Aaser and Reid, 1975). Similar adverse effects have been reported for an oxidoreductase, a transferase, and a hydrolase (Anderson, 1967).

In an elaborate procedure, Baudhuin *et al.* (1967) did use glutaraldehyde prefixation for the study of acid phosphatase, Unger and Buchwalow (1975) prefixed isolated nuclei with formaldehyde (2%) for 15 min before incubation for the study of NAD-pyrophosphorylase, and Trelease (1975) prefixed both the cotyledons and isolated glyoxysomes with glutaraldehyde (2%) for the study of malate synthase; however, this practice does not seem to be necessary except in certain cases and is not recommended for routine studies. Generally, glutaraldehyde is used both to terminate the reaction after incubation has been completed and to impart light fixation to the tissue fraction. Studies of the effect of duration of fixation with glutaraldehyde and formaldehyde, prior to incubation, on the activity of glucose-6-phosphatase in the liver homogenate, indicate that the maximum enzyme activity is preserved when the duration of fixation does not exceed 30 min. The pellet is obtained by centrifugation and is postfixed with OsO_4. Reported fixation and other conditions for studying enzyme activity in subcellular fractions are given in Table 14.2.

In some studies it might be necessary to fix tissue fractions prior to incubation. A case in point is that of unfixed bovine heart mitochondria, which, when incubated with endotoxin, undergo transition from condensed state to orthodox configuration and simultaneously become uncoupled (Harris *et al.*, 1968). This transition may change enzyme activities within the mitochondria. It has been well documented that very low concentrations of glutaraldehyde stabilize unstable configurational states of mitochondria (Packer and Greville, 1969; Wakabayashi *et al.*, 1970).

If tissue fractions require fixation, it can be accomplished by using very low concentrations of glutaraldehyde (0.01–0.05%). These concentrations of glutaraldehyde seem to preserve both the activity of some enzymes and the structure of cell organelles. The usefulness of this approach in preserving the activity of individual enzymes needs further study. Formaldehyde is not recommended for fixing cell fractions. Since prolonged contact between the fixative and the cell fraction results in diminished preservation of enzyme activity, centrifugation at a high speed can be used to shorten the duration of fixation. It has been shown that monoamine oxidase activity in mitochondria is drastically decreased when the duration of fixation exceeds 6 min (Yoo *et al.*, 1974). In addition, a sucrose density gradient method can be used for rapid and controlled fixation of mitochondria by glutaraldehyde (Utsumi and Packer, 1967).

It should be noted that fixative concentration, as well as temperature and duration of homogenization, affect the enzyme activity that remains after fixation. By using purified 0.01% glutaraldehyde for 45 sec at 25°C followed by centrifugation (12,000 rpm) for 2 min (for mitochondrial fraction) or 6 min (for lysosomal fraction), ~70% of monoamine oxidase, ATPase, and cytochrome oxidase activities has been preserved (Asano *et al.*, 1976). In this study glucose-6-phosphatase in microsomal fractions was well preserved with 0.01–0.05% glutaraldehyde, but the activity was considerably decreased with glutaraldehyde concentrations higher than 0.05%. The activity of acid phosphatase was increased when the fraction was fixed with 0.1–1.0% glutaraldehyde; the effect of these higher concentrations of glutaraldehyde on lysosomal membranes is probably similar to that resulting from freezing and thawing.

An agreement on the extent of retention of enzyme activity in intact tissues versus that in homogenates is lacking. Various enzymes differ with respect to the degree to which their activity is affected by homogenization. In the case of hamster renal acid phosphatase, the activity is more or less the same in the two situations (Christie and Stoward, 1974). Magnesium-activated ATPase is significantly more active in intact sections than in homogenates (Stoward, quoted by Christie and Stoward, 1974). On the other hand, in the homogenate renal aminopeptidase loses half the activity it has in the intact, unfixed sections (Felgenhauer and Glenner, 1966). Comparative studies on the retention of enzyme activity in fixed sections versus that in fixed homogenates are needed. Such studies are also needed in unfixed sections versus unfixed homogenates.

The activity of acid phosphatase and other acid hydrolases (which show latency) has been increased by the addition of a lysosome labilizer (e.g., Triton X-100) to homogenates (e.g., Meany *et al.,* 1967). However, according to Christie and Stoward (1974), the activity of acid phosphatase in hamster kidney is depressed in the presence of Triton X-100. Although the reasons for this discrepancy are not known, it can be assumed that, if the homogenate contains a large number of intact lysosomes, then the addition of the labilizer should result in an increased activity of acid hydrolases. On the other hand, if the homogenate contains a significant number of ruptured lysosomes and free acid hydrolases, the labilizer will inhibit some of the enzyme activity. Thus, the efficiency of the homogenizer determines whether a labilizer is needed or not. This assumption, of course, can be tested by comparing the homogenates by electron microscopy.

CRYOULTRAMICROTOMY

Cryoultramicrotomy is a technique for obtaining ultrathin sections without using dehydration and embedding either prior to or after incubation. Another advantage of this technique is that enzyme-containing sites are available to substrates without the problem of membrane barrier. Although unfixed or fixed specimens can be sectioned, the use of the latter is recommended. The use of this technique in unfixed tissues is limited to enzymes that are insoluble and that therefore will not be displaced during incubation. The best approach is to use briefly fixed tissues, although conventionally fixed specimens have been used. The reason for using fixed tissues is that when ultrathin sections of unfixed tissues are brought into contact during incubation with the liquid medium con-

TABLE 14.2

Reported Conditions for Enzyme Cytochemistry on Subcellular Fractions

Authors	Material studied	Enzyme activity	Pretreatment of the fraction
Tice and Engel (1966)	Rabbit muscle microsomal fraction	Mg^{2+}-ATPase	Suspend in 0.25 M sucrose
Marchesi and Palade (1967)	Guinea pig red cell ghosts	ATPases, Mg^{2+} and Na^+/K^+	Make concentrated suspension in pH 7.0 Tris-HCl
Baudhuin et al. (1967)	Rat liver fractions, e.g., lysosome-enriched	"Acid phosphatase"	Treat sucrose suspension with 1.5% glutaraldehyde in pH 7.4 cacodylate; → Millipore filter; particles (~10 μm layer) washed with pH 7.4 buffer
Leskes et al. (1971)	Rat liver rough microsomes from gradient	Glucose-6-phosphatase (G-6-Pase)	Microsomes (4°C) in suspension gently sonicated (to disperse aggregates)
Cohen et al. (1971)	Pea seed zymogen granule fraction	An acidic ATPase, not cation-activated	Pellet suspended in pH 5.4 acetate buffer, 50 mM
Widnell (1972)	Rat liver microsomes and derived subfractions	5'-nucleotidase	Make concentrated suspension in 0.25 M sucrose use 1 ml in the incubation medium

(Usually as Suspensions)[a]

Incubation medium	Incubation conditions and workup	Remarks
mM Tris-ATP, 4 mM Mg²⁺; 125 mM Tris-maleate pH 7.5; 0.25 M sucrose; 1 mM Pb²⁺	17°C, '20 min, finally aldehyde fixation and negative staining	Enzyme half-inhibited by 1 mM Pb²⁺, Pb phosphate deposition (heavier if 37°C incubation) *inside* vesicles
mM Tris-maleate pH 7.0; 4 mM Mg²⁺; if desired, 100 mM Na⁺, 20 mM K⁺. 0.1 mM ouabam; 4 mM ATP; 0.5 mM Pb²⁺; added last with agitation	30°C (or 37°C) say 15 min; then → 4°C Add 2 vol glutaraldehyde, pH 7.0, 4% in 100 mM cacodylate; keep a few hr, then centrifuge hard; wash pellet. 1% OsO₄, pH 7.0	Red cell ghosts served as model system. Na⁺/K⁺ (not Mg²⁺) enzyme half-inhibited by 0.5 mM Pb²⁺; Pb phosphate deposition was on inner face of membrane; prefixation undesirable
crose-free medium containing β-glycerophosphate, presumably pH 7.4, and 2 mM Pb²⁺	20 min at room temp or 1 hr at 0°C pH 7.4 washing (5 min) and postfixation (1% OsO₄). Use *soft* Epon	Commercial filtration tube ('Filterfuge,' IEC) was modified for the centrifugal filtration step; heaviest deposits in large lysosomes
mM G-6-P; 50 mM cacodylate, pH 6.6; 2 mM Pb²⁺	25°C, 60 min; then → 4°C. Dialyze and sonicate (to disaggregate); then gradient centrifugation with final OsO₄	Amylase step, before microsomes isolated, useful to remove glycogen; deposition study was incidental to fractionation of vesicles and deposits
5.4 buffer; 10 mM ATP; 0.5 mM Pb²⁺ added with agitation	37°C, 60 min; centrifuge (20,000 g, 10 min); 2% glutaraldehyde for 1½ min, pH 7.2; recentrifuge then wash and OsO₄ fix	Deposits on membranes; some deposition in controls; poor results with β-glycerophosphate "etched" granules later used (Shain et al., 1974)
s-acetate pH 7.5, 5'-AMP, Mg²⁺, Pb²⁺, each 1 mM; 3–5 ml	22–25°C, 30 min; then →0°C. Stepped gradient centrifugation, 12 hr, 22,500 rev/min; band or pellet harvested and suspended in pH 7.2 cacodylate + Mg²⁺ OsO₄ 1% 0–4°C. 4 hr. Recentrifuge	Special care was taken over cleanliness and choice of controls; enzyme 55–60% inhibited by 1 mM Pb²⁺; deposition pattern heterogeneous in the vesicle population

(Continued)

TABLE 14.2

Authors	Material studied	Enzyme activity	Pretreatment of the fraction
El-Aaser and Reid (1969)	Rat liver fractions, especially membranes	5′-nucleotid-ase; acid phosphatase; G-6-Pase; ATPase	Fraction suspended in 0.25 M sucrose; use 1 ml $\approx$0.1 g liver (or, if plasma membrane fraction, $\approx$10 g
Morré *et al.* (1974)	Rat liver fractions, especially Golgi	NADH-terri-cyanide reduc-tase [→ppt. of cupric terrocyanide]	Pellet of <1 mm diam. fixed (intact) at 0–4°C for 10 min in 50 mM pH 7.2 phosphate/0.2 M sucrose containing 0.1% glutaral-dehyde, then wash for 1–3 hr, 0–4°C

[a] From El-Aaser and Reid (1975).
[b] Pb²⁺ signifies a lead salt, usually lead nitrate. Conditions for dehydration and subsequent steps were

taining the substrates, thawing causes morphological distortion, as well as displacement and loss of cellular substances, including enzymes.

Several enzymes have been localized in ultrathin frozen sections of tissues fixed in glutaraldehyde. Hydrolytic enzymes (Leduc *et al.*, 1967), nuclear phosphatases (Vorbrodt and Bernhard, 1968), nucleases (Zotikov and Bernhard, 1970), and 5′-nucleotidase (Leung and Babaï, 1974) have been localized by the use of this technique. A detailed discussion of cryoultramicrotomy is outside the scope of this book; a comprehensive discussion of its methodology for enzyme studies has been presented by Babaï (1977). The use of this technique in conjunction with X-ray microanalysis of enzyme activity has been discussed by Ryder and Bowen (1974) and Bowen and Ryder (1977). For its use in enzyme immunocytochemistry, the reader is referred to Scott and Avrameas (1968), Kuhlmann and Miller (1971), and Painter *et al.* (1973).

It should be remembered that cryoultramicrotomy presents many problems,

(Continued)

Incubation medium	Incubation conditions and workup	Remarks
mM Tris-maleate pH 7.2; 2 mM Pb^{2+}; 5 mM Mg^{2+}; 5 mM 5'-UMP (not AMP; also serves for acid phosphatase, pH 5.0) or G-6-Pase or ATP	2°C, say 15 min. for nucleotidase; but 1 hr or 20°C for microsomal G-6-Pase. Add glutaraldehyde, pH 7.2, to 1%; 15 min at 2°C. Spin; wash and cut up pellet; 1% OsO$_4$, 1 hr, 2°C	Little enzyme inhibition with 2 mM Pb^{2+}. Reagents filtered; for nucleotidase, fluoride can replace glutaraldehyde to terminate reaction; *don't* use OsO$_4$; deposition results discussed
10 ml; 6 mM Na, K-tartrate; 3 mM Cu sulfate; 85 mM (pH 7.2) K phosphate; 2% dimethyl sulfoxide; 0.5 mM K ferricyanide; 0.075 mM NADH; 0.2 M sucrose. *Fresh. Add dropwise in that order*	Pellet; 30°C, 40 min; wash 1 hr at 0–4°C in buffered sucrose; fix for 1 hr in buffered sucrose containing 1% glutaraldehyde-2% formaldehyde *or* 2% OsO$_4$	Morré *et al.* cite previous authors for the cupric ferrocyanide approach (M. J. Karnovsky and M. Roots; A. Lukaszyk); "quantitative morphometry" performed on "stained" membranes by counting with aid of a grid

nerally as commonly used in morphological studies. For controls, the usual practice is to omit the substrate.

and at best it can be considered as a complementary technique in the study of enzyme activity. The problems include the relatively less precise localization of enzyme activity, the loss of soluble and unbound enzymes during incubation, and the need for considerable skill and experience for preparation of ultrathin frozen sections.

15

Fixation
for Immunoelectron
Microscopy

An ideal fixative for immunoelectron microscopy would be that which preserves cellular structure and at the same time prevents loss and secondary location of antigens. In addition, it should not interfere seriously with the ability of antigens to bind with antibodies. Also, it should minimize background staining. If the antigenic determinant is an oligosaccharide, there is little problem in preserving its activity and cellular structure, because commonly used fixatives do not modify carbohydrate structure. However, most antigens are proteins or glycoproteins, the molecules of which undergo conformational changes when treated with standard fixatives. This is undesirable because secondary and tertiary structures of proteins are important in antigen-antibody reaction. A fixative capable of stabilizing the antigens without destroying their antigenicity and simultaneously satisfactorily preserving the cellular structure is not available.

Currently, it is therefore necessary to strike a compromise between satisfactory preservation of the fine structure and immunoreactivity, since these two goals appear opposed to each other. In other words, fixatives that improve the preservation of fine structure bring about a decrease in immunoreactivity. This is exemplified by glutaraldehyde, which is unsurpassed in preserving the cellular structure but which destroys antigenicity by denaturation of the antigens. The loss of antigenicity is considered to be caused primarily by the change in tertiary structure of proteins. This is to be expected, because the dialdehyde is an extremely efficient protein cross-linking agent. Formaldehyde is relatively less destructive to antigenicity but is also less efficient in preserving cellular structure. Treatment of bovine serum albumin with formaldehyde causes little inhibition of its ability to bind with the corresponding antibody (Habeeb, 1969).

Even the mode of fixation may affect the preservation of immunoreactivity. For example, immersion fixation with formaldehyde resulted in better immunolocalization of neurohypophyseal hormones (vasopressin) than in the case of specimens fixed by vascular perfusion with a mixture of formaldehyde and glutaraldehyde (Van Leeuwen, 1977). In the absence of an ideal fixative, generally solutions of low concentrations of glutaraldehyde or paraformaldehyde, or a mixture of these two aldehydes, are used for brief periods of time for fixing cells and tissue specimens.

In general, for a given low concentration of glutaraldehyde, the loss of antigenicity increases with time of exposure, but the loss does not continue indefinitely after a certain period. The loss of antigenicity also increases with increasing concentrations of the fixative, although the response of individual antigens varies quantitatively (Kraehenbuhl and Jamieson, 1974a,b). It has been suggested that 0.5% glutaraldehyde is desirable for surface localization procedures, whereas 4% formaldehyde is useful for diffusion localization procedures.

Different specimens differ in their susceptibility to a given aldehyde with respect to destruction of their antigenicities. A few examples will suffice. When 0.5% glutaraldehyde is used for 16 hr at room temperature, 40% of the capacity of bovine trypsinogen to bind to its antibodies is lost (Kraehenbuhl and Jamieson, 1972). On the other hand, 2% glutaraldehyde used for 1 hr at 0°C has little effect on the antibody binding capacity of bovine ribonuclease (Painter *et al.*, 1973). These and other examples indicate that the best fixation method for one antigen may not necessarily be suitable for another antigen. A dilute solution (1–2%) of paraformaldehyde can preserve antigenicities in some specimens. There are a few cases in which even a mixture of glutaraldehyde and paraformaldehyde can allow the retention of sufficient antigenicity. One such example is the antigenicity of vasopressin in the posterior pituitary gland (Rufener and Nakane, 1973).

A mixture of formaldehyde and picric acid (Zamboni and De Martino, 1967) has also been used. The use of picric acid in immunoelectron microscopy is justified, because it presumably reacts with proteins, precipitating them *in situ* without appreciable denaturation (De Martino *et al.*, 1972). The fine structure of both tissues and free cells is preserved satisfactorily by this mixture (Stefanini *et al.*, 1967). The mixture has also been used for studying proteinaceous hormones by peroxidase-labeled antibody technique (Mazurkiewicz and Nakane, 1972;

Moriarity and Halmi, 1972) and, with ferritin-conjugated antibody, for detecting antigens localized in renal structures (Accinni *et al.*, 1974). Duration of fixation should be kept to a minimum (~15 min).

Penetration of antibodies through plasma membrane is a problem because of the stabilizing effects of aldehyde fixatives. As a result, only surface reaction is observed. Another problem is impenetrability of the fixed cytosol matrix. Several approaches have been used to facilitate the penetration of the plasma membrane by antibodies. These include enzymatic digestion of the cell surface (Kuhlman *et al.*, 1974), detergents such as Triton X-100, saponin, and digitonin (Dales *et al.*, 1965); solvents, freezing, and thawing (Leduc *et al.*, 1969); and mechanical sectioning with a Vibratome (after primary fixation) (Pickel *et al.*, 1975). These treatments have not been very successful, since either they cause excessive damage to the cell structure or antibody penetration is too slow. Treatment with detergents, for example, disrupts the structure of membranes.

Saponin is considered superior to Triton, since the former is less damaging to membrane morphology (Ohtsuki *et al.*, 1978). Saponin in combination with aldehydes was used as a membrane-attacking agent. Although the effect of saponin on cell membranes has been studied thoroughly (Seeman, 1967; Graham *et al.*, 1967), the actual mechanism of saponin interaction with the membrane is still not well understood (Assa *et al.*, 1973). The mixture used for BHK-21 cells infected with the Shope fibroma virus is composed of 0.05% saponin, 0.0125–0.05% glutaraldehyde, and 1% formaldehyde (Bohn, 1978). Cell monolayers were treated with the mixture for 5 min at 4°C, followed by the aldehyde mixture without saponin for another 45 min in the cold. After washing in the buffer, specimens were incubated and postfixed in 1% OsO_4. This mixture allows antibodies to penetrate to intracellular sites, does not alter the antigenic structure excessively, and provides an adequate cell preservation. The concentration of the three ingredients in the mixture is critical, and the optimal concentration for each cell system must be found by trial and error. It should be noted that increasing concentrations of glutaraldehyde depreciate the efficiency of saponin.

In those cases in which even a mild treatment with an aldehyde results in a substantial inactivation of an antigen, there are at least two alternatives. One solution is to use bifunctional diimidoesters, such as diethylmalonimidate (DEM) and dimethylsuberimidate (DMS), and carbodiimides as fixatives rather than aldehydes. These reagents are known to preserve protein activities. Indeed, DEM has been used as a fixative for immunoelectron microscopy of erythrocytes (McLean and Singer, 1970). The chemistry and use of DMS and carbodiimides have been discussed on pp. 195 and 198.

Since the preceding bifunctional reagents, when used alone, fail to preserve the fine structure satisfactorily, they have been commonly used in combination with aldehydes. A mixture of glutaraldehyde and water-soluble carbodiimide (Yamamoto and Yasuda, 1977) has proved effective in preserving ultrastructure, as well as in localizing intracellular antigens in cultured cells (Willingham and Yamada, 1979) (Fig. 15.1). These authors indicate that this mixture obviates the problems of membrane impermeability, nonspecific globulin binding, and impermeability of fixed cytosol protein matrix. This procedure produces a permeable cytosol matrix so that antibodies can gain access to fixed proteins therein.

Another solution may be to devise a fixative that would preserve cellular structure by cross-linking primarily carbohydrates and lipids rather than proteins. This approach may prove useful because many antigens contain some carbohydrate moieties that could be fixed with little loss of the antigenicity that resides in the accompanying protein moiety. In such a procedure, the hydroxyl groups of carbohydrates are oxidized with periodate to produce aldehydes, which in turn can be made to react with diamino compounds (e.g., lysine), resulting in the cross-linking of carbohydrates (Nakane, 1973). Paraformaldehyde can be added to the periodate–lysine mixture, since lysine does not interact with the monoaldehyde at neutral pH. Such a mixture containing periodate, lysine, and paraformaldehyde was used by McLean and Nakane (1977). Periodate is employed to oxidize carbohydrates to form aldehyde groups, lysine is used to cross-link the newly formed aldehyde groups, and paraformaldehyde is added to achieve some stabilization of proteins and lipids. This mixture was developed primarily for fixing the carbohydrate moiety in order to stabilize the molecule without significantly denaturing the antigenic sites. Approximately 45% of the antigenicity is retained by this procedure. The quality of ultrastructural preservation is better than that obtained with most other fixative solutions used for immunoelectron microscopy.

However, studies with Novikoff tumor cells indicate that the mechanism of periodate–lysine–paraformaldehyde fixation, at least for cell surface-glycoproteins, does not involve extensive cross-links between oxidized carbohydrate moities and the free base lysine (Hixson *et al.,* 1981). It is suggested that a complex interaction occurs between periodate oxidized plasma membrane glycoproteins and polymeric complexes of lysine and formaldehyde. The fixatives containing only formaldehyde and lysine may be of greater value for immunocytochemical studies.

In conclusion, it is emphasized that various fixatives have different effects on the antigenicity of different antigens. Optimal localization of different antigens will require a systematic evaluation of different fixatives and fixation times depending on the antigen–antibody system utilized and the tissue studied (Sisson and Vernier, 1980). Since prefixation is essential for stabilization of cellular structure and prevention of extraction and dislocation of antigens, the best approach currently available is to fix very small tissue slices with a mixture of formaldehyde (1%) and glutaraldehyde (0.05–0.5%) for relatively short durations. Nonspecific staining due to cross-linking of antibodies with glutaraldehyde can be prevented by incubating the tissue specimens in dilute solutions of lysine

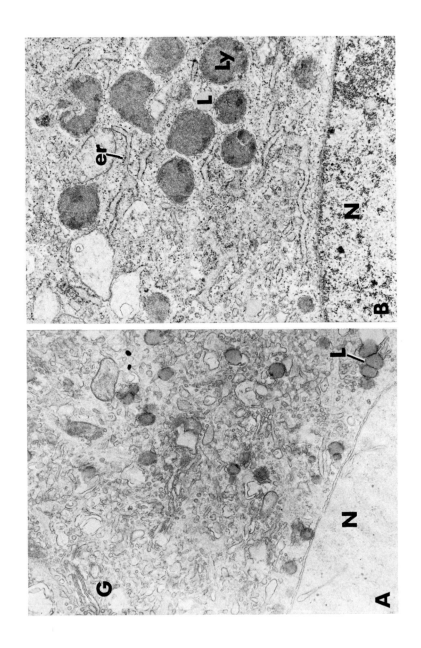

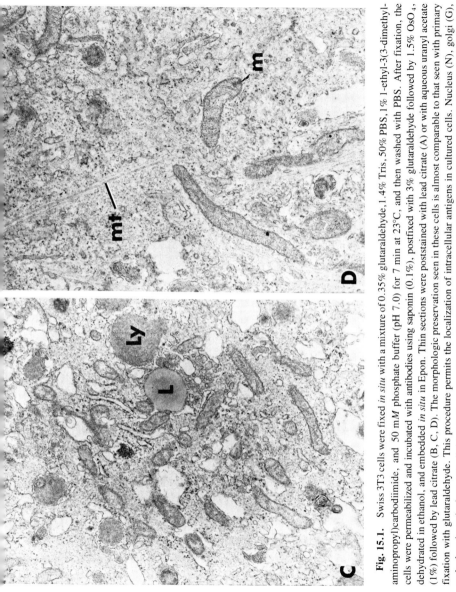

Fig. 15.1. Swiss 3T3 cells were fixed *in situ* with a mixture of 0.35% glutaraldehyde, 1.4% Tris, 50% PBS, 1% 1-ethyl-3(3-dimethyl-aminopropyl)carbodiimide, and 50 mM phosphate buffer (pH 7.0) for 7 min at 23°C, and then washed with PBS. After fixation, the cells were permeabilized and incubated with antibodies using saponin (0.1%), postfixed with 3% glutaraldehyde followed by 1.5% OsO$_4$, dehydrated in ethanol, and embedded *in situ* in Epon. Thin sections were poststained with lead citrate (A) or with aqueous uranyl acetate (1%) followed by lead citrate (B, C, D). The morphologic preservation seen in these cells is almost comparable to that seen with primary fixation with glutaraldehyde. This procedure permits the localization of intracellular antigens in cultured cells. Nucleus (N), golgi (G), endoplasmic reticulum (er), lipid (L), lysosome (Ly), mitochondrion (m), and microtubules (mt). ×15,600 (A, C); ×13,000 (B); ×19,500 (D). (Willingham and Yamada, 1979.)

(0.1 ml/l) in serum (10%) in Tris/saline. Freezing methods with fixed specimens have also been attempted (Kraehenbuhl and Jamieson, 1974b; Sisson and Vernier, 1980).

FIXATIVES

Method I. Periodate–lysine–paraformaldehyde. This mixture was introduced by McLean and Nakane (1977), and the following method of its preparation is essentially the same as given by these workers. Sufficient dibasic sodium phosphate (0.1 mol/l) is added to lysine–HCl (0.2 mol/l) until the pH has reached 7.4. The solution is diluted to 0.1 mol/l lysine with 0.1 mol/l sodium phosphate buffer (pH 7.4). The resulting phosphate concentration is 0.05 mol/l. Aqueous 2% paraformaldehyde (1 part) is added to the lysine–phosphate buffer (3 parts). To this solution is added sufficient sodium m-periodate to attain periodate concentration of $\sim$0.1 mol/l. The pH of the mixture is $\sim$6.2.

After fixation, tissue specimens are washed in the following 0.05 mol/l sodium phosphate buffer (pH 7.2) series: buffer containing 7% sucrose for 4 hr, buffer containing 15% sucrose for 4 hr, and buffer containing 25% sucrose and 10% glycerol for 2 hr. Specimens are embedded and quick-frozen.

Method II. Glutaraldehyde–formaldehyde–parabenzoquinone. A mixture of glutaraldehyde (1%), formaldehyde (1.5%), and parabenzoquinone (2%) in 0.1 mol/l cacodylate buffer is used for 2–4 hr. The mixture was employed for localizing peptide antigens in neurons (Larsson, 1977).

Method III. Formaldehyde–picric acid

Saturated aqueous solution of picric acid	150 ml
Paraformaldehyde powder	20 g
2.52% NaOH	a few drops

A clear solution is obtained by heating at 60°C and alkalinization. After cooling and filtering, the mixture is made up to 1000 ml by adding phosphate buffer (pH 7.3, 320 mosmols). Tissue specimens are fixed for 15 min, thoroughly washed, and then stained with specific ferritin conjugated antisera. After washing thoroughly in cold 0.01 mol/l phosphate buffered saline (PBS), pH 7.2, the specimens are postfixed in 1% buffered OsO_4. For further details, see Stefanini *et al.* (1967) and Accinni *et al.* (1974).

Method IV. Glutaraldehyde–carbodiimide. The stock phosphate buffer is made as follows:

Tris base	1.4 g
Dibasic $Na_2HPO_4 \cdot 7H_2O$	0.67 g
Monobasic $NaH_2PO_4 \cdot H_2O$	0.345 g

Dulbecco's PBS   50 ml
Distilled water   50 ml

This buffer is stored at 4°C, and brought to room temperature before use.

Fixative mixture:

Stock buffer   10 ml
Carbodiimide (EDC)   100 mg
50% glutaraldehyde   10-100 μl

The final concentration of glutaraldehyde is 0.05–0.5%, and the pH is adjusted to 7.0 with 1 mol/l NaOH (requiring 1–8 drops). At precisely 4 min after initial mixing, the fixative solution is added to the cell culture dish that had been prewashed with PBS. Fixation time is 7 min, followed by washing with PBS (Willingham and Yamada, 1979) (Fig. 15.1).

16

Use of Dimethylsulfoxide in Fixation

To avoid confusion, the well-known designation for dimethylsulfoxide, DMSO, will be used instead of its correct abbreviation, Me_2SO. Dimethylsulfoxide is a dipolar aprotic solvent, is completely miscible with water, and is commonly used as a solvent for water-insoluble substances. It is a mildly toxic electrolyte and has the property of penetrating living tissues without causing significant damage. This characteristic of DMSO is probably related to its polar nature, its ability to accept hydrogen bonds, and its relatively small and compact structure. It has a tendency to accept rather than to donate protons. These and other properties of DMSO facilitate its ready association with water, proteins, nucleic acids, carbohydrates, ionic substances, and many other cellular constituents. Perhaps the most relevant properties of DMSO to fixation are its ability to replace some of the water molecules associated with cellular substances and to facilitate the movement of molecules, including those of aldehydes, through biological membranes. The solvent action of DMSO weakens or breaks down the permeability barrier of tissues during fixation. The substitution of DMSO for water in membranous proteins is likely to alter their configurations, which may account for easy penetration of ions and molecules through cell membranes. Possible mechanisms involved in the effect of DMSO on the permeability of biological membranes have been discussed by Jacob *et al.* (1964) and Franz and van Bruggen (1967).

The effects of DMSO on biological systems are numerous; only a few of these are mentioned here. Dimethylsulfoxide is thought to form hydrogen bonds with proton donor groups on biopolymers, and these bonds are stronger than those formed with water. Many electron transfer reactions proceed relatively slowly in water because of its high dielectric constant, which acts as an insulator and keeps charged molecules apart (Hanker, 1975). Dimethylsulfoxide, on the other hand, has a relatively low dielectric constant, has less tendency to keep charged molecules apart, and facilitates electron transfer. Dimethylsulfoxide also forms hydrate structures in which the hydrogen bonds between DMSO and water or other proton donor groups are stronger than the hydrogen bonds between water molecules. Hanker (1975) has explained the intercalation of DMSO between associated water molecules. Dimethylsulfoxide affects hydrophilic bonding, and at higher concentrations may also affect hydrophobic bonding in proteins. Since DMSO contains methyl groups, the latter effect is likely.

The previous mentioned effects are reversible as long as DMSO is used at low concentrations (less than 20%) and low temperatures (lower than 25°C). At higher concentrations or temperatures, DMSO may rupture hydrophobic bonds, which may result in an irreversible loss of protein activity (Henderson *et al.*, 1975). This effect can be considered to result from protein denaturation, because hydrophobic bonding is important in the maintenance of the structure of certain proteins.

Dimethylsulfoxide may also affect the structure of omnipresent water. Since the activity of water is not necessarily the same in the different states of water, DMSO may exert an indirect effect on biological systems by virtue of the changes that it causes in the liquid structure of water (Szmant, 1975).

On the other hand, it has been proposed that DMSO, when used as a cryo-protectant, prevents denaturation of cell membranes by acting as a buffer against the high concentration of damaging electrolytes (e.g., NaCl) in the bathing medium during freezing (Elford and Walter, 1972). In addition, DMSO may act as an intracellular pH buffer, and thus may help to prevent denaturation of intracellular proteins. Indeed, it has been suggested that DMSO stabilizes labile proteins or enzyme preparations at low temperatures, and thus may be useful for the storage of tissues in the cold (but unfrozen) state. That DMSO postpones both the cell shrinkage and the onset of cation leakage to higher osmolarities has been demonstrated (Farrant, 1972). In this case, the onset of sodium or potassium leaks may be linked to cell shrinkage. It has been suggested that the efficacy of DMSO may result from its ability to postpone cation leakage to higher os-molarities.

Dimethylsulfoxide is known or claimed to have a wide spectrum of biological action (Jacob and Herschler, 1975) such as cryoprotection (Karow, 1974; Hayat,

1973, has reviewed its role as a cryoprotectant), radioprotection (Ashwood-Smith, 1967, 1975; Lappenbuschand and Willis, 1970), and analgesia (Becker *et al.*, 1969), and percutaneous absorption (Elfbaum and Laden, 1968). We are interested here only in its role in the preservation of ultrastructure and enzyme activity.

EFFECT ON ENZYMES

Various enzymes differ with respect to their activity after exposure to DMSO. There are enzymes that seem to be unaffected by high concentration of DMSO, whereas other enzymes are potentiated in catalytic activity, and still others are inhibited in relatively dilute DMSO solutions. These responses may reflect pH changes or modifications of enzyme conformation in terms of changes in functional groups. These possibilities have been discussed in detail by Rammler (1967, 1971).

There is some evidence to indicate that better preservation and staining of certain enzymes are obtained when specimens are treated with DMSO. Several possible explanations can be offered to explain the improved localization of enzyme activity. Dimethylsulfoxide increases the permeability of cell and/or organelle membranes. It penetrates cells rapidly, and thus exerts its effect on intracellular sites, as well as on the limiting cell membranes. Dimethylsulfoxide, for example, may cause the release of lactate dehydrogenase from cardiac muscle mitochondria (Feuvray and de Leiris, 1971). It appears that substrate molecules are made more available to an enzyme in the presence of DMSO. The increase in permeability of the lysosomal membrane to enzyme substrate in the presence of DMSO has been demonstrated (Misch and Misch, 1967, 1968, 1969). Similarly, 10% DMSO caused rapid penetration by succinate and isocitrate dehydrogenase media into the mucilage covering the hyphae of the fungus *Cercosporella herpotrichoides* (Reiss, 1971). A more homogeneous activity of succinic dehydrogenase has been obtained in the mitochondria of cardiac and skeletal muscles of the mouse by adding 3–5% DMSO to the copper ferrocyanide incubation medium (Makita and Sandborn, 1970, 1971). This improvement is also related to an increase in osmolality of the medium (to 800–900 mosmols) because of the presence of DMSO.

Improvement in the preservation of enzyme activity is also related to electron transfer facilitated by DMSO. Dimethylsulfoxide, for instance, facilitates electron transfer from cysteine, a number of coenzyme-linked dehydrogenases, and monoamine oxidase (MAO) to a variety of artificial acceptors such as ferricyanide (Hanker *et al.*, 1970, 1973; Hanker, 1975).

Several approaches can be utilized to obtain improved preservation of enzyme activity in the presence of DMSO. The rate of penetration by the incubation media for certain enzymes can be increased by adding DMSO. The majority of incubation media penetrate the tissue at a rather slow rate; various components of the incubation medium probably penetrate at different rates. Copper ferrocyanide medium for succinic dehydrogenase, for instance, does not penetrate more than 20 μm in 30 min at room temperature. Furthermore, the extent of reaction between the medium and the enzyme varies, depending on the distance of the reaction site from the surface of the tissue block. It is well documented that the addition of DMSO to the incubation media used for the demonstration of several coenzyme-linked dehydrogenases and MAO results in increased activity. No reaction is obtained when DMSO is omitted from the incubation medium for the localization of MAO (Bloom *et al.*, 1972; Hanker *et al.*, 1972). In this method, ferricyanide is used as a suitable artificial electron acceptor and tryptamine or tyramine as the substrate.

The prolonged incubation necessary for adequate penetration and interaction may result in the formation of artifacts and damage to the fine structure of the tissue. The incubation time can be shortened by accelerating the rate of penetration of the incubation medium by the addition to it of DMSO. Enhanced staining of enzymes has been demonstrated by the addition of DMSO to the incubation medium (Reiss, 1971; Makita and Sandborn, 1971). Dimethylsulfoxide was added to naphthol AS BI phosphate medium for the localization of acid phosphatase activity in rat and slug liver (Bowen, 1971).

An alternate approach is to treat the tissue specimen with DMSO prior to incubation. Such an approach has been utilized by Göthlin and Ericsson (1971) to obtain localization of acid phosphomonoesterase activity in brush border osteoclasts. They immersed the fracture callus in 20% DMSO for 24–48 hr prior to incubation. Another approach is to fix the tissue in the presence of DMSO during immersion or vascular perfusion. The addition of DMSO to the fixative has been shown to result in improved preservation of frozen tissues (Etherton and Botham, 1970).

As mentioned earlier, exposure of the tissue to DMSO does not result in increased activity of all enzymes. In fact, the addition of DMSO to the fixative, incubation media, or pre- or postfixation rinsing solutions inhibits the activity of certain enzymes. Furthermore, DMSO can cause greater diffusion of the cytochemical reaction product (Davies and Garrett, 1972).

EFFECT ON ULTRASTRUCTURE

The solvent properties of DMSO and its influence on membrane permeability have prompted the use of this reagent during fixation. The presumed advantage of this approach is that, with the DMSO treatment, the fixative is able to enter the cell without traversing a membrane barrier because of the solvent action of this reagent. This approach has also been claimed to be useful for preserving the fine structure when tissue specimens are fixed for a brief period of time in order to avoid excessive enzyme inhibition. According to Sandborn *et al.* (1969, 1975), DMSO has proved useful in the preservation of ultrastructure as well as enzyme activity. These workers have indicated an improvement in membrane preservation, an increase in the continuity between organelles, and greater precision in enzyme localization. Kalt and Tandler (1971) also indicate that a larger amount of membranes is retained in the cytoplasm of embryonic liver cells fixed in the presence of DMSO. The implication is that in the absence of DMSO, a certain amount of membrane is lost during the prolonged durations required for the fixative to traverse the plasma membrane of the cell. Dimethylsulfoxide blocks certain membrane receptors (Wincek and Sweat, 1976), although many of these effects can be reversed by washing (Shlafer and Karow, 1975).

One should be aware of the fact that DMSO imposes ionization on the constituents of an aqueous buffer system to which it is added. Dilution with DMSO will change not only the pH but also the concentration of the buffer. Therefore, the pH and osmolarity of the buffer should be checked after adding DMSO; otherwise the results of the experiment may relate simply to a change in the pH of the buffer rather than to any effect of DMSO. It should be emphasized that the addition of DMSO results in higher osmolarity of the incubation medium.

Caution is warranted in the use of DMSO, for the production of artifacts by this reagent cannot be overlooked. A few examples of the effect of low concentrations of DMSO on the fine structure follow. It has been demonstrated that the action of DMSO at the cell surface may produce blebbing in hepatocytes (Shilkin *et al.*, 1971). Even a brief (2 min) perfusion of isolated rat heart with DMSO resulted in ultrastructural alterations; the effect on the T system and mitochondria was immediate (Feuvray and de Leiris, 1971). Isolated rabbit kidneys perfused with 10% DMSO solution showed relatively high sensitivity: proximal tubular cells, including their microvilli, were disrupted, whereas vascular endothelial cells remained intact (Jeske *et al.*, 1974). These workers suggested that ultrastructural effects of DMSO must be considered in assessment of damage done in whole kidneys during freeze-preservation studies. Malinin (1973) found marked cytological alterations in cultured rhesus kidney cells after incubation with as low as 7.5% (0.05 mol/l) DMSO for 10 min at 4°C.

Dimethylsulfoxide at a concentration of 10% in Ringer's solution within 30 min at room temperature transformed the native myelin into a new, highly ordered structure with a repeat period approximately two-thirds that of the original structure (Kirschner and Caspar, 1975). This effect may be accounted for by the loss of water from the spaces between the membrane units without significant modification of the bilayer. The transformation is reversible when DMSO solution is replaced by Ringer's solution. Dimethylsulfoxide at low concentrations (>2.5%) inhibits vacuole-forming capacity of *Tetrahymena* cells, whereas cell motility and cell division remain unaffected (Nilsson, 1974). These and most other alterations caused by low concentrations of DMSO are reversible. That DMSO, when used during perfusion fixation, causes shrinkage and argentophilia in neurons and damaged neurons has been demonstrated (Cammermeyer, 1980).

Dimethylsulfoxide used in as high as 10% concentration has been reported to stabilize microtubules *in vitro* by preventing their depolymerization under certain conditions of cooling, pH, and ionic strength (Dulak and Crist, 1974). According to Brunk and Ericsson (1972), treatment with DMSO does not significantly alter the fine structure of tissues such as rat cerebral cortex. However, in spite of the advantages claimed by some workers, DMSO is a toxic drug and so its damaging effect on the tissue ultrastructure cannot be overlooked. It should be remembered that DMSO acts on the cell fine structure prior to its preservation with the fixative. Therefore, its use in routine immersion or perfusion fixation cannot be recommended.

Criteria for Satisfactory Specimen Preservation

It is difficult to set up absolute criteria for determining satisfactory preservation of biological specimens for two main reasons: (1) components of a typical plant or animal cell in the living state have not been viewed in the electron microscope and (2) incomplete information on the molecular basis of fixation. However, there are several reliable procedures and criteria that can be used to test the quality of preservation. The process of fixation in the living cells can be observed under a phase contrast or Nomarski interference contrast microscope and can be recorded by cinematography. Plant cells (e.g., hair cells) are well suited for this purpose. Thick sections (0.5–2.0 μm) of tissue blocks processed for electron microscopy can be examined with the light microscope or with the high-voltage electron microscope. Ordered structures such as myelin sheath or myofibrils can be used to follow dimensional changes within a cell. Mitochondria, since they are extremely sensitive to changes in the osmotic pressure, can be used as test sites to detect shrinkage or swelling. Microvilli also act as osmometers, since they are more osmotically sensitive than is the remainder of the cell surface. Comparison may also be made between the dimensions and orientation of the ultrastructure of fixed cells and data collected with the aid of other techniques such as polarization microscope, X-ray diffraction, or freeze-etching (this procedure also causes some dimensional distortion because of fixation and coating). The X-ray diffraction technique has been effectively used in the case of highly ordered structures such as myelin sheath and collagen fibers (McPherson, 1976), and the freeze-etching technique has provided extremely useful data on the structure of cell components, especially membranes (Koehler, 1972; Bullivant, 1973).

One of the criteria for satisfactory fixation is the preservation of the continuity of membranes, without distortion or break, and in some cases their trilaminar configuration. Although such continuity may not be the proof that other structures have also been preserved well, the presence of breaks is an obvious indication of defective preservation. The removal of enzymes affects the appearance of membranes in that the granular substructure on the membrane surface disappears.

It is generally recognized that in satisfactory preservation, the spaces between the membranes are filled with granular material, although some workers consider such material to be the dissolution of substances from various cell components (Lehman and Mancuso, 1957). Although some dissolution may occur, it is apparent that the spaces between the membranes in living cells are not empty. In fact, no truly empty spaces are found in the ground substance or within organel-

TABLE 17.1

Appearance of Well-Fixed Cellular Components

Components	Appearance
Cell wall	Dense and essentially layered; no breaks
Cytoplasmic ground substance	Fine granular precipitate showing no empty spaces
Endoplasmic reticulum (rough)	Flattened cisternae with attached ribosomes; usually arranged in long profiles
Endoplasmic reticulum (smooth)	Single or branching tubules with intact membranes not associated with ribosomes; tubular appearance instead of vesicular
Glycogen	Dark individual granules or clumps
Golgi	Intact membranes
Lipid	Uniformly dark
Lysosomes	Dense matrix bound by an unbroken single membrane
Microbodies	Dense matrix bound by an unbroken single membrane
Mitochondria	Neither swollen nor shrunken; outer double membrane and cristae intact; dense matrix
Nuclear envelope	Double membranes intact and essentially parallel to each other; two membranes different in width, often showing pores
Nuclear contents	Uniformly dense with masses of chromatin scattered adjacent to nuclear membrane
Plasma membrane	Single and intact; apposed to cell wall in plants
Plastids	Outer double membrane and lamellae intact; dense stroma
Vacuoles	Bound by unbroken single membrane; absence of intravacuolar membranes

les in living cells. Although the absence of enlarged intercellular spaces is a sign of satisfactory fixation, the presence of relatively large intercellular (extracellular) spaces is not uncommon in the well-fixed CNS. It should be noted that the presence of widened and clear spaces is common in pathological tissues.

In the case of kidney, the criteria of satisfactory fixation include open tubular lumina, uniform tubular cytoplasmic density without any swelling, absence of prominent extracellular compartments, presence of an interstitial space between tubules, and uncompressed capillaries. In the living kidney, proximal tubules have large open lumina that close within 5–20 sec after clamping the renal artery or interrupting the blood supply.

Subjectively, a satisfactory image is distinct, orderly, and overall "grayish" in electron opacity (as visualized on the electron micrograph), especially within mitochondria, microbodies, plastids, and lumina of blood and lymph vessels. In other words, every component of the tissue (including the extracellular entities) should possess a certain degree of opacity. However, there are certain exceptions (e.g., pituitary secretory granules). Extracellular materials and cell walls, which are almost electron lucent under some conditions of fixation, can assume a distinct opacity under other conditions of fixation. Other examples are elastin and collagen, which are electron lucent under certain conditions, and highly electron opaque under other conditions of fixation, but neither state denoting necessarily a poor fixation. The characteristic ultrastructural appearance of cellular components in well-preserved specimens is summarized in Table 17.1.

Artifacts

The preparation of living specimens for electron microscopy inevitably causes alterations in specimen structure. These artifacts are many and varied, but consist primarily of distortions of existing structures, loss or displacement of diffusible components, and addition of foreign substances. Artifacts can be introduced prior to, during, and after fixation. It is not always easy to determine precisely which stage or procedure was responsible for a given artifact. Furthermore, universal agreement on whether a given structure is an artifact or represents an *in vivo* condition does not exist. Artifacts introduced prior to and during fixation will be discussed; alterations caused by dehydration, embedding, sectioning, staining, and exposure to the electron beam have been presented elsewhere (Hayat, 1981). Artifacts resulting from critical point drying or freeze-drying and metal coating for scanning electron microscopy have also been discussed by Hayat (1978). Detailed discussions of artifacts encountered in light microscopy have been presented by Thompson and Luna (1978) and Wallington (1979).

All fixation procedures introduce artifacts in the specimen, and it is impossible to eliminate artifacts in a fixed specimen. The best one can accomplish is to minimize their introduction as much as possible. By using the utmost care, the production of artifacts can be substantially reduced. It should be remembered that once an artifact has been introduced, it cannot be removed. (Dimensional changes caused by osmolarity are reversible.) Obviously, avoidable artifacts must be prevented; however, more importantly, the process that causes artifacts should be understood, and the range of effects produced by various fixation procedures should be recognized. It is necessary that the appearance of the specimen structure be interpreted in the context of the fixation and other treatments that it has undergone. This understanding is essential in order to relate what is shown by electron microscopy to what may have existed *in vivo*. As long as unavoidable artifacts are interpretable and are acknowledged, they can be accepted. The major goal of improved fixation methodology is, in fact, the elimination of uninterpretable artifacts.

The magnitude of artifacts in a specimen depends on manifold fixation parameters such as mode of fixation (immersion or vascular perfusion), type of the fixative and its concentration, pH, temperature, type and osmolarity of the vehicle, addition of electrolytes or nonelectrolytes, duration of fixation, specimen type, and pre- and postfixation treatments. Less than optimal conditions of fixation obviously cause known and unknown artifacts.

Specimens may be damaged in a variety of ways prior to fixation. Such damage can occur during their collection, including removal at dissection or surgical operation. Soft tissues (e.g., tonsils and lymph nodes) are especially sensitive to pressure from forceps and knives, razor blades, or scissors. Brain, spinal cord, and lung are also very susceptible to damage on their removal from the body. Abnormal distribution of nuclear chromatin can result in the brain subjected to postmortem concussion prior to fixation (Cammermeyer, 1960). Artifacts may sometimes be directly related to a particular surgical procedure. Specimens obtained by cautery or aspiration biopsy can be damaged; the former may cause condensation of fibers. Percutaneous needle biopsies of liver and kidney may also include fragments of skin or skeletal muscle in the sample. An inadequately cooled turbine hand-piece can produce enough heat to damage dentine structure (Kramer, 1961). Reorientation of collagen fibers is common in skin that had been stretched (Craik and McNeil, 1965). Artifacts resulting from immersion fixation are discussed on p. 256. The following discussion presents some other known fixation and fixation-related artifacts that have been observed and described in specimens fixed for the TEM and SEM.

TRANSMISSION ELECTRON MICROSCOPY

Fixation can bring about diffusion and/or precipitation of cellular materials. A case in point is the production of a coarse artifactual network in the nuclear sap during fixation with glutaraldehyde; the formation of this artifact can be observed while living cells are subjected to fixation (Skaer and Whytock, 1976, 1977). This network could be mistaken for chromatin fibers, since the two stain alike. Moreover, the threads of the network might adhere to chromatin fibers, resulting in artifactual thickening. Fixation with formaldehyde may cause the nuclear sap to appear as granules or beads approximately the same size as histone nucleosomes ($\sim$10 nm in diameter); this is another potential source of confusion in the understanding of the fine structure of intranuclear chromosomes and chromatin.

Every osmiophilic droplet observed in the fixed cell is not necessarily present in the living cell. The appearance of certain types of osmiophilic droplets could very well be artifacts created by inadequate fixation with glutaraldehyde. Artifactual osmiophilic droplets have been observed on membranes of erythrocytes, endothelial and epithelial cells, within pinocytotic vesicles, mitochondria, and plastids. Artifactual precipitates caused by various staining procedures have been discussed by Hayat (1975, 1981).

Fixed specimens, however well preserved, do not necessarily reveal any spatial displacement of cell constituents that might take place during the process of fixation. It has been suggested that, during fixation, a contraction or molecular rearrangement within the cytonet material results in the deformation of the surface membrane and consequent formation of pinocytotic vesicles and condensation of cytonet around the plain synaptic vesicles. The implication is therefore that the "coats" of coated vesicles could be artifactual condensations of proteinaceous material. An example of this type of fixation artifact may be the presence of a greater number of pinocytotic vesicles and coated pinocytotic vesicles in the CNS (Gray, 1972; Paula-Barbosa et al., 1977).

The previously mentioned and certain other types of vesicles can be formed in a wide variety of cells. Certain vesicles can be formed as a result of unnatural fusion of membranes. Although membrane fusion in cells is not an uncommon natural phenomenon occurring during certain kinds of metabolic activity, every membrane fusion observed in the electron microscope is not necessarily an expression of an event taking place in the living cell. The presence of small, closed vesicles in the fractions of outer and inner mitochondrial membranes isolated by homogenization and centrifugation (Sottocasa, 1967) is a well-known example of fusion of membranes during preparatory procedure.

Other examples of nonphysiological membrane fusion include demembranated ciliary axonemes within the pellicular regions of the ciliate *Pseudomicrothorax* (Hausmann, 1977). The membranes of the cilia are thought to fuse with the plasma membrane and are in part incorporated in it. The presence of bridgelike connections or even the fusion of the outer leaflets of adjacent ciliary membranes in the ciliate *Paramecium* seems to be a fixation artifact (Hausmann, 1977). The artifactual fusion of membranes in these two ciliates appears to be derived from the collidine buffer used as a vehicle. Even the membranes of two different organisms may fuse during the preparatory procedure. The fusion of plasma membranes in *Aphelidium* (the parasite) with those in *Scenedesmus* cells (the host) has been reported (Schnepf, 1972). These plasma membranes not only fuse, but also occasionally form vesicles. Membranous structures inside a plant cell vacuole are considered to be fixation artifacts.

The appearance of light and dark cells in various organs has been explained as an artifact of poor fixative penetration by immersion. For example, light and dark cells of the adrenal cortex are thought to represent fixation artifacts rather than different functional states of individual cells. Another example is liver, which shows light and dark cells after fixation by immersion. The larger the size of the tissue block, the greater the number of light and dark cells, particularly in the core of the block. No such cell types are seen in the periphery of the liver tissue block. Osmotic phenomena may also in some way be responsible for the occurrence of such cells. However, there are exceptions—light and dark cells in the epithelium of mammalian choroid plexus fixed by vascular perfusion might represent varying states of cellular hydration at the time of fixation (Dohrmann, 1970).

Most published reports indicate that only 15–20% of all hepatocytes contain albumin and that they are randomly distributed throughout the liver. These data were obtained from liver filled with blood before fixation. It appears that during fixation, most or all hepatocytes lose their albumin, whereas an artifactual passive movement of plasma albumin occurs into the cytosol of some liver cells. Fixation penetration into blood-filled liver tissue is slow. However, when all blood is flushed from the liver (with saline containing 0.04% sodium nitrite) prior to fixation, albumin is found in all liver cells (Lebouton and Masse, 1980a,b). Only when all plasma protein is removed from the space of Disse before fixation can the presence of albumin in liver cells be considered valid.

Caution is warranted when glutaraldehyde fixation is used to avoid certain artifacts (e.g., dissociation of conformational changes) during sedimentation analysis of mixtures containing different kinds of ribosomal particles; dimerization or aggregation of ribosomal particles may be induced by this aldehyde. The dimerization is especially noticeable with 30S subunits (Garcia-Patrone and Algranati, 1976). The addition of albumin to the mixture decreases the aggregation, but does not prevent it. The fixation does not seem to induce association between 30S and 50S subunits.

Many biopsy specimens are fixed with formaldehyde, resulting in the formation of various types of artifacts. The presence of septate junctions between digestive organelles in phagocytic cells from human lymphatic ganglia is considered artifactual (Nistal *et al.*, 1978). Such junctions are absent in tissue samples fixed with a mixture of glutaraldehyde and formaldehyde. Intramitochondrial and intermitochondrial septate junctions reported in certain mammalian cells (Pearse and Welsch, 1968) may also be formaldehyde-induced artifacts. Only certain cells and organelles have the capability of developing septate junctions when fixed under certain conditions. It is possible that such organelles possess materials that react with formaldehyde and produce septate junctions.

The production of artifactual contraction bands and hypercontraction of sarcomeres in fresh cardiac tissue obtained by the biopsy procedure has been reported (Adomian *et al.*, 1977). Probably the biopsy procedure activates the contractile mechanism in the muscle. Such artifacts should be distinguished from the contraction bands associated with various cardiac pathological states. Artifactual contraction bands do not occur in specimens obtained after death or vascular perfusion. In addition, the distortion of cells and organelles is inevitable when the biopsy needle cuts through the beating myocardium.

SCANNING ELECTRON MICROSCOPY

It is impossible to prevent deformation of internal and external surfaces of a specimen completely during fixation. The extent of artifacts introduced during fixation depends on the method used, the type of specimen under study, and its pre- and postfixation treatments. Air drying, for example, causes drastic alterations in the specimen in contrast to those caused by critical point drying. Even this last treatment causes considerable shrinkage in the specimen (see Hayat, 1978). Soft specimens are much more vulnerable to damage during processing than are hard specimens. Furthermore, the difficulty in interpreting the artifacts is far greater in the case of soft specimens than that encountered in hard specimens. In all cases, however, it is necessary that the appearance of the specimen surface be interpreted with respect to the treatments that a specimen has undergone. The following discussion gives some of the factors that introduce artifacts and thus make correct interpretation difficult.

Environmental conditions readily affect surface morphology and surface forms (monolayer, colony, spherical, elongate, spindle-shaped, and stellate) of cells grown *in vitro*. The smooth or villous nature of the cell surface, ruffling activity, pinocytosis, and cell proliferation are known to be influenced by culture conditions. Both the composition of the growth medium and the nature of the substratum influence the surface morphology of cells. Components such as serum of the medium have been reported to affect cell morphology (Taylor, 1961; Westbrook *et al.*, 1975). It has been shown, for instance, that different serum concentrations in the medium influence ruffling activity in normal human glial cells (Schellens *et al.*, 1976). A decrease in the serum concentration from 10 to 0.5% resulted in pronounced reduction in the ruffling activity. Cell morphology may also be affected by the ambient temperature, pH, cell density, consistency of the medium, and nutrient levels. Westbrook *et al.* (1975) have shown that HBT-3 (human breast tumor line) assume very different forms in response to specific changes in the culture environment. The need of standardization of culture conditions for each type of experiment cannot be overemphasized, in order that meaningful comparative data be obtained from various laboratories.

As stated previously, interaction between cells and the underlying substratum can cause profound changes in the form of the cells. The two techniques used to prepare cells, aspiration filtration and suspension, differ in their effect on the surface architecture. The density of cells applied to the filters affects the ratio of smooth to villous cells in a given population (Alexander *et al.*, 1976). The surface morphology and size of those cells situated on top of the substrate or atop other cells may differ from those in direct contact with the substrate. Centrifugation on a Ficoll–Hypaque gradient stimulates the lymphocytes to increase their interaction with the substrate, which leads to alterations in cell morphology.

When rabbit PBLs are in suspension, they exhibit smooth surfaces, but on adhering to glass they display microvilli and pseudopodia (Linthicum *et al.*, 1974). Different cell types within peritoneal cell suspensions interact in different ways with artificial substrates such as Millipore filters and glass. Various types of cells differ in their response to the same type of substrate. Orenstein and Shelton (1976) have shown that macrophages and lymphocytes differ with respect to their interaction with the Millipore filter. Macrophage flattening is extremely variable among the individual cells; lymphocytes do not flatten, and their only visible interaction with the substrate is by means of short, fine processes that arise from the underside of the cell. The rate of flattening and morphology are subsequently different when these cells are allowed to settle on glass; the morphology of lymphocytes is not visibly affected.

Cleaning or washing may cause swelling or shrinkage of the cell. The use of solutions having nonphysiological pH or osmolarity prior to, during, or even after fixation may introduce artifacts. Vehicles used for preparing fixative solutions and buffers can produce small or large crystals or depositions. Treatments with digestive enzymes, acids, and other reagents prior to fixation can easily alter the surface of the specimen. Mechanical treatments (e.g., stripping, dissection, and sectioning) prior to and/or after fixation can also change the topography of the specimen. These treatments may cause deformation of the topography of either the fixed or unfixed specimen. An example of artifacts caused by mechanical treatments is the presence of randomly oriented fibers (indifferent fiber plexus) in the periodontal ligament of rat incisor. Although this artifact was accepted in the past as a normal structure, it is caused by the use of a rotating abrasive disc (because of its grinding action). On the other hand, freeze-fractured surfaces and sections of demineralized specimens of similar tissue show a highly organized collagen fiber architecture (Sloan *et al.*, 1976).

As stated earlier, the appearance of surface morphology may be affected by temperature. The temperature employed during fixation may affect, for example, the presence or absence of hairlike processes, folds, and cratering on the luminal surface of the aorta (Edanaga, 1974). When sinus aorta is washed with cold (2°C) physiological saline, hairlike processes of endothelial cells become obscure. Temperature is also known to affect the appearance and density of microvilli of cultured lymphocytes (Lin *et al.*, 1973). Cold temperatures tend to smoothen the surface architecture of lymphocytes.

Even the mode of fixation affects the surface morphology of certain tissues. As already discussed, fixation by vascular perfusion causes far fewer artifacts than does immersion fixation. The characteristic ridged pattern, for example, seen on the surface of luminal arterial wall after fixation by immersion is not seen after fixation by vascular perfusion. Not all types of tissues, however, can be fixed by vascular perfusion.

Artifactual cellular blebbing is not uncommon when corneal tissue is fixed by immersion. When the lens is removed before fixation, while the tissue is alive, endothelial cells adhere to the stroma rather than to the lens, and the cells exposed to the external milieu show pronounced surface blebbing (Bard et al., 1975). On the other hand, when the lens is removed after fixation, endothelial cells exhibit very little surface blebbing, and many of the endothelial cells adhere to the lens surface instead of to the corneal surface. It is likely that manipulation of the lens, as performed prior to fixation, disrupts the fibrous matrix and that its removal causes the blebbing. Cells protected in part by fibrous matrix show much less blebbing than cells on the surface.

The plasma membrane of a variety of free cells and cells in culture fixed with glutaraldehyde followed by OsO_4 show blebs (blisters) when examined either in the SEM or in the TEM. Similar artifacts are present in the glutaraldehyde-fixed, freeze-fractured cells. Such blebs are membrane bound, virtually empty looking, and their diameter ranges from 0.2 to 0.6 μm (Fig. 7.1). They seem to be an extension of plasma membrane and are quite different from the zeiotic blebs described by Price (1967), Godman et al. (1975), and Wollman (1972). It seems that lipid moiety of the membrane is destabilized by glutaraldehyde, resulting in its release and expansion into blebs. Phase microscopy has shown that blebs do not appear instantaneously on contact with glutaraldehyde, but appear gradually during the first 2 min after initial exposure to this fixative (Shelton and Mowczko, 1978).

Primary fixation with glutaraldehyde is responsible for the loss of previously mentioned membrane integrity. This artifact is produced by glutaraldehyde and subsequently preserved by OsO_4. Fixation with glutaraldehyde alone may not produce the artifact, since the artifact is lost during dehydration and critical point drying. Fixation with OsO_4 alone or with a mixture of glutaraldehyde and OsO_4 also does not show the artifact. The formation of this artifact is independent of other fixation parameters such as osmolarity of the fixation solution and temperature at which fixation is carried out.

Different cell types and cells grown in different manners are not equally influenced by the postosmication phenomenon. For example, plated mouse peritoneal cells are more sensitive to the effects of fixation compared with cultured fibroblasts and kidney cells (Shelton and Mowczko, 1978). It is apparent that, before assigning any functional specificity to cellular blebs present on the surface of cells in the SEM, the possible role of preparatory methods in their production should be considered.

Chemical dehydration can cause serious alterations in cell morphology. Distortions of cell and tissue surface details can result from partial or total removal or disruption of underlying intracellular structures during solvent exchange procedures (Clark and Glagov, 1976). The removal of free water, changes in pH, and the action of dehydration solvents can bring about diffusion and/or precipitation of cellular materials, which constitutes a problem in high-resolution scanning electron microscopy. Time schedules of dehydration after fixation can alter surface details of the specimen. For example, protoplasts dehydrated at an accelerated rate showed a characteristic polygonal outline resulting from the collapse of the membrane onto the underlying peripheral chloroplast (Burgess *et al.*, 1978). Rates of dehydration as being a potential source of poor preservation for scanning electron microscopy have been discussed by Hughes *et al.* (1976).

It should be pointed out that preparatory methods are not responsible for all morphological differences among the population of a given cell type; surface morphology of a particular cell may differ, depending on its stage of development. The cell cycle is known to affect the surface morphology (Fraumeni *et al.*, 1973). Wetzel *et al.* (1974) have aptly suggested that the *in vivo* microenvironment affects lymphocyte morphology.

The likelihood of interpretative pitfalls diminishes when studies with the SEM are substantiated by corresponding transmission electron micrographs and/or photomicrographs of similar specimens prepared separately. This is especially desirable in the case of structures not previously identified with the SEM, for a considerable consensus already exists on the appearance of a wide variety of well-fixed specimens viewed with the TEM. Furthermore, it is desirable that specimens be processed by more than one method so that the validity of the results obtained by one method can be checked against those gained by the use of another method. For example, similar specimens can be processed by the critical point drying and by the freeze-drying methods. With this approach, it is possible to gauge the relationship between the method used and the appearance of the real surface structure. This approach does require more time, but its value should not be overlooked, since presently the understanding about the effect of fixation and other procedures on cell morphology is inadequate. Finally, it is helpful that, when electron micrographs are published, they should be accompanied by a detailed explanation of the methodology used. The need of such information for correct interpretation becomes clear when one considers that the appearance of similar specimens may differ according not only to the type of fixation method used but also to certain parameters within the method itself.

Specific Fixation Methods

ALGAE (green)
  Solution A:
    25% glutaraldehyde   10 ml
    0.025 mol/l phosphate buffer (pH 7.2)   37 ml
    Distilled water to make   100 ml
  Solution B:
    OsO$_4$ 1 g
    0.025 mol/l phosphate buffer (pH 7.2)   50 ml
Fix specimens in solution A for 1 hr at room temperature. Wash with the buffer for 5 min, and postfix with solution B for 3 hr at 4°C.

ALGAE (green, unicellular marine)
  Solution A:
    25% glutaraldehyde   24 ml
    Distilled water to make   100 ml
  Solution B:
    Solution A   50 ml
    Seawater   50 ml
  Solution C:
    Distilled water   50 ml
    Seawater   50 ml

Solution D:
2% OsO$_4$	50 ml
Seawater	50 ml

Centrifuge and then fix the cells in solution B for 90 min at room temperature. Wash three times in solution C, and postfix in solution D for 1 hr at 4°C.

ALGAE (red)
  Method 1
  Solution A:
Sodium cacodylate (0.1 mol/l)	2.14 g
Sucrose (0.25 mol/l)	8.558 g
Distilled water to make	100 ml

  Solution B:
Solution A	80 ml
25% glutaraldehyde	20 ml

The addition of buffer is for the purpose of adjusting the osmotic pressure of the buffer to that of sea water. Fix the cells for 3 hr in solution B at 4°C, followed by washing in a buffered series of sucrose concentrations (0.25, 0.15, and 0.05 mol/l sucrose) (Ramus, 1969). Postfix in 2% OsO$_4$ in the same buffer without sucrose for 3 hr at room temperature. Figure 19.1 shows a well-fixed red alga cell.

Method 2. Fix the cells with 3% glutaraldehyde in 0.1 mol/l Millonig's phosphate buffer (pH 6.6) containing 0.25 mol/l sucrose and 50 mg/μl CaCl$_2$ for 1 hr at 4°C. Rinse twice in the buffer containing CaCl$_2$, followed by rinses in a graded series of buffer containing decreasing amounts of CaCl$_2$ until only buffer remains. Postfix in 1% OsO$_4$ for 1 hr at 4°C. Fresh specimens yield better results. For additional details, see Lin and Sommerfeld (1978).

AMOEBA
  Cell suspension in growth medium is prefixed by dilution with an equal volume of 2.5% glutaraldehyde in 0.05 mol/l cacodylate buffer (pH 6.8) containing 2 mM CaCl$_2$ and 0.2 mol/l sucrose for 1 hr at room temperature. After centrifugation, the cells are fixed in the same fixative solution for an additional period of ~12 hr at 4°C. The cells are washed twice for 30 min at 4°C in the buffer and centrifuged, and a dense suspension of cells is suspended in a drop of 3% agar. After cooling, the agar is cut into small cubes and postfixed in 1% OsO$_4$ in the same buffer for 1 hr at room temperature.

BACTERIA (Ryter and Kellenberger method)
  Solution A: tryptone medium
Bacto-tryptone	1.0 g
NaCl	0.5 g
Distilled water to make	100 ml

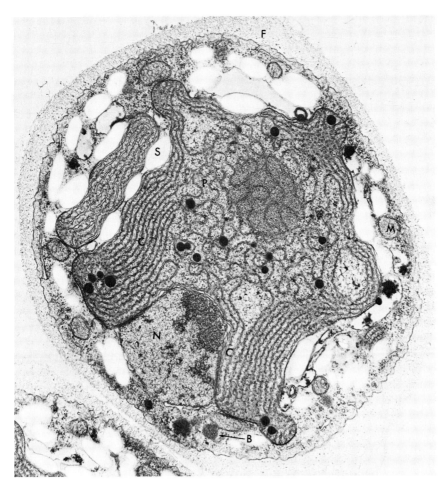

Fig. 19.1. Red algae (*Porphyridium purpureum*) in 5 ml of growth medium were fixed by adding 5 ml of 2% glutaraldehyde in 0.1 mol/l phosphate buffer (pH 6.6) containing 0.15 mol/l sucrose for 1 hr at room temperature. Cells were gently pelleted with a clinical centrifuge, rinsed three times in the same buffer, and postfixed with 1% OsO_4 in the same buffer for 2½ hr at room temperature. After again rinsing in the buffer, cells were embedded in 2% agar and stained *en bloc* with 2% uranyl acetate. Nucleus (N), pyrenoid (P), chloroplast (C), starch (S), mitochondrion (M), microbody (B), and fibrous sheath (F). ×26,600. (K. Schornstein, unpublished.)

Solution B: veronal acetate buffer
Sodium veronal	2.94 g
Sodium acetate	1.94 g
Sodium chloride	3.40 g
Distilled water to make	100 ml

Solution C: Kellenberger buffer
Veronal acetate buffer	5.0 ml
Distilled water	13.0 ml
0.1 mol/l HCl	7.0 ml
1 mol/l CaCl$_2$	0.25 ml

Adjust the pH to 6.0 with HCl. Freshly prepared buffer is recommended.

Solution D: fixative
Kellenberger buffer	100 ml
OsO$_4$	0.5 g

Solution E:
Kellenberger buffer	100 ml
Uranyl acetate	0.5 g

Prefixation is carried out by suspending bacteria in solution A and mixing 30 ml aliquot with 6 ml of solution D in a centrifuge tube. Centrifuge for 5 min at 1800 g, and resuspend the pellet in 1 ml of solution D and 0.1 ml of solution A for 16 hr at room temperature. Dilute the suspension with 8 ml of solution C and centrifuge for 5 min at 1800 g. Resuspend the pellet in 2% warm agar in solution A (~0.03 ml), transfer it as a drop on a glass slide, and allow it to solidify. Cut the hardened agar into small cubes (1 mm³) and treat with solution E for 2 hr at room temperature. Dehydrate and embed. Figure 19.2 shows a well-fixed bacterium.

BACTERIA (glutaraldehyde and OsO$_4$)

Add an equal volume of 6% glutaraldehyde in 0.2 mol/l cacodylate buffer to bacterial cells suspended in growth medium containing 0.01% CaCl$_2$ · H$_2$0 and fix for 1 hr at 4°C. Centrifuge at 1800 g and suspend the pellet in 0.1 mol/l cacodylate buffer and wash in three changes for a total duration of ~2 hr. Further processing is same as given in the preceding entry except that the prefixation step is omitted.

Fig. 19.2. Bacterium *Klebsiella pneumoniae* Biover d fixed under Ryter–Kellenberger conditions. Capsule, cell wall, ribosomes, and fine fibrillar cytoplasmic structure are visible. Prefixation with 0.2% OsO$_4$ in Michaelis buffer, fixation with 1% OsO$_4$ in Michaelis buffer containing 0.1% amino acid, embedding in 0.2% agar in Michaelis buffer, postfixation with 2% uranyl acetate in the same buffer, dehydration in acetone, final embedding in Epon, and section staining with lead citrate for 20 sec; section thickness is ~25 nm. (E. N. Schmid, B. Menge, and K. G. Lickfield, unpublished.)

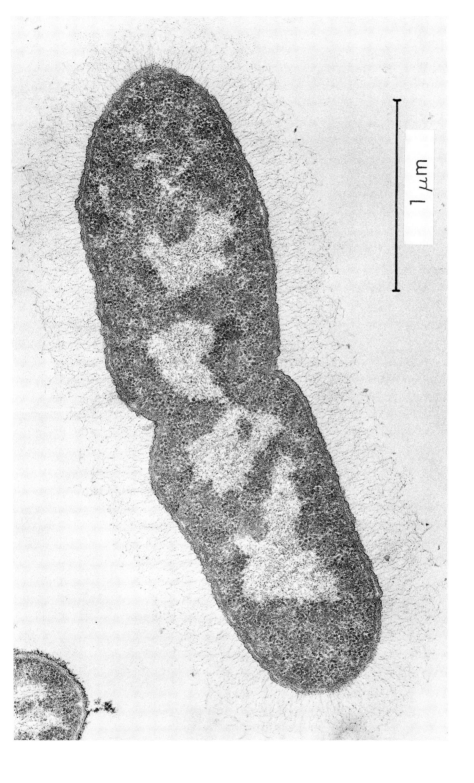

1 μm

BACTERIA (infected with phage) (Séchaud and Kellenberger, 1972)

Bacteria are sedimented and the well-drained pellet is resuspended into a mixture of 0.1% uranyl acetate and 5% glutaraldehyde in Michaelis buffer at a final pH of 5.4 (the initial pH of the buffer is 5.9, which upon addition of a concentrated aqueous solution of uranyl acetate drops to 5.4). Fixation is completed overnight at room temperature.

About 1 ml of the fixed cells are spun down in a Micro centrifuge tube used in a small swinging bucket-type tabletop centrifuge. The pellet ($2-4 \times 10^8$ bacteria) is suspended in 2% agar in Michaelis buffer (pH 5.7). This is done by pouring 3–4 mm of agar in the tube followed by mixing with a microsyringe, while holding the tube in a bath at 45°C. The agar–bacteria mixture pulled back into the syringe makes a cylinder ~6 mm long. The syringe is transferred into the refrigerator for some minutes; then the solidified cylinder is pushed out of the tubing onto a glass slide and cut into small blocks, which contain a sufficiently high concentration of bacteria. These blocks are treated in a saturated aqueous solution of uranyl acetate for 2 hr at room temperature, followed by dehydration and embedding.

BONE (undecalcified)

Solution A:

Formaldehyde (2%)	1 part
Glutaraldehyde (2%)	1 part

Solution B:

OsO_4	1 g
0.1 mol/l phosphate buffer (pH 7.2)	50 ml

Fix very small pieces of bone in 0.1 mol/l phosphate-buffered solution A for 2 hr at room temperature. After washing with the same buffer for 30 min at 4°C, the specimens are postfixed with solution B for 2 hr at 4°C. Dehydrate and infiltrate with graded series of acetone and Spurr's embedding medium, respectively. Each step of infiltration should last 2 hr with continuous agitation. The specimens are left in the pure embedding medium overnight with agitation, followed by evacuation, preferably 10^{-3} torr at 60°C for 30 min.

CARTILAGE (calcifying, e.g., epiphyseal plate) (Thyberg and Friberg, 1970)

Solution A:

Glutaraldehyde (25%)	24 ml
Cetylpyridinium chloride	0.5 g
0.071 mol/l cacodylate buffer (pH 7.4) to make	100 ml

Solution B:

OsO_4	0.5 g
Krebs–Ringer phosphate buffer (pH 7.4)	50 ml

Excise proximal end of the tibia from an anesthetized animal and immediately immerse it in solution A. Insert a syringe into the diaphysis, up to the metaphysis, and inject solution A to rinse out bone marrow (this will give the fixative access to the epiphyseal plate from below and thus increase the rate of initial fixation). Under a dissection microscope, cut the cartilage free from surrounding bone and slice into pieces of 1 mm on a side with a clean razor blade. Fix in fresh solution A for 2–6 hr at 4°C. Transfer the specimens to 0.1 mol/l cacodylate buffer (pH 7.4) containing 7.5% sucrose and store for 2–24 hr at 4°C. Postfix in solution B for 2 hr at 4°C. Rinse the specimens in Krebs-Ringer.

CELL FRACTIONS (plasma membrane) (Meldolesi *et al.*, 1971)

Mince the tissue, pass through a stainless steel tissue press, and homogenize in 10 volumes of 0.3 mol/l sucrose at 4°C in a glass homogenizer fitted with a rubber pestle which permits a very gentle homogenization (a loose-fitting Dounce homogenizer would also be adequate). Filter the homogenate through 110-mesh nylon cloth at 4°C. Centrifuge 6 ml of filtered homogenate in 12 ml glass conical tubes at 1000 g for 12 min. Resuspend the pellet in 15 ml of 0.3 mol/l sucrose by one stroke of a loose-fitting ground glass homogenizer, Dilute with 48 ml of 2 mol/l sucrose added dropwise with continuous stirring to obtain a final concentration of 1.58 mol/l sucrose. Transfer 20 ml of the suspension to Spinco SW 25 tubes and cover by a 5 ml layer of 0.3 mol/l sucrose. Centrifuge at 25,000 rpm for 60 min to obtain a pellet and a band at the interface. Collect the band and dilute with distilled water to 0.3 mol/l sucrose. Centrifuge at 1000 g for 12 min. Discard the supernatant and wash the pellet twice with 0.3 mol/l sucrose and once with 1 ml of 0.17 mol/l NaCl plus 3 ml of 0.2 mol/l $NaHCO_3$ at pH 7.8 (this treatment reduces the contaminants such as microsomes). Fix the pellet in 2% glutaraldehyde in 0.1 mol/l cacodylate buffer (pH 7.4) for 1 hr at 4°C. Rinse and postfix in 1% OsO_4 in the same buffer for 30 min at 4°C. Stain the pellet *en bloc* with 0.5% Mg uranyl acetate in 0.9% NaCl for 15 min, and dehydrate and embed. The preparation of other cell fractions is presented by Hayat (1972).

CELL MONOLAYERS

Method 1. Clean a 2.5 × 7.5 cm microscope slide in a 1:5 mixture of nitric acid and sulfuric acid. Evaporate a rather heavy coat of carbon onto the slide in a conventional vacuum evaporator. Stabilize the carbon film by placing the slide in an oven at 180°C for 24 hr. Seal ~5 mm deep steel or glass ring into the slide to form a shallow well. Grow cell monolayers in the well. Fix in 5.5% glutaraldehyde, prepared in Tyrode's solution at pH 7.3, for 5 min at room temperature. Postfix in 1% OsO_4, prepared in one-half isotonic Tyrode's solution at pH 7.3 for 30 min at room temperature. Dehydrate with acetone and infiltrate with Epon embedding mixture. Fill predried gelatin capsules with Epon and invert onto the monolayer. After polymerization, pull the capsule off the slide.

Method 2. Cells are grown on plastic cover slips (made of polymethylpentene), briefly rinsed in Ca^{2+}- and Mg^{2+}-free phosphate-buffered saline (pH 7.4) and fixed immediately with 2.5% glutaraldehyde in 0.1 mol/l cacodylate buffer (pH 7.4) containing 0.12 mol/l sucrose for 30 min at room temperature. After rinsing for 10 min in the same buffer, the cells are postfixed with 1% OsO_4 in 0.04 cacodylate buffer (pH 7.4) containing 0.14 mol/l sucrose for 1 hr at 4°C. After dehydration with acetone and infiltration with Epon embedding mixture, pieces of the cover slip are placed in the caps of BEEM capsules for polymerization. Cover slips can easily be separated from the polymerized resin by immersion in liquid nitrogen.

Polymethylpentene substrate is superior to glass in that the former does not require prior coatings (carbon) to assure resin separation (Trusal *et al.*, 1979). If desired, a portion of the same cover slip, after the dehydration step, can be critical point dried for scanning electron microscopy.

CELLS IN CULTURE

Method I (e.g., chick embryo spinal cord cells). The culture is fixed for 10 min at 4°C in 3% glutaraldehyde containing 3% dextran and 3% glucose adjusted to pH 7.2 with 1 mol/l NaOH. After washing in 0.25 mol/l sucrose, the culture is postfixed for 1 hr at 4°C in 1% OsO_4 in 0.1 mol/l phosphate buffer (pH 7.6).

Method II (e.g., rat liver parenchyma and human glia cells). Glutaraldehyde (2%) in 0.1 mol/l cacodylate buffer (pH 7.2) containing 0.1 mol/l sucrose is recommended. The vehicle osmolality is 300 mosmols and the total osmolality is 510 mosmols.

Method III (e.g., limpet blood cells). Cells are fixed in 2.5% vacuum-distilled glutaraldehyde (2.5%) in 0.1 mol/l phosphate buffer (pH 8.2) for 3 min, spun down at 300 g for 5 min, and resuspended in fresh fixative for 20 min at room temperature. After washing in artificial seawater, the cells are postfixed in 1% OsO_4 in the same buffer for 15 min at room temperature. The cells are washed in artificial seawater and then suspended in 0.25% uranyl acetate in 0.1 mol/l acetate buffer (pH 6.3) for 15 min. After centrifugation, two washings in the seawater, and a further centrifugation, the pellet is warmed to 55°C and suspended in 2% agar at 55°C. The cell suspension is centrifuged for 7 min at 55°C, and the pellet is solidified by cooling with ice. For further details see Gillett *et al.* (1975).

CILIATES

Solution A:

0.15 mol/l phosphate buffer (pH 7.0)	4.8 ml
1 mM $MgSO_4$	0.1 ml
0.1 mol/l sucrose	0.1 ml
2% OsO_4	5.0 ml

Solution B: As solution A except that 25% glutaraldehyde replaces OsO_4. Ciliates are fixed in solution A for ½–3 min at room temperature. To the suspension is added an equal volume of solution B and fixation is allowed to continue for 20-30 min. For embedding, Spurr's medium is recommended.

Alternate Method

Solution A:

0.15 mol/l phosphate buffer (pH 7.0)	7.2 ml
1 mM sucrose	0.2 ml
2% glutaraldehyde	0.2 ml

Solution A is mixed with 10 ml of 2% OsO_4 immediately prior to use. Ciliates are fixed for 20-30 min at room temperature. The culture medium should be pipetted off before fixation. For additional details see Shigenaka *et al.* (1973).

CYSTS

Pelleted organisms are fixed with 5% glutaraldehyde in 0.5 mol/l cacodylate buffer (pH 6.8) containing 1 mM $CaCl_2$ for 6 hr at room temperature. The specimens are centrifuged at 400 g for 10 min in 12 ml conical centrifuge tubes. After decanting the supernatant, the pellet is embedded in agar. This can be accomplished by layering 2 ml of 2% agar at 45°C over the pellet. With a dental explorer the pellet is lifted into the molten agar, and the centrifuge tube is plunged into an ice bath for 30 sec to solidify the agar.

The agar block containing the specimens is cut into small pieces and washed in 0.05 mol/l cacodylate buffer (pH 6.8) containing 7.5% sucrose for 16 hr. Postfixation is carried out with 2% OsO_4 in 0.05 mol/l cacodylate containing sucrose for 6 hr at 4°C. After washing in distilled water for 2 hr, the specimens are treated with 1% uranyl acetate for 12-16 hr at 4°C. For embedding, Spurr's mixture is recommended. Chiovetti (1978) has given an alternate method for embedding.

ECHINODERMS

Specimens are fixed in 3% glutaraldehyde in phosphate buffer (pH 7.8) for 3-4 hr at 4°C, followed by postfixation in 2% OsO_4 in the same buffer. The glutaraldehyde fixative should have an osmolality of 1000 mosmols (by adding sucrose).

EGGS AND ZYGOTES (Longo and Anderson, 1970)

Solution A: Dissolve 2.5 g bovine serum albumin in 30 ml of seawater. Dissolve 2 g of paraformaldehyde powder in 20 ml of seawater; mix the two solutions. Add 7 g of sucrose and 2 ml of acrolein to 30 ml of 0.2 mol/l collidine buffer (pH 7.4). Mix this solution with the previous mixture and then add slowly 6 ml of 50% glutaraldehyde. Bring the mixture to 100 ml with 0.2 mol/l collidine buffer.

After insemination of oocytes, collect fertilized eggs at intervals of a few minutes. Fix in solution A for 1 hr at room temperature. Wash in seawater for 2 hr at room temperature and postfix in 1% OsO_4 in seawater for 90 min at 4°C.

EMBRYOS (e.g., chick)
Solution A:

Glutaraldehyde (50%)	5 ml
Formaldehyde (8%)	25 ml
0.2 mol/l cacodylate buffer (pH 7.6)	50 ml
Calcium chloride (anhydrous)	25 mg

Fix *in situ* embryos by pouring chilled solution A after removing the amnion. After about 5 min, transfer the embryo or a part of it to fresh solution A for 30 min at 4°C. Wash briefly in cold buffer and postfix in 1% OsO_4 in 0.2 mol/l cacodylate buffer (pH 7.6) for 2 hr at 4°C.

ERYTHROCYTES (e.g., avian)

Glutaraldehyde (2%) in 0.1 mol/l cacodylate buffer (pH 6.85) containing sucrose is recommended. The vehicle osmolality is 320 mosmols. A slightly acidic pH enhances the stability of microtubules and microfilaments. The osmolality of avian blood plasma is 323 mosmols. The addition of 12% hexylene glycol to the buffer improves the preservation but is not essential. The addition of $MgCl_2$ helps to preserve cell integrity, although it may cause fusion of the two nuclear membranes when used in a hypertonic buffer. For additional details see Brown (1975).

EYES
Solution A:

Glutaraldehyde (25%)	3 ml
Formaldehyde (10%)	50 ml
NaCl	3.0 g
Sucrose	4.5 g
0.1 mol/l phosphate buffer (pH 7.2) to make	100 ml

Fix the specimens in solution A for 1–2 hr at room temperature. Wash briefly in 0.1 mol/l phosphate buffer containing 8% sucrose and then postfix in 1% OsO_4 in phosphate buffer for 1 hr at 4°C.

FLAGELLATES
Solution A:

Glutaraldehyde (25%)	20 ml
Acrolein	2 ml
0.1 mol/l cacodylate buffer (pH 7.4) to make	100 ml

Fix the specimens with solution A for 2 hr at 4°C. Wash the specimens with 0.1 mol/l cacodylate buffer (pH 7.4) containing 0.2 mol/l sucrose, and postfix in 2% OsO_4 in 0.1 mol/l cacodylate buffer for 2 hr at 4°C.

FETAL TISSUE (e.g., pig liver and tooth germs)

A mixture of 1.25% glutaraldehyde and 2% formaldehyde in 0.1 mol/l cacodylate buffer (pH 7.0) having an osmolality of 950 mosmols is recommended. The fixation is carried out for 90 min at ~20°C. Postfixation is accomplished in 2% OsO_4 in 0.1 mol/l cacodylate buffer for 2 hr. For additional details see Rømert and Matthiessen (1975).

FREE CELLS (e.g., cerebrospinal fluid, urine, ascites, and pleural effusion) (Ito and Inaba, 1970)

Centrifuge the fluid at 2000 rpm for 20 min. Using a pipette, remove all but the last drop of the supernatant (do not disturb the sedimented cells). Add 3–5 ml of serum of the same patient and shake the test tube gently to resuspend the cells. Centrifuge at 3000 rpm for 20–30 min. Using a pipette, remove the supernatant and place the test tube at an angle in a beaker containing ice. Using a pipette, gently introduce 3% chilled glutaraldehyde into the test tube and fix for 30 min at 4°C (do not remove the test tube during fixation). Remove the fixed and so-lidified serum containing the cells from the test tube and cut into small slices. Postfix in 1% OsO_4 for 1–2 hr at 4°C.

FUNGI

Hyphae and spores are fixed with 4% glutaraldehyde in 0.1 mol/l cacodylate buffer (pH 7.0) containing 0.01 mol/l $CaCl_2$ and 3% Triton X100 for 3 hr. After pelleting, the specimens are further fixed in a fresh solution of glutaraldehyde without detergent for a period of 15 hr (overnight) in the refrigerator. The specimens are washed twice in collidine buffer (pH 7.4) and postfixed with 1% OsO_4 in collidine buffer for 12 hr at 4°C.

FUNGI (*Neurospora*)

Form a compact pellet of hyphae by removing the water in a Buchner funnel under vacuum. Cut segments of ~2 mm on a side. Fix the specimens in 3% glutaraldehyde in cacodylate buffer (pH 7.2) for 1 hr at room temperature. A brief vacuum treatment in the beginning of fixation removes air bubbles trapped in the specimens. Rinse the specimens with the buffer and postfix with 1% OsO_4 for 1 hr at 4°C. A well-fixed cell is shown in Fig. 19.3.

HeLa CELLS
Method I
Solution A:
Glutaraldehyde (8%)   10 ml
OsO_4 (2%)   10 ml
Solutions are prepared in 0.1 mol/l cacodylate buffer (pH 7.25), mixed im-mediately prior to use at 4°C, and the mixture has an osmolality of 607 mosmols. Cultures are grown in Falcon T-30 flasks at 37°C using Eagle's minimal essential media with 10% fetal calf serum as the fluid phase and 5% CO_2 and air as the gas phase. The cultures are allowed to grow for 2 weeks until a monolayer has formed. Fixation is carried out in solution A for 1 hr at 4°C. Cells are prestained

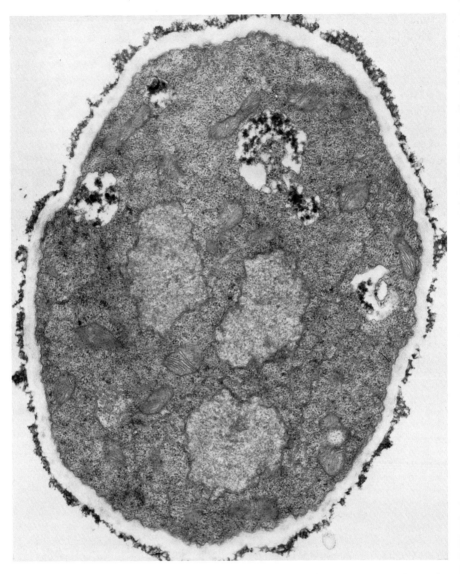

Fig. 19.3.   Fungus *Neurospora crassa* fixed with glutaraldehyde followed by OsO$_4$. Note multinucleate condition.

with 0.5% uranyl acetate in veronal acetate buffer prior to dehydration. For additional details, see Chang (1972).

Method II

Solution A:

Glutaraldehyde (25%)	4 ml
Acrolein (25%)	4 ml
0.1 mol/l phosphate buffer (pH 7.3) to make	100 ml
CaCl$_2$	Trace

The osmolality should be 393 mosmols.

Solution B:

OsO$_4$	1 g
0.1 mol/l phosphate buffer	50 ml

The osmolality should be adjusted to 366 mosmols with glucose.

Fix cells in solution A for 10 min. Pellet the cells and continue fixation for another 15 min. Postfix in solution B for 1 hr at 4°C.

LEAVES

Solution A:

Formaldehyde (8%)	25 ml
Glutaraldehyde (25%)	15 ml
0.2 mol/l cacodylate buffer (pH 7.2)	18 ml
CaCl$_2$	25 mg

Seal a metal ring (3 mm high and 12 mm in diameter) with lanolin onto the upper surface of an intact leaf. No lanolin should be present within the ring. Fill the ring with solution A using a hypodermic syringe and cover with a glass cover slip. Change this solution after 45 min and leave it for another 45 min at room temperature. Cut out the leaf tissue in contact with the fixative and slice 0.5 mm strips. Place the tissue strips into fresh solution A for 30 min at 4°C. Rinse rapidly in the buffer and postfix with 2% OsO$_4$ in the same buffer for 2 hr at 4°C.

MICROTUBULES AND MICROFILAMENTS

Fix root tissue with 1% glutaraldehyde in 50 mM phosphate buffer (pH 6.8) containing 0.2% tannic acid for 2 hr at 22°C. After a brief wash in the buffer, the specimens are postfixed with 1% OsO$_4$ in the same buffer for 1 hr at the same temperature. The diameter of microfilaments increases with increased concentration of tannic acid. The addition of various ions, use of low temperatures, and excessive hydrostatic pressure are undersirable for satisfactory preservation of these structures.

Animal tissues are fixed with 2% glutaraldehyde in 0.05 mol/l phosphate buffer (pH 7.2) containing 4–8% tannic acid and 0.015 mol/l CaCl$_2$ for 1 hr at room temperature. After a brief rinse in the buffer, the specimens are postfixed with OsO$_4$. Tannic acid is dissolved in a buffer with moderate heating; on cooling, a brown precipitate is formed, which can be cleared by centrifugation.

The resultant solution is clear, brownish. Glutaraldehyde is added just before fixation.

NEMATODES
Solution A:

Paraformaldehyde powder	1.5 g
Distilled water	25 ml
1 mol/l NaOH	5 drops
Glutaraldehyde (25%)	2.5 ml
0.1 mol/l phosphate buffer (pH 7.3) containing 0.2 mol/l sucrose	23 ml

The paraformaldehyde is dissolved in water at 60°C. The solution is cleared by adding NaOH. After cooling the solution, glutaraldehyde and buffer are added. The final pH is 7.2, and the concentrations of formaldehyde and glutaraldehyde are 3% and 1%, respectively.

Solution B:

Glutaraldehyde (25%)	5 ml
DMSO	2.5 ml
Distilled water	17.5 ml
0.1 mol/l phosphate buffer (pH 7.3)	25.0 ml

The fixative solution contains 2% glutaraldehyde and 5% DMSO. Immediately after removing the specimens from the soil, they are placed in a drop of water and then solution A preheated to 70°C is poured over the specimens and allowed to stay for 1 hr at room temperature. The container with the specimen is placed for 1 hr at 4°C. The specimens (one at a time) are transferred to a drop of solution B on a plastic slide or Petri dish and severed with a surgical eye knife. The desired parts are transferred to the fresh solution B and kept for 16–24 hr at 4°C. After washing, the specimens are postfixed with 2% OsO_4 for 2 hr at room temperature. If the specimens need to be oriented, they can be embedded in agar prior to postfixation. For further details see Zeikus and Aldrich (1975).

OOCYTES (see also eggs)

Relative impermeability of the oocyte wall poses a problem in obtaining satisfactory fixation and embedding. The following double-sectioning technique overcomes some of these problems. Oocytes are washed in deionized water and densely packed by centrifugation into a solution consisting of 20% (w/v) bovine serum albumin (BSA) and 15% (w/v) sucrose in phosphate-buffered saline (PBS) (pH 7.4). The supernatant is discarded and the pellet solidified by cross-linking the BSA with glutaraldehyde. This is accomplished by exposing 1 ml of the pellet with three drops of 25% glutaraldehyde. The pellet is soaked in 50% sucrose for 1 hr and then placed on a cryostat chuck and rapidly frozen either in Freon 22 cooled to its melting point by liquid nitrogen or in supercooled liquid nitrogen (Umrath, 1974).

The pellet is sectioned at the 20 μm setting of a cryostat with knife and pellet temperatures adjusted to between $-20°$ and $-30°$C. The sections are rapidly transferred and allowed to thaw in a mixture of glutaraldehyde and formaldehyde (Karnovsky, 1965) in cacodylate buffer (pH 7.2) and 0.01 mol/l $CaCl_2$ at least 3 hr at room temperature. The sections are washed by repeated centrifugation in PBS and finally soaked overnight in the same solution.

The sections are embedded in 20% BSA (without sucrose) and solidified by adding glutaraldehyde as before. The block thus obtained is cut into 1 mm^3 cubes, postfixed in 1% OsO_4 in veronal acetate buffer (pH 7.3) for 1 hr, and treated *en bloc* with 2% uranyl acetate for another hour. For further details see Birch-Andersen *et al.* (1976).

PARAMECIUM

Add 5 ml of 4% glutaraldehyde to 5 ml of paramecium suspension and allow to fix for 1 hr. Cells are further fixed with 4% glutaraldehyde in 0.075 mol/l cacodylate buffer (pH 7.2) containing 1 mM $CaCl_2$ for 1 hr at room temperature. Equal volumes of paramecium suspension and glutaraldehyde fixative are recommended. After washing in the buffer for 30 min, the cells are postfixed with 1% OsO_4 for 2–4 hr at room temperature.

PLANT CELLS (mature)

Specimens should be prefixed for 1 hr with 5% glutaraldehyde in 0.08 mol/l PIPES buffer (pH 8.0) having an osmolality of 800 mosmols, rinsed for 1 hr in 0.2 mol/l PIPES buffer (pH 6.8) having an osmolality of 600 mosmols, and postfixed with 2% OsO_4 in 0.18 mol/l PIPES buffer having an osmolality of 680 mosmols for 1 hr. All treatments are carried out at room temperature with gentle agitation. Osmium tetroxide solution in PIPES should be prepared immediately prior to use, since a brownish color develops after ~1 hr. For further details see Salema and Brandão (1973). Figure 19.4 shows excellent preservation of meristematic cells.

PLATELETS

Blood is collected in a plastic test tube containing 3.8% trisodium citrate anticoagulant in a ratio of nine parts to one part coagulant. The sample is mixed by gentle inversion, and then centrifuged at 750 rpm for 8 min. Citrated platelet-rich plasma (CPRP) is transferred to a clean plastic test tube and kept at room temperature to avoid cold-induced morphological changes (White and Krivit, 1967).

One volume of CPRP is poured in a slow, steady stream into 10 volumes of freshly prepared 0.2% glutaraldehyde in 0.1 mol/l cacodylate buffer, which is continuously agitated. Equal volumes of fixative and CPRP are used. The addition of the fixative stops immediately the physiological reaction. This initial fixation is accomplished in 30 min at room temperature.

Preparation for the TEM. CPRP in glutaraldehyde is centrifuged at 3100 rpm for 10 min, and the glutaraldehyde is decanted. The intact pellet is fixed with 3% glutaraldehyde for 2 hr at room temperature; care should be taken not to break the

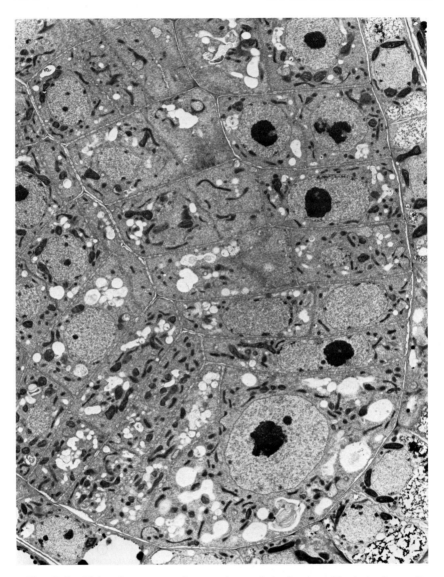

Fig. 19.4. High voltage survey electron micrograph (at low magnification) of a median-longitudinal section (0.4 μm thick) of an Azolla root apex. The curved outline around the periphery of the micrograph is the junction between the root cap and the root proper. Specimens were conventionally fixed and embedded in Spurr's resin. Sections were stained by immersion in saturated uranyl acetate dissolved in 50% ethanol, followed by 20 min in lead citrate. Sections were carbon-coated on both sides just prior to examination at one million volts. Note mitochondria (dark) and vacuoles (light) scattered throughout the cytoplasm. ×2080. (Gunning, 1980.)

pellet. The pellet is gently dislodged and transferred to a glass Petri dish containing 3% glutaraldehyde. It is cut into 1 mm pieces which are placed into a vial, washed twice in the buffer, and postfixed in 1% OsO_4 for 1 hr at 4°C.

Preparation for the SEM. An equal volume of 6% glutaraldehyde is added to the CPRP in 0.2% glutaraldehyde and fixed for 24 hr at room temperature. The platelets are centrifuged at 3100 rpm for 10 min and the supernatent is discarded. The platelets are gently resuspended in 5 ml of distilled water, washed for 10 min, and centrifuged at 3100 rpm for 10 min, and the supernatant wash is discarded. This is repeated three more times, followed by dehydration.

ROOTS

Cut segments less than 1 mm on a side in 3% glutaraldehyde in 0.1 mol/l PIPES buffer (pH 8.0) having an osmolality of 750 mosmols, and allow it to be fixed for 1 hr. After rinsing in the same buffer for 30 min, the specimens are postfixed in 2% OsO_4 in the same buffer for 1 hr. All treatments are carried out at room temperature.

SEEDS

Solution A:

Glutaraldehyde (8%)	25 ml
Formaldehyde (8%)	25 ml
0.05 mol/l collidine buffer (pH 7.3) to make	100 ml

Fix the specimens in solution A for 5 min at room temperature, and then transfer to 4°C for another 2 hr. After rinsing in the same buffer for 1–2 hr, postfix specimens with 1% OsO_4 in the same buffer for 2–12 hr at 4°C.

SKIN (e.g., frog)

Solution A:

Formaldehyde (8%)	25 ml
Glutaraldehyde (50%)	5 ml
0.1 mol/l cacodylate buffer (pH 7.3) to make	50 ml

Remove skin biopsy and cut into very small pieces. Fix in solution A for 90 min at room temperature. After rinsing in the buffer for 10 min, postfix in 1% OsO_4 in the same buffer at 4°C.

SLIME MOLD

Solution A:

Formaldehyde (2%)	25 ml
Glutaraldehyde (12%)	5 ml
0.2 mol/l phosphate buffer (pH 7.2) to make	50 ml

Expose agar-grown cells to OsO_4 vapors for 30 min at room temperature. Wash cells and spores from the agar plate with the same buffer, and centrifuge at low speeds. Embed in 1.5% agar, and fix small cubes of agar in solution A for 45 min at room temperature. After washing in the same buffer, postfix in 1% OsO_4 in the same buffer for 45 min at room temperature. Treat with aqueous 0.5% uranyl acetate for 5 min.

SMOOTH MUSCLE

Specimens are bathed with hyperosmolar Ringer's (normal Ringer's made 1.5 times hyperosmolar by adding sucrose (0.17 mol/l) to yield an osmolality of 500 mosmols for at least 2 hr at 4°C. This treatment tends to relax or stabilize the muscle. Specimens are fixed in a mixture of acrolein (1.5%) and potassium dichromate (1%) containing 6% sucrose at pH 7.35 for 4 hr at 4°C. The specimens are transferred to the bathing solution for 12 hr and then postfixed with 1% OsO_4 containing sucrose (7%) at pH 7.0 for 1 hr at 4°C. Figure 19.5 shows well-fixed smooth muscle cells.

SPERMATIDS

Solution A:
Glutaraldehyde (25%)	25 ml
0.1 mol/l cacodylate buffer (pH 7.2)	75 ml

Solution B:
Sucrose	7.5 g
0.1 mol/l cacodylate buffer to make	100 ml

Solution C:
Potassium bichromate (5%)	15 ml
2.5 mol/l KOH	2 ml
Distilled water	2 ml
0.1 mol/l cacodylate buffer (pH 7.2)	10 ml

Solution D:
OsO_4	1 g
0.1 mol/l cacodylate buffer	100 ml
Sucrose	0.4 mol/l

Fix small pieces of the testis in solution A for 1 hr at 4°C. After a rinse in the buffer, store the specimens in solution B for 30 hr at 4°C. Transfer the specimens to solution C for 30 min at 4°C. Rinse in solution B and postfix in solution D for 80 min at 4°C.

SPERMATOZOA (human or animal ejaculate)

Solution A:
Glutaraldehyde (8%)	25 ml
Formaldehyde (10%)	25 ml
Picric acid	0.02 g
0.1 mol/l phosphate buffer (pH 7.2) to make	100 ml

Collect first two drops directly into 20 ml of solution A at room temperature. Discard proteinaceous seminal plasma clumps which are formed rapidly due to coagulation. Centrifuge the remaining nearly pure cellular suspension at 1000

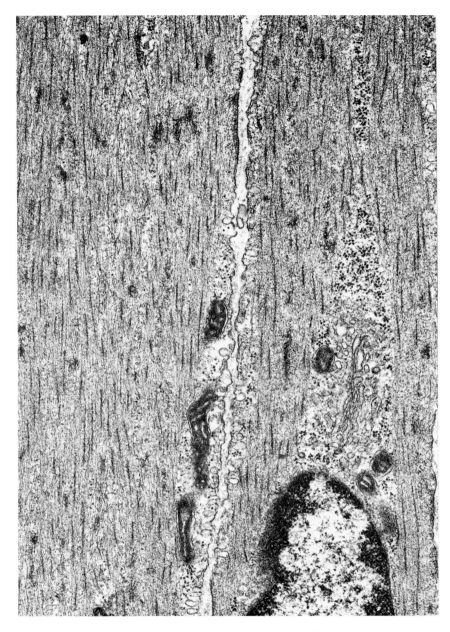

Fig. 19.5. Longitudinal-oblique section of two smooth muscle fibers of rabbit portal-anterior mesentric vein. Specimens were incubated for 30 min in Krebs solution under physiological stretch, and then fixed with a mixture of 3% glutaraldehyde, 2% formaldehyde, and 1% acrolein. Postfixation was carried out with 2% OsO_4 and *en bloc* staining with saturated aqueous uranyl acetate for 1–2 hr. Most of the thick myofilaments are sectioned rather short because of their oblique orientation relative to the plane of the section; the correct length of the whole filament is 2.2 μm. Thin, intermediate, and thick filaments have diameters of 6.4 nm, 10 nm, and 15.5 nm, respectively. (Somlyo *et al.,* 1973.)

rpm for 10 min (the total duration of fixation is ~25 min). Discard the supernatant and wash the pellet of highly concentrated spermatozoa in phosphate buffer for 15 min. Postfix the pellet in 1% OsO_4 in 0.1 mol/l phosphate buffer (pH 7.2) for 15 min at room temperature.

SPORES

Method I. Harvest the spores by centrifugation at 3000 g and wash twice with distilled water. Resuspend the spores in molten 2% agar and solidify in the refrigerator. Cut the agar into 1 mm cubes, and fix with 15% formaldehyde in 0.1 mol/l phosphate buffer (pH 7.2) for 2 hr at room temperature. Transfer the cubes to 6% glutaraldehyde in the phosphate buffer for 2 hr. After a rinse in the buffer, postfix with 1% OsO_4 in the phosphate buffer for 2 hr. If fixative penetration is a problem, use Method II.

Method II. Fix the spores with 4% glutaraldehyde in 0.1 mol/l cacodylate buffer (pH 7.1) containing 0.01 mol/l $CaCl_2$ and 3% Triton X100 for 2–4 hr at room temperature. Pellet the spores and refix overnight in the above fixative without detergent at 4°C. Wash in the cacodylate buffer (0.2 mol/l) and in collidine buffer (pH 7.2), and postfix in 1% OsO_4 in collidine buffer for 12 hr at 4°C. Spores of different species and developmental stages may require changes in the duration of fixation.

TETRAHYMENA

Pelleted specimens are fixed in a mixture of glutaraldehyde (0.5%) and OsO_4 (1%) buffered with 0.5 mol/l cacodylate buffer (pH 7.2) having an osmolality of 120 mosmols for 2 hr at 4°C. After a rinse in the buffer, the pellet is postfixed with 1% OsO_4 for 1 hr at 4°C.

WOOD

Solution A:
    Glutaraldehyde (25%)   12 ml
    0.025 mol/l phosphate buffer (pH 6.9) to make   100 ml
Solution B:
    0.05 mol/l phosphate buffer (pH 6.9)
    0.2 mol/l sucrose
    0.001 mol/l $CaCl_2$

Fix small pieces (less than 1 mm on a side) of the tissue in solution A for 2 hr at room temperature, and continue fixation in the same solution for another 22 hr at 4°C. Wash the specimens in solution B for 4 hr, and postfix with 2% OsO_4 in 0.05 mol/l phosphate buffer (pH 6.9) for 2 hr at 4°C. Dehydrate after washing in solution B for 10 min.

YEAST

Method I. Cells are pelleted by centrifugation at 1000 g and then washed twice in 0.02 mol/l collidine buffer (pH 7.4) containing 0.2 mol/l sucrose and

0.001 mol/l CaCl$_2$; this is done by resuspension and recentrifugation. Cells are resuspended in 50% DMSO containing 2% acrolein and 2% glutaraldehyde buffered with 0.02 mol/l collidine buffer (pH 7.4), and allowed to be fixed for 7 hr at 4°C. Cells are washed twice (30 min each) in the first solution, and postfixed with 2% OsO$_4$ in veronal acetate buffer (pH 7.4) for 7 hr at 4°C. Cells are washed twice (10 min each) in the first solution.

Method II. Cells in agar are placed in 0.1–0.5 mol/l cysteine solution in phosphate buffer (pH 7.4) for 15–30 min. After a rinse in the same buffer, the cells are fixed in 6% glutaraldehyde in the same buffer for 2 hr at room temperature. Postfix in 2% OsO$_4$ in the same buffer for 1–2 hr, followed by in 3% uranyl acetate in 30% ethanol for 2 hr at room temperature. Cysteine treatment facilitates the penetration of the fixatives into the cells, presumably by weakening the cell wall.

Appendix

BALANCED SALT SOLUTIONS

TYRODE
Solution A:

Distilled water	70 ml
NaCl	20 g
KCl	0.5 g
$CaCl_2 \cdot 6H_2O$	0.5 g
$MgCl_2 \cdot 6H_2O$	0.25 g
Distilled water to make	100 ml

Solution B:

Distilled water	90 ml
$NaHCO_3$	5.0 g
NaH_2PO_4	0.25 g
Distilled water to make	100 ml

Final solution:

Distilled water	940 ml
Solution A	40 ml
Solution B	20 ml
Glucose	1 g

AMPHIBIAN RINGER

Distilled water	180 ml
NaCl	1.3 g

KCl	0.04 g
CaCl$_2 \cdot$6H$_2$O	0.04 g
Distilled water to make	200 ml

MAMMALIAN RINGER

Distilled water	180 ml
NaCl	1.8 g
KCl	0.84 g
CaCl$_2 \cdot$6H$_2$O	0.48 g
NaHCO$_3$	0.1 g
Glucose	0.1 g
MgCl$_2 \cdot$6H$_2$O	0.05 g

PHOSPHATE-BUFFERED PHYSIOLOGICAL SALINE

1.0 mol/l NaCl	12 ml
0.2 mol/l KCl	2 ml
0.2 mol/l CaCl$_2$	1 ml
0.1 mol/l MgSO$_4 \cdot$7H$_2$O	1 ml
0.2 mol/l phosphate buffer (pH 6.9)	10 ml
Distilled water to make	100 ml

STANDARD CACODYLATE WASHING SOLUTION

0.2 mol/l sodium cacodylate	24 ml
0.2 mol/l cacodylic acid	10 ml
0.2 mol/l calcium acetate	1 ml
Isotonic sodium sulfate to make	100 ml

If cacodylic acid is not readily available, the following solution will suffice:

0.2 mol/l	34 ml
0.1 mol/l nitric or sulfuric acid	20 ml
0.2 mol/l calcium acetate or nitrate	1 ml
0.2 mol/l sodium sulfate	38 ml
Distilled water to make	100 ml

COMMONLY USED SALTS AND THEIR PHYSICOCHEMICAL PROPERTIES[a]

Sodium cacodylate:	
Anhydrous Na(CH$_3$)$_2$AsO$_2$	Unstable, readily absorbs three molecules of water
Trihydrate Na(CH$_3$)$_2$AsO$_2 \cdot$3H$_2$O	Stable
Disodium hydrogen phosphate:	
Anhydrous Na$_2$HPO$_4$	Unstable, absorbs 2–7 mols of water
Dihydrate Na$_2$HPO$_4 \cdot$2H$_2$O	Stable

(Continued)

Heptahydrate $Na_2HPO_4 \cdot 7H_2O$	Relatively stable
Dodecahydrate $Na_2HPO_4 \cdot 12H_2O$	Unstable, loses water
Sodium dihydrogen phosphate:	
Anhydrous NaH_2PO_4	Unstable, absorbs water
Monohydrate $NaH_2PO_4 \cdot H_2O$	Unstable, absorbs water
Dihydrate $NaH_2PO_4 \cdot 2H_2O$	Stable
Sodium acetate:	
Anhydrous CH_3COONa	Unstable, readily absorbs three molecules of water
Trihydrate $CH_3COONa \cdot 3H_2O$	Relatively stable, water content may vary
Calcium chloride ($CaCl_2$):	
Anhydrous	Unstable, very hygroscopic
Dihydrate $CaCl_2 \cdot 2H_2O$	Unstable, absorbs water
Hexahydrate $CaCl_2 \cdot 6H_2O$	Unstable, water content may vary

[a] Kalimo and Pelliniemi (1977).

COMMONLY USED CHEMICALS

Chemical	Formula	Mol. wt.
Calcium chloride	$CaCl_2$	110.99
s-Collidine	$2,4,6\text{-}(CH_3)_3(C_5H_2N)$	121.18
Hydrochloric acid	HCl	36.465
Glucose	$C_6H_{12}O_6$	180.16
Sodium acetate (crystals)	$CH_3COONa \cdot 3H_2O$	136.09
Sodium cacodylate (trihydrate)	$Na(CH_3)_2AsO_2 \cdot 3H_2O$	214.03
Sodium hydroxide	NaOH	40.005
Sodium phosphate (monobasic)	$NaH_2PO_4 \cdot H_2O$	137.99
Sodium phosphate (anhydrous, dibasic)	NaH_2PO_4	141.98
Tris acid maleate	$H_2NC(CH_2OH)_3$	121.14

References

Abramczuk, J. (1972). Effect of formalin fixation on the dry mass of isolated rat liver nuclei. *Histochemie* **29**, 207.

Abrunhosa, R. (1972). Microperfusion fixation of embryos for ultrastructural studies. *J. Ultrastruct. Res.* **41**, 176.

Accinni, L., Hsu, K. C., Spiele, H., and De Martino, C. (1974). Picric acid-formaldehyde fixation for immunoferritin studies. *Histochemistry* **42**, 257.

Adams, C. W. M. (1958). Histochemical mechanisms of the Marchi reaction for degenerating myelin. *J. Neurochem.* **2**, 178.

Adams, C. W. M. (1960). Osmium tetroxide and the Marchi method: Reactions with polar and nonpolar lipids, protein, and polysaccharide. *J. Histochem. Cytochem.* **8**, 262.

Adams, C. W. M., and Bayliss, O. B. (1968). Reappraisal of osmium tetroxide and OTAN histochemical reactions. *Histochemie* **16**, 162.

Adams, C. W. M., Abdulla, Y. H., and Bayliss, O. B. (1967). Osmium tetroxide as a histochemical and histological reagent. *Histochemie* **9**, 68.

Adamson, I. Y. R., and Bowden, D. H. (1970). The surface complexes of the lung. A cytochemical partition of phospholipid surfactant and mucopolysaccharides. *Am. J. Pathol.* **61**, 359.

Adomian, G. E., Laks, M. M., and Billingham, M. E. (1977). Contraction bands in human hearts: Pathology or artifact? *Proc. 35th Annu. Meet., Electron. Microsc. Soc. Am.*, p. 578.

Afzelius, B. A. (1962). Chemical fixatives for electron microscopy. *Symp. Int. Soc. Cell Biol.* **1**, 1.

Aihara, K., Nomizo, K., Nagata, K., Nishikawa, H., and Suzuki, K. (1967). An attempt to incorporate the automated tissue processing in electron microscopy. *J. Electron Microsc.* **16**, 285.

Aihara, K., Saito, H., Yajima, G., Suzuki, K., Inoue, K., Nishikawa, H., and Nakamura, S. (1972). A revised device for automated tissue processing in electron microscopy. *Proc. 30th Annu. Meet., Electron Microsc. Soc. Am.*, p. 302.

Aihara, K., Suzuki, K., Hirohata, Y., Yajima, M., Yajima, G., Hata, T., Goro, A., Matsumoto, K., Kurokawa, T., Takahashi, K., Mori, H., and Tsuzuki, T. (1978). Incorporation of the automated tissue processing system in biological electron microscopy. *Proc. Int. Congr. Electron Microsc., 9th*, Vol. 2, 76. Microscopical Soc. of Canada.

Al-Azzawi, H. T., and Stoward, P. J. (1970). The effects of various forms of euthanasia on the histochemically demonstrable activities of lysosomal enzymes. *Proc. R. Microsc. Soc.* **5,** 13.

Albert, E. N., and Rucker, R. D. (1975). Electron microscopic demonstration of cholesterol in atheromatous aortae. *Histochem. J.* **7,** 517.

Albin, T. B. (1962). Handling and toxicology. *In* "Acrolein" (C. W. Smith, ed.). p. 234. Wiley, New York.

Alderson, T. (1964). Crosslinking of fibrous protein by formaldehyde. *Nature (London)* **187,** 485.

Alexander, E., Sanders, S., and Braylan, R. (1976). Purported difference between human T- and B-cell surface morphology is an artifact. *Nature (London)* **261,** 239.

Allman, D. W., Wakabashi, T., Korman, F., and Green, D. E. (1970). Studies on the transition of the cristal membrane from the orthodox to the aggregated configuration. I. Topology of bovine adrenal mitochondria in the orthodox configuration. *Bioenergetics* **1,** 331.

Altman, F. P., and Barnett, R. J. (1975). The ultrastructural localization of enzyme activity in unfixed sections. *Histochemistry* **44,** 179.

Altman, P. L., and Dittmer, D. S., eds. (1973) "Biological Data Handbook," 2nd ed., Vols. I–III. Fed. Am. Soc. Exp. Biol., Bethesda, Maryland.

Amsterdam, A., and Schramm, M. (1966). Rapid release of the zymogen granule protein by osmium tetroxide and its retention during fixation by glutaraldehyde. *J. Cell Biol.* **29,** 199.

Andersen, C. A. (1967). An introduction to the electron probe microanalyzer and its application to biochemistry. *In* "Methods of Biochemical Analysis" (D. Glick, ed.), Vol. 15. Wiley (Interscience), New York.

Anderson, T. F. (1951). Techniques for the preservation of three-dimensional structure in preparing specimens for the electron microscope. *Trans. N.Y. Acad. Sci.* **13,** 130.

Anderson, P. J. (1967). Purification and quantitation of glutaraldehyde and its effects on several enzyme activities in skeletal muscle. *J. Histochem. Cytochem.* **15,** 652.

Andrews, P. M., and Porter, K. R. (1974). A scanning electron microscopic study of the nephron. *Am. J. Anat.* **140,** 81.

Anniko, M., and Bagger-Sjöbäck, D. (1977). Early postmortem change of the crista ampullaris. *Virchows Arch. B* **25,** 137.

Anniko, M., and Lundquist, P.-G. (1980). Temporal bone morphology after systemic arterial perfusion or intralabyrinthine *in situ* immersion. I. Hair cells of the vestibular organs and the cochlea. *Micron* **11,** 73.

Aoki, A., Einstein, J., and Fawcett, D. W. (1976). Tannic acid as an electron-dense probe in the testis. *Am. J. Anat.* **146,** 449.

Arborgh, B., Ericsson, J. L. E., and Helminen, H. (1971). Inhibition of renal acid phosphatase and aryl sulfatase activity by glutaraldehyde fixation. *J. Histochem. Cytochem.* **19,** 449.

Arborgh, B., Bell, P., Brunk, U., and Collins, V. P. (1976). The osmotic effect of glutaraldehyde during fixation. A transmission electron microscopy, scanning electron microscopy and cytochemical study. *J. Ultrastruc. Res.* **56,** 339.

Arnold, J. D., Berger, A. E., and Allison, O. L. (1971). Some problems of fixation of selected biological samples for SEM examination. *Proc. SEM.,* p. 249.

Arstila, A. U., Hirsimaki, P., and Trump, B. F. (1974). Studies on the subcellular pathophysiology of sublethal chronic injury. *Beitr. Pathol.* **152,** 211.

Artvinli, S. (1975). Biochemical aspects of aldehyde fixation and a new formaldehyde fixative. *Histochem. J.* **7,** 435.

Asano, M., Kurono, C., Wakabayashi, T., and Kimura, H. (1976). Stabilization of configurational states and enzyme activities in subcellular fractions after fixation with extremely low concentrations of glutaraldehyde. *Histochem. J.* **8,** 113.

Ashwood-Smith, M. J. (1967). Radioprotective and cryoprotective properties of dimethyl sulfoxide in cellular systems. *Ann. N. Y. Acad. Sci.* **141,** 45.

Ashwood-Smith, M. J. (1975). Current concepts concerning radioprotective and cryoprotective properties of dimethyl sulfoxide in cellular systems. *Ann. N. Y. Acad. Sci.* **243**, 246.

Ashwood-Smith, M. J., and Warby, C. (1971). Studies on the molecular weight and cryoprotective properties of polyvinyl pyrrolidone and dextran with bacteria and erythrocytes. *Cryobiology* **8**, 453.

Ashworth, C. T., Leonard, J. S., Eigenbrodt, E. H., and Wrightsman, F. J. (1966). Hepatic intracellular osmiophilic droplets: Effect of lipid solvents during tissue preparation. *J. Cell Biol.* **31**, 301.

Aso, C., and Aito, Y. (1962). Polymerization of bifunctional monomers. II. Polymerization of glutaraldehyde. *Makromol. Chem.* **58**, 195.

Assa, Y., Shany, S., Gestetner, B., Tencer, Y., Birk, Y., and Bondi, A. (1973). Interaction of alfalfa saponins with components of the erythrocyte membrane in hemolysis. *Biochim. Biophys. Acta* **307**, 83.

Assimacopoulos, A., and Kapanci, Y. (1974). Lung fixation by perfusion—a simple method to control the pressure in the perfusion circuit. *J. Microsc. (Oxford)* **100**, 227.

Atwood, D. G., Crawford, B. J., and Braybrook, G. B. (1975). A technique for processing mucus-coated marine invertebrate spermatozoa for scanning electron microscopy. *J. Microsc. (Oxford)* **103**, 259.

Atwood, H. L., Lang, F., and Morin, W. A. (1972). Synaptic vesicles: Selective depletion in crayfish excitatory and inhibitory axons. *Science* **176**, 1353.

Aukland, K. (1973). Autoregulation of interstitial fluid volume. *Scand. J. Clin. Lab. Invest.* **31**, 247.

Babaï, F. (1977). Application of cryoultramicrotomy to enzyme cytochemistry. *In* "Electron Microscopy of Enzymes: Principles and Methods" (M. A. Hayat, ed.), Vol. 5, p. 162. Van Nostrand-Reinhold, New York.

Baccetti, B. (1975). Spermatozoa. *In* "Principles and Techniques of Scanning Electron Microscopy: Biological Applications" (M. A. Hayat, ed.), Vol. 4, Van Nostrand-Reinhold, New York.

Bach, D., and Miller, I. R. (1967). Polarographic investigation of binding of Cu^{++} and Cd^{++} by DNA. *Biopolymers* **5**, 161.

Bachofen, M., Weibel, E. R., and Roos, B. (1975). Postmortem fixation of human lungs for electron microscopy. *Am. Rev. Respir. Dis.* **111**, 247.

Bahr, G. F. (1954). Osmium tetroxide and ruthenium tetroxide and their reactions with biologically important substances. *Exp. Cell Res.* **7**, 457.

Bahr, G. F. (1955). Continued studies about the fixation with osmium tetroxide. Electron stains. IV. *Exp. Cell Res.* **9**, 277.

Bahr, G. F., Bloom, G., and Friberg, U. (1957). Volume changes of tissues in physiological fluids during fixation in osmium tetroxide or formaldehyde and during subsequent treatment. *Exp. Cell Res.* **12**, 342.

Bairati, A., Petruccioli, M. G., and Tarelli, L. T. (1972). Birefringence of collagen fibrils after glutaraldehyde treatment. *J. Submicrosc. Cytol.* **4**, 89.

Bajer, A., and Molé-Bajer, J. (1971). Architecture and function of the mitotic spindle. *Adv. Cell Mol. Biol.* **1**, 213.

Baker, J. R. (1950). "Cytological Technique." Methuen, London.

Baker, J. R. (1958). "Principles of Biological Microtechnique: A Study of Fixation and Dyeing." Wiley, New York.

Baker, J. R., and McCrae, J. M. (1966). The fine structure resulting from fixation by formaldehyde: The effects of concentration, duration, and temperature. *J. R. Microsc. Soc.* **58**, 391.

Bakst, M., and Haworth, B. (1975). SEM preparation and observations of the hen's oviduct. *Anat. Rec.* **181**, 211.

Baldwin, K. M. (1970). The fine structure and electrophysiology of heart muscle cell injury. *J. Cell Biol.* **46**, 455.

Banfield, W. G. (1970). Automation in tissue processing for electron microscopy. *In* "Biological Techniques in Electron Microscopy" (D. F. Parsons, ed.), p. 165. Academic Press, New York.

Barber, T., and Burkholder, P. (1975). Relation of surface and internal ultrastructure of thymus and bone marrow derived lymphocytes to specimen preparatory technique. *Proc. SEM.*, p. 369.

Bard, J. B. L., Hay, E. D., and Meller, S. M. (1975). Formation of the endothelium of the avian cornea: A study of cell movement *in vivo. Dev. Biol.* **42**, 334.

Bernard, T. (1976). An empirical relationship for the formulation of glutaraldehyde-based fixatives. *J. Ultrastruct. Res.* **54**, 478.

Barnes, C. D., and Etherington, L. D. (1973). "Drug Doses and Laboratory Animals." Univ. of California Press, Berkeley.

Barnicot, N. A. (1967). An electron microscopic study of newt mitotic chromosomes by negative staining. *J. Cell Biol.* **32**, 585.

Barnett, R. J., Perney, D. P., and Hagström, P. E. (1964). Additional new aldehyde fixatives for histochemistry and electron microscopy. *J. Histochem. Cytochem.* **12**, 36.

Barrow, P. C., and Holt, S. J. (1971). Differences in distribution of esterase between cell fractions of rat liver homogenates prepared in various media. *Biochem. J.* **125**, 545.

Barton, R. (1966). Fine structure of mesophyll cells in senescing leaves of *Phaseolus. Planta* **71**, 314.

Battaglia, E., and Maggini, F. (1968). Use of osmium for staining nucleolus in squash technique. *Caryologia* **21**, 287.

Battistoni, C., Furlani, C., Mattogno, G., and Tom, G. (1977). ESCA spectra of some N-bonded osmium complexes. *Inorg. Chim. Acta* **21**, L25.

Baudhuin, P. (1969). Liver peroxisomes: Cytology and function. *Ann. N. Y. Acad. Sci.* **168**, 214.

Baudhuin, P., Evrard, P., and Berthet, J. (1967). Electron microscopic analysis of subcellular fractions. I. The preparation of representative samples from suspensions of particles. *J. Cell Biol.* **32**, 181.

Baur, P. S., and Stacy, T. R. (1977). The use of PIPES buffer in the fixation of mammalian and marine tissues for electron microscopy. *J. Microsc. (Oxford)* **109**, 315.

Baur, P. S., and Walkinshaw, C. H. (1974). Fine structure of tannin accumulations in callus cultures of *Pinus elliotii. Can. J. Bot.* **52**, 615.

Beaufay, H., Hers, H. G., Berthet, J., and de Duve, C. (1954). Le système hexosphosphatasique. V. Influence de divers agents sur l'activité et la stabilité de la glucose-6-phosphatase. *Bull. Soc. Chim. Biol.* **36**, 1539.

Beaulaton, J., and Gras, R. (1980). Influence of aldehyde fixatives on the ultrastructure of the prothoracic gland cells in *Rhodnius prolixus* (Insecta, Heteroptera) with special reference to their osmotic effects. *Microsc. Acta* **82**, 351.

Becker, D. P., Young, H. F., Nulsen, F. E., and Jane, J. A. (1960). Physiological effects of dimethyl sulfoxide on peripheral nerves: Possible role in pain relief. *Exp. Neurol.* **24**, 272.

Becker, R. (1959). Diplom. Arbeit Germany, Technische Hochschule Karlsruhe.

Beer, M., and Thomas, C. A. (1961). The electron microscopy of phage DNA molecules with denatured regions. *J. Mol. Biol.* **3**, 699.

Beer, M., Stern, S., Carmalt, D., and Mohlenrich, K. H. (1966). Determination of base sequence in nucleic acids with the electron microscope. V. The thymine specific reactions of osmium tetroxide with deoxyribonucleic acid and its components. *Biochemistry* **5**, 2283.

Begg, D., and Rebhun, L. I. (1978). Visualization of actin filament polarity in thin sections. *Biophys. J.* **21**, 23a.

Behbehani, B. I., Nayak, R. K., and McCoy, R. E. (1977). Scanning electron microscopy of the male nematode *Physaloptera*. *Proc. 35th Annu. Meet.,Electron Microsc. Soc. Am.,* p. 674.

Behnke, O., and Zelander, T. (1970). Preservation of intercellular substances by the cationic dye alcian blue in preparative procedures for electron microscopy. *J. Ultrastruct. Res.* **31**, 424.

Behrman, E. J. (1980a). On the oxidation state of osmium in fixed tissues. *J. Histochem. Cytochem.* **28**, 285.

Behrman, E. J. (1980b). On the oxidation state of osmium in fixed tissues: An addendum. *J. Histochem. Cytochem.* **28**, 1032.

Bell, P. B., Brunk, U., Collins, P., Forsby, N., and Fredricksson, B.-A. (1975). SEM of cells in culture: Osmotic effects during fixation. *Proc. SEM.,* p. 380.

Bello, J., Haas, D., and Bello, H. R. (1966). Interactions of protein-denaturing salts with model amides. *Biochemistry* **5**, 2539.

Berger, N. A., and Eichhorn, G. L. (1971). Interaction of metal ions with polynucleotides and related compounds. XIV. Nuclear magnetic resonance studies of the binding of copper II to adenine nucleotides. *Biochemistry* **10**, 1847.

Berjak, P. 1973). Ultrastructure of a functional senescence process. *Proc—Electron Microsc. Soc., South Afr.* **3**, 51.

Bernhard, W. (1955). Appareil de deshydration continue. *Exp. Cell Res.* **8**, 248.

Bernheim, F. (1971). The effects of alcohol and sugars on the swelling rate of cells of *Pseudomonas aeruginosa* in various salts. *Microbios* **4**, 49.

Berteloot, A., and Hugon, J. S. (1975). Effect of glutaraldehyde and lead on the activity of hepatic glucose-6-phosphatase. *Histochemistry* **43**, 197.

Bessis, M., and Weed, R. I. (1972). Preparation of red blood cells (RBC) for SEM—a survey of various artifacts. *Proc. SEM.,* p. 9.

Bhisey, A. N., and Freed, J. J. (1971). Cross-bridges on the microtubules of cooled interphase Hela cells. *J. Cell Biol.* **50**, 557.

Birch-Andersen, A., Ferguson, D. J. P., and Pontefract, R. D. (1976). A technique for obtaining thin sections of coccidian oocytes. *Acta Pathol. Microbiol. Scand.* **84**, 235.

Birks, R. I., and Davey, D. F. (1972). An analysis of volume changes in the T-tubes of frog skeletal muscle exposed to sucrose. *J. Physiol. (London)* **222**, 95.

Blass, J., Verriest, C., Leau, A., and Weiss, M. (1976). Monomeric glutaraldehyde as an effective crosslinking reagent for proteins. *J. Leather Chem. Assoc.* **71**, 121.

Bloom, F. E. (1970). The fine structural localization of biogenic monoamines in nervous tissue. *Int. Rev. Neurobiol.* **13**, 27.

Bloom, F. E. (1973). Ultrastructural identification of catecholamine containing central synaptic terminals. *J. Histochem. Cytochem.* **21**, 333.

Bloom, F. E., Sims, K. L., Weitsen, H. A., Davis, G. A., and Hanker, J. S. (1972). Cytochemical differentiation between monoamine oxidase and other neuronal oxidases. *Adv. Biochem. Psychopharmacol.* **5**, 244.

Bloom, G. D., and Haegermark, O. (1965). Studies on morphological changes and histamine release induced by compound 48/80 in rat peritoneal mast cells. *Exp. Cell Res.* **40**, 637.

Blough, H. A. (1966). Selective inactivation of biological activity of Myxoviruses by glutaraldehyde. *J. Bacteriol.* **92**, 266.

Blouin, A., Bolender, R. P., and Weibel, E. R. (1977). Distribution of organelles and membranes between hepatocytes and nonhepatocytes in the rat liver parenchyma. *J. Cell Biol.* **72**, 441.

Bluemink, J. G. (1970). The first cleavage of the amphibian egg. An electron microscope study of the onset of cytokinesis in the egg of *Ambystoma mexicanum. J. Ultrastruct. Res.* **32**, 142.

Bluemink, J. G. (1972). Cortical wound healing in the amphibian egg: an electron microscopical study. *J. Ultrastruct. Res.* **41**, 95.

Bodian, D. (1970). An electron microscopic characterization of classes of synaptic vesicles by means of controlled aldehyde fixation. *J. Cell Biol.* **44**, 115.

Bohman, S.-O. (1974). The ultrastructure of the rat renal medulla as observed after improved fixation methods. *J. Ultrustruct. Res.* **47**, 329.

Bohman, S.-O., and Maunsbach, A. B. (1970). Effects on tissue fine structure of variations in colloid osmotic pressure of glutaraldehyde fixatives. *J. Ultrastruct. Res.* **30**, 195.

Bohn, W. (1978). A fixation method for improved antibody penetration in electron microscopical immunoperoxidase studies. *J. Histochem. Cytochem.* **26**, 293.

Bone, Q., and Denton, E. J. (1971). The osmotic effects of electron microscope fixatives. *J. Cell Biol.* **49**, 571.

Bone, Q., and Ryan, K. P. (1972). Osmolarity of osmium tetroxide and glutaraldehyde fixatives. *Histochem. J.* **4**, 331.

Bonilla, E. (1977). Staining of transverse tubular system of skeletal muscle by tannic acid-glutaraldehyde fixation. *J. Ultrastruct. Res.* **58**, 162.

Boom, A., Daems, W. T., and Luft, J. H. (1974). On the fixation of intestinal absorptive cells. *J. Ultrastruct. Res.* **48**, 350.

Boucher, R. M. G. (1972). Advances in sterilization techniques. *Am. J. Hosp. Pharm.* **29**, 661.

Boucher, R. M. G. (1974). Potentiated acid 1,5-pentanedial solution—a new chemical sterilizing and disinfecting agent. *Am. J. Hosp. Pharm.* **31**, 546.

Boucher, R. M. G., Last, A. J., and Smith, D. K. (1973). Biocidal mechanisms of saturated dialdehydes. *Proc. West. Pharmacol. Soc.* **16**, 282.

Bowen, I. D. (1971). A high resolution technique for the fine structural localization of acid hydrolases. *J. Microsc. (Oxford)* **94**, 25.

Bowen, I. D., and Ryder, T. A. (1977). The application of x-ray microanalysis to enzyme cytochemistry. *In* "Electron Microscopy of Enzymes: Principles and Methods" (M. A. Hayat, ed.), Van Nostrand Reinhold, New York.

Bower, A. J. (1981). A simple, inexpensive apparatus to allow sequential perfusion of an animal with several different fluids without an intervening drop in perfusion pressure. *J. Microsc.* **122**, 103.

Bowes, D., Bullock, G. R., and Winsey, N. J. P. (1970). A method for fixing rabbit and rat hind limb skeletal muscle by perfusion. *Proc. 7th Ann. Int. Cong. Electron Microsc.* **1**, 397. Grenoble, France.

Bowes, J. H., and Cater, C. W. (1966). The reaction of glutaraldehyde with proteins and other biological materials. *J. R. Microsc. Soc.* **85**, 193.

Bowes, J. H., and Kenton, R. H. (1949). The effect of deamination and esterification on the reactivity of collagen. *J. Biochem. (Tokyo),* **44**, 142.

Bowes, J. H., and Raistrick, A. S. (1964). The action of heat and moisture on leather. V. Chemical changes in collagen and tanned collagen. *J. Am. Leather Chem. Assoc.* **59**, 201.

Bowes, J. H., Cater, C. W., and Ellis, M. J. (1965). Determination of formaldehyde and glutaraldehyde bound to collagen by carbon[14] assay. *J. Am. Leather Chem. Assoc.* **60**, 275.

Boyde, A., and Echlin, P. 1974). Freezing and freeze-drying—a preparative technique for SEM. *Proc. SEM.,* p. 759.

Boyne, A. F., Bohan, T. P., and Williams, T. H. (1975). Effects of calcium-containing fixation solutions on cholinergic synaptic vesicles. *J. Cell Biol.* **63**, 780.

Bradbury, S., and Meek, G. A. (1960). A study of potassium permanganate 'fixation' for electron microscopy. *Q. J. Microsc. Sci.* [N.S.] **101**, 241.

Bradbury, S., and Stoward, P. J. (1967). The specific cytochemical demonstration in the electron microscope of periodate-reactive mucosubstances and polysaccharides containing *vic*-glycol groups. *Histochemie* **11**, 71.

Brand, E., Kassell, B., and Saidel, L. (1944). Chemical, clinical, and immunological studies on the products of human plasma fractionation. III. Amino acid composition of plasma proteins. *J. Clin. Invest.* **23**, 437.

Brightman, M. W., and Reese, T. S. (1969). Junctions between intimately apposed cell membranes in the vertebrate brain. *J. Cell Biol.* **40**, 648.

Brinkley, B. R., and Cartwright, J. (1975). Cold-labile and cold-stable microtubules in the mitotic spindle of mammalian cells. *Ann. N. Y. Acad. Sci.* **253**, 428.

Brissie, R. M., Spicer, S. S., Hall, B. J., and Thompson, N. T. (1974). Ultrastructural staining of thin sections with iron hematoxylin. *J. Histochem. Cytochem.* **22**, 895.

Brody, A. R., and Craighead, J. E. (1973). A simple perfusion apparatus for lung fixation. *Proc. Soc. Exp. Biol. Med.* **143**, 388.

Brody, A. R., and Craighead, J. E. (1975). Preparation of human lung biopsy specimens by perfusion fixation. *Am. Rev. Respir. Dis.* **112**, 645.

Brooks, B. R., and Klamerth, O. L. (1968). Interaction of DNA with bifunctional aldehydes. *Eur. J. Biochem.* **5**, 178.

Brown, G. L., and Locke, M. (1978). Nucleoprotein localization by bismuth staining. *Tissue Cell* **10**, 365.

Brown, J. N. (1975). The avian erythrocyte: A study of fixation for electron microscopy. *J. Microsc. (Oxford)* **104**, 293.

Brown, J. N. (1977). A simple low-cost critical point dryer with continuous flow dehydration attachment. *J. Microsc. (Oxford)* **111**, 351.

Bruin, A., Fasulo, M. P., Tosi, B., Dall'Olio, G., and Vannini, G. L. (1976). Fluorogenic detection of primary amines in plant histochemistry with fluorescamine: A comparative study on the effects of coagulant and non-coagulant fixatives. *Histochemistry* **48**, 269.

Brunings, E. A., and de Priester, W. (1971). Effect of mode of fixation on formation of extrusions in the midgut epithelium of Calliphora. *Cytobiologie* **4**, 487.

Brunk, U. T., and Ericsson, J. L. E. (1972). Electron microscopical studies on rat brain neurons. Localization of acid phosphatase and mode of formation of lipofuschin bodies. *J. Ultrastruct. Res.* **38**, 1.

Buckley, I. K. (1973a). Studies in fixation for electron microscopy using cultured cells. *Lab. Invest.* **29**, 398.

Buckley, I. K. (1973b). The lysosomes of cultured chick embryo cells. A correlated light and electron microscopic study. *Lab. Invest.* **29**, 411.

Bulger, R. E., Siegel, F. L., and Pendergrass, R. (1976). Proximal tubule tendrils: Fact or artifact. *Am. J. Anat.* **146**, 323.

Bullivant, S. (1973). Freeze-etching and freeze-fracturing. *In* "Advanced Techniques in Biological Electron Microscopy" (J. K. Koehler, ed.). Springer-Verlag, Berlin and New York.

Bullivant, S., Rayns, D. G., Bertaud, W. S., Chalcroft, J. P., and Grayston, G. F. (1972). Freeze-fractured myosin filaments. *J. Cell Biol.* **55**, 520.

Bunge, M. B., Bunge, R. P., and Ris, H. (1961). Ultrastructural study of remyelination in an experimental lesion in cat spinal cord. *J. Biophys. Biochem. Cytol.* **10**, 67.

Burdett, I. D. J., and Rogers, H. J. (1970). Modification of the appearance of mesosomes in sections of *Bacillus licheniformis* according to the fixation procedures. *J. Ultrastruct. Res.* **30**, 354.

Burgess, J., Linstead, P. J., and Harnden, J. M. (1978). The interpretation of scanning electron micrographs. *Micron* **8**, 181.

Burkl, W., and Schiechl, H. (1968). A study of osmium tetroxide fixation. *J. Histochem. Cytochem.* **16**, 157.

Burnham, J. C., and Hageage, G. J., Jr. (1973). Effect of aldehyde fixatives on the adenosine phosphatases and ultrastructure of *Vitreoscilla*. *Can. J. Microbiol.* **19**, 1059.

Burton, K. (1967). Oxidation of pyrimidine nucleosides and nucleotides by osmium tetroxide. *Biochem. J.* **104**, 686.

Burton, K., and Riley, W. T. (1966). Selective degradation of thymidine and thymine deoxynucleotides. *Biochem. J.* **98**, 70.

Buss, H., Klose, J. P., and Hollweg, H. C. (1976). Endothelial surfaces of renal, coronary, and cerebral arteries. *Proc. SEM* **2**, 217.

Busson-Mabillot, S. (1971). Influence de la fixation Chimique sur les ultrastructures I. Etude sur les organites du follicule ovarien d'un poisson téléostéen. *J. Microsc. (Paris)* **12**, 317.

Butler, R. D., and Simon, E. W. (1971). Ultrastructural aspects of senescence in plants. *Adv. Gerontol. Res.* **3**, 73.

Butler, T. C., Waddell, W. J., and Poole, D. T. (1967). Intracellular pH based on the distribution of weak electrodes. *Fed. Proc.* **26**, 1327.

Bylock, A., Björkerud, S., Brattsand, R., Hansson, G. R., Hansson, H. A., and Bondjers, G. (1977). Endothelial structure in rabbits with moderate hypercholesterolaemia. A SEM study. *Acta Pathol. Microbiol. Scand., Sect. A* **85A**, 671.

Caldwell, J. B., and Milligan, B. (1972). The sites of reaction of wool with formaldehyde. *Text. Res. J.* **42**, 122.

Callas, G. (1974). A new fixation technique for the electron microscopic study of pulmonary surfactant. *Anat. Rec.* **180**, 457.

Cammermeyer, J. (1960). The postmortem origin and mechanism of neuronal hyperchromatosis and nuclear pykonosis. *Exp. Neurol.* **3**, 379.

Cammermeyer, J. (1968). Peripheral vasoconstriction with epinephrine for selective fixation of the central nervous system by perfusion. *Acta Neuropathol.* **11**, 368.

Cammermeyer, J. (1978). Is the solitary dark neuron a manifestation of postmortem trauma to the brain inadequately fixed by perfusion? *Histochemistry* **56**, 97.

Cammermeyer, J. (1980). Segmental shrinkage and argentophilia of dendrons after fixation by perfusion with dimethyl sulfoxide (DMSO)- containing solutions. *Exp. Neurol.* **67**, 621.

Carling, D. E., White, J. A., and Brown, M. F. (1977). The influence of fixation procedure on the ultrastructure of the host-endophyte interface of vesicular-arbuscular mycorrhizae. *Can. J. Bot.* **55**, 48.

Carpenter, D. C., and Nebel, B. R. (1931). Ruthenium tetroxide as a fixative in cytology. *Science* **74**, 154.

Carraway, K. L., and Koshland, D. E. (1968). Reaction of tyrosine residues in proteins with carbodiimide reagent. *Biochim. Biophys. Acta* **160**, 272.

Carraway, K. L., and Triplett, R. B. (1970). Reaction of carbodiimides with protein sulfhydryl groups. *Biochim. Biophys. Acta* **200**, 564.

Carson, F., Lynn, J. A., and Martin, J. H. (1972). Ultrastructural effect of various buffers, osmolality, and temperature on paraformaldehyde fixation of the formed elements of blood and bone marrow. *Tex. Rep. Biol. Med.* **30**, 125.

Carson, J. H., Martin, J. H., and Lynn, J. A. (1973). Formalin fixation for electron microscopy: A reevaluation. *Am. J. Clin. Pathol.* **59**, 365.

Carstensen, E. L., Aldridge, W. G., Child, S. Z., Sullivan, P., and Brown, H. H. (1971). Stability of cells fixed with glutaraldehyde and acrolein. *J. Cell Biol.* **50**, 529.

Casanova, S., Marchetti, M., Bovina, C., and Laschi, R. (1972). A study of the effects of fixation on liver glucose-6-phosphatase activity for electron microscope cytochemistry. *J. Submicrosc. Cytol.* **4**, 261.

Casciani, R. V. and Behrman, E. J. (1978). Reduction of osmium (VII) to osmium (II) by reaction with an olefin and EDTA. Formation of an Os II-EDTA complex. *Inorg. Chim. Acta* **28**, 69.

Casley-Smith, J. R. (1967). Some observations on the electron microscopy of lipids. *J. R. Microsc. Soc.* **81**, 235.

Cater, C. W. (1963). The evaluation of aldehydes and other bifunctional compounds as crosslinking agents for collagen. *J. Soc. Leather Traders' Chem.* **47**, 259.

Cater, C. W. (1965). Further investigations into the efficiency of dialdehydes and other compounds as crosslinking agents for collagen. *J. Soc. Leather Trades' Chem.* **49**, 455.

Catesson, A. M. (1974). Cambial cells. *In* "Dynamic Aspects of Plant Ultrastructure" (A. W. Robards, ed.), p. 358. McGraw-Hill, New York.

Ceccarini, C., Campo, M. S., and Andronico, F. (1970). Biosynthesis and distribution of ribosomal particles in the cellular slime mould. *J. Mol. Biol.* **54**, 33.

Chalkley, R., and Hunter, C. (1975). Histone-histone propinquity by aldehyde fixation of chromatin. *Proc. Natl. Acad. Sci. U.S.A.* **72**, 1304.

Chambers, R. W., Bowling, M. C., and Grimley, P. M. (1968). Glutaraldehyde fixation in routine histopathology. *Arch. Pathol.* **85**, 18.

Chang, C.-H., Beer, M., and Marzilli, L. G. (1977). Osmium-labeled polynucleotides. The reaction of osmium tetroxide with deoxyribonucleic acid and synthetic polynucleotides in the presence of tertiary nitrogen donor ligands. *Biochemistry* **16**, 33.

Chang, J. H. T. (1972). Fixation and embedding, *in situ,* of tissue culture cells for electron microscopy. *Tissue Cell* **4**, 561.

Chapman, D., and Fluck, D. J. (1966). Physical studies of phospholipids. III. Electron microscope studies of some pure fully saturated 2,3-diacyl-diphosphatidyl-ethanolamines and phosphatidyl-cholines. *J. Cell Biol.* **30**, 1.

Chédid, A., and Nair, V. (1972). Diurnal rhythum in endoplasmic reticulum of rat liver: electron microscopic study. *Science* **175**, 179.

Cheng, K.-J., and Costerton, J. W. (1973). Localization of alkaline phosphatase in three gram-negative rumen bacteria. *J. Bacteriol.* **116**, 424.

Cheng, K.-J., Ingram, J. M., and Costerton, J. W. (1970). Alkaline phosphatase localization and spheroplast formation of *Pseudomonas aeruginosa. Can. J. Microbiol.* **16**, 1319.

Chiovetti, R. (1978). Encystment of *Naegleria gruberi.* I. Preparation of cysts for electron microscopy. *Trans. Am. Microsc. Soc.* **97**, 244.

Christensen, A., and Fawcett, D. (1961). The normal fine structure of opossum testicular interstitial cells. *J. Biophys. Biochem. Cytol.* **9**, 653.

Christie, K. N., and Stoward, P. J. (1974). A quantitative study of the fixation of acid phosphatase by formaldehyde and its relevance to histochemistry. *Proc. R. Soc. London, Ser. B* **186**, 137.

Clark, A. W. (1976). Changes in the structure of neuromuscular junctions caused by variations in osmotic pressure. *J. Cell Biol.* **69**, 521.

Clark, J. M., and Glagov, S. (1976). Luminal surface of distended arteries by scanning electron microscopy: Eliminating configurational and technical artifacts. *Br. J. Exp. Pathol.* **57**, 129.

Clark, R. L., and Behrman, E. J. (1976). On the question of phosphate complexes of osmium tetroxide: kinetics of reactions with nucleotides and dinucleoside monophosphates. *Bioinorg. Chem.* **5**, 359.

Clark, R. V. (1976). Three-dimensional organization of testicular interstitial tissue and lympatic space in the rat. *Anat. Rec.* **184**, 203.

Claude, A. (1962). Fixation of nuclear structures by unbuffered solutions of osmium tetroxide in slightly acid distilled water. *Proc. Int. Congr. Electron Microsc., 5th, 1962 Vol. 2, Artic.* L-14.

Clement, R. M., Sturm, J., and Daune, M. P. (1973). Interaction of metallic cations with DNA. VI. Specific binding of Mg^{++} and Mn^{++}. *Biopolymers* **12**, 405.

Cliff, W. J. (1971). The ultrastructure of aortic elastica as revealed by prolonged treatment with OsO_4. *Exp. Mol. Pathol.* **15**, 220.

Clift, F. F., and Cook, R. P. (1932). A method of determination of biologically important aldehydes and ketones, with special reference to pyruvic acid and methylgloxal. *Biochem. J.* **26**, 1788.

Coggeshall, R. E. (1979). A fine structural analysis of the myelin sheath in rat spinal roots. *Anat. Rec.* **194**, 201.

Cohen, A. L. (1974). Critical point drying. *In* "Principles and Techniques of Scanning Electron Microscopy: Biological Applications" (M. A. Hayat, ed.), Vol. 1, Van Nostrand-Reinhold, New York.

Cohen, A. L. (1977). A critical look at critical point drying-theory, practice, and artifacts. *Proc. SEM.*, p. 525.

Cohen, A. L. (1979). Critical point drying—principles and procedures. *Proc. SEM* **2**, 303.

Cohen, E., Shain, Y., Ben-Shaul, Y., and Mayer, A. M. (1971). Structural and enzymatic characterization of plant zygomen bodies. *Can. J. Bot.* **49**, 2053.

Coimbra, A., and Lopes-Vaz, A. (1971). The presence of lipid droplets and the absence of stable Sudanophilia in osmium-fixed human leukocytes. *J. Histochem. Cytochem.* **19**, 551.

Cole, M. D., Wiggins, J. W., and Beer, M. (1977). Molecular microscopy of labeled polynucleotides; stability of osmium atoms. *J. Mol. Biol.* **117**, 387.

Cole, R. S. (1975). Psoralen monoadducts and interstrand crosslinks in DNA. *Biochim. Biophys. Acta* **254**, 30.

Coleman, J. E., Ricciutti, C., and Severn, D. (1956). Improved preparation of ketohydroxystearic acids. *J. Am. Chem. Soc.* **78**, 5342.

Coleman, J. R., and Terepka, A. R. (1974). Preparatory methods for electron probe analysis. *In* "Principles and Techniques of Electron Microscopy: Biological Applications" (M. A. Hayat, ed.), Vol. 4, Van Nostrand-Reinhold, New York.

Coleman, S. E., Duggan, J., and Hackett, R. L. (1975). Changes in freeze-fractured nuclei following renal ischemia and reflow in the rat. *Exp. Mol. Pathol.* **23**, 59.

Coleman, S. E., Duggan, J., and Hackett, L. (1978). Quantitation particles in the freeze-fractured nuclear membrane after renal ischemia. *Virchows Arch. B* **28**, 119.

Collin, R. J., and Griffith, W. P. (1974). Mechanism of tissue component staining by osmium tetroxide. *J. Histochem. Cytochem.* **22**, 992.

Collin, R. J., Griffith, W. P., Phillips, F. L., and Skapski, A. C. (1973). Staining and fixation of unsaturated membrane lipids by osmium tetroxide: Crystal structure of a model osmium (VI) intermediate. *Biochim. Biophys. Acta* **320**, 745.

Collin, R. J., Griffith, W. P., Phillips, F. L., and Skapski, A. C. (1974). Staining and fixation of unsaturated membrane lipids by osmium tetroxide: Crystal structure of a model osmium (VI) di-ester. *Biochim. Biophys. Acta* **354**, 152.

Collins, C. J., and Guild, W. R. (1969). Irreversible effects of formaldehyde by DNA. *Biochim. Biophys. Acta.* **157**, 107.

Colquhoun, W. R., and Rieder, C. L. (1980). Contrast enhancement based on rapid dehydration in the presence of phosphate buffer. *J. Ultrastruct. Res.* **73**, 1.

Conger, K. A., Garcia, J. H., Lossinsky, A. S., and Kauffman, F. C. (1978). The effect of aldehyde fixation on selected substrates for energy metabolism and amino acids in mouse brain. *J. Histochem. Cytochem.* **26**, 423.

Conn, J. F., Kim, J. J., Suddath, F. L., Blattmann, P., and Rich, A. (1974). Crystal and molecular structure of an osmium bispyridine ester adenosine. *J. Am. Chem. Soc.* **96**, 7152.

Cope, G. H. (1968). Low temperature embedding in water miscible methacrylates after treatment with antifreezes. *J. R. Microsc. Soc.* **88**, 259.

Cope, G. H., and Williams, M. A. (1968). Quantitative studies on neutral lipid preservation in electron microscopy. *J. R. Microsc. Soc.* **88**, 259.

Corrodi, H., and Jonsson, G. (1967). The formaldehyde fluorescence method for the histochemical demonstration of biogenic monoamines. A review on the methodology. *J:. Histochem. Cytochem.* **15**, 65.

Costerton, J. W., and Marks, I. (1977). Localization of enzymes in prokaryotic cells. *In* "Electron Microscopy of Enzymes: Principles and Methods" (M. A. Hayat, ed.), Vol. 5, Van Nostrand-Reinhold, New York.

Costerton, J. W., Geesey, G. G., and Cheng, K. J. (1978). "How bacteria stick"? *Sci. Am.* **238**, 86.

Cotran, R. S., and Karnovsky, M. J. (1968). Ultrastructural studies on the permeability of the mesothelium to horseradish peroxidase. *J. Cell Biol.* **37**, 123.

Cotta-Pereira, G., Guerra-Rodrigo, F., and David-Ferreira, J. F. (1976). The use of tannic acid-glutaraldehyde in the study of elastic and elastic-related fibers. *Stain Technol.* **51**, 7.

Coulter, H. D., and Terracio, L. (1977). Preparation of biological tissues for electron microscopy by freeze-drying. *Anat. Rec.* **187**, 477.

Coupland, R. E. (1965). "The Natural History of the Chromaffin Tissue." Longmans, Green, New York.

Coupland, R. E., and Hopwood, D. (1966). The mechanism of the differential staining reaction for adrenalin and noradrenalin-storing granules in tissues fixed in glutaraldehyde. *J. Anat.* **100**, 227.

Cragg, B. (1979). Overcoming the failure of electron microscopy to preserve the brain's extracellular space. *Trends Neurosci.* **2**, 159.

Cragg, B. (1980). Preservation of extracellular space during fixation of the brain for electron microscopy. *Tissue Cell* **12**, 63.

Craik, J. E., and McNeil, I. R. R. (1965). Histological studies of stressed skin. *In* "Biomechanics and Related Bioengineering Topics" (R. M. Kenedi, ed.). Pergamon, Oxford.

Crawford, R. M. (1972). The structure and formation of siliceous wall of diatom *Melosira nummuloides. In* "Recent and Fossil Marine Diatoms" (R. Simonsen, ed.), p. 131. Cramer, Weinheim.

Criegee, R. (1936). Osmiumsäure-ester als Zwischenprodukte bei Oxydationen. *Justus Liebigs Ann. Chem.* **522**, 75.

Criegee, R. (1938). Organische Osmiumverbindungen. *Angew. Chem.* **51**, 519.

Criegee, R., Marchand, B., and Wannowius, A. (1942). Zur Kenntnis der organischen Osmiumverbindungen. *Justus Liebigs Ann. Chem.* **550**, 99.

Crowe, J. H., and Cooper, A. F. (1971). Cryptobiosis. *Sci. Am.* **225**, 30.

Curgy, J.-J. (1968). Influence du mode de fixation sur la possibilité d'observer des structures myeliniques dans les hépatocytes d'embryons de poulet. *J. Microsc. (Paris)* **7**, 63.

Cutler, R.W.P., and Dudzinsky, D. S. (1974). Effect of pentobarbitol on uptake and release of ^{3}H-GABA and ^{14}C-glutamate by brain slices. *Brain Res.* **67**, 546.

Daems, W. T., and Persijn, J. P. (1963). Sections staining with heavy metals of osmium-fixed and formol-fixed mouse liver tissue. *J. R. Microsc. Soc.* **81**, 199.

Dahl, H. A., and From, S. Hj. (1971). Some effects of polyvinyl alcohol and polyvinyl pyrrolidone on the activity of lactate dehydrogenase and its isoenzymes. *Histochemie* **25**, 182.

Dalen, A. B. (1976). The effect of glutaraldehyde on the stability of erythrocytes and on virus receptor substances. *Acta Pathol. Microbiol. Scand., Sect. B* **84B**, 196.

Dales, S., Gomatos, P. J., and Hsu, K. C. (1965). The uptake and development of reovirus in strain L-cells followed with labeled viral ribonucleic acid and ferritin-antibody conjugates. *Virology* **25**, 193.

Dallam, R. D. (1957). Determination of protein and lipid lost during osmic acid fixation of tissues and cellular particulates. *J. Histochem. Cytochem.* **5**, 178.

Daniel, F. B., and Behrman, E. J. (1975). Reactions of osmium ligand complexes with nucleosides. *J. Am. Chem. Soc.* **97**, 7352.

Daniel, F. B., and Behrman, E. J. (1976). Osmium (VI) complexes of the 3,5-dinucleoside monophosphates, ApU and UpA. *Biochemistry* **15**, 565.

Dankelman, W., and Daemen, J. M. H. (1976). Gas chromatographic and nuclear magnetic resonance determination of linear formaldehyde oligomers in formalin. *Anal. Chem.* **48**, 401.

Darrah, H. K., Hedley Whyte, J., and Hedley Whyte, E. T. (1971). Radioautography of cholesterol in lung. An assessment of different tissue processing techniques. *J. Cell Biol.* **49**, 345.

Dauwalder, M., Whaley, W. G., and Starr, R. C. (1980). Differentiation and secretion in *Volvox*. *J. Ultrastruct. Res.* **70**, 318.

Davey, D. F. (1972). The effect of fixative tonicity on the myosin filament lattice volume of frog muscle fixed following exposure to normal or hypertonic Ringer. *In* "Fixation in Histochemistry" (P. J. Stoward, ed.), p. 103. Chapman & Hall, London.

Davies, G. E., and Stark, G. R. (1970). Use of dimethylsuberimidate, a crosslinking reagent in studying the subunit structure of oligomeric proteins. *Proc. Natl. Acad. Sci. U.S.A.* **66**, 651.

Davies, H. G., and Spencer, M. (1962). The variation in the structure of erythrocyte nuclei with fixation. *J. Cell Biol.* **14**, 445.

Davies, K. J., and Garrett, J. R. (1972). Improved preservation of alkaline phosphatase in salivary glands of the cat. *In* "Fixation in Histochemistry" (P. J. Stoward, ed.), p. 151. Chapman & Hall, London.

Davies, P. F., and Bowyer, D. E. (1975). Scanning electron microscopy: Arterial endothelial integrity after fixation at physiological pressure. *Atherosclerosis* **21**, 464.

Davis, J. M., and Himwich, W. A. (1971). The amino acid, norepinephrine, and serotonin content of rat brain fixed with glutaraldehyde. *Brain Res.* **33**, 568.

Dawes, C. J. (1969). *Saprochaete saccharophilis:* Ultrastructure, x-ray diffraction, and chitin essay of cell walls as aids in evaluating taxonomic position. *Trans. Am. Microsc. Soc.* **88**, 572.

Dawes, C. J. (1979). "Biological Techniques for Transmission and Scanning Electron Microscopy." Barnes & Noble, New York.

Deamer, D. W., and Baskin, R. J. (1969). Ultrastructure of sarcoplasmic reticulum preparations. *J. Cell Biol.* **42**, 296.

Deamer, D. W., and Crofts, A. (1967). Action of Triton X-100 on chloroplast membranes. *J. Cell Biol.* **33**, 395.

de Bruijn, W. C. (1968). A modified OsO_4- (double) fixation procedure which selectively contrasts glycogen. *Proc. 4th Eur. Reg. Conf. Electron Microsc.*, p. 65.

de Bruijn, W. C. (1973). Glycogen, its chemistry and morphologic appearance in the electron microscope. I. A modified OsO_4 fixative which selectively contrasts glycogen. *J. Ultrastruct. Res.* **42**, 29.

de Bruijn, W. C., and Den Breejen, P. (1975). Glycogen, its chemistry and morphological appearance in the electron microscope. II. The complex formed in the selective contrast staining of glycogen. *Histochem. J.* **7**, 205.

de Bruijn, W. C., and Den Breejen, P. (1976). Glycogen, its chemistry and morphological appearance in the electron microscope. III. Identification of the tissue ligands involved in the glycogen contrast staining reaction with the osmium (VI)-iron (II) complex. *Histochem. J.* **8**, 121.

de Duve, C., and Baudhuin, P. 1966). Peroxisomes (microbodies and related particles). *Physiol. Rev.* **46**, 323.

DeFilippis, L. F., and Pallaghy, C. K. (1975). Localization of zinc and mercury in plant cells. *Micron* **6**, 111.

deF Webster, H. (1971). The geometry of peripheral myelin sheaths during their formation and growth in rat sciatic nerves. *J. Cell Biol.* **48**, 348.

deF Webster, H., and Ames, A. (1965). Reversible and irreversible changes in the fine structure of nervous tissue during oxygen and glucose deprivation. *J. Cell Biol.* **26**, 885.

de Harven, E., Lampen, N., Polliack, A., Warfel, A., and Fogh, J. (1975). New observations on methods for preparing cell suspensions for scanning electron microscopy. *Proc. SEM.*, p. 361.

De Jong, D. W., Olson, A. C., and Jansen, E. F. (1967). Glutaraldehyde activation of nuclear phosphatase in cultured plant cells. *Science* **155**, 1672.

De Martino, C., Stefanini, M., Bellocci, M., and Quintarelli, G. (1972). Osmium tetroxide-picric acid: A new fixation technique in electron microscopy. *J. Submicrosc. Cytol.* **4**, 111.

Dempsey, G. P., Bullivant, S., and Watkins, W. B. (1973). Endothelial cell membranes: Polarity of particles as seen by freeze-fracturing. *Science* **179**, 190.

Dempster, W. T. (1960). Rates of penetration of fixing fluids. *Am. J. Anat.* **107**, 59.

Dermer, G. B. (1968). An autoradiographic and biochemical study of oleic acid absorption by intestinal slices including determinations of lipid loss during preparation for electron microscopy. *J. Ultrastruct. Res.* **22**, 312.

Dermer, G. B. (1970). The fixation of pulmonary surfactant for electron microscopy. *J. Ultrastruct. Res.* **31**, 229.

Desmet, V. J. (1962). The hazard of acid differentiation in Gomori's method for acid phosphatase. *Stain Technol.* **37**, 373.

Devon, R. M., and Jones, D. G. (1979). Synaptic terminal parameters in unanesthetized rat cerebral cortex. *Cell Tissue Res.* **203**, 189.

Dewar, M. K., Johns, R. B., Kelly, D. P., and Yates, J. F. (1975). Crosslinking of amino acids by formaldehyde. ^{13}C.N.M.R. spectra of model compounds. *Aust. J. Chem.* **28**, 917.

Diamond, J. M., and Wright, E. (1969). Biological membranes: The physical basis of ion and nonelectrolyte selectivity. *Annu. Rev. Physiol.* **31**, 581.

Diers, L., and Schieren, M. T. (1972). Der Einfluss einiger elektronenmikroskopischer Fixierungs- und Einbettungsmittel auf die Chloroplasten- und Zellgrösse bei *Elodea. Protoplasma* **74**, 321.

Diers, L., and Schötz, F. (1966). Uber die dreidmensionale Gestaltung des Thylakoidsystems in den Chloroplasten. *Planta* **70**, 322.

Di Giamberardino, L., Koller, T., and Beer, M. (1969). Electron microscopic study of the base sequence in nucleic acids. IX. Absence of fragmentation and of crosslinking during reaction with osmium tetroxide and cyanide. *Biochim. Biophys. Acta* **182**, 523.

Dimmock, E. (1970). The surface structure of cultured rabbit kidney cells as revealed by electron microscopy. *J. Cell Sci.* **7**, 719.

Dodd, R. E. (1959). Infrared spectra of ruthenium and osmium tetroxide. *Trans. Faraday Soc.* **55**, 1480.

Doggenweiler, C. F., and Frenk, S. (1965). Staining properties of lanthanum on cell membranes. *Proc. Natl. Acad. Sci. U.S.A.* **53**, 425.

Doggenweiler, C. F., and Heuser, J. E. (1967). Ultrastructure of the prawn nerve sheaths: role of fixative and osmotic pressure in vesiculation of thin cytoplasmic laminae. *J. Cell Biol.* **34**, 407.

Dohrmann, G. J. (1970). Dark and light epithelial cells in the choroid plexus of mammals. *J. Ultrastruct. Res.* **32**, 268.

Donaldson, R. P., Tolbert, N. E., and Schnarrenberger, C. (1972). A comparison of microbody membranes with microsomes and mitochondria from plant and animal tissue. *Arch. Biochem. Biophys.* **152**, 199.

Done, J., Shorey, C. D., Loke, J. P., and Pollak, J. K. (1965). The cytochemical localization of alkaline phosphatase in *Escherichia coli* at the electron microscopy level. *Biochem. J.* **96**, 270.

Donnelly, E. H., and Goldstein, I. J. (1970). Glutaraldehyde-insolubilized concanavalin A: an absorbant for the specific isolation of polysaccharides and glycoproteins. *Biochem. J.* **118**, 679.

Doty, P., Boedtker, H., Fresco, J. R., and Haselkorn, M. (1959). Secondary structure in ribonucleic acids. *Proc. Natl. Acad. Sci. U.S.A.* **45**, 482.

Drawert, H. (1955). Der pH-Wert des Zellsaftes. *Encycl. Plant Physiol. New. Ser.* **1**.

Dreher, K. D., Schulman, J. H., Anderson, O. R., and Roels, O. A. (1967). The stability and structure of mixed lipid monolayers and bilayers. I. Properties of lipid and lipoprotein monolayers on OsO_4 solutions and the role of cholesterol, retinal, and tocopherol in stabilizing lecithin monolayers. *J. Ultrastruct. Res.* **19**, 586.

428

Dulak, L., and Crist, R. D. (1974). Microtubule properties in dimethyl sulfoxide (DMSO). *J. Cell Biol.* **63**, 900.

Dunbar, A. (1978). The use of some compound fixatives in the study of Campanulaceae. *J. Ultrustruct. Res.* **63**, 108.

Eagle, H. (1971). Buffer combinations for mammalian cell culture. *Science* **174**, 500.

Edanaga, M. (1974). A SEM study on the endothelium of the vessel. I. Fine structure of endothelium of the aorta and some other arteries in normal rabbit. *Arch. Histol. Jpn.* **37**, 1.

Eddy, A. A., and Johns, P. (1965). Preliminary observations on the proteins of the erythrocyte membranes with special reference to their cytological behavior. *SCI Monog.* **19**, 24.

Edwardson, J. R., Purcifull, D. E., and Christie, R. G. (1966). Electron microscopy of two small spherical plant viruses in thin sections. *Can. J. Bot.* **44**, 821.

Eisenberg, B. R., and Mobley, B. A. (1975). Size changes in single muscle fibers during fixation and embedding. *Tissue Cell* **7**, 383.

Eisenman, G. (1962). Cation selective glass electrodes and their mode of operation. *Biophys. J.* **2**, 259.

Eisenstat, L. F., Levin, B., Golomb, H. M., and Riddell, R. H. (1976). A technique for removing mucus and debris from mucosal surfaces. *Proc. SEM.*, p. 263.

El-Aaser, A. A., and Reid, E. (1969). Rat liver 5′-nucleotidase. *Histochem. J.* **I**, 417.

El-Aaser, A. A., and Reid, E. (1975). Localization of enzymatic activity in subcellular fractions. *In* "Electron Microscopy of Enzymes: Principles and Methods" (M. A. Hayat, ed.), Vol. 4, Van Nostrand-Reinhold, New York.

Elbers, P. F. (1966). Ion permeability of the egg *Limnaea stagnalis* L. on fixation for electron microscopy. *Biochim. Biophys. Acta* **112**, 318.

Elbers, P. F., Ververgaert, P. H. J. T., and Demel, R. (1965). Tricomplex fixation of phospholipids. *J. Cell Biol.* **24**, 23.

Elfbaum, S. G., and Laden, K. (1968). The effect of dimethyl sulfoxide on percutaneous absorption: a mechanistic study. Part III. *J. Soc. Cosmet. Chem.* **19**, 841.

Elford, B. C., and Walter, C. A. (1972). Effects of electrolyte composition and pH on the structure and function of smooth muscle cooled at $-79°C$ in unfrozen media. *Cryobiology* **9**, 82.

Elgin, S. C. R., Froehner, S. C., Stuart, J. E., and Bonner, J. (1971). The biology and chemistry of chromosomal proteins. *Adv. Cell Mol. Biol.* **1**, 1.

Elgjo, R. F. (1976). Platelets, endothelial cells, and macrophages in the spleen. An ultrastructural study on perfusion-fixed organs. *Am. J. Anat.* **145**, 101.

Elias, P. M., Goerke, J., Friend, D. S., and Brown, B. E. (1978). Freeze-fracture identification of sterol-digitonin complexes in cell and liposome membranes. *J. Cell Biol.* **78**, 577.

Ellar, D. J., Muñoz, E., and Salton, M. R. J. (1971). The effect of low concentrations of glutaraldehyde in *Micrococcus lysodeikticus*. Changes in the release of membrane associated enzymes and membrane structure. *Biochim. Biophys. Acta.* **225**, 140.

Elleder, M., and Lojda, Z. (1968a). Remarks on the detection of osmium derivatives in tissue sections. *Histochemie* **13**, 276.

Elleder, M., and Lojda, Z. (1968b). Remarks on the "OTAN" reaction. *Histochemie* **14**, 47.

Elling, F., Hasselager, E., and Friis, C. (1977). Perfusion fixation of kidneys in adult pigs for electron microscopy. *Acta Anat.* **98**, 340.

Eränkö, O. (1972). Light and electron microscopic evidence of granular and non-granular storage of catecholamines. *Histochem. J.* **4**, 213.

Erasmus, D. A., ed. (1978). "Electron Probe Microanalysis in Biology." Chapman & Hall, London.

Ericsson, J. L. E. (1966). On the fine structural demonstration of glucose-6-phosphatase. *J. Histochem. Cytochem.* **14**, 301.

Ericsson, J. L. E., and Biberfeld, P. (1967). Studies on aldehyde fixation: Fixation rates and their relation to fine structure and some histochemical reactions in liver. *Lab. Invest.* **17**, 281.

Ericsson, J. L. E., Saldino, A. J., and Trump, B. F. (1965). Electron microscopic observations of the influence of different fixatives on the appearance of cellular ultrastructure. *Z. Zellforsch. Mikrosk. Anat.* **66**, 161.

Ernst, S. A., and Philpott, C. W. (1970). Preservation of Na-K-activated and Mg-activated adenosine triphosphatase activities of avian salt gland. *J. Histochem. Cytochem.* **20**, 23.

Essner, E. (1973). Phosphatases. *In* "Electron Microscopy of Enzymes: Principles and Methods" (M. A. Hayat, ed.), Vol. 1, Van Nostrand-Reinhold, New York.

Etherton, J. E., and Botham, C. M. (1970). Factors affecting lead capture methods for the fine localization of rat lung acid phosphatase. *Histochem. J.* **2**, 507.

Evan, A. P., Dail, W. G., Dammrose, D., and Palmer, C. (1976). Scanning electron microscopy of cell surfaces following removal of extracellular material. *Anat. Rec.* **185**, 433.

Evans, L. V., and Holligan, M. S. (1972). Correlated light and electron microscopic studies on brown algae. II. Physical production in Dictyota. *New Phytol.* **71**, 1173.

Evans, P. S. (1966). Preparation of specimens for microtomy using continuous dehydration. *J. R. Microsc. Soc.* **85**, 375.

Eyring, E. J., and Ofengand, J. (1967). Reaction of formaldehyde with heterocyclic imino nitrogen of purine and pyrimidine nucleosides. *Biochemistry* **6**, 2500.

Fahimi, H. D. (1974). Effect of buffer storage on fine structure and catalase cytochemistry of peroxisomes. *J. Cell Biol.* **63**, 675.

Fahimi, H. D., and Amarasingham, C. R. (1964). Cytochemical localization of lactic dehydrogenase in white skeletal muscle. *J. Cell Biol.* **22**, 291.

Fahimi, H. D., and Drochmans, P. (1965a). Essais de standardisation de la fixation au glutaraldéhyde. I. Purification et détermination de la concentration du glutaraldéhyde. *J. Microsc. (Paris)* **4**, 725.

Fahimi, H. D., and Drochmans, P. (1965b). Essais de standardisation de la fixation au glutaraldéhyde. II. Influence des concentrations en aldehyde et de l'osmolalite. *J. Microsc. (Paris)* **4**, 737.

Falck, B., Hillarp, N.-Å., Thieme, G., and Torp, A. (1962). Fluorescence of catecholamines and related compounds condensed with formaldehyde. *J. Histochem. Cytochem.* **10**, 348.

Falk, H. (1969). Rough thylakoids: Polysomes attached to chloroplast membranes. *J. Cell Biol.* **42**, 582.

Fallah, E., Schuman, B. M., Watson, J. H. L., and Goodwin, J. (1976). Scanning electron microscopy of gastroscopic biopsies. *Gast. Endoscopy* **22**, 137.

Farrant, J. (1972). Human red cells under hypertonic conditions: A model system for investigating freezing damage. *Cryobiology* **9**, 131.

Favali, M. A., and Conti, G. G. (1971). Fine structure of healthy and virus-infected leaves following ultracentrifugation. *Cytobiologie* **3**, 153.

Fedorko, M. E., and Levine, R. F. (1976). Tannic acid effect on membrane of cell surface origin in guinea pig megakaryocytes and platelets. *J. Histochem. Cytochem.* **24**, 601.

Fedorko, M. E., Hirsch, J. G., and Cohen, Z. A. (1968). Autographic vacuoles produced *in vitro. J. Cell Biol.* **38**, 377.

Fein, M. L., and Harris, E. H. (1962). Quantitative analytical procedure for determining glutaraldehyde and chrome in tanning solutions. *U. S., Agric. Res. Serv., ARS* **ARS -73-37.**

Fein, M. L., Harris, E. H., Naghski, J., and Filachione, E. M. (1959). Tanning with glutaraldehyde. I. Rate studies. *J. Am. Leather Chem. Assoc.* **54**, 488.

Feldman, M. Ya. (1962). The condensation of adenine and adenosine with formaldehyde. *Biokhimiya* **27**, 321.

Felgenhauer, K., and Glenner, G. G. (1966). Quantitation of tissue-bound renal aminopeptidase by a microdensitometric technique. *J. Histochem. Cytochem.* **18**, 251.

Ferguson, C. C., and Richardson, J. B. (1978). A simple technique for the utilization of postmortem tracheal and bronchial tissues for ultrastructural studies. *Hum. Pathol.* **9**, 463.

Ferguson, W. J., Braunschweiger, K. I., Braunschweiger, W. R., Smith, J. R., McCormick, J. J.,

Wasman, C. C., Jarvis, N. P., Bell, D. H., and Good, N. E. (1980). Hydrogen ion buffers for biological research II. *Anal. Biochem.* (in press).

Ferrier, L. K., Richardson, T., and Olson, N. F. (1972). Crystalline catalase insolubilized with glutaraldehyde. *Enzymologia* **42**, 273.

Feuvray, D., and de Leiris, J. (1971). Effect of dimethyl sulfoxide on isolated rat heart and lacticodehydrogenase release. *Eur. J. Pharmacol.* **16**, 8.

Fieser, L. F., and Fieser, M. (1956). "Organic Chemistry," p. 207. Heath, Boston, Massachusetts.

Fieser, L. F., and Fieser, M. (1961). "Advanced Organic Chemistry." Van Nostrand-Reinhold, New York.

Finlay, J. B., Hunter, J. A. A., and Steven, F. S. (1971). Preparation of human skin for high resolution scanning electron microscopy using phosphate buffered crude bacterial α-amylase. *J. Microsc. (Oxford)* **93**, 73.

Finlay-Jones, J.-M., and Papadimitriou, J. M. (1972). Demonstration of pulmonary surfactant by tracheal injection of tricomplex salt mixture: Electron microscopy. *Stain Technol.* **47**, 59.

Finlay-Jones, J.-M., and Papadimitriou, J. M. (1973). The autolysis of neonatal pulmonary cells *in vitro:* An ultrastructural study. *J. Pathol.* **111**, 125.

Fischer, A. (1899). "Fixierung, Färbung und Bau des Protoplasmas." Fischer, Jena.

Fishman, J. A., Ryan, G. B., and Karnovsky, M. B. (1975). Endothelial regeneration in the rat carotid artery and the significance of endothelial denudation in the pathogenesis of myointimal thickening. *Lab. Invest.* **32**, 339.

Flasch, A. (1955). Die Wirkung von coffein auf vitalgefärbte pffanzenzellen. *Protoplasma* **44**, 412.

Fleischer, S., Fleischer, B., and Stoeckenius, W. (1967). Fine structure of lipid depleted mitochondria. *J. Cell Biol.* **32**, 193.

Flitney, F. W. (1966). The time course of the fixation of albumin by formaldehyde, glutaraldehyde, acrolein, and other higher aldehydes. *J. R. Microsc. Soc.* **85**, 353.

Fonkalsrud, E. W., Sanchez, M. Zerubavel, R., Lassaletta, L., Smeesters, C., and Mahoney, A. (1976). Arterial endothelial changes after ischemia and perfusion. *Surg., Gynecol. Obster.* **142**, 715.

Fooke-Achterrath, M., Lickfeld, K. G., Reusch, V. M., Aebi, U., Tschöpe, U., and Menge, B. (1974). Close-to-life preservation of *Staphylococcus aureus* mesosomes for transmission electron microscopy. *J. Ultrastruct. Res.* **49**, 270.

Forbes, M. S., Rubio, R., and Sperelakis, N. (1972). Tubular systems of *Limulus* myocardial cells investigated by use of electron opaque tracers and hypertonicity. *J. Ultrastruct. Res.* **39**, 580.

Forer, A. (1978). Electron microscopy of actin. *In* "Principles and Techniques of Electron Microscopy: Biological Applications" (M. A. Hayat, ed.), Vol. 9, Van Nostrand-Reinhold, New York.

Forssmann, W. G. (1969). "Präparation biologischer Gewebe für die Elektronenmikro-skopie. I. Fixation." Seminar für Elektronenmikroskopie, Technische Akademie, Esslingen.

Forssmann, W. G., Siegrist, G., Orci, L., Girardier, L., Pictet, R., and Rouiller, C. (1967). Fixation par perfusion pour le microscope électronique essai de généralisation. *J. Microsc. (Paris)* **6**, 279.

Forssmann, W. G., Ito, S., Weihe, E., Aoki, A., Dym, M., and Fawcett, D. W. (1977). An improved perfusion fixation method for the testis. *Anat. Rec.* **188**, 307.

Fowke, L. C., and Setterfield, G. (1968). Cytological responses in Jerusalem artichoke tuber slices during aging and subsequent auxin treatment. *Biochem. Physiol. Plant Growth Subst., Proc. Int. Conf. Plant Growth Subst.,* **6**, 581.

Fozzard, H. A., and Dominguez, C. (1969). Effect of formaldehyde and glutaraldehyde on electrical properties of cardiac Purkinje fibers. *J. Gen. Physiol.* **53**, 530.

Franke, W. W., Krien, S., and Brown, J. R. M. (1969). Simultaneous glutaraldehyde-osmium tetroxide fixation with postosmication. *Histochemie* **19**, 162.

Franz, T. J., and van Bruggen, J. T. (1967). A possible mechanism of action of DMSO. *Ann. N. Y. Acad. Sci.* **141**, 302.

Franzini-Armstrong, C., and Porter, K. R. (1964). Sarcolemmal invaginations constituting the T system in fish muscle fibers. *J. Cell Biol.* **22**, 675.

Fraumeni, J. F., Li., F. P., and Dalager, N. (1973). Teratomas in children: Epidemiologic features. *JNCI J. Natl. Cancer Inst.* **51**, 1425.

Frederiksen, O., and Rostgaard, J. (1974). Absence of dilated lateral intercellular spaces in fluid-transporting frog gallbladder epithelium. *J. Cell Biol.* **61**, 830.

Fredrick, S. E., and Newcomb, E. H. (1969). Cytochemical localization of catalase in leaf microbodies (peroxisomes). *J. Cell Biol.* **43**, 343.

Freinkel, N., Pedley, K. C., Wooding, P., and Dawson, R.M.C. (1978). Localization of inorganic phosphate in the pancreatic B cell and its loss on glucose stimulation. *Science* **201**, 1124.

Frenzel, H., Kremer, B., Richter, I.-E., and Hücker, H. (1976). The fine structure of liver sinusoids after perfusion fixation with various pressures. A transmission and scanning electron microscopic study. *Res. Exp. Med.* **168**, 229.

Frigerio, N. A., and Nebel, B. R. (1962). The quantitative determination of osmium tetroxide in fixatives. *Stain Technol.* **37**, 347.

Frigerio, N. A., and Shaw, M. J. (1969). A simple method for determination of glutaraldehyde. *J. Histochem. Cytochem.* **17**, 176.

Frühling, J., Penasse, W., Sand, G., Morena, E., and Claude, A. (1970). Etude comparative par microscopie électronique des reactions cytochimiques de la digitonine avec le cholestérol et d'autres lipides presents dans les cellules de la corticosurrénale. *Arch. Int. Physiol. Biochim.* **78**, 997.

Fujime, S., and Ishiwata, S. (1971). Dynamic study of F-actin by quasielastic scattering of laser light. *J. Mol. Biol.* **62**, 251.

Fujimoto, S., Yamamoto, K., and Takeshige, Y. (1975). Electron microscopy of endothelial microvilli of large arteries. *Anat. Rec.* **183**, 259.

Fulcher, R. G., and McCully, M. E. (1971). Histological studies on the genus *Fucus*. V. An autoradiographic and electron microscopic study of the early stages of regeneration. *Can. J. Bot.* **49**, 161.

Furness, J. B., Costa, M., and Blessing, W. W. (1977). Simultaneous fixation and production of catecholamine fluorescence in central nervous tissue by perfusion with aldehydes. *Histochem. J.* **9**, 745.

Furness, J. B., Heath, J. W., and Costa, M. (1978). Aqueous aldehyde (Faglu) methods for the fluorescence histochemical localization of catecholamines and for ultrastructural studies of central nervous tissue. *Histochemistry* **57**, 285.

Futaesaku, Y., and Mizuhira, U. (1974). Fine structure of the microtubules by means of the tannic acid fixation. *Proc. 8th Int. Congr. Electron Microsc.*, II, p. 340. Aust. Acad. Sci., Canberra.

Galigher, A. E., and Kosloff, E. N. (1971). "Essentials of Practical Microtechnique," 2nd ed., p. 114. Lea & Febiger, Philadelphia, Pennsylvania.

Gander, E. S., and Moppert, J. M. (1969). Der Einfluss von Dimethyl sulfoxid auf die Permeabilitat der Lysosomenmembrane bei quantitativer Darstellung der sauren Phosphatse. *Histochemie* **20**, 211.

Garbarsch, C., Tranum-Jensen, J., and van Deurs, B. (1980). SEM and TEM of the normal rabbit aortic endothelium after controlled perfusion fixation. *J. Ultrastruct. Res.* **73**, 95.

Garcia-Patrone, M., and Algranati, I. D. (1976). Artifacts induced by glutaraldehyde fixation of ribosomal particles. *Mol. Biol. Rep.* **2**, 507.

Garrett, J. R., Davies, K. J., and Parsons, P. A. (1972). Consumer's guide to glutaraldehyde. *Proc. R. Microsc. Soc.* **7**, 116.

Gaylarde, P., and Sarkany, I. (1968). Ruthenium tetroxide for fixing and staining cytoplasmic membranes. *Science* **161,** 1157.

Gertz, S. D., Rennels, M. L., Forbes, M. S., and Nelson, E. (1975). Preparation of vascular endothelium for scanning electron microscopy: A comparison of the effects of perfusion and immersion fixation. *J. Microsc. (Oxford)* **105,** 309.

Ghosh, B. K. (1971). Grooves in the plasmalemma of *Saccharomyces cerevisiae* seen in glancing sections of double aldehyde-fixed cells. *J. Cell Biol.* **48,** 192.

Ghosh, B. K., and Nanninga, N. (1976). Polymorphism of the mesosome in *Bacillus licheniformis* (749/c and 749). Influence of chemical fixation monitored by freeze-etching. *J. Ultrastruct. Res.* **56,** 107.

Ghosh, B. K., Wouters, J. T. M., and Lampen, J. O. (1971). Distribution of the sites of alkaline phosphatase(s) activity in vegetative cells of *Bacillus subtilis. J. Bacteriol.* **108,** 928.

Gibbons, I. R. (1959). An embedding resin miscible with water for electron microscopy. *Nature (London)* **184,** 375.

Giese, A. C. (1968). "Cell Physiology," 4th ed. Saunders, Philadelphia, Pennsylvania.

Gigg, R., and Payne, S. (1969). The reaction of glutaraldehyde with tissue lipids. *Chem. Phys. Lipids* **3,** 292.

Gil, J. (1972). Effect of tricomplex fixation on lung tissue. *J. Ultrastruct. Res.* **40,** 122.

Gil, J., and Weibel, E. R. (1968). The role of buffers in lung fixation with glutaraldehyde and osmium tetroxide. *J. Ultrastruct. Res.* **25,** 331.

Gil, J., and Weibel, E. R. (1969-1970). Improvements in demonstration of lining layer of lung alveoli by electron microscopy. *Respir. Physiol.* **8,** 13.

Gil, J., and Weibel, E. R. (1971). An improved apparatus for perfusion fixation with automatic pressure control. *J. Microsc. (Oxford)* **94,** 241.

Gillett, R., and Gull, K. (1972). Glutaraldehyde—its purity and stability. *Histochemie* **30,** 162.

Gillett, R., Jones, G. E., and Partridge, T. (1975). Distilled glutaraldehyde: Its use in an improved fixation regime for cell suspensions. *J. Microsc. (Oxford)* **105,** 325.

Glassman, T. A., Cooper, C., Harrison, L. W., and Swift, T. J. (1971). A proton magnetic study of metal ion-adenine ring interactions in metal ion complexes with adenosine triphosphate. *Biochemistry* **10,** 843.

Glassman, T. A., Klopman, G., and Cooper, C. (1973). Use of the generalized perturbation theory to predict the interaction of purine nucleotides with metal ions. *Biochemistry* **12,** 5013.

Glauert, A. M. (1974). Fixation, dehydration, and embedding of biological specimens. *In* "Practical Methods in Electron Microscopy" (A. M. Glauert, ed.). North-Holland Publ., Amsterdam.

Glaumann, H. (1975). Ultrastructural demonstration of phosphatases by perfusion fixation followed by perfusion incubation of rat liver. *Histochemistry* **44,** 169.

Godman, G. C., Miranda, A. F., Deitch, A. D., and Tanenbaum, S. W. (1975). Action of cytochalasin D on cells of established lines. III. Zeiosis and movements at the cell surface. *J. Cell Biol.* **64,** 644.

Goff, C. W., and Oster, M. O. (1974). Formation of 235-nanometer absorbing substance during glutaraldehyde fixation. *J. Histochem. Cytochem.* **22,** 913.

Goland, L. P., Grand, N. G., Green, F. J., and Booker, B. F. (1969). Immunofluorescence microscopy of cyanurated tissues. *Stain Technol.* **44,** 227.

Goldfarb, D., Miyai, K., and Hegenauer, J. (1977). A hydrostatic device for tissue dehydration. *Stain Technol.* **52,** 171.

Goldfischer, S., Essner, E., and Schiller, B. (1971). Nucleoside diphosphatase and thiamine pyrophosphatase activities in the endoplasmic reticulum and Golgi apparatus. *J. Histochem. Cytochem.* **19,** 349.

Goldman, R. D., Larzarides, E., Pollack, R., and Weber, K. (1975). The distribution of actin in non-muscle cells. *Exp. Cell Res.* **90,** 333.

Goldstein, M. A., and Entman, M. L. (1979). Microtubules in mammalian heart muscle. *J. Cell Biol.* **80**, 183.

Gomez Dumm, C. L., and Echave Llanos, J. M. (1970). Variations in the Golgi complex of mouse somatotrops at different times in a 24-hour period. *Experientia* **26**, 177.

Good, N. E., Winget, G. D., Winster, W., Connolly, T. N., Izawa, S., and Singh, R. M. M. (1966), Hydrogen ion buffers for biological research. *Biochemistry* **5**, 467.

Goodenough, D. A., and Revel, J.-P. (1971). The permeability of isolated and *in situ* mouse hepatic gap juntions studied with enzymatic tracers. *J. Cell Biol.* **50**, 81.

Goodfriend, T. L., Levine, L., and Fasman, G. D. (1964). Antibodies to Bradykinin and Angiotensin. A use of carbodiimides in immunology. *Science* **144**, 1344.

Goodman, T., and Moore, A. (1971). A simple flow indicator for use in perfusion fixation. *Stain Technol.* **46**, 36.

Göthlin, G., and Ericsson, J. L. E. (1971). Fine structural localization of acid phosphomonoesterase in the brush border region of osteoclasts. *Histochemie* **28**, 337.

Graham, R. C., Karnovsky, M. J., Schafer, A. W., Glass, E. A., and Karnovsky, M. L. (1967). Metabolic and morphological observations on the effect of surface-active agents on leukocytes. *J. Cell Biol.* **32**, 629.

Grantham, J., Cuppage, F. E., and Fanestil, D. (1971). Direct observation on toad bladder response to vasopression. *J. Cell Biol.* **48**, 695.

Gray, E. G. (1972). Are the coats of coated vesicles artifacts? *J. Neurocytol.* **1**, 363.

Greenberg, D. M. (1944). The interaction between the alkali earth cations, particularly calcium and proteins. *Adv. Protein Chem.* **1**, 121.

Gregorius, F. K., and Rand, R. W. (1975). Scanning electron microscopic observations of common carotid artery endothelium in the rat. I. Crater artifacts. *Surg. Neurol.* **4**, 252.

Griffin, J. L. (1963). Motion picture analysis of fixation for electron microscopy: *Amoeba proteus.* *J. Cell Biol.* **19**, 77A.

Griffiths, G. W., and Beck, S. D. (1977). Effect of dietary cholesterol on the pattern of osmium deposition in the symbiote-containing cells of the pea aphid. *Cell Tissue Res.* **176**, 191.

Griffith, W. P. (1965). Osmium and its compounds. *Q. Rev., Chem. Soc.* **19**, 254.

Griffith, W. P. (1967) "The Chemistry of the Rarer Platinum Metals." Wiley (Interscience), New York.

Griffith, W. P., and Rossetti, R. (1972). Oxo complexes of osmium and ruthenium with organic ligands. *J. Chem. Soc., Dalton Trans.* **14**, 1449.

Grinnell, F., Anderson, R. G. W., and Hackenbrock, C. R. (1976). Glutaraldehyde induced alterations of membrane anionic sites. *Biochim. Biophys. Acta* **426**, 772.

Grube, H. L. (1965). Section 29. The platinum metals. *In* "Handbook of Preparative Inorganic Chemistry" (G. Brauer, ed.), Vol. 2, pp. 1560–1605. Academic Press, New York.

Gunning, B. E. S. (1980). Spatial and temporal regulation of nucleating sites for arrays of cortical microtubules in root tip cells of the water fern *Azella pinnata. Eur. J. Cell Biol.* **23**, 53.

Gusnard, D., and Kirschner, R. H. (1977). Cell and organelle shrinkage during preparation for scanning electron microscopy: Effects of fixation, dehydration, and critical point drying. *J. Microsc. (Oxford)* **110**, 51.

Gustavson, K. H. (1949). Some protein chemical aspects of tanning process. *Adv. Protein Chem.* **5**, 353.

Guth, L., and Watson, P. K. (1968). A correlated histochemical and quantitative study on cerebral glycogen after brain injury in the rat. *Exp. Neurol.* **22**, 590.

Habeeb, A. F. S. A. (1969). A study of the antigenicity of formaldehyde and glutaraldehyde-treated bovine serum albumin and ovalbumin-bovine serum albumin conjugate. *J. Immunol.* **102**, 457.

Habeeb, A. F. S. A., and Hiramoto, R. (1968). Reaction of proteins with glutaraldehyde. *Arch. Biochem. Biophys.* **126**, 16.

Hackenbrock, C. R. (1966). Ultrastructural bases for metabolically linked mechanical activity in mitochondria. I. Reversible ultrastructural changes with change in metabolic steady state in isolated liver mitochondria. *J. Cell Biol.* **30**, 269.

Hackenbrock, C. R. (1968). Ultrastructural bases for metabolically linked mechanical activity in mitochondria. II. Electron transport-linked ultrastructural transformations in mitochondria. *J. Cell Biol.* **37**, 345.

Hackenbrock, C. R. (1972). Energy-linked ultrastructural transformations in isolated liver mitochondria and mitoplasts. Preservation of configurations by freeze-cleaving compared to chemical fixation. *J. Cell Biol.* **53**, 450.

Hagström, L., and Bahr, G. F. (1960). Penetration rates of osmium tetroxide with different fixation vehicles. *Histochemie* **2**, 1.

Hajdu, J., Solti, M., and Friedrich, P. (1975). Crosslinking and coupling of rabbit muscle aldolase and glyceraldehyde-3-phosphate dehydrogenase by glutaraldehyde. *Acta Biochim. Biophys. Sci. Hung.* **10**, 7.

Hajós, F., and Kerpel-Fronius, S. (1970). The incubation of unfixed tissues for electron histochemistry. *Histochemie* **23**, 120.

Håkanson, R., and Sundler, F. (1971). Formaldehyde condensation. A method for the fluorescence microscopic demonstration of peptides with NH_2-terminal tryptophan residues. *J. Histochem. Cytochem.* **19**, 477.

Hake, T. (1965). Studies on the reaction of OsO_4 and $KMnO_4$ with amino acids, peptides, and proteins. *Lab. Invest.* **14**, 470.

Hall, J. L., Yeo, A. R., and Flowers, T. J. (1974). Uptake and localization of rubidium in the halophyte *Suaeda maritima. Z. Pflanzenphysiologie* **71**, 200.

Hall, R. H., and Stern, E. S. (1955). Unsaturated aldehydes and related compounds. Part VII. Thermal fission of 1,1,3-trialkoxypropanes. *J. Chem. Soc.* p. 2657.

Hallam, N. D. (1976). Anhydrous fixation of dry plant tissue using non-aqueous fixatives. *J. Microsc. (Oxford)* **106**, 337.

Hampton, J. C. (1965). Effects of fixation on the morphology of Paneth cell granules. *Stain Technol.* **40**, 283.

Hand, A. R., and Hassell, J. R. (1976). Tissue fixation with diimidoesters as an alternative to aldehydes. II. Cytochemical and biochemical studies of rat liver fixed with dimethylsuberimidate. *J. Histochem. Cytochem.* **24**, 1000.

Handley, D., and Ghosh, B. K. (1980). Subcellular distribution of marker enzymes in cells of a minute fungus. *Fusidium* sp. 100-3. *J. Bacteriol.* **141**, 946.

Hanker, J. S. (1975). Oxidoreductases. *In* "Electron Microscopy of Enzymes: Principles and Methods" (M. A. Hayat, ed.), Vol. 4, Van Nostrand-Reinhold, New York.

Hanker, J. S. Seaman, A. R., Weiss, L. P., Ueno, H., Bergman, R. A., and Seligman, A. M. (1964). Osmiophilic reagents: New cytochemical principle for light and electron microscopy. *Science* **146**, 1039.

Hanker, J. S., Deb, C., and Seligman, A. M. (1966). Staining tissue for light and electron microscopy by bridging metals with multidentate ligands. *Science* **152**, 1631.

Hanker, J. S., Kasler, F., Bloom, M. G., Copeland, J. S., and Seligman, A. M. (1967). Coordination polymers of osmium: The nature of osmium black. *Science* **156**, 1737.

Hanker, J. S., Kusyk, C. J., Clapp, D. H., and Yates, P. E. (1970). Effect of dimethyl sulfoxide (DMSO) on the histochemical demonstration of dehydrogenases. *J. Histochem. Cytochem.* **18**, 673.

Hanker, J. S., Anderson, W. A., and Bloom, F. E. (1972). Osmiophilic polymer generation: Catalysis by transition metal compounds in ultrastructural cytochemistry. *Science* **175**, 991.

Hanker, J. S., Kusyk, C. J., Bloom, F. E., and Pearse, A.G.E. (1973). The demonstration of dehydrogenases and monoamine oxidase by the formation of osmium blacks at the site of Hatchett's brown. *Histochemie* **33**, 205.

Hanker, J. S., Romanovicz, D. K., and Padykula, H. A. (1976). Tissue fixation and osmium black formation with nonvolatile octavalent osmium compounds. *Histochemistry* **49**, 263.

Hanks, J. H., and Wallace, J. H. (1958). Determination of cell viability. *Proc. Soc. Exp. Biol. Med.* **98**, 188.

Hanson, C. V., Shen, C. K. J., and Hearst, J. E. (1976). Crosslinking of DNA *in situ* as a probe for chromatin structure. *Science* **193**, 62.

Hansson, H. P. J. (1967). Histochemical demonstration of carbonic anhydrase activity. *Histochemie* **11**, 112.

Hanzon, V. (1959). An apparatus for continuous rinsing, dehydration, and plastic impregnation of electron microscope specimens. *Sci. Tools* **6**, 18.

Happich, W. F., Taylor, M. M., and Feairheller, S. H. (1970). Amino acid composition of glutaraldehyde-stabilized wool. *Text. Res. J.* **40**, 768.

Hardman, A. R., and Gunning, B. E. S. (1978). Structure of cortical microtubule arrays in plant cells. *J. Cell Biol.* **77**, 14.

Hardonk, M. J., Haarsma, T. J., Dijkhuis, F. W. J., Poel, M., and Koudstaal, J. (1977). Influence of fixation and buffer treatment on the release of enzymes from the plasma membrane. *Histochemistry* **54**, 57.

Hardy, P. M., Nicholls, A. C., and Rydon, H. N. (1969). The nature of glutaraldehyde in aqueous solution. *J. Chem. Soc. D* **1**, 565.

Hardy, P. M., Nicholls, A. C., and Rydon, H. N. (1972). The hydration and polymerization of succinaldehyde, glutaraldehyde, and adipaldehyde. *J. Chem. Soc.* **15**, 2270.

Hardy, P. M., Nicholls, A. C., and Rydon, H. N. (1976a). The nature of the crosslinking of porteins by glutaraldehyde. Part I. Interaction of glutaraldehyde with the amino-groups of 6-aminohexanoic acid and of α-N-acetyl-lysine. *J. Chem. Soc., Perkin Trans.* **1**, 958.

Hardy, P. M., Hughes, G. J., and Rydon, H. N. (1976b). Formation of quarternary pyridinium compounds by the action of glutaraldehyde on proteins. *J. Chem. Soc., Chem. Commun.* p. 157.

Hardy. P. M., Hughes, G. J., and Rydon, H. N. (1977). Identification of a 3-(2-piperidyl)pyridinium derivative ('anabilysine') as a crosslinking entity in a glutaraldehyde-treated protein. *J. Chem. Soc., Chem. Commun.* p. 759.

Hargraves, P. E., and Guillard, R. R. L. (1974). Structural and physiological observations on some small marine diatoms. *Phycologia* **13**, 163.

Harris, E. D., and Farrell, M. E. (1972). Resistance to collagenase: A characteristic of collagen fibrils crosslinked by formaldehyde. *Biochim. Biophys. Acta* **278**, 133.

Harris, P. 1962). Some structural and functional aspects of the mitotic apparatus in sea urchin embryos. *J. Cell Biol.* **14**, 475.

Harris, R. A., Harris, D. L., and Green, D. E. (1968). Effect of *Bordetella* endotoxin upon mitochondrial respiration and energized process. *Arch. Biochem. Biophys.* **128**, 219.

Hartman, F. C., and Wold, F. (1967). Crosslinking of bovine pancreatic ribonuclease A with dimethyl adipimate. *Biochemistry* **8**, 2439.

Hartman, H., and Hayes, T. L. (1971). Scanning electron microscopy of *Drosophila*. *J. Hered.* **62**, 41.

Harvey, D. M. R. (1980). The preparation of botanical samples for ion localization studies at the subcellular level. *Proc. SEM.,* p. 409.

Haselkorn, R., and Doty, P. (1961). The reaction of formaldehyde with polynucleotides. *J. Biol. Chem.* **236**, 2738.

Haslam, E. (1966). "Chemistry of Vegetable Tannins." Academic Press, New York.

Hasle, G. R., and Fryxell, G. A. (1970). Diatoms: Cleaning and mounting for light and electron microscopy. *Trans. Am. Microsc. Soc.* **89,** 469.

Hassell, J., and Hand, A. R. (1974). Tissue fixation with diimidoesters as an alternative to aldehydes. I. Comparison of crosslinking and ultrastructure obtained with dimethylsuberimidate and glutaraldehyde. *J. Histochem. Cytochem.* **22,** 223.

Hasty, D. L., and Hay, E. D. (1978). Freeze-fracture studies of the developing cell surface. II. Particle-free membrane blisters on glutaraldehyde-fixed corneal fibroblasts are artifacts. *J. Cell Biol.* **78,** 756.

Hatta, T. (1976). Recognition and measurement of small isometric virus particles in thin sections. *Virology* **69,** 237.

Hatta, T., and Mathews, R. E. F. (1976). Sites of coat protein accumulation in turnip yellow mosaic virus-infected cells. *Virology* **73,** 1.

Haudenschild, C., Baumgartner, H. R., and Studer, A. (1972). Significance of fixation procedure for preservation of arteries. *Experientia* **29,** 828.

Hauser, M. (1978). Demonstration of membrane-associated and oriented microfilaments in *Amoeba proteus* by means of a Schiff base-glutaraldehyde fixative. *Cytobiologie* **18,** 95.

Hausmann, K. (1977). Artifactual fusion of membranes during preparation of material for electron microscopy. *Naturwissenschaften* **64,** 95.

Hayashi, M., and Freiman, D. G. (1966). An improved method of fixation fo formalin-sensitive enzymes with special reference to myosin adenosine triphosphatase. *J. Histochem. Cytochem.* **14,** 577.

Hayat, M. A. (1968a). A study of glutaraldehyde-potassium permanganate fixation of plant embryos for electron microscopy. *Proc. Eur. Reg. Conf. Electron Microsc. 4th.,* Vol. 2, p. 379.

Hayat, M. A. (1968b). Triple fixation for electron microscopy. *Proc. 26th Annu. Meet., Electron Microsc. Soc. Am.* p. 90.

Hayat, M. A. (1973). Specimen preparation. *In* "Electron Microscopy of Enzymes: Principles and Methods" (M. A. Hayat, ed.), Vol. 1, Van Nostrand-Reinhold, New York.

Hayat, M. A. (1975). "Positive Staining for Electron Microscopy." Van Nostrand-Reinhold, New York.

Hayat, M. A. (1978). "Introduction to Biological Scanning Electron Microscopy." Univ. Park Press, Baltimore, Maryland.

Hayat, M. A., ed. (1980). "X-ray Microanalysis in Biology." Univ. Park Press, Baltimore, Maryland.

Hayat, M. A. (1981). "Principles and Techniques of Electron Microscopy: Biological Applications," 2nd ed., Vol. 1. Univ. Park Press, Baltimore, Maryland.

Hayat, M. A. (1982). "Basic Electron Microscopy Techniques." (in press).

Hayat, M. A., and Guiquinta, R. (1970). Vapor fixation prior to fixation by immersion for electron microscopy. *Proc. 7th Int. Congr. Electron Microsc.* p. 391. Grenoble.

Hayat, M. A., and Zirkin, B. R. (1973). Critical point drying method. *In* "Principles and Techniques of Electron Microscopy: Biological Applications" (M. A. Hayat, ed.), Vol. 3, Van Nostrand-Reinhold, New York.

Hayes, T. L., Lindgren, F. T., and Gofman, J. W. (1963). A quantitative determination of the osmium tetroxide-lipoprotein interaction. *J. Cell Biol.* **19,** 251.

Heath, I. B., and Heath, M. C. (1978). Microtubules and organelle movements in the rust fungus *Uromyces phaseoli.* *Cytobiologie* **16,** 393.

Heckman, C. A., and Barnett, R. J. (1973). GACH: A water-miscible, lipid-retaining embedding polymer for electron microscopy. *J. Ultrastruct. Res.* **42,** 156.

Hegenauer, J. C., Tartof, K. D., and Nace, G. W. (1965). Mixing device for generating simple chromatographic gradients. *Anal. Biochem.* **13,** 6.

Helander, H. F. (1962). On the preservation of gastric mucosa. *Proc. Int. Congr. Electron Microsc., 5th, 1962* Vol. 2, Artic. P-7.

Heller, J., Ostwald, T. J., and Bok, D. (1971). The osmotic behavior of rod photoreceptor outer segment discs. *J. Cell Biol.* **48,** 633.

Henderson, T. R., Henderson, R. F., and York, J. L. (1975). Effects of dimethyl sulfoxide on subunit proteins. *Ann. N. Y. Acad. Sci.* **243,** 38.

Henle, J. (1865). Ueber das Gewebe der Nebenniere und der Hypophyse. *Z. Rat. Med.* **24,** 143.

Herzog, V., and Fahimi, H. D. (1974). The effect of glutaraldehyde on catalase. Biochemical and cytochemical studies with beef liver catalase and rat liver peroxisomes. *J. Cell Biol.* **60,** 303.

Heuser, J. E., Reese, T. S., and Landis, D.M.D. (1976). Preservation of synaptic structure by rapid freezing. *Cold Spring Harbor Symp. Quant. Biol.* **40,** 17.

Heuser, J. E., Reese, T. S., Dennis, M. J., Jan, Y., Jan, L., and Evans, L. (1979). Synaptic vesicle exocytosis captured by quick freezing and correlated with quantal transmitter release. *J. Cell Biol.* **81,** 275.

Higgins, M. L., and Daneo-Moore, L. (1974). Factors influencing the frequency of mesosomes observed in fixed and unfixed cells of *Streptococcus faecalis*. *J. Cell Biol.* **61,** 288.

Hillman, H., and Deutsch, K. (1978). Area changes in slices of rat brain during preparation for histology or electron microscopy. *J. Microsc. (Oxford)* **114,** 77.

Hinton, D. E. (1975). Perfusion fixation of whole fish for electron microscopy. *J. Fish. Res. Board Can.* **32,** 416.

Hinton, R. H., Burge, M. L. E., and Hartman, G. C. (1969). Sucrose interference in the assay of enzymes and proteins. *Anal. Biochem.* **29,** 248.

Hirsch, J. G., and Fedorko, M. E. (1968). Ultrastructure of human leukocytes after simultaneous fixation with glutaraldehyde and osmium tetroxide and postfixation in uranyl acetate. *J. Cell Biol.* **38,** 615.

Hirsch, J. G., Fedorko, M. E., and Cohn, Z. A. (1968). Vesicle fusion and formation at the surface of pinocytic vacuoles in macrophages. *J. Cell Biol.* **38,** 629.

Hixson, D. C., Yep, J. M., Glenney, J. R., Hayes, T., and Walborg, E. F. (1981). Evaluation of periodate-lysine-paraformaldehyde fixation as a method for crosslinking plasma membrane glycoproteins. *J. Histochem. Cytochem.* **29,** 561.

Hobbs, M. J. (1969). Fixation of microscopic fresh-water green algae by 31% OsO_4 in CCl_4 in an unbuffered two-phase fixative system. *Stain Technol.* **44,** 217.

Hökfelt, T., and Jonsson, G. (1968). Studies on reaction and binding of monoamines after fixation and processing for electron microscopy with special reference to fixation with potassium permanganate. *Histochemistry* **16,** 45.

Hollweg, H. G., and Buss, H. (1980). Problems with the preparation of blood vessels for scanning electron microscopy: A critical review. *Scanning* **3,** 3.

Holt, S. J., and Hicks, R. M. (1961). Use of veronal buffers in formalin fixatives. *Nature (London)* **191,** 832.

Holt, S. J., and Hicks, R. M. (1966). The importance of osmiophilia in the production of stable azoindoxyl complexes of high contrast for combined enzyme cytochemistry and electron microscopy. *J. Cell Biol.* **29,** 361.

Homès, J. (1974). Scanning electron microscopy of carrot cells and embroids growing *in vitro* under various culture conditions. *Proc. SEM.*, p. 335.

Hopwood, D. (1967a). Some aspects of fixation with glutaraldehyde: A biochemical and histochemical comparison of the effects of formaldehyde and glutaraldehyde fixation on various enzymes and glycogen with a note on penetration of glutaraldehyde into liver. *J. Anat.* **101,** 83.

Hopwood, D. (1967b). The behavior of various glutaraldehyde on Sephadex G-10 and some implications for fixation. *Histochemie* **11**, 289.

Hopwood, D. (1968a). The effect of pH and various fixatives on isolated ox chromaffin granules with respect to the chromaffin reaction. *J. Anat.* **102**, 415.

Hopwood, D. (1968b). Some aspects of fixation by glutaraldehyde and formaldehyde. *J. Anat.* **103**, 581.

Hopwood, D. (1969a). Fixation of proteins by osmium tetroxide, potassium dichromate and potassium permanganate. *Histochemie* **18**, 250.

Hopwood, D. (1969b). Fixatives and fixation: A review. *Histochem. J.* **1**, 323.

Hopwood, D. (1970). The reactions between formaldehyde, glutaraldehyde, and osmium tetroxide, and their fixation effects on bovine serum albumin and on tissue blocks. *Histochemie* **24**, 50.

Hopwood, D. (1972). Theoretical and practical aspects of glutaraldehyde fixation. *In* "Fixation in Histochemistry" (P. J. Stoward, ed.), p. 47. Chapman & Hall, London.

Hopwood, D. (1975). The reactions of glutaraldehyde with nucleic acids. *Histochem. J.* **7**, 267.

Hopwood, D., Allen, C. R., and McCabe, M. (1970). The reactions between glutaraldehyde and various proteins. An investigation of their kinetics. *Histochem. J.* **2**, 137.

Horridge, G. A., and Tamm, S. L. (1969). Critical point drying for scanning electron microscopic study of ciliary motion. *Science* **163**, 817.

Hossman, K. A., Sakaki, S., and Zimmermann, V. (1977). Cation activities in reversible ischemia of the cat brain. *Stroke* **8**, 77.

Howell, S. L., and Tyhurst, M. (1976). 45-Calcium localization in islets of Langerhans, a study by electron microscope autoradiography. *J. Cell Sci.* **21**, 415.

Howell, S. L., Young, D. A., and Lacy, P. E. (1969). Isolation and properties of secretory granules from rat islets of Langerhans. III. Studies on the stability of the isolated beta granule. *J. Cell Biol.* **41**, 167.

Hughes, B. G., White, F. G., and Smith, M. A. (1976). Scanning electron microscopy of barley protoplasts. *Protoplasma* **90**, 399.

Hunter, M. J., and Ludwig, M. L. (1972). Amidination. *In* "Methods in Enzymology" (C. H. W. Hirs and S. N. Timasheff, eds.), Vol. 25, Part B, p. 585. Academic Press, New York.

Hunter-Duvar, I. M. (1977). Morphology of the normal and the acoustically damaged cochlea. *Proc. SEM.*, p. 421.

Huxley, H. E. (1957). The double array of filaments in cross-striated muscle. *J. Biophys. Biochem. Cytol.* **3**, 631.

Huxley, H. E. (1971). The structural basis of muscular contraction. *Proc. R. Soc. London, Ser. B* **178**, 131.

Hyde, J. M., and Peters, D. (1970). The influence of pH and osmolality of fixation on the fowlpox virus core. *J. Cell Biol.* **46**, 179.

Idelman, S. (1964). Modification de la technique de Luft en vue de la conservation des lipides en microscopie électronique. *J. Microsc. (Paris)* **3**, 715.

Idelman, S. (1965). Conservation des lipides par les techniques utilises en microscopie electronique. *Histochemie* **5**, 18.

Inman, R. B. (1966). A denaturation map of the λ phage DNA molecule determined by electron microscopy. *J. Mol. Biol.* **18**, 464.

Inman, R. B. (1967). Denaturation maps of the left and right sides of the lambda DNA molecule determined by electron microscopy. *J. Mol. Biol.* **28**, 103.

Irino, S., Murakami, T., Fujita, T., Nagatani, T., and Kaneshige, T. (1978). Microdissection of tannin-osmium impregnated specimens in the scanning electron microscope: Demonstration of arterial terminals in human spleen. *Proc. SEM.* **2**, 111.

Ishii, S. (1974). An electron microscopic study of the autonomic nerve in the human uterus by lithium permanganate fixation. *J. Electron Microsc.* **23**, 185.

Ishikawa, H., Bischoff, R., and Holtzer, H. (1969). Formation of arrowhead complexes with heavy meromyosin in a variety of cell types. *J. Cell Biol.* **43**, 312.

Itakura, T., Maeda, T., Tohyama, M., Kashiba, A., and Shimizu, N. (1975). Electron microscopic alteration of synaptic vesicles after α-methyl-p-tyrosine administration in sympathetic nerve terminals demonstrated by potassium permanganate fixation. *Ann. Histochim.* **20**, 173.

Ito, S. (1961). The endoplasmic reticulum of gastric parietal cells. *J. Biophys. Biochem. Cytol.* **11**, 333.

Ito, S. (1974). Form and function of the glycocalyx on free cell surfaces. *Philos. Trans. R. Soc. London, Ser. B* **268**, 55.

Ito, S., and Karnovsky, M. J. (1968). Formaldehyde-glutaraldehyde fixatives containing trinitro compounds. *J. Cell Biol.* **39**, 168a.

Ito, S., and Winchester, R. J. (1963). The fine structure of the gastric mucosa in the rat. *J. Cell Biol.* **16**, 541.

Ito, U., and Inaba, Y. (1977). Electron microscopic observation of cerebrospinal fluid cells: A new method for embedding of CSF cells. *J. Electron Microsc.* **19**, 265.

Izard, C., and Libermann, C. (1978). Acrolein. *Mut. Res.* **47**, 115.

Jacob, S. W., and Herschler, R., eds. (1975). Biological actions of dimethyl sulfoxide. *Ann. N. Y. Acad. Sci.* **243**, 1.

Jacob, S. W., Bischel, M. B., and Herschler, R. J. (1964). Dimethyl sulfoxide: Effects on the permeability of biologic membranes (preliminary report). *Curr. Ther. Res.* **6**, 193.

Jacobs, G. F., and Liggett, S. J. (1971). An oxidative-distillation procedure for reclaiming osmium tetroxide from used fixative solutions. *Stain Technol.* **46**, 207.

Jaim-Etcheverry, G., and Zieher, L. M. (1968). Cytochemistry of 5-hydroxytryptamine at the electron microscope level. II. Localization in the autonomic nerves of the rat pineal gland. *Z. Zellforsch. Mikrosk. Annat.* **86**, 393.

James, R., and Branton, D. (1971). The correlation between the saturation of membrane fatty acids and the presence of membrane fracture faces after osmium fixation. *Biochim. Biophys. Acta* **233**, 504.

Janigan, D. T. (1964). The effects of aldehyde fixation on β-glucuronidase, β-galactosidase, N-acetyl- β-glucusaminidase, and β-glucosidase in tissue blocks. *Lab. Invest.* **13**, 1038.

Janigan, D. T. (1965). The effects of aldehyde fixation on acid phosphatase activity in tissue blocks. *J. Histochem. Cytochem.* **13**, 473.

Jansen, E. F., and Olson, A. C. (1969). Properties and enzymatic activities of papain insolubilized with glutaraldehyde. *Arch. Biochem. Biophys.* **129**, 221.

Jansen, E. F., Tomimatsu, Y., and Olsen, A. C. (1971). Crosslinking of α-chymotrypsin and other proteins by reaction with glutaraldehyde. *Arch. Biochem. Biophys.* **144**, 394.

Jard, S., Bourget, J., Carasso, N., and Favard, P. (1966). Action des fixatives sen la perméabilité et l'ultrastructure de la véssie de grenouille. *J. Microsc. (Paris)* **5**, 31.

Jeske, A. H., Fonteles, M. C., and Karow, A. M. (1974). Functional preservation of the mammalian kidney. III. Ultrastructural effects of perfusion with dimethyl sulfoxide (DMSO). *Cryobiologie* **11**, 170.

Johannessen, J. V., ed. (1978). "Electron Microscopy in Human Medicine," Vol. 1. McGraw-Hill, New York.

Johnson, D. M. (1977). Some observations on the effects of glutaraldehyde on horse liver alcohol dehydrogenase. *Int. J. Biochem.* **8**, 473.

Johnson, H. M., Brenner, K., and Hall, H. E. (1966). The use of water-soluble carboiimide as a coupling reagent in the passive hemagglutination test. *J. Immunol.* **97**, 791.

Johnson, T. J. A., and Rash, J. E. (1980). Glutaraldehyde chemistry: Fixation reactions consume O_2 and are inhibited by tissue anoxia. *J. Cell Biol.* **87,** 234a.

Johnston, W. H., Latta, H., and Osvaldo, L. (1973). Variation in glomerular ultrastructure in rat kidneys fixed by perfusion. *J. Ultrastruct. Res.* **45,** 149.

Joly, M. (1965). "A Physico-Chemical Approach to the Denaturation of Proteins." Academic Press, New York.

Jones, D. (1969a). Acrolein as a histological fixative. *J. Microsc. (Oxford)* **90,** 75.

Jones, D. (1969b). The reaction of formaldehyde with unsaturated fatty acids during histological fixation. *Histochem. J.* **1,** 459.

Jones, D. (1972). Reactions of aldehydes with unsaturated fatty acids during histological fixation. *In* "Fixation in Histochemistry" (P. J. Stoward, ed.), p. 1. Chapman & Hall, London.

Jones, D., and Gresham, G. A. (1966). Reaction of formaldehyde with unsaturated fatty acids during histological fixation. *Nature (London)* **210,** 1386.

Jones, D. G., and Devon, R. M. (1978). An ultrastructural study into the effects of pentobarbitone on synaptic organization. *Brain Res.* **147,** 47.

Jones, G., Gallant, P., and Butler, W. H. (1977). Improved techniques in light and electron microscopy. *J. Pathol.* **121,** 141.

Jones, G. J. (1974). Polymerization of glutaraldehyde at fixative pH. *J. Histochem. Cytochem.* **22,** 911.

Jones, R. T., and Trump, B. F. (1975). Cellular and subcellular effects on ischemia on the pancreatic acinar cell. *In vitro* studies of rat tissue. *Virchows Arch. B* **19,** 325.

Jost, P. C., and Griffith, O. H. (1973). The molecular reorganization of lipid bilayers by osmium tetroxide. A spin-label study of orientation and restricted Y-axis anisotropic motion in model membrane systems. *Arch. Biochem. Biophys.* **159,** 70.

Jost, P. C., Brooks, U. J., and Griffith, O. H. (1973). Fluidity of phospholipid bilayers and membranes after exposure to osmium tetroxide and glutaraldehyde. *J. Mol. Biol.* **76,** 313.

Kadar, A., Bush, V., and Gardner, D. L. (1971). Direct elastase treatment of ultrathin sections embedded in water-soluble Durcupan. *J. Pathol.* **103,** 64.

Kahan, L., and Kaltschmidt, E. (1972). Glutaraldehyde reactivity of the proteins of *Echerichia coli* ribosomes. *Biochemistry* **11,** 2691.

Kaibara, M., and Kickkawa, Y. (1971). Osmiophilia of the saturated phospholipid, dipalmitoyl lecithin, and its relationship to the alveolar lining layer of the mammalian lung. *Am. J. Anat.* **132,** 61.

Kalderon, N., and Gilula, N. B. (1979). Membrane events involved in myoblast fusion. *J. Cell Biol.* **81,** 411.

Kalimo, H. (1974). Ultrastructural studies on the hypothalmic neurosecretory neurons of the rat. Ph.D. Dissertation, Medical Faculty, University of Turku, Finland.

Kalimo, H. (1976). The role of the blood-brain barrier in perfusion fixation of the brain for electron microscopy. *Histochem. J.* **8,** 1.

Kalimo, H., Garcia, J. H., Kamijyo, Y., Tanaka, J., Viloria, J. E., Valigorsky, J. M., Jones, R. T., Kim, K. M., Mergner, W. J., Pendergrass, R. E., and Trump, B. F. (1974). Cellular and subcellular alterations of human CNS. Studies utilizing *in situ* perfusion fixation at immediate autopsy. *Arch. Pathol.* **97,** 352.

Kalimo, H., and Pelliniemi, L. J. (1977). Pitfalls in the preparation of buffers for electron microscopy. *Histochem. J.* **9,** 241.

Kalina, M., and Pease, D. C. (1977a). The preservation of ultrastructure in saturated phosphatidyl cholines by tannic acid in model systems and type II pneumocytes. *J. Cell. Biol.* **74,** 726.

Kalina, M., and Pease, D. C. (1977b). The probable role of phosphatidyl cholines in the tannic acid enhancement of cytomembrane electron contrast. *J. Cell Biol.* **74,** 742.

Kalt, M. R. (1971). The relationship between cleavage and blastocoel formation in *Xenopus laevis*. II. Electron microscopic observations. *J. Embryol. Exp. Morphol.* **26,** 51.

Kalt, M. R., and Tandler, B. (1971). A study of fixation of early amphibian embryos for electron microscopy. *J. Ultrastruct. Res.* **36,** 633.

Kanamura, S. (1971). Demonstration of glucose-6-phosphatase activity in hepatocytes following transparenchymal perfusion fixation with glutaraldehyde. *J. Histochem. Cytochem.* **19,** 386.

Kanazawa, K., Hamano, M., and Akahori, H. (1972). Application of scanning electron microscope to gastro-intestinal research. *Jpn. J. Clin. Electron Microsc.* **5,** 307.

Kanerva, L., Hervonen, A., and Rechardt, L. (1977). Permanganate fixation demonstrates the monoamine-containing granular vesicles in the SIF cells but not in the adrenal medulla of mast cells. *Histochemistry* **52,** 61.

Karlsson, U. L., and Schultz, R. L. (1965). Fixation of the central nervous system for electron microscopy by aldehyde perfusion. I. Preservation with aldehyde perfusates versus direct perfusion with osmium tetroxide with special reference to membranes and the extracellular space. *J. Ultrastruct. Res.* **12,** 160.

Karlsson, U. L., Schultz, R. L., and Hooker, W. M. (1975). Cation-dependent structures associated with membranes in the rat central nervous system. *J. Neurocytol.* **4,** 537.

Karnovsky, M. J. (1965). A formaldehyde-glutaraldehyde fixative of high osmolarity for use in electron microscopy. *J. Cell Biol.* **27,** 137A.

Karnovsky, M. J. (1967). The ultrastructural basis of capillary permeability studied with peroxidase as a tracer. *J. Cell Biol.* **35,** 213.

Karnovsky, M. J. (1971). Use of ferrocyanide-reduced osmium tetroxide in electron microscopy. *Proc. 14th Annu. Meet. Am. Soc. Cell Biol.* p. 146.

Karow, A. M. (1974). Cryopreservation: Pharmacological considerations. *In* "Organ Preservation for Transplantation" (A. M. Karow, G. M. Abouna, and A. L. Humphries, eds.), Little, Brown, Boston, Massachusetts.

Kasten, F. H., and Lala, R. (1975). The Feulgen reaction after glutaraldehyde fixation. *Stain Technol.* **50,** 197.

Katora, M. E., and Hollis, T. M. (1976). Regional variation in rat aortic endothelial surface morphology: Relationship to regional aortic permeability. *Exp. Mol. Pathol.* **24,** 23.

Kawarai, Y., and Nakane, P. K. (1970). Localization of tissue antigens on the ultrathin sections with peroxidase-labeled antibody method. *J. Histochem. Cytochem.* **18,** 161.

Kawata, T., Masuda, K., and Nakasone, N. (1978). Mesosomes observed in unfixed cells of *Staphylococcus aurens* by negative staining. *J. Electron Microsc.* **27,** 317.

Kellenberger, E., Ryter, A., and Sechaud, J. (1958). Electron microscope study of DNA-containing plasms. II. Vegetative and mature phage DNA as compared with normal bacterial nucleoids in different physiological states. *J. Biophys. Biochem. Cytol.* **4,** 671.

Kelley, R. O., Dekker, R. A. F., and Bluemink, J. G. (1975). Thiocarbohydrazide-mediated osmium binding: a technique for protecting soft biological specimens in the scanning electron microscope. *In* "Principles and Techniques of Scanning Electron Microscopy: Biological Applications" (M. A. Hayat, ed.), Vol. 4, Van Nostrand-Reinhold, New York.

Kendall, P. A., Polak, J. M., and Pearse, A. G. E. (1971). Carbodiimide fixation for immuno-histochemistry: Observations on the fixation of polypeptide hormones. *Experientia* **27,** 1104.

Kessel, R. G. (1969). The effect of glutaraldehyde fixation on the elucidation of the morphogenesis of annulate lamellae in oocytes of *Rana pipiens*. *Z. Zellforsch. Mikrosk. Anat.* **94,** 454.

Khan, A. A., Riemersma, J. C., and Booij, H. L. (1961). The reactions of osmium tetroxide with lipids and other compounds. *J. Histochem. Cytochem.* **9,** 560.

Kiernan, J. A. (1978). Recovery of osmium tetroxide from used fixative solutions. *J. Microsc. (Oxford)* **113,** 77.

Kirschner, D. A., and Caspar, D. L. D. (1975). Myelin structure transformed by dimethyl sulfoxide. *Proc. Natl. Acad. Sci. U.S.A.* **72,** 3513.

Kirschner, R. H. (1978). High resolution scanning electron microscopy of isolated cell organelles. *In* "Principles and Techniques of Scanning Electron Microscopy: Biological Applications" (M. A. Hayat, ed.), Vol. 6, Van Nostrand-Reinhold, New York.

Klein, S., and Shochat, M. (1971). The electron microscopic image of chloroplasts after osmium tetroxide fixation in the presence of sorbitol. *J. Microsc. (Paris)* **10,** 117.

Koehler, J. K. (1972). The freeze-etching technique. *In* "Principles and Techniques of Electron Microscopy: Biological Applications" (M. A. Hayat, ed.), Vol. 2, Van Nostrand-Reinhold, New York.

Koelle, G. B., Davis, R., Smyrl, E. G., and Fine, A. V. (1974). Refinement of the bis-(thioacetoxy) aurate (I) method for the electron microscopic localization of acetylcholinesterase and nonspecific cholinesterase. *J. Histochem. Cytochem.* **22,** 252.

Koelle, W. A., Hossaine, K. S., Akbarzadeh, P., and Koelle, G. B. (1970). Histochemical evidence and consequences of the occurence of isoenzymes of acetylcholinesterase. *J. Histochem. Cytochem.* **18,** 812.

Kolb-Bachofen, V. (1977). Electron microscopic localization of acid phosphatase in *Tetrahymena pyriformis:* The influence of activities of lysosomal enzymes on fixation and structural preservation. *Cytobiologie* **15,** 135.

Kölbel, H. K. (1978). Specimen course-automated EM specimen processing. *Proc. Int. Congr. Electron Microsc.,* 2, p. 74. Miscroscopical Society of Canada, Ontario.

Konwinski, M., Abramczuk, J., Barańska, W., and Szymkowiak, W. (1974). Size changes of mouse ova during preparation for morphometric studies in the electron microscope. *Histochemistry* **42,** 315.

Korn, A. H., Feairheller, S. H., and Filachione, E. M. (1972). Glutaraldehyde: Nature of the reagent. *J. Mol. Biol.* **65,** 525.

Korn, E. D. (1966a). II. Synthesis of bis(methyl 9,10-dihydroxystearte)osmate from methyl oleate and osmium tetroxide under conditions used for fixation of biological material. *Biochim. Biophys. Acta* **116,** 317.

Korn, E. D. (1966b). III. Modification of oleic acid during fixation of amoebae by osmium tetroxide. *Biochim. Biophys. Acta* **116,** 325.

Korn, E. D. (1967). A chromatographic and spectrophotometric study of the products of the reaction of osmium tetroxide with unsaturated lipids. *J. Cell Biol.* **34,** 627.

Korn, E. D. (1969). Current concepts of membrane structure and function. *Fed. Proc.* **28,** 6.

Korn, E. D., and Weisman, R. A. (1966). I. Loss of lipids during preparation of amoebae for electron microscopy. *Biochim. Biophys. Acta* **116,** 309.

Korneliussen, H. (1972). Elongated profiles of synaptic vesicles in motor endplates, morphological effects of fixative variations. *J. Neurocytol.* **1,** 279.

Kotowycz, C., and Suzuki, O. (1973). A carbon-13 nuclear magnetic resonance study of binding of manganese (II) to purine and pyrimidine nucleosides and nucleotides. *Biochemistry* **12,** 3434.

Kraehenbuhl, J. P., and Jamieson, J. D. (1972). Solid phase conjugation of ferritin to Fab-fragments of immunoglobulin G for use in antigen localization on thin sections. *Proc. Natl. Acad. U.S.A.* **69,** 1771.

Kraehenbuhl, J. P., and Jamieson, J. D. (1974a). Localization of intracellular antigens using immunoelectron microscopy. *In* "Electron Microscopy and Cytochemistry" (E. Wisse, W. T. Daems, I. Molenaar, and P. van Duijn, eds.). North-Holland Publ., Amsterdam.

Kraehenbuhl, J. P., and Jamieson, J. D. (1974b). Localization of intracellular antigens by immunoelectron microscopy. *Int. Rev. Exp. Pathol.* **13,** 1.

Kramer, I. R. H. (1961). Changes in the dental tissues due to cavity prepartion using a turbine hand-piece. *Proc. R. Soc. Med.* **54,** 239.

Kramer, P. J. (1955). Physical chemistry of the vacuoles. *Encycl. Plant Physiol., New Sev.* **1.**

Krames, B., and Page, E. (1968). Effects of electron-microscopic fixatives on cell membranes of the perfused rat heart. *Biochim. Biophys. Acta* **150,** 24.

Kreibich, G., Freienstein, C. M., Pereyra, B. N., Ulrich, B. L., and Sabatini, D. D. (1978). Proteins of rough microsomal membranes related to ribosome binding. II. Crosslinking of bound ribosomes to specific membrane proteins exposed at the binding sites. *J. Cell Biol.* **77,** 488.

Ksiezak, H., Zaleska, M., and Gromek, A. (1974). Level of total and bound acetylcholine in conditions of hypoxia, ischemia, and barbiturate anesthesia in guinea-pig brain. *Bull. Acad. Pol. Sci.* **22,** 649.

Kuczmarski, E. R., and Rosenbaum, J. L. (1976). Microfilament arrangement in nerve cells. *J. Cell Biol.* **70,** 247a.

Kuhlman, W. D., and Miller, H. R. P. (1971). A comparative study of the techniques for ultrastructural localization of antienzyme antibodies. *J. Ultrstruct. Res.* **35,** 370.

Kuhlman, W. D., Avrameas, S., and Ternynck, T. (1974). A comparative study for the ultrastructural localization of intracellular immunoglobulins using peroxidase conjugates. *Immunol. Methods* **5,** 33.

Kuran, H., and Olszewska, M. J. (1974). Effect of some buffers on the ultrastructure dry mass content and radioactivity of nuclei of *Haemanthus katharinae*. *Folia Histochem. Cytochem.* **12,** 173.

Kushida, H. (1962). A study of cellular swelling and shrinkage during fixation, dehydration, and embedding in various standard media. *Proc. Int. Congr. Electron Microsc, 5th. 1962* Vol. 2, Academic Press, New York.

Kushnarev, V. M., and Smirnova, T. A. (1966). Electron microscopy of alkaline phosphatase of *Escherichia coli*. *Can. J. Microbiol.* **12,** 605.

Kushnaryev, V. M., Dunne, W. M., and Buckmire, F. L. A. (1980). Electron microscopy of malachite green-glutaraldehyde fixed bacteria. *Stain Technol.* **54,** 331.

Kuthy, E., and Csapó, Z. (1976). Peculiar artifacts after fixation with glutaraldehyde and osmium tetroxide. *J. Microsc. (Oxford)* **107,** 177.

Kuwabara, T. (1969). The scanning electron microscopic study of the cell surface *Proc. 27th Annu. Meet., Electron Microsc. Soc. Am.,* p. 36.

Kuwabara, T. (1970). Surface structure of the eye tissue. *Proc. SEM.,* p. 185.

Kuypers, G. A. J., and Roomans, G. M. (1980). Postmortem elemental redistribution in rat studied by x-ray microanalysis and electron microscopy. *Histochemistry* **69,** 145.

LaFountain, J. R., Zobel, C. R., Thomas, H. R., and Galbreath, C. (1977). Fixation and staining of F-actin and microfilaments using tannic acid. *J. Ultrastruct. Res.* **58,** 78.

Laiho, K. U., Shelburne, J. D., and Trump, B. F. (1971). Observations on cell volume, ultrastructure, mitochondrial conformation and vital-dye uptake in Ehrlich ascites tumor cells. *Am. J. Pathol.* **65,** 203.

Landboe-Christiansen, E., and Parapat, S. (1972). The gastrointestinal mucosa in man, its surface architecture: Some observations by the scanning electron microscope. *JEOL News* **9E,** 12.

Landis, W. J., Paine, M. C., and Glimcher, M. J. (1980). Use of acrolein vapors for the anhydrous preparation of bone tissue for electron microscopy. *J. Ultrastruct. Res.* **70,** 171.

Langenberg, W. G. (1979). Chilling of tissue before glutaraldehyde fixation preserves fragile inclusions of several plant viruses. *J. Ultrastruct. Res.* **66,** 120.

Langenberg, W. G. (1980). Glutaraldehyde nonfixation of isolated viral and yeast RNAs. *J. Histochem. Cytochem.* **28,** 311.

Langenberg, W. G., and Schroeder, H. F. (1973). Effects of preparative procedures on the preservation of tobacco mosaic virus inclusions. *Phytopathology* **63,** 1003.

Langenberg, W. G., and Sharpee, R. L. (1978). Chromic acid-formaldehyde fixation of nucleic acids of bacteriophage $\phi6$ and infectious bovine rhinotracheitis virus. *J. Gen. Virol.* **39,** 377.

Langford, L. A., and Coggeshall, R. E. (1980). The use of potassium ferricyanide in neural fixation. *Anat. Rec.* **197**, 297.

Lappenbusch, W. L., and Willis, D. L. (1970). The effect of dimethyl sulfoxide on the radiation response of the rough-skinned newt (*Taricha tranulosa*). *Int. J. Radiat. Biol. Relat. Stud. Phys.* **18**, 217.

Larsson, L.-I. (1975). Effects of different fixatives on the ultrastructure of the developing proximal tubule in the rat kidney. *J. Ultrastruct. Res.* **51**, 140.

Larsson, L.-I. (1977). Ultrastructural localization of a new neuronal peptide (VIP). *Histochemistry* **54**, 173.

Läuchli, A. (1972). Electron probe analysis. *In* "Microautoradiography and Electron Probe Analysis," p. 191. Springer-Verlag, Berlin and New York.

Lawton, J. R., and Harris, P. J. (1978). Fixation of senescing plant tissues: Sclerenchymatous fiber cells from the flowering stem of grass. *J. Microsc. (Oxford)* **112**, 307.

Lebouton, A. V., and Masse, J. P. (1980a). A random arrangement of albumin-containing hepatocytes seen with histoimmunologic methods. I. Verification of the artifact. *Anat. Rec.* **197**, 183.

Lebouton, A. V., and Masse, J. P. (1980b). A random arrangement of albumin-containing hepatocytes seen with histoimmunological methods. II. Conditions that produce the artifact. *Anat. Rec.* **197**, 195.

Leduc, E. H., Bernhard, W., Holt, S. J., and Tranzer, J. P. (1967). Ultrathin sections. II. Demonstration of enzymatic activity. *J. Cell Biol.* **34**, 757.

Leduc, E. H., Scott, G. B., and Avrameas, S. (1969). Ultrastructural localization of intracellular immune globulins in plasma cells and lymphoblasts by enzyme-labeled antibodies. *J. Histochem. Cytochem.* **17**, 211.

Lee, C. S. (1978). Chromatin structure of *Drosophila nasutoides* satellites as probed by cross-linking DNA *in situ. Chromosoma* **65**, 103.

Lee, R. M. K. W., Garfield, R. E., Forrest, J. B., and Daniel, E. E. (1979). The effects of fixation, dehydration, and critical point drying on the size of cultured smooth muscle cells. *Proc. SEM* **3**, 439.

Lee, R. M. K. W., Garfield, R. E., Forrest, J. B., and Daniel, E. E. (1980). Dimensional changes of cultured smooth muscle cells due to preparatory process for transmission electron microscopy. *J. Microsc. (Oxford)* **120**, 85.

Lee, S. H. (1972). Isolation of parietal cells from glutaraldehyde-fixed rabbit stomach. *J. Histochem. Cytochem.* **20**, 634.

Lee, S. H., Dusek, J., and Rona, G. (1971). Electron microscopic cytochemical study of glutamic oxalacetic transaminase activity in ischemic myocardium. *J. Mol. Cell. Cardiol.* **3**, 103.

Leffingwell, H. A. (1974). Fossil palynomorphs. *In* "Principles and Techniques of Scanning Electron Microscopy: Biological Applications" (M. A. Hayat, ed.), Vol. 2, Van Nostrand-Reinhold, New York.

Lehman, F. E., and Mancuso, V. (1957). Improved fixative for astral rays and nuclear membrane of Tubifex embryos. *Exp. Cell Res.* **13**, 161.

Lenard, J., and Singer, S. J. (1968). Alterations of the conformation of proteins in red blood cell membranes and in solution by fixatives used in electron microscopy. *J. Cell Biol.* **37**, 117.

Lenn, N. J., and Beebe, B. (1977). A simple apparatus for controlled pressure perfusion fixation. *Microsc. Acta* **79**, 139.

Leskes, A., Siekevitz, P., and Palade, G. E. (1971). Differentiation of endoplasmic reticulum in hepatocytes. I. Glucose-6-phosphatase distribution *in situ. J. Cell Biol.* **49**, 264.

Lesseps, R. J. (1967). The removal by phospholipase C of a layer of lanthanum-staining material external to the cell membrane in embryonic chick cells. *J. Cell Biol.* **34**, 173.

Leung, T. K., and Babaï, F. (1974). Détection ultracytochimique de la 5'-nucléotidase sur les coupes à congelation ultrafines. *J. Microsc. (Paris)* **21**, 111.

Levy, M., Toury, R., and André, J. (1967). Séparation des membranes mitochondriales. Purification et caracterisation enzymatique de la membrane externe. *Biochim. Biophy. Acta* **135**, 599.

Levy, W. A., Herzog, I., Suzuki, K., Katzman, R., and Scheinberg, L. (1965). Method for combined ultrastructural and biochemical analysis of neural tissue. *J. Cell Biol.* **27**, 119.

Lewin, S. (1966). Reaction of salmon sperm deoxyribonucleic acid with formaldehyde. *Arch. Biochem. Biophys.* **113**, 584.

Lewis, E. R. (1971). Syudying neural architecture and organization with the scanning electron microscope. *Proc. SEM.*, p. 281.

Lewis, P. R., and Knight, D. P. (1977). Staining Methods for Sectioned Material. *In* "Practical Methods in Electron Microscopy" (A. M. Glauert, ed.), North-Holland Publ., Amsterdam.

L'Hermite, P., and Israel, M. (1969). Action du glutaraldéhyde sur les lipide et les protéines myéliniques. *Ann. Histochim.* **14**, 1.

Li, H. J. (1972). Thermal denaturation of nucleohistones—effects of formaldehyde reaction. *Biopolymers* **11**, 835.

Libanati, C. M., and Tandler, C. J. (1969). The distribution of the water-soluble inorganic orthophosphate ions within the cell: Accumulation in the nucleus. Electron probe microanalysis. *J. Cell Biol.* **42**, 754.

Lillie, R. D. (1977). "Conn's Biological Stains," 9th ed. Williams & Wilkins, Baltimore, Maryland.

Lillie, R. D., and Pizzolato, P. (1972). Histochemical use of borohydrides as aldehyde blocking reagents. *Stain Technol.* **47**, 13.

Lim, B. S., and Solomon, J. D.(1975). Electron microscopic study of freshly fixed and time-delayed fixed tissue. *Proc. 33rd Annu. Meet., Electron Microsc. Soc. Am.*, p. 618.

Lim, B. S., and Solomon, J. D. (1976). Electron microscopic study of time-delay fixed tissues in rats. *EMSA Bull.* **6**, 7.

Lim, B. S., Veech, R., Slovis, R., and Solomon, J. D. (1976). Preparation of freeze-stopped rat liver tissue for transmission electron microscopy. *Proc. 34th Annu. Meet., Electron Microsc. Soc. Am.*, p. 332.

Lim, D. J. (1976). Morphological and physiological correlates in cochlear and vestibular sensory epithelia. *Proc. SEM.*, p. 269.

Lim, D. J., and Lane, W. C. (1969). The scanning electron microscopic observation of the vestibular sensory epithelia. *Proc. 27th Annu. Meet., Electron Microsc. Soc. Am.*, p. 40.

Lima-de-Faria, A. (1978). Scanning electron microscopy of an isolated gene complex. *In* "Principles and Techniques of Scanning Electron Microscopy: Biological Applications" (M. A. Hayat, ed.), Vol. 6, Van Nostrand-Reinhold, New York.

Lin, H.-P., and Sommerfeld, M. R. (1978). Stabilization of cytomembranes in red algae during ultrastructural fixation with calcium chloride. *Trans. Am. Microsc. Soc.* **97**, 94.

Lin, P. S. D., Wallach, D. F. H., and Tsai, S. (1973). Temperature-induced variations in the surface topology of cultured lymphocytes are revealed by scanning electron microscopy. *Proc. Natl. Acad. Sci. U.S.A.* **70**, 2492.

Linscott, W. D., Faulk, W. P., and Perucca, P. J. (1969). Complement, chemical coupling of a functionally active component to erythrocytes in the absence of antibody. *J. Immunol.* **103**, 474.

Linthicum, D. S., Sell, S., Wagner, R. M., and Trefts, P. (1974). Scanning immunoelectron microscopy of mouse B and T lymphocytes. *Nature (London)* **252**, 173.

Lisak, J. C., Kaufman, H. W., Maupin-Szamier, P., and Pollard, T. D. (1976). The action of osmium tetroxide on proteins and amino acids. *Biol. Bull. (Woods Hole, Mass.)* **151**, 418.

Lisý, V., Kovárů, H., and Lodin, Z. (1971). *In vitro* effects of polyvinyl pyrrolidone and sucrose on the acetylcholinesterase, succinic dehydrogenase, and lactate dehydrogenase activity in the brain. *Histochemie* **26**, 205.

Litke, L. L., and Low, F. N. (1977). Fixative tonicity for scanning electron microscopy of delicate chick embryos. *Am. J. Anat.* **148,** 121.

Litman, R. B., and Barrnett, R. J. (1972). The mechanism of the fixation of tissue components by osmium tetroxide via hydrogen bonding. *J. Ultrastruct. Res.* **38,** 63.

Ljubešić, N. (1970). Osmiophile substanz in Blattzellen der Brombeere (*Rubus fructicosus*). *Protoplasma* **169,** 49.

Löhr, G. W., and Walker, H. D. (1963). *In* "Methods of Enzymatic Analysis" (H. U. Bergmeyer, ed.), Academic Press, New York.

Lojda, A. (1965). Fixation in histochemistry. *Folia Morphol.* **13,** 65.

Lollar, R. M. (1958). The mechanism of vegetable tannage. *In* "The Chemistry and Technology of Leather" (F. O'Flaherty, W. T. Roddy, and R. M. Lollar, eds.), Vol. 2, p. 201. Van Nostrand-Reinhold, New York.

Longo, F. J., and Anderson, E. (1970). An ultrastructural analysis of fertilization in the surf clam, *Spisula solidissima:* Polar body formation and development of the female pronucleus. *J. Ultrastruct. Res.* **33,** 495.

Loomis, W. D. (1974). *In* "Methods in Enzymology" (A. S. Fleischer and L. Packer, eds.), Vol. 31, pp. 528–544. Academic Press, New York.

Lotan, R., and Sharon, N. (1976). Modification of the biological properties of plant lectins by chemical crosslinking. *In* "Protein Crosslinking" (M. Friedman, ed.), p. 149. Plenum, New York.

Lowry, O. H., Passonneau, J. V., Hasselberger, F. X., and Schulz, D. W. (1964). Effect of ischemia on known substrates and co-factors of the glycolytic pathway in brain. *J. Biol. Chem.* **239,** 18.

Lubig, R. (1972). Ein Beitrag zur Umsetzung von Wolle mit Glutaraldehyde. Dissertation, Technische Hochschule, Aachen.

Luft, J. H. (1956). Permanganate—a new fixative for electron microscopy. *J. Biophys. Biochem. Cytol.* **2,** 799.

Luft, J. H. (1959). The use of acrolein as a fixative for light and electron microscopy. *Anat. Rec.* **133,** 305.

Luft, J. H., and Wood, R. L. (1963). The extraction of tissue protein during and after fixation with osmium tetroxide in various buffer systems. *J. Cell Biol.* **19,** 46A.

Luftig, R. B., McMillan, P. N., Weatherbee, J. A., and Weihing, R. R. (1977). Increased visualization of microtubules by an improved fixation procedure. *J. Histochem. Cytochem.* **25,** 175.

Lumb, W. V. (1963). "Small Animal Anesthesia." Lea & Febiger, Philadelphia, Pennsylvania.

Luzardo-Baptista, M. (1972). Correlation between molecular structure and glycogen ultrastructure. *Ann. Histochem.* **17,** 141.

Lynn, J. A., Martin, J. G., and Race, G. J. (1966). Recent improvements of histologic techniques for the combined light and electron microscopic examination of surgical specimens. *Am. J. Clin. Pathol.* **45,** 704.

Ma, J. C. W., and Jeffrey, L. M. (1978). Description and comparison of a new cleaning method of diatom frustules for light and electron microscope studies. *J. Microsc. (Oxford)* **112,** 235.

McDowell, E. M., and Trump, B. F. (1976). Histologic fixatives suitable for diagnostic light and electron microscopy. *Arch. Pathol. Lab. Med.* **100,** 405.

McDuffie, N. G. (1974). Lithium permanganate fixation of mammalian cells and viruses for electron microscopy. *J. Microsc. (Paris)* **19,** 197.

McFarland, W. N., and Klontz, G. W. (1969). Anesthesia in fishes. *Fed. Proc.* **28,** 1535.

McGarvey, K. A., Reidy, M. A., and Roach, M. R. (1980). A quantitative study of the preparation of rabbit aortic endothelial cells for scanning electron microscopy. *J. Microsc. (Oxford)* **118,** 229.

McGhee, J. D., and von Hippel, P. H. (1977). Formaldehyde as a probe of DNA structure. 4. Mechanism of the initial reaction of formaldehyde with DNA. *Biochemistry* **16**, 3276.

Machado, A. B. M. (1967). Straight OsO_4 versus glutaraldehyde-OsO_4 in sequence as fixatives for the granular vesicles in sympathetic axons of the rat pineal body. *Stain Technol.* **42**, 293.

McIlwain, H., and Buddle, H. L. (1953). Techniques in tissue metabolism. I. A mechanical chopper. *Biochem. J.* **53**, 412.

McIntyre, J. A., Gilula, N. B., and Karnovsky, M. J. (1974). Cryoprotectant - induced redistribution of intramembranous particles in mouse lymphocytes. *J. Cell Biol.* **60**, 192.

McLean, I. W., and Nakane, P. K. (1977). Periodate-lysine-paraformaldehyde fixative. A new fixative for immunoelectron microscopy. *J. Histochem. Cytochem.* **22**, 1077.

McLean, J. D., and Singer, S. J. (1970). A general method for the specific staining of intracellular antigens with ferritin antibody conjugates. *Proc. Natl. Acad. Sci. U.S.A.* **65**, 122.

McManus, J. F. A., and Mowry, R. W. (1958). Effects of fixation on carbohydrate histochemistry. *J. Histochem. Cytochem.* **6**, 309.

McMillan, P. N., and Luftig, R. B. (1973). Preservation of erythrocyte ultrastructure achieved by various fixatives. *Proc. Natl. Acad. Sci. U.S.A.* **70**, 3060.

McMillan, P. N., and Luftig, R. B. (1975). Preservation of membrane ultrastructure with aldehyde or imidate fixative. *J. Ultrastruct. Res.* **52**, 243.

McNicholas, J. M., and Hulett, F. M. (1977). Electron microscope histochemical localization of alkaline phosphatase(s) in *Bacillus licheniformis*. *J. Bacteriol.* **129**, 501.

McPherson, A. (1976). The analysis of biological structure with x-ray diffraction techniques. *In* "Principles and Techniques of Electron Microscopy: Biological Applications" (M. A. Hayat, ed.), Vol. 6, Van Nostrand-Reinhold, New York.

Magnusson, M. O., Osborne, D. G., Shimoji, T., Kiser, W. S., and Hawk, W. A. (1975). Ultrastructural changes of canine kidneys during 24-hour perfusion on a LI-400 preservation system. *Proc. 29th Annu. Meet., Electron Microsc. Soc. Am.*, p. 316.

Makita, T., and Sandborn, E. B. (1970). Fine structural localization of reductases and transferases in muscle mitochondria and the effect of dimethyl sulfoxide (DMSO) in the incubation medium. *J. Histochem. Cytochem.* **18**, 686.

Makita, T., and Sandborn, E. B. (1971). The effect of dimethyl sulfoxide (DMSO) in the incubation medium for the cytochemical localization of succinate dehydrogenase. *Histochemie* **26**, 305.

Malick, L. E., and Wilson, R. B. (1975). Evaluation of a modified technique for SEM examination of vertebrate specimens without evaporated metal layers. *Proc. SEM.*, p. 259.

Malinin, G. I. (1973). Cytotoxic effect of dimethyl sulfoxide on the ultrastructure of cultured Rhesus kidney cells. *Cryobiology* **10**, 22.

Mangum, C. P., and Shick, J. M. (1972). The pH of body fluids of marine invertebrates. *Comp. Biochem. Physiol. A* **42A**, 693.

Mann, D. M. A., Barton, C. M., and Davies, J. S. (1978). Postmortem changes in human central nervous tissue and the effects on quantitation of nucleic acids and enzymes. *Histochem. J.* **10**, 127.

March, J. (1968). "Advanced Organic Chemistry: Reactions, Mechanisms, and Structure." MdGraw-Hill, New York.

Marchant, H. J. (1978). Scanning electron microscopy of microtubules associated with the plasma membrane of protoplasts of the alga *Mougeotia*. *Proc. SEM* **2**, 1071.

Marchesi, V. T., and Palade, G. E. (1967). The localization of Mg-Na-K-activated adenosine triphosphatase on red cell ghost membranes. *J. Cell Biol.* **35**, 385.

Marinozzi, V. (1963). The role of fixation in electron staining. *J. R. Microsc. Soc.* **81**, 141.

Mariscal, R. N., Conklin, E. J., and Bigger, C. H. (1978). The putative sensory receptors associated with the cnidae of cnidarians. *Proc. SEM.* **2**, 959.

Marquardt, M. D., and Gordon, J. A. (1975). Glutaraldehyde fixation and the mechanism of erythrocyte agglutination by concanavalin A and soybean agglutinin. *Exp. Cell Res.* **91**, 310.

Marshall, A. T. (1980). Frozen-hydrated bulk specimens. *In* "X-Ray Microanalysis in Biology" (M. A. Hayat, ed.), Univ. Park Press, Baltimore, Maryland.

Marszalek, D. A., and Small, E. B. (1969). Preparation of soft biological materials for scanning electron microscopy. *Proc. SEM.,* p. 231.

Martin, C. J., Lam, D. P., and Marini, M. A. (1975). Reaction of formaldehyde with the histidine residues of proteins. *Bioorg. Chem.* **4**, 22.

Maser, M. D., Powell, T. E., III, and Philpott, C. W. (1967). Relationships among pH, osmolality, and concentration of fixative solutions. *Stain Technol.* **42**, 175.

Massie, H. R., Samis, H. V., and Baird, M. B. (1972). The effects of the buffer HEPES on the division potential of WI-38 cells. *In vitro* **7**, 191.

Mathieu, O., Claassen, H., and Weibel, E. R. (1978). Differential effect of glutaraldehyde and buffer osmolarity on cell dimensions: A study on lung tissue. *J. Ultrastruct. Res.* **63**, 20.

Matile, P. (1975). "The Lytic Compartment of Plant Cells." Springer-Verlag, Berlin and New York.

Maugel, T. K., Bonar, D. B., Creegan, W. J., and Small, E. B. (1980). Specimen preparation techniques for aquatic organisms. *Proc. SEM* **2**, 57.

Maul, G. G., Maul, H. M., Slogma, J. E., Lieberman, M. W., Stein, G. S., Hsu, B. Y. L., and Borum, T. W. (1972). Time sequence of nuclear pore formation in phytohemagglutinin-stimulated lymphocytes and in Hela cells during cell cycle. *J. Cell Biol.* **55**, 433.

Maunsbach, A. B. (1966). The influence of different fixatives and fixation methods on the ultrastructure of rat kidney proximal tubule cells. II. Effects of varying osmolality, ionic strength, buffer system and fixative concentration of glutaraldehyde solutions. *J. Ultrastruct. Res.* **15**, 283.

Maupin-Szamier, P., and Pollard, T. D. (1977). Destruction of actin filaments by osmium tetroxide. *Proc. 35th Annu. Meet., Electron Microsc. Soc. Am.,* p. 466.

Maupin-Szamier, P., and Pollard, T. D. (1978). Actin filament destruction by osmium tetroxide. *J. Cell Biol.* **77**, 837.

Maupin-Szamier, P., Pollard, T. D., and Fujiwara, K. (1975). Tropomyosin prevents the destruction of actin filaments by osmium tetroxide. *J. Cell Biol.* **67**, 424a.

Mayor, H. D., and Jordan, L. E. (1963). Acrolein - a fine structure fixative for viral cytochemistry. *J. Cell Biol.* **18**, 207.

Mazhul, L. A., Dobrov, E. N., and Tikhonenko, T. I. (1978). Tight binding of RNA to protein in particles of rod-like virus under the action of formaldehyde. *Biokhimiya* **43**, 138.

Mazurkiewicz, J. E., and Nakane, P. K. (1972). Light and electron microscopic localization of antigens in tissues embedded in polyethylene glycol with a peroxidase-labeled antibody method. *J. Histochem. Cytochem.* **20**, 969.

Meany, A., Gahan, G. B., and Maggi, V. (1967). Effects of Triton X-100 on acid phosphatases with different substrate specificities. *Histochemie* **11**, 280.

Mehler, A. H. (1954). Potential errors in spectrophotometry with optically dense solutions. *Science* **120**, 1043.

Meldolesi, J., Jamieson, J. D., and Palade, G. E. (1971). Composition of cellular membranes in the pancreas of the guinea pig. I. Isolation of membrane fractions. *J. Cell Biol.* **49**, 109.

Mellor, J. W. (1932). "A Comprehensive Treatise on Inorganic and Theoretical Chemistry," p. 265. Longmans, Green, New York.

Merck, W., Sparwald, E., and Cürten, I. (1974). Über den Einfluss verschiedener Fixationsmethoden und Fixationsmittel auf die electronenmikroskopische Struktur der Stria vascularis. Prominentia spiralis und des Ligamentum spirale beim Meerschweinchen. *Arch. Oto-Rhino-Laryngol.* **206**, 299.

Mersey, B., and McCully, M. E. (1978). Monitoring of the course of fixation of plant cells. *J. Microsc. (Oxford)* **114**, 49.

Meszler, R. M., and Gennaro, J. F. (1970). In "Biology of the Reptilia" (C. Grans, ed.), p. 305. Academic Press, New York.

Meves, H. (1970). The ionic permeability of nerve membranes. In "Permeability and Function of Biological Membranes" (L. Bolis, ed.), p. 261. North-Holland Publ., Amsterdam.

Meyers, S. (1978). Development of the embryonic otic placode. II. Electron microscopic analysis. *Anat. Rec.* **191**, 459.

Milch, R. A., Frisco, L. J., and Szymkowiak, E. A. (1965). Solid-state dielectric properties of aldehyde-treated goatskin collagen. *Biorheology* **3**, 9.

Millonig, G. (1961). Advantages of a phosphate buffer for OsO_4 solutions in fixation. *J. Appl. Phys.* **32**, 1637.

Millonig, G. (1964). Study on the factors which influence preservation of fine structure. In "Electron Microscopy" (P. Buffa, ed.), p. 347. Consiglio Nazionale delle Ricerche, Roma.

Millonig, G. (1966). Model experiments on fixation and dehydration. *Proc. Int. Congr. Electron Microsc., 6th, 1966* Vol. 2, p. 21.

Millonig, G., and Marinozzi, V. (1968). Fixation and embedding in electron microscopy. *Adv. Opt. Electron Microsc.* **2**, 251.

Milne, R. G. (1967). Electron microscopy of leaves infected with sowbane mosaic virus and other small polyhedral viruses. *Virology* **32**, 589.

Minássian, H., and Huang, S.-N. (1979). Effect of sodium azide on the ultrastructural preservation of tissues. *J. Microsc. (Oxford)* **117**, 243.

Minio, F., Lombardi, L., and Gautier, A. (1966). Mise en évidence et ultrastructure du glycogène hépatique influence des techniques de préparation. *J. Ultrastruct. Res.* **16**, 339.

Minns, R. J., and Steven, F. S. (1975). Observations on the fine structure of the human aorta. *Micron* **6**, 25.

Misch, D. W., and Misch, M. S. (1967). Dimethyl sulfoxide: activation of lysosomes *in vitro*. *Proc. Natl. Acad. Sci. U.S.A.* **58**, 2462.

Misch, D. W., and Misch, M. S. (1968). Lysosomes: Histochemical demonstration of latency using dimethyl sulfoxide. *Proc. Int. Congr. Histochem. Cytochem., 3rd, 1968*. p. 179.

Misch, D. W., and Misch, M. S. (1969). Reversible activation of lysosomes in dimethyl sulfoxide. *Nature (London)* **221**, 862.

Mitchell, C. D. (1969). Preservation of the lipids of the human erythrocyte stroma during fixation and dehydration for electron microscopy. *J. Cell Biol.* **40**, 869.

Mizuhira, V., and Futaesaku, Y. (1971). On the new approach of tannic acid and digitonine to the biological fixatives. *Proc. 29th Annu. Meet., Electron Microsc. Soc. Am.*, p. 494.

Moffat, D. B., and Creasey, M. (1971). A technique for the fixation for electron microscopy of the vascular bundles of the medulla of the kidney. *J. Microsc. (Paris)* **12**, 293.

Molin, S.-O, Nygren, H., and Dolonius, L. (1978). A new method for the study of glutaraldehyde-induced crosslinking properties in proteins with special reference to the reaction with amino groups. *J. Histochem. Cytochem.* **26**, 412.

Mollenhauer, H. H., and Totten, C. (1971). Studies on seeds. I. Fixation of seeds. *J. Cell Biol.* **48**, 387.

Møller, J. C., Skriver, E., SteenOlsen, T., and Maunsbach, A. B. (1980). Perfusion fixation of human kidneys for ultrastructural analysis. *J. Ultrastruct. Res.* **73**, 94.

Molnar, J., and Szepessy, G. L. (1972). Electron microscopic localization of alkaline phosphatase in an adenine-dependent mutant of *Bacillus antracis*. *Acta Microbiol. Acad. Sci. Hung.* **19**, 7.

Monsan, P., Puzo, G., and Mazarguil, H. (1975). Etude du mécanisme d'etablissement des liaisons glutaraldéhyde-proteines. *Biochimie* **57**, 1281.

Mooseker, M. S., and Tilney, L. G. (1975). The organization of an actin filament-membrane complex: Filament polarity and membrane attachment in the microvilli of intestinal epithelial cells. *J. Cell Biol.* **67**, 725.

Morel, F. M. M., Baker, R. F., and Wayland, H. (1971). Quantitation of human red blood cell fixation by glutaraldehyde. *J. Cell Biol.* **48**, 91.

Morest, D. K., and Morest, R. R. (1966). Perfusion fixation of the brain with chrom-osmium solutions for the rapid Golgi method. *Am. J. Anat.* **118**, 811.

Moretz, R. C., Akers, C. K., and Parsons, D. F. (1969). Use of small angle x-ray diffraction to investigate disordering of membranes during preparation for electron microscopy. I. Osmium tetroxide and potassium permanganate. *Biochim. Biophys. Acta* **193**, 1.

Morgan, A. J. (1980). Preparation of specimens: Changes in chemical integrity. *In* "X-ray Microanalysis in Biology" (M. A. Hayat, ed.), Univ. Park Press, Baltimore, Maryland.

Morgan, A. J., Davies, T. W., and Erasmus, D. A. (1975). Changes in the concentration and distribution of elements during electron microscope preparative procedures. *Micron.* **6**, 11.

Morgan, T. E., and Huber, G. L. (1967). Loss of lipid during fixation for electron microscopy. *J. Cell Biol.* **32**, 757.

Moriarty, G. C., and Halmi, N. S. (1972). Electron microscopic study of the adrenocorticotropin-producing cell with the use of unlabeled antibody and the soluble peroxidase-antiperoxidase complex. *J. Histochem. Cytochem.* **20**, 590.

Morré, D. J., and Mollenhauer, H. H. (1969). Studies on the mechanisms of glutaraldehyde stabilization of cytomembranes. *Proc. Indiana Acad. Sci.* **78**, 167.

Morré, D. J., and Junghans, W. N., and Vigil, E. L., and Kennan, T. W. (1974). Isolation of organelles and endomembrane components from rat liver: Biochemical markers and quantitative morphometry. *In* "Methodological Developments in Biochemistry" (E. Reid, ed.), Vol. 4, pp. 195–236. Longmans, Green, New York.

Morris, J. F., and Cannata, M. A. (1973). Ultrastructural preservation of the dense core of posterior pituitary neurosecretory granules and its implications for hormone release. *J. Endocrinol.* **57**, 517.

Mortensen, N. J. McC., and Morris, J. F. (1977). The effect of fixation conditions on the ultrastructural appearance of gastrin cell granules in the rat gastric pyloric antrum. *Cell Tissue Res.* **176**, 251.

Moss, G. I. (1966). Glutaraldehyde as a fixative for plant tissues. *Protoplasma* **22**, 194.

Motta, P., and Porter, K. R. (1974). Structure of rat liver sinusoids and associated tissue spaces as revealed by scanning electron microscopy. *Cell Tissue Res.* **148**, 111.

Moyer, W. W., and Grev, D. A. (1963). Linear polyglutaraldehyde. *Polym. Lett.* **1**, 29.

Mueller, W. C., and Beckman, C. H. (1974). Ultrastructure of the phenol-storing cells in the roots of banana. *Physiol. Plant Pathol.* **4**, 187.

Mueller, W. C., and Beckman, C. H. (1976). Ultrastructure and development of phenolic-storing cells in cotton roots. *Can. J. Bot.* **54**, 2074.

Mueller, W. C., and Rodehorst, E. (1977). The effect of some alkaloids on the ultrastructure of phenolic-containing cells in the endodermis of cotton roots. *Proc. 35th Annu. Meet., Electron Microsc. Soc. Am.,* p. 544.

Munger, B. L. (1958). Polarization optical properties of the pancreatic acinar cells of the mouse. *J. Biophys. Biochem. Cytol.* **4**, 177.

Munkres, M., and Wachtel, A. (1967). Histochemical localization of phosphatases in *Mycoplasma gallisepticum*. *J. Bacteriol.* **93**, 1096.

Munton, T. J., and Russell, A. D. (1970). Aspects of the action of glutaraldehyde on *Escherichia coli*. *J. Appl. Bacteriol.* **33**, 410.

Murakami, Y. (1976). Puncture perfusion of small tissue pieces for scanning electron microscopy. *Arch. Histol. Jpn.* **39**, 99.

Murphy, J. A., Pappelis, A. J., and Thompson, M. R. (1974). Morphological aspects of *Diplodia maydis* and its role in the stalk rot of corn. *Proc. SEM.,* p. 405.

Muscatello, U., Guarriera-Bobyleva, V., and Buffa, P. (1972). Configurational changes in isolated rat liver mitochondria as revealed by negative staining. II. Modifications caused by changes in respiratory states. *J. Ultrastruct. Res.* **40**, 235.

Nachmias, V. T. (1963). Fibrillar structures in the cytoplasm of *Chaos chaos*. *J. Cell Biol.* **19**, 51A.

Nachmias, V. T. (1968). Inhibition of streaming in amoeba by pinocytosis inducers. *Exp. Cell Res.* **51**, 347.

Nagata, T., and Murata, F. (1972). Supplemental studies on the method for electron microscopic demonstration of lipase in the pancreatic acinar cells of mice and rats. *Histochemie* **29**, 8.

Nakai, Y. (1971). On different types of nerve endings in the frog median eminence after fixation with permanganate and glutaraldehyde-osmium. *Z. Zellforsch. Mikrosk. Anat.* **119**, 164.

Nakane, P. K. (1973). Ultrastructural localization of tissue antigens with the peroxidase-labeled antibody method. *In* "Electron Microscopy and Cytochemistry" (E. Wisse, W. T. Daems, I. Molenaar, and P. van Duijn, eds.). North-Holland Publ., Amsterdam.

Nanninga, N. (1971). The mesosome of *Bacillus subtilis* as affected by chemical and physical fixation. *J. Cell Biol.* **48**, 219.

Napolitano, L. M., Lebaron, F., and Scaletti, J. (1967). Preservation of myelin lamellar structure in the absence of lipid. *J. Cell Biol.* **34**, 817.

Napolitano, L. M., Saland, L., Lopex, J., Sterzing, P. R., and Kelley, R. O. (1972). Localization of cholesterol in peripheral nerve: Use of (3H) digitonin for electron microscopic autoradiography. *Anat. Rec.* **174**, 157.

Neale, E. K., and Chapman, G. B. (1970). Effect of low temperature on the growth and fine structure of *Bacillus subtilis*. *J. Bacteriol.* **104**, 518.

Needles, H. L. (1967). Crosslinking of gelatin by aqueous peroxydisulfate. *J. Polym. Sci.* **A5**, 1.

Neidle, S., and Stuart, D. I. (1976). The crystal and molecular structure of an osmium bispyridine adduct of thymine. *Biochim. Biophys. Acta* **418**, 226.

Nemanic, M. K., and Pitelka, D. R. (1971). Scanning electron microscope study of the lactating mammaary gland. *J. Cell Biol.* **48**, 410.

Nermut, M. V. (1977). Freeze-drying for electron microscopy. *In* "Principles and Techniques of Electron Microscopy: Biological Applications" (M. A. Hayat, ed.), Vol. 7, Van Nostrand-Reinhold, New York.

Nermut, M. V., and Ward, B. J. (1974). Effect of fixatives on the fracture plane in red blood cells. *J. Microsc. (Oxford)* **102**, 29.

Nevalainen, T. J., and Anttinen, J. (1977). Ultrastructural and functional changes in pancreatic acinar cells during autolysis. *Virchows Arch. B* **24**, 197.

Nichols, D. B., Cheng, H., and Leblond, C. P. (1974). Variability of the shape and argentaffinity of the granules in the enteroendocrine cells of the mouse duodenum. *J. Histochem. Cytochem.* **22**, 929.

Nielson, A. J., and Griffith, W. P. (1978). Tissue fixation and staining with osmium tetroxide: The role of phenolic compounds. *J. Histochem. Cytochem.* **26**, 138.

Niimi, M., Edanaga, M., Murakami, H., and Tokunaga, M. (1975). Several problems in critical point drying method for SEM- especially shrinkage and distortion of specimens. *J. Electron Microsc.* **24**, 201.

Nilsson, J. R. (1974). Effects of DMSO on vacuole formation, contractile vacuole function, and nuclear division in *Tetrahymena pyriformis* GL. *J. Cell Sci.* **16**, 39.

Nir, I., and Hall, M. O. (1974). The ultrastructure of lipid-depleted rod photoreceptor membranes. *J. Cell Biol.* **63**, 587.

Nistal, M., Rodriguez-Echandia, E. L., and Paniagua, R. (1978). Formaldehyde-induced appearance of septate junctions between digestive vacuoles. *Tissue Cell* **10**, 735.

Noel, A. R. A. (1972). Rapid continuous flow method for tissue processing. *Lab. Pract.* **21**, 815.

Nonomura, Y. (1971). Structure of liver ribosomes studied with negative staining. *J. Mol. Biol.* **60**, 303.

Nordmann, J. J. (1977). Ultrastructural appearance of neurosecretory granules in the sinus gland of the crab after different fixation procedures. *Cell Tissue Res.* **185,** 557.

Norris, G., Banfield, W., and Chalifoux, H. (1967). Tissue processor for automatic dehydration and infiltration of small specimens. *Sci. Tools* **14,** 13.

Norton, T. N., Gelfand, M., and Brotz, M. (1962). Studies in the histochemistry of plasmalogens. I. The effect of formalin and acrolein fixation on the plasmalogens of adrenal and brain. *J. Histochem. Cytochem.* **10,** 375.

O'Brien, T. P., Kuo, J., McCully, M. E., and Zee, S.-Y. (1973). Coagulant and non-coagulant fixation of plant cells. *Aust. J. Biol. Sci.* **26,** 1231.

Ockleford, C. D., and Tucker, J. B. (1973). Growth, breakdown, repair, and rapid contraction of microtubular axopodia in the heliozoan *Actinophrys* sol. *J. Ultrastruct. Res.* **44,** 396.

Ogata, K. M., Otteson, M., and Svendsen, I. (1968). Preparation of water-insoluble, enzymatically active derivatives of subtilisin type novo by crosslinking with glutaraldehyde. *Biochim. Biophys. Acta* **159,** 403.

Ogawa, K., and Barrnett, R. J. (1965). Electron cytochemical studies of succinic dehydrogenase and dihydronicotinamide-adenine dinucleotide diaphorase activities. *J. Ultrastruct. Res.* **12,** 488.

Ohkura, T. (1966). Electron microscopic demonstration of acid mucopolysaccharides in the synovial membrane of an adult dog. *Proc. Int. Congr. Electron Microsc., 6th, 1966* Vol. 2, p. 67.

Ohmiya, K., Tanimura, S., Kobayashi, T., and Shimizu, S. (1978). Preparation and properties of proteases immobilized on anion exchange resin with glutaraldehyde. *Biotechnol. Bioeng.* **20,** 1.

Ohnishi, A., Offord, K., and Dyck, P. J. (1974). Studies to improve fixation of human nerves. I. Effect of duration of glutaraldehyde fixation on peripheral nerve morphometry. *J. Neurol. Sci.* **23,** 223.

Ohtsuki, I., Manzi, R. M., Palade, G. E., and Jamieson, J. D. (1978). Entry of macromolecular tracers into cells fixed with low concentrations of aldehydes. *Biol. Cell.* **31,** 119.

Okabayashi, K., Futai, M., and Mizuno, D. (1974). Localization of acid and alkaline phosphatases in *Staphylococcus aureus. J. Microbiol. (Jpn)* **18,** 287.

Ökrös, I. (1968). Digitonin reaction in electron microscopy. *Histochemie* **13,** 91.

Olah, J., and Rohlich, P. (1966). Peculiar membrane configurations after fixation in glutaraldehyde. *Acta Biol. Acad. Sci. Hung.* **17,** 65.

Olins, D. E., and Wright, E. B. (1973). Glutaraldehyde fixation of isolated eukaryotic nuclei. *J. Cell Biol.* **59,** 304.

Olson, L. W., and Heath, I. B. (1971). Observations on the ultrastructure of microtubules. *Z. Zellforsch. Mikrosk. Anat.* **115,** 388.

Ongun, A., Thomson, W. W., and Mudd, J. B. (1968). Lipid fixation during preparation of chloroplasts for electron microscopy. *J. Lipid Res.* **9,** 416.

Orenstein, J. M., and Shelton, E. (1976). Surface topography and interactions between mouse peritoneal cells allowed to settle on an artificial substrate: Observations by scanning electron microscopy. *Exp. Mol. Pathol.* **24,** 201.

Oschman, J. L., and Wall, B. J. (1972). Calcium binding to intestinal membranes. *J. Cell Biol.* **55,** 58.

Oschman, J. L., Hall, T. A., Peters, P. D., and Wall, B. J. (1974). Association of calcium with membranes of squid giant axon: ultrastructure and microprobe analysis. *J. Cell Biol.* **61,** 156.

Paavola, L. G. (1977). The corpus luteum of the guinea pig. Fine structure at the time of maximum progesterone secretion and during regression. *Am. J. Anat.* **150,** 565.

Packer, L., and Greville, G. D. (1969). Energy linked oxidation of glutaraldehyde by rat liver mitochondria. *FEBS Lett.* **3,** 112.

Page, S. G. (1964). Filament lengths in resting and excited muscles. *Proc. R. Soc. London, Ser. B* **160,** 460.

Page, S. G., and Huxley, H. E. (1963a). Filament lengths in striated muscle. *J. Cell Biol.* **19**, 369.

Page, S. G., and Huxley, H. E. (1963b). Filament lengths in resting and excited muscles. *Proc. R. Soc. London, Ser. B* **160**, 460.

Painter, R. G., Tokuyasu, K. T., and Singer, S. J. (1973). Immunoferritin localization of intracellular antigens. The use of ultracryotomy to obtain ultrathin sections suitable for direct immunoferritin staining. *Proc. Natl. Acad. Sci. U.S.A.* **70**, 1649.

Palade, G. (1952). A study of fixation for electron microscopy. *J. Exp. Med.* **95**, 285.

Palay, S. L., McGee-Russell, S. M., Gordon, S., and Grillo, M. A. (1962). Fixation of neural tissue for electron microscopy by perfusion with solutions of osmium tetroxide. *J. Cell Biol.* **12**, 385.

Palfrey, A. J., and Davies, D. V. (1966). The fine structure of chondrocytes. *J. Anat.* **100**, 213.

Paliwal, Y. C. (1970). Electron microscopy of brome grass mosaic virus in infected leaves. *J. Ultrastruct. Res.* **30**, 491.

Paolilli, D. J., Falk, R., and Reighard, J. A. (1967). The effect of chemical fixation on the fretwork of chloroplasts. *Trans. Am. Microsc. Soc.* **86**, 225.

Papadimitriou, J. M., and van Duijn, P. (1970). Effects of fixation and substrate protection on the isozymes of asparate aminotransferase studied in a quantitative cytochemical model system. *J. Cell Biol.* **47**, 71.

Parducz, B. (1967). Ciliary movement and coordination in ciliates. *Int. Rev. Cytol.* **21**, 91.

Parish, G. R. (1975). Changes of particle frequency in freeze-etched erythrocyte membranes after fixation. *J. Microsc. (Oxford)* **104**, 245.

Park, R. B., Kelley, J., Drury, S., and Sauer, K. (1966). The Hill reaction of chloroplasts isolated from glutaraldehyde-fixed spinach leaves. *Proc. Natl. Acad. Sci. U.S.A.* **55**, 1056.

Parker, F., and Odland, G. F. (1973). Ultrastructural and lipid biochemical comparisons of human eruptive tuberous and planar xanthomas. *Isr. J. Med. Sci.* **9**, 395.

Pathak, M. A., and Kramer, D. M. (1969). Photosensitization of skin *in vivo* by furocoumarins (Psoralens). *Biochim. Biophys. Acta* **195**, 197.

Patrick, R., and Reimer, C. W. (1966). "The Diatoms of the United States," Vol. 1. Monographs of the Acad. Nat. Sci., Philadelphia, Pennsylvania.

Pattle, R. E., Schock, C., and Creasey, J. M. (1972). Electron microscopy of the lung surfactant. *Experientia* **28**, 286.

Pattle, R. E., Schock, C., and Creasey, J. M. (1974). Postmortem changes at electron microscope level in the Type II cells of the lung. *Br. J. Exp. Pathol.* **55**, 221.

Paula-Barbosa, M. (1975). The duration of aldehyde fixation as a "flattening factor" of synaptic vesicles. *Cell Tissue Res.* **164**, 63.

Paula-Barbosa, M., and Gray, E. G. (1974). The effects of various fixatives at different pH on synaptic coated vesicles, reticulosomes, and cytonet. *J. Neurocytol.* **3**, 471.

Paula-Barbosa, M., Sobrinho-Simões, M. A., and Gray, E. G. (1977). The effects of different methods of fixation on central nervous system synaptic pinocytotic vesicles. *Cell Tissue Res.* **178**, 323.

Pawley, J. B., Hayes, T. L., and Nowell, J. A. (1975). Microdissection. *In* "Principles and Techniques of Scanning Electron Microscopy: Biological Applications" (M. A. Hayat, ed.), Vol. 3, Van Nostrand-Reinhold, New York.

Payne, J. W. (1973). Polymerization of proteins with glutaraldehyde. *Biochem. J.* **135**, 867.

Pearse, A. G. E. (1963). Some aspects of the localization of enzyme activity with the electron microscope. *J. R. Microsc. Soc.* **81**, 107.

Pearse, A. G. E., and Polak, J. M. (1975). Bifunctional reagents as vapor- and liquid-phase fixatives for immunohistochemistry. *Histochem. J.* **7**, 179.

Pearse, A. G. E., and Welsch, V. (1968). Ultrastructural characteristic of the thyroid C cells in the summer, autumn and winter states of the hedgehog (*Erinaceus europaeus*), with some reference to other mammalian species. *Z. Zellforsch. Mikrosk. Anat.* **92**, 596.

Pearson, G. R., and Logan, E. F. (1978). Scanning electron microscopy of early postmortem artifacts in the small intestine of a neonatal calf. *Br. J. Exp. Pathol.* **59**, 499.

Pease, D. C. (1966). The preservation of unfixed cytological detail by dehydration with 'inert' agents. *J. Ultrastruct. Res.* **14**, 356.

Pease, D. C. (1973a). Substitution techniques. *In* "Advanced Techniques in Biological Electron Microscopy" (J. K. Koehler, ed.), Springer-Verlag, Berlin and New York.

Pease, D. C. (1973b). Glycol methacrylate copolymerized with glutaraldehyde and urea as an embedment retaining lipids. *J. Ultrastruct. Res.* **45**, 124.

Pease, D. C., and Peterson, R. G. (1972). Polymerizable glutaraldehyde-urea mixtures as polar, water-containing embedding media. *J. Ultrastruct. Res.* **41**, 133.

Pelttari, A., and Helminen, H. J. (1979). The effects of various fixatives on the relative thickness of cellular membranes in the ventral lobe of the rat prostrate. *Histochem. J.* **11**, 599.

Penttilä, A., and Ahonen, A. (1976). Electron microscopical and enzyme histochemical changes in the rat myocardium during prolonged autolysis. *Beitr. Pathol.* **157**, 126.

Penttilä, A., Kalimo, H., and Trump, B. F. (1974). Influence of glutaraldehyde and/or osmium tetroxide on cell volume, ion content, mechanical stability, and membrane permeability of Ehrlich ascites tumor cells. *J. Cell Biol.* **63**, 197.

Peracchia, C., and Mittler, B. S. (1972a). New glutaraldehyde fixation procedures. *J. Ultrastruct. Res.* **39**, 57.

Peracchia, C., and Mittler, B. S. (1972b). Fixation by means of glutaraldehyde-hydrogen peroxide reaction products. *J. Cell Biol.* **53**, 234.

Peracchia, C., and Robertson, J. D. (1971). Increase in osmiophilia of axonal membranes of crayfish as a result of electrical stimulation, asphyxia, or treatment with reducing agents. *J. Cell Biol.* **51**, 223.

Perlin, M., and Hallum, J. V. (1971). Effect of acid pH on macromolecular synthesis in L cells. *J. Cell Biol.* **49**, 66.

Perner, E. (1965). Electron mikroscopische untersuchungen an zellen von embryonen im zustand volliger Samenruhe. I. Die zellulare strukturordnung in den radicula Lufttrockner samen von *Pisum sativum*. *Planta* **65**, 334.

Peters. A., and Palay, S. L. (1966). The morphology of laminae A and A_1 of the dorsal nucleus of the lateral geniculate body of the cat. *J. Anat.* **100**, 451.

Peters, A., Proskauer, C. C., and Kaiserman-Abramof, I. R. (1968). The small pyramidal neuron of the rat cerebral cortex. *J. Cell Biol.* **39**, 604.

Peters, K., and Richards, F. M. (1977). Chemical crosslinking: Reagents and problems in studies of membrane structure. *Annu. Rev. Biochem.* **46**, 523.

Peters, T., and Ashley, C. A. (1967). An artifact in autoradiography due to binding of free amino acids to tissues by fixatives. *J. Cell Biol.* **33**, 53.

Peterson, E. A., and Sober, H. A. (1962). Chromatography of proteins: Substituted celluloses. *In* "Methods in Enzymology" (S. P. Colwick, and N. O. Kaplan, eds.), Vol. 5, pp. 15–18. Academic Press, New York.

Pexieder, T. (1977). The role of buffer osmolarity in fixation for SEM and TEM. *Experientia* **32**, 806.

Pfenninger, K. H. (1979). Subplasmalemmal vesicle clusters: Real or artifact? *In* "Freeze-Fracture: Methods, Artifacts, and Interpretation" (J. E. Rash and C. S. Hudson, eds.), Raven, New York.

Phillips, E. R., Kletzien, R. F., and Perdue, J. F. (1977). A supravital polyaldehyde fixative for external cell surfaces. *Exp. Cell Res.* **105**, 51.

Pickel, V. M., Joh, T. H., and Reis, D. J. (1975). Ultrastructural localization of tyrosine hydroxylase in noradrenergic neurons of brain. *Proc. Natl. Acad. Sci. U.S.A.* **72**, 659.

Pickett-Heaps, J. D., and Fowke, L. C. (1969). Cell division in *Oedogonium*. I. Mitosis, cytokinesis and cell elongation. *Aust. J. Biol. Sci.* **22**, 857.

Pilström, L., and Nordlund, U. (1975). The effect of temperature and concentration of the fixative on morphometry of rat liver mitochondria and rough endoplasmic reticulum. *J. Ultrastruct. Res.* **50**, 33.

Polacow, I., Cabasso, L., and Li, H. J. (1976). Histone redistribution and conformational effect on chromatin induced by formaldehyde. *Biochemistry* **15**, 4559.

Pollard, T. D. (1976). Cytoskeletal functions of cytoplasmic contractile proteins. *J. Supramol. Struct.* **5**, 317.

Pollard, T. D., and Korn, E. D. (1973). Electron microscopic identification of actin associated with isolated amoeba membranes. *J. Biol. Chem.* **248**, 448.

Pollard, T. D., and Maupin, P. (1978). Electron microscopy of cytoplasmic contractile proteins. *Proc. Int. Congr. Electron Microsc., 9th, 1978* Vol. 3, p. 606.

Porter, K. R., and Kallman, F. (1953). The properties and effects of osmium tetroxide as a tissue fixative with special reference to its use for electron microscopy. *Exp. Cell Res.* **4**, 127.

Poste, G., and Papahadjopoulos, D. (1976). Lipid vesicles as carriers for introducing materials into cultured cells: Influence of vesicle lipid composition on mechanism(s) of vesicle incorporation into cells. *Proc. Natl. Acad. Sci. U.S.A.* **73**, 1603.

Poste, G., Porter, C. W., and Papahadjopoulos, D. (1978). Identification of a potential artifact in the use of electron microscope autoradiography to localize saturated phospholipids in cells. *Biochim. Biophys. Acta* **510**, 256.

Pourcho, R. G., Bernstein, M. H., and Gould, S. F. (1978). Malachite green: Applications in electron microscopy. *Stain Technol.* **53**, 29.

Powell, P. C. (1975). Immunity to Marek's disease induced by glutaraldehyde-treated cells of Marek's disease lymphoblastoid cell lines. *Nature (London)* **257**, 684.

Prescott, T. D., Rao, M. V. N., Evenson, D. P., Stone, G. E., and Thrasher, J. D. (1966). Isolation of single nuclei from various protozoans. *Methods Cell Physiol.* **2**, 131.

Pricam, C., Fisher, K. A., and Friend, D. S. (1977). Intramembranous particle distribution in human erythrocytes: Effects of lysis, glutaraldehyde, and poly-L-lysine. *Anat. Rec.* **189**, 595.

Price, Z. H. (1967). The micromorphology of zeiotic blebs in cultured human epithelial (HEP) cells. *Exp. Cell Res.* **48**, 82.

Puchtler, H., Waldrop, F. S., Meloan, S., Terry, M. S., and Conner, H. M. (1970). Methacarn (methenol-Carnoy) fixation: Practical and theoretical considerations. *Histochemie* **21**, 97.

Quick, D. C., and Waxman, S. G. (1977). Ferric ion, ferrocyanide, and inorganic phosphate as cytochemical reactants at peripheral nodes of Ranvier. *J. Neurocytol.* **6**, 555.

Quiocho, F. A., and Richards, F. M. (1964). Intermolecular crosslinking of a protein in the crystaline state: Carboxypeptidase-A. *Proc. Natl. Acad. Sci. U.S.A.* **52**, 833.

Quiocho, F. A., and Richards, F. M. (1966). The enzyme behavior of carboxypeptidase-A in the solid state. *Biochemistry* **5**, 4062.

Quiocho, F. A., Bishop, W. H., and Richards, F. M. (1967). Effects of changes in some solvent parameters on carboxypeptidase-A in solution and in crosslinked crystals. *Proc. Natl. Acad. Sci. U.S.A.* **57**, 525.

Radda, G. K. (1975). Dynamic aspects of membrane structure. *In* "Biological Membranes. Twelve Essays on their Organization, Properties, and Functions" (D. S. Parsons, ed.), pp. 81–95. Oxford Univ. Press (Clarendon), London and New York.

Ragazzo, J. A., and Behrman, E. J. (1976). The reaction of oxo-osmium ligand complexes with isopentenyl adenine and its nucleoside. *Bioinorg. Chem.* **5**, 343.

Rammler, D. H. (1967). The effect of DMSO on several enzyme systems. *Ann. N. Y. Acad. Sci.* **141**, 291.

Rammler, D. H. (1971). Use of DMSO in enzyme-catalyzed reactions. *In* "Dimethyl Sulfoxide" (S. W. Jacob, E. E. Rosenbaum, and D. C. Wood, eds.), Vol. 1, p. 189. Dekker, New York.

Ramus, J. (1969). Pit connection formation in the red alga *Pseudogloiophloea*. *J. Phycol.* **5,** 57.

Ranvier, M. L. (1887). De l'emploi de l'acide perruthénique dans les recherches histologiques et de l'application de ce réactif à l'étude des vacuoles des cellules calciformes. *C. R. Hebd. Seances Acad. Sci.* **105,** 145.

Rash, J. E., and Hudson, C. S., eds. (1979). "Freeze-Fracture: Methods, Artifacts, and Interpretations." Raven, New York.

Rash, J. E., Biesele, J., and Gey, G. O. (1970). Three classes of filaments in cardiac differentiation. *J. Ultrastruct. Res.* **33,** 408.

Rasmussen, K. E., and Albrechtsen, J. (1974). Glutaraldehyde. The influence of pH, temperature, and buffering on the polymerization rate. *Histochemistry* **38,** 19.

Reale, E., and Luciano, L. (1964). A probable source of errors in electron histochemistry. *J. Histochem. Cytochem.* **12,** 713.

Rebhun, L. I. (1972). Freeze-substitution and freeze-drying. *In* "Principles and Techniques of Electron Microscopy: Biological Applications" (M. A. Hayat, ed.), Vol 2, Van Nostrand-Reinhold, New York.

Reboud, A.-M., Buisson, M., Madjar, J-J., and Reboud, J.-P. (1975). Study of mammalian ribosomal protein reactivity *in situ.* II. Effect of glutaraldehyde and salts. *Biochimie* **57,** 295.

Rechardt, L., Kanerva, L., and Hervonen, A. (1977). Ultrastructural demonstration of amine granules in the adrenal medullary cells of the rat using acid permanganate fixation. *Histochemistry* **54,** 339.

Reedy, M. K. (1971). Electron microscope observations concerning the behavior of the cross-bridge in striated muscle. *In* "Contractility of Muscle Cells and Related Processes" (R. J. Podolsky, ed.), p. 229. Prentice-Hall, Englewood Cliffs, New Jersey.

Reese, T. S., and Karnovsky, M. J. (1967). Fine structural localization of a blood-brain barrier to exogenous peroxidase. *J. Cell Biol.* **34,** 207.

Reichen, J., and Paumgartner, G. (1976). Uptake of bile acids by the perfused rat liver. *Am. J. Physiol.* **231,** 734.

Reimann, B. E., Lewin, J. M., and Guillard, R. L. (1963). *Cyclotella cryptica,* a new brackish-water diatom species. *Phycologia* **3,** 75.

Reimann, V. B. (1960). Bildung, Bau und Zuzammenhang der Bacillariophyceenschalen. *Nova Hedwigia* **2,** 349.

Reiss, J. (1971). Dimethyl sulfoxide as carrier in enzyme cytochemistry. *Histochemie* **26,** 93.

Richards, F. M., and Knowles, J. R. (1968). Glutaraldehyde as a protein crosslinking reagent. *J. Mol. Biol.* **37,** 231.

Richardson, K. C. (1966). Electron microscopic identification of autonomic nerve endings. *Nature (London)* **210,** 756.

Riecken, E. O., Goebell, H., and Bode, C. (1969). Untersuchungen zum Einfluss von tiefen Temperaturen und Speicherdauer auf Speicherdauer auf einige histochemisch nachweisbare Enzymaktivitaten in Leber, Niere und Jejunum der Ratte. *Histochemie* **20,** 225.

Riede, U. N., Lobingen, A., Grünholz, P., Steimer, R., and Sandritter, W. (1976). Einfluss einer einstündigen Autolyse auf die quantitative Zytoarchitektur der Rattenleberzelle (Eine ultrastrukturellmorphometrische). *Bietr. Pathol.* **157,** 391.

Riemersma, J. C. (1963). Osmium tetroxide fixation of lipids: Nature of the reaction products. *J. Histochem. Cytochem.* **11,** 436.

Riemersma, J. C. (1968). Osmium tetroxide fixation of lipids for electron microscopy: A possible reaction mechanism. *Biochim. Biophys. Acta* **152,** 718.

Riemersma, J. C. (1970). Chemical effects of fixation in biological systems. *In* "Biological Techniques in Electron Microscopy" (D. F. Parsons, ed.), Academic Press, New York.

Riemersma, J. C., and Booij, H. L. (1962). The reaction of osmium tetroxide with lecithin: application of staining procedures. *J. Histochem. Cytochem.* **10,** 89.

Robards, A. W. (1968). On the ultrastructure of differentiating secondary xylem in willow. *Protoplasma* **65,** 449.

Robbins, E. (1961). Some theoretical aspects of osmium tetroxide fixation with special reference to the metaphase chromosomes of cell cultures. *J. Biophys. Biochem. Cytol.* **11,** 449.

Roberts, K. (1974). Crustalline glycoprotein cell walls of algae: Their structure, composition, and assembly. *Philos. Trans. R. Soc. London, Ser. B* **268,** 129.

Robertson, J. D. (1959). The ultrastructure of cell membranes and their derivatives. *Biochem. Soc. Symp.* **16,** 3.

Robertson, J. D. (1963). The occurrence of a subunit pattern in the unit membranes of club endings in Maunther cell synapses in goldfish brain. *J. Cell Biol.* **19,** 201.

Robertson, E. A., and Schultz, R. L. (1970). The impurities in commercial glutaraldehyde and their effect on the fixation of brain. *J. Ultrastruct. Res.* **30,** 275.

Robertson, J. G., Lyttleton, P., Williamson, K. I., and Batt, R. D. (1975). The effect of fixation procedures on the electron density of polysaccharide granules in *Nocardia corallina*. *J. Ultrastruct. Res.* **50,** 321.

Robinson, D. R., and Grant, M. E. (1966). The effects of aqueous salt solutions on the activity coefficients of purine and pyrimidine bases and their relation to the denaturation of deoxyribonucleic acid by salts. *J. Biol. Chem.* **241,** 4030.

Rodewald, R., and Karnovsky, M. J. (1974). Porous structure of the glomerular slit diaphragm in the rat and mouse. *J. Cell Biol.* **60,** 423.

Rodriguez, E. M. (1969). Fixation of the central nervous system by perfusion of the cerebral ventricles with a threefold aldehyde mixture. *Brain Res.* **15,** 395.

Rømert, P., and Matthiessen, M. E. (1975). Fixation of foetal pig liver for electron microscopy. I. The effect of various aldehydes and of delayed fixation. *Anat. Embryol.* **147,** 243.

Roomans, G. M. (1975). Calcium binding to the acrosomal membrane of human spermatozoa. *Exp. Cell Res.* **96,** 23.

Roozemond, R. C. (1969). The effect of fixation with formaldehyde and glutaraldehyde on the composition of phospholipids extractable from rat hypothalamus. *J. Histochem. Cytochem.* **17,** 482.

Rosa, J. J., and Sigler, P. B. (1974). The site of covalent attachment in the crystalline osmium-tRNA^{fMet} isomorphous derivative. *Biochemistry* **13,** 5102.

Rosenbluth, J. (1963). Contrast between osmium-fixed and permanganate-fixed toad spinal ganglia. *J. Cell Biol.* **16,** 143.

Rosene, D. L., and Mesulam, M.-M. (1978). Fixation variables in horseradish peroxidase neurohistochemistry. I. The effects of fixation time and perfusion procedures upon enzyme activity. *J. Histochem. Cytochem.* **26,** 28.

Rosenfeld, A., Stevens, C. L., and Printz, M. P. (1970). Studies on the secondary structure of phenylalanyl transfer ribonucleic acid. *Biochemistry* **9,** 4971.

Rossi, G. L. (1975). Simple apparatus for perfusion fixation for electron microscopy. *Experientia* **31,** 998.

Rostgaard, J., and Tranum-Jensen, J. (1980). A procedure for minimizing cellular shrinkage in electron microscope preparation: A quantitative study on frog gallbladder. *J. Microsc. (Oxford)* **119,** 213.

Roth, L. E., Jenkins, R. A., Johnson, C. W., and Robinson, R. W. (1963). Additional stabilizing conditions for electron microscopy of the mitotic apparatus of giant amoebae. *J. Cell Biol.* **19,** 62A.

Rowsell, H. (1969). ''A Textbook of Veterinary Clinical Pathology.'' Williams & Wilkins, Baltimore, Maryland.

Rubbo, S. D., Gardner, J. F., and Webb, R. L. (1967). Biocidal activities of glutaraldehyde and related compounds. *J. Appl. Bacteriol.* **30**, 78.

Rufener, C., and Nakane, P. K. (1973). Light and electron microscopic localization of vasopression in the supraoptic nucleus of hypothalamus and posterior pituitary gland of the rat. *Anat. Rec.* **175**, 432.

Russell, A. D., and Haque, H. (1975). Inhibition of EDTA-lysozyme lysis of *Pseudomonas aeruginosa* by glutaraldehyde. *Microbios* **13**, 151.

Russell, A. D., and Vernon, G. N. (1975). Inhibition by glutaraldehyde of lysostaphin-induced lysis of *Staphylococcus aureus*. *Microbios.* **13**, 147.

Russell, L., and Burguet, S. (1977). Ultrastructure of Leydig cells as revealed by secondary tissue treatment with a ferrocyanide-osmium mixture. *Tissue Cell* **9**, 751.

Rybicka, K. (1977). Morphology, histochemistry, and staining of glycogen in the dog heart. *Proc. 35th Annu. Meet., Electron Microsc. Soc. Am.,* p. 448.

Ryder, T. A., and Bowen, I. D. (1974). The use of x-ray microanalysis to investigate problems encountered in enzyme cytochemistry. *J. Microsc. (Oxford)* **101**, 143.

Ryter, A. (1968). Association of the nucleus and the membrane of bacteria: A morphological study. *Bacteriol. Rev.* **32**, 39.

Ryter, A., and Kellenberger, E. (1958). Etude en microscopie électronique de plasmas contenant de l'acide désoxyribonucléique. *Z. Naturforsch* **13**, 597.

Sabatini, D. D., Bensch, K., and Barrnett, R. J. (1962). New means of fixation for electron microscopy and histochemistry. *Anat. Rec.* **142**, 274.

Sabatini, D. D., Bensch, K., and Barrnett, R. J. (1963). Cytochemistry and electron microscopy. The preservation of cellular structures and enzymatic activity by aldehyde fixation. *J. Cell Biol.* **17**, 19.

Sabatini, D. D., Miller, F., and Barrnett, R. J. (1964). Aldehyde fixation for morphological and enzyme histochemical studies with the electron microscope. *J. Histochem. Cytochem.* **12**, 57.

Sachs, D. H., and Winn, H. J. (1970). The use of glutaraldehyde as a coupling agent for ribonuclease and bovine serum albumin. *Immunochemistry* **7**, 581.

Sahyoun, N., Hock, R. A., and Hollenberg, M. D. (1978). Insulin and epidermal growth factor-urogastrone: Affinity crosslinking to specific binding sites in rat liver membranes. *Proc. Natl. Acad. Sci. U.S.A.* **75**, 1675.

Saito, A., Wang, C.-T., and Fleischer, S. (1978). Membrane asymmetry and enhanced ultrastructural detail of sarcoplasmic reticulum revealed with use of tannic acid. *J. Cell Biol.* **79**, 601.

Saito, T., and Keino, H. (1976). Acrolein as a fixative for enzyme cytochemistry. *J. Histochem. Cytochem.* **24**, 1258.

Saito, T., Iwata, K., and Yamazaki, Y. (1973). Ultracytochemical study of the blastocyst development. I. Acid phosphatase. *J. Electron Microsc.* **22**, 289.

Saito, Y., and Tanaka, Y. (1980). Glutaraldehyde fixation of fish tissue for electron microscopy. *J. Electron Microsc.* **29**, 1.

Salema, R., and Brandão, I. (1973). The use of PIPES buffer in the fixation of plant cells for electron microscopy. *J. Submicrosc. Cytol.* **5**, 79.

Salisbury, F. B., and Ross, C. (1969). "Plant Physiology." Wadsworth, Belmont, California.

Sandberg, P. A. (1970). SEM of freeze-dried ostracoda (Crustacea). *Trans. Am. Microsc. Soc.* **89**, 113.

Sandborn, E. B. (1966). Electron microscopy of the neuron membrane systems and filaments. *Can. J. Physiol. Pharmacol.* **44**, 329.

Sandborn, E. B., and Makita, T. (1972). Scanning and transmission electron microscopy of intracellular organelles. *J. Cell Biol.* **55**, 71a.

Sandborn, E. B., Makita, B. T., and Lin, K. (1969). The use of dimethyl sulfoxide as an accelerator in the fixation of tissues for ultrastructural and cytochemical studies and in freeze-etching of cells. *Anat. Rec.* **163**, 255.

Sandborn, E. B., Stephens, H., and Bendayan, M. (1975). The influence of dimethyl sulfoxide on cellular ultrastructure and cytochemistry. *Ann. N. Y. Acad. Sci.* **243**, 122.

Sanders, H. J., Walker, G. O., Edwards, H. S., Jr., and Hall, T. J. (1958). Derivatives of acrolein and peracetic acid. *Ind. Eng. Chem.* **50**, 854.

Sanderson, C. J., and Frost, P. (1974). The induction of tumor immunity in mice using glutaraldehyde-treated tumor cells. *Nature (London)* **248**, 690.

Sandström, B. (1970). Liver fixation for electron microscopy by means of transparenchymal perfusion with glutaraldehyde. *Lab. Invest.* **23**, 71.

Sanger, F. (1949). Fractionation of oxidized insulin. *Biochem. J.* **44**, 126.

Sannes, P. L., Katsuyama, T., and Spicer, S. S. (1978). Tannic acid-metal salt sequences for light and electron microscopic localization of complex carbohydrates. *J. Histochem. Cytochem.* **26**, 55.

Santos-Sacchi, J. (1978). Differential effects of primary fixation with glutaraldehyde and osmium upon the membranous systems of the strial and external sulcus cells. *Acta Otolaryngol.* **86**, 56.

Sasse, D. (1965). Untersuchungen zum cytochemischen Glykogennachweis. VII. Mitteilung versuch zur OsO$_4$-bedingten Diastaseresistenz. *Histochemie* **5**, 378.

Satir, B. H., and Satir, P. (1979). Partitioning of intramembrane particles during the freeze-fracture procedure. *In* "Freeze-Fracture: Methods, Artifacts, and Interpretations" (J. E. Rash, and C. S. Hudson, eds.), Raven, New York.

Sauerbier, W. (1960). Reaction of formaldehyde with deoxyribonucleic acid in phage T 1. *Nature (London)* **188**, 327.

Saunders, D. R., Wilson, J., and Rubin, C. E. (1968). Loss of absorbed lipid during fixation and dehydration of jejunal mucosa. *J. Cell Biol.* **37**, 183.

Sauer, W. (1969). Effect of fixation on histochemical determination of deoxyribonucleic acid in plant material. II. Ultraviolet microspectrophotometry. *Mikroskopie* **24**, 291.

Scallen, T. J., and Dietert, S. E. (1969). The quantitative retention of cholesterol in mouse liver prepared for electron microscopy by fixation in a digitonin containing aldehyde solution. *J. Cell Biol.* **40**, 802.

Schäfer, D., and Hündgen, M. (1971). Der Einfluss acht verschiedener Aldehyde und des pH-Wertes auf die Dartstellbarkeit der Glukose-6-phosphatase in Kultivierten Zellen. *Histochemie* **26**, 362.

Scheehan, J. C., and Hlavka, J. J. (1956). The use of water-soluble basic carbodiimides in peptide synthesis. *J. Org. Chem.* **21**, 439.

Scheehan, J. C., and Hlavka, J. J. (1957). The crosslinking of gelatin using a water soluble carbodiimide. *J. Am. Chem. Soc.* **79**, 4528.

Schejter, A., and Bar-Eli, A. (1970). Preparation and properties of crosslinked water-insoluble catalase. *Arch. Biochem. Biophys.* **136**, 325.

Schellenberg, D. A., and Pangborn, J. (1980). Preservation of the microbial association to intestinal mucosal epithelium in rats for scanning electron microscopy. *Scanning* **3**, 237.

Schellens, J. P. M., Brunk, U. T., and Lindgren, A. (1976). Influence of serum on ruffling activity, pinocytosis, and proliferation of *in vitro* cultivated human glia cells. *Cytobiologie* **13**, 93.

Schidlovsky, G. (1965). Contrast in multilayer systems after various fixations. *Lab. Invest.* **14**, 1213.

Schiff, R. I., and Gennaro, J. F. (1979). The role of the buffer in the fixation of biological specimens for transmission and scanning electron microscopy. *Scanning* **2**, 135.

Schlatter, C., and Schlatter-Lanz, I. (1971). A simple method for the regeneration of used osmium tetroxide fixative solution. *J. Microsc. (Oxford)* **93**, 85.

Schliwa, M. (1977). Influence of calcium on intermicrotubule bridges within heliozoan axonemes. *J. Submicrosc. Cytol.* **9**, 221.

Schmalbruch, H. (1980). Delayed fixation alters the pattern of intramembrane particles in mammalian muscle fibers. *J. Ultrastruct. Res.* **70**, 15.

Schmid, H. H. O., and Takahashi, T. (1968). The condensation of long-chain aldehydes to 2,-3,dialkylacroleins in the presence of ethanolamine phosphatides. *Hoppe-Seyler's Z. Physiol. Chem.* **349**, 1673.

Schneider, G., and Schneider, G. (1967). Liber Stoffverluste bei Formal-Fixierung von menschlichen Gehirngewebe. *Histochemie* **4**, 348.

Schneider, G. B. (1976). The effects of preparative procedures for scanning electron microscopy on the size of isolated lymphocytes. *Am. J. Anat.* **146**, 93.

Schnepf, E. (1972). Structural modifications in the plasmalemma of *Aphelidium*-infected *Scenedesmus* cells. *Protoplasma* **75**, 155.

Schook, P. (1980). The effective osmotic pressure of the fixative for transmission and scanning electron microscopy. *Acta Morphol. Neerl.-Scand.* **18**, 31.

Schreil, W. H. (1964). Studies on the fixation of artificial and bacterial DNA plasms for the electron microscopy of thin sections. *J. Cell Biol.* **22**, 1.

Schultz, R. L., and Case, N. M. (1968). Microtubule loss with acrolein and bicarbonate-containing fixatives. *J. Cell Biol.* **38**, 633.

Schultz, R. L., and Case, N. M. (1970). A modified aldehyde perfusion technique for preventing certain artifacts in electron microscopy of the central nervous system. *J. Microsc. (Oxford)* **92**, 69.

Schultz, R. L., and Karlsson, U. L. (1972). Brain extracellular space and membrane morphology variations with preparative procedures. *J. Cell Sci.* **10**, 181.

Schultze, M., and Rudneff, M. (1865). Weitere mittheilungen über die einwirkung der ueberosmiumsaüre auf thierische gewebe. *Arch. Mikrosk. Anat.* **1**, 299.

Schurer, J. W., Hoedemaeker, J., and Molenaar, I. (1977). Polyethyleneimine as tracer particle for (immuno) electron microscopy. *J. Histochem. Cytochem.* **25**, 384.

Schurer, J. W., Kalicharan, D., Hoedemaeker, J., and Molenaar, I. (1978). The use of polyethyleneimine for demonstration of anionic sites in basement membranes and collagen fibrils. *J. Histochem. Cytochem.* **26**, 688.

Schwab, D. W., Janney, A. H., and Scala, J. (1970). Preservation of fine structure in yeast by fixation in a dimethyl sulfoxide-acrolein-glutaraldehyde solution. *Stain Technol.* **45**, 143.

Scott, G. B., and Avrameas, S. (1968). Intracellular antibody formation demonstrated on ultrathin frozen sections with alkaline phosphatase used as an antigen and as a marker. *Proc. Eur. Reg. Conf. Electron Microsc., 4th, 1968* Vol. 2, p. 201.

Seagull, R. W., and Heath, I. B. (1979). The effects of tannic acid on the *in vivo* preservation of microfilaments. *Eur. J. Cell Biol.* **20**, 184.

Seagull, R. W., and Heath, I. B. (1980). The organization of cortical microtubule arrays in the radish root hair. *Protoplasma* **103**, 205.

Séchaud, J., and Kellenberger, E. (1972). Electron microscopy of DNA-containing plasms. IV. Glutaraldehyde-uranyl acetate fixation of virus-infected bacteria for thin sectioning. *J. Ultrastruct. Res.* **39**, 598.

Seeman, P. (1967). Transient holes in the erythrocyte membrane during hypotonic hemolysis and stable holes in the membrane after lysis by saponin and lysolecithin. *J. Cell Biol.* **32**, 55.

Seigel, N. J., Spargo, B. H., and Kashgarian, M. (1973). An evaluation of routine electron microscopy in the examination of renal biopsies. *Nephron* **10**, 209.

Seip, H. M., and Stølevik, R. (1966). Studies on the failure of the first born approximation in electron diffraction. *Acta Chem. Scand.* **20**, 385.

Seligman, A. M., Hanker, J. S., Wasserkrug, H., Dimochowski, H., and Katzoff, L. (1965).

Histochemical demonstration of some oxidized macromolecules with thiocarbohydrazide (TCH) and osmium tetroxide. *J. Histochem. Cytochem.* **13**, 629.

Seligman, A. M., Wasserkrug, H. L., and Hanker, J. S. (1966). A new staining method (OTO) for enhancing contrast of lipid-containing membranes and droplets in osmium tetroxide fixed tissue with osmiophilic thiocarbohydrazide (TCH). *J. Cell Biol.* **30**, 424.

Seligman, A. M., Wasserkrug, H. L., Deb, C., and Hanker, J. S. (1968). Osmium-containing compounds with multiple basic or acidic groups as stains for ultrastructure. *J. Histochem. Cytochem.* **16**, 87.

Seligman, A. M., Wasserkrug, H. L., and Plapinger, R. E. (1970). Comparison of the ultrastructural demonstration of cytochrome oxidase activity with three bis(phenylenediamines). *Histochemie* **23**, 63.

Sentein, P. (1975). Action of glutaraldehyde and formaldehyde on segmentation mitosis. *Exp. Cell Res.* **95**, 233.

Shah, D. O. (1970). The effect of potassium permanganate on lecithin and cholesterol monolayers. *Biochim. Biophys. Acta* **211**, 358.

Shain, Y., Cohen, E., and Ben-Shaul, Y. (1974). Separation and characterization of plant zymogen bodies. *In* "Methodological Developments in Biochemistry" Vol. 4: Subcellular Studies (E. Reid, ed.), pp. 367–376. Longman, London.

Shannon, W. A. (1974). A simplified agar-polyethylene disc method for the Sorvall TC-2 tissue sectioner. *Stain Technol.* **49**, 109.

Sharpe, V., and Denny, P. (1976). Electron microscope studies on the absorption and localization of lead in the leaf tissue of *Potamogeton pectinatus* L. *J. Exp. Bot.* **27**, 1155.

Shaw, J. (1960). The mechanisms of osmoregulation. *In* "Comparative Biochemistry" (M. Florkin and H. S. Mason, eds.), Vol. 2, p. 471. Academic Press, New York.

Shaw, M., and Manocha, M. S. (1965). Fine structure in detached, senescing wheat leaves. *Can. J. Bot.* **43**, 747.

Shea, S. M. (1971). Lanthanum staining of the surface coat of cells. Its enhancement by the use of fixatives containing alcian blue or cetylpyridinium chloride. *J. Cell Biol.* **51**, 611.

Shea, S. M., and Karnovsky, M. J. (1969). The cell surface and intercellular junctions in liver as revealed by lanthanum staining after fixation with glutaraldehyde with added alcian blue. *J. Cell Biol.* **43**, 128a.

Shelton, E., and Mowczko, W. E. (1977). Membrane blebs: A fixation artifact. *J. Cell Biol.* **75**, 206a.

Shelton, E., and Mowczko, W. E. (1978). Membrane blisters: A fixation artifact. A study in fixation for scanning electron microscopy. *Scanning* **1**, 166.

Shienvold, F. L., and Kelly, D. E. (1974). Desmosome structure revealed by freeze-fracturing and tannic acid staining. *J. Cell Biol.* **63**, 313a.

Shigenaka, Y., Watanabe, K., and Kaneda, M. (1973). Effects of glutaraldehyde and osmium tetroxide on hypotrichous ciliates, and determination of the most satisfactory fixation methods for electron microscopy. *J. Protozool.* **20**, 414.

Shigenaka, Y., Watanabe, K., and Kaneda, M. (1974). Degrading and stabilizing effects of Mg^{2+} ions on microtubule containing axopodia. *Exp. Cell Res.* **85**, 391.

Shih, C. Y. (1974). SEM studies of the internal organization of plant organs. *Proc. SEM.*, p. 344.

Shilkin, K. B., Papadimitriou, J. M., and Walters, M. N.-I. (1971). The effect of dimethyl sulfoxide on hepatic cells of rats. *Aust. J. Exp. Biol. Med. Sci.* **44**, 581.

Shimamura, A., and Tokunaga, J. (1970). Scanning electron microscopy of sensory (fungiform) papillae in the frog tongue. *Proc. SEM.*, p. 225.

Shlafer, M., and Karow, A. M. (1975). Pharmacological effects of dimethyl sulfoxide on the mammalian myocardium. *Ann. N. Y. Acad. Sci.* **243**, 110.

Shnitka, T. K., Youngman, M. M., and Jewell, L. D. (1968). Simple mechanical tissue chopper for the preparation of specimens for electron microscopy and ultrastructural cytochemistry. *Lab. Pract.* **17**, 918.

Shrager, P. G., Strickholm, A., and Macey, R. I. (1969). Chemical modification of crayfish axons by protein crosslinking aldehydes. *J. Cell. Physiol.* **74**, 91.

Sidwell, R. W., Westbrook, L. Dixon, G. J., and Happich, W. F. (1970). Potentially infectious agents associated with shearling bedpads. Part I. *J. Appl. Microbiol.* **19**, 53.

Silva, M. T. (1971). Changes included in the ultrastructure of the cytoplasmic and intracytoplasmic membranes of several gram-positive bacteria by variations in OsO_4 fixation. *J. Microsc. (Oxford)* **93**, 227.

Silva, M. T. (1975). The ultrastructure of the membranes of gram-positive bacteria as influenced by fixatives and membrane-damaging treatments. *In* "Biomembranes—Lipids, Proteins and Receptors" (R. M. Burton and L. Packer, eds.), p. 255. BI-Science Publ. Div., Webster Groves, Missouri.

Silva, M. T., Sousa, J. C. F., Polónia, J. J., Macedo, M. A. E., and Parente, A. M. (1976). Bacterial mesosomes: Real structures or artifacts. *Biochim. Biophys. Acta* **443**, 92.

Simionescu, M., Simionescu, N., and Palade, G. E. (1976). Segmental differentiations of cell junctions in the vascular endothelium. *J. Cell Biol.* **68**, 705.

Simionescu, N., and Simionescu, M. (1976a). Galloylglucoses of low molecular weight as mordants in electron microscopy. I. Procedure and evidence for mordanting effect. *J. Cell Biol.* **70**, 608.

Simionescu, N., and Simionescu, M. (1976b). Galloylglucoses of low molecular weight as mordants in electron microscopy. II. The moiety and functional groups possibly involved in the mordanting effect. *J. Cell Biol.* **70**, 622.

Simson, J. A. V. (1977). The influence of fixation on the carbohydrate cytochemistry of rat salivary gland secretory granules. *Histochem. J.* **9**, 645.

Simson, J.A.V., and Spicer, S. S. (1975). Selective subcellular localization of cations with variants of potassium (pyro) antimonate technique. *J. Histochem. Cytochem.* **23**, 575.

Simson, J. A. V., Spicer, S. S., and Hall, B. J. (1974). Morphology and cytochemistry of rat salivary gland acinar secretory granules and their alteration by isoproterenol. I. Parotid gland. *J. Ultrastruct. Res.* **48**, 465.

Simson, J. A. V., Dom, R. M., and Sannes, P. L. (1978). Morphology and cytochemistry of acinar secretory granules in normal and isoproterenol-treated rat submandibular glands. *J. Microsc. (Oxford)* **113**, 185.

Singal, P. K., and Sanders, E. J. (1974). An ultrastructural study of the first cleavage of *Xenopus* embryos. *J. Ultrastruct. Res.* **47**, 433.

Singley, C. T., and Solursh, M. (1980). The use of tannic acid for the ultrastructural visualization of hyaluronic acid. *Histochemistry* **65**, 93.

Sisson, S. P., and Vernier, R. L. (1980). Methods for immunoelectron microscopy: Localization of antigens in rat kidney. *J. Histochem. Cytochem.* **28**, 441.

Sitte, V. P. (1960). Daseindringen von osmiumtetroxyd, kaliumpermanganat und formaldehyd-lösungen in pflanzengewege. *Histochemie* **2**, 76.

Sjöstrand, F. S. (1976). The problems of preserving molecular structure of cellular components in connection with electron microscopic analysis. *J. Ultrastruct. Res.* **55**, 271.

Sjöstrand, F. S., and Barajas, L. (1968). Effect of modifications on conformation of protein molecules on structure of mitochondrial membranes. *J. Ultrastruct. Res.* **25**, 121.

Sjöstrand, F. S., and Barajas, L. (1970). A new model for mitochondrial membranes based on structural and on biochemical information. *J. Ultrastruct. Res.* **32**, 293.

Sjöstrand, F. S., and Elfvin, L.-G. (1962). The layered asymmetric structure of the plasma membrane in the exocrine pancreas cells of the cat. *J. Ultrastruct. Res.* **7**, 504.

Skaer, R. J., and Whytock, S. (1976). The fixation of nuclei and chromosomes. *J. Cell Sci.* **20**, 221.

Skaer, R. J., and Whytock, S. (1977). Chromatin-like artifacts from nuclear sap. *J. Cell Sci.* **26**, 301.

Skehan, P. (1975). The mechanisms of glutaraldehyde-fixed sarcoma 180 ascites cell aggregation. *J. Memb. Biol.* **24**, 87.

Sloan, P., Shellis, R. P., and Berkovitz, B. K. B. (1976). Effect of specimen preparation on the appearance of the rat periodontal ligament in the scanning electron microscope. *Arch. Oral Biol.* **21**, 633.

Smith, J. T., Funckes, A. J., Barak, A. J., and Thomas, L. E. (1957). Cellular lipoproteins. I. The insoluble lipoprotein of whole liver cells. *Exp. Cell Res.* **13**, 96.

Smith, R. E. (1970). Comparative evaluation of two instruments and procedures to cut nonfrozen sections. *J. Histochem. Cytochem.* **18**, 590.

Smith, R. E., and Farquhar, M. G. (1965). Preparation of nonfrozen sections for electron microscope cytochemistry. *Sci. Instrum. News, R.C.A.* **10**, 13.

Smith, R. E., and Farquhar, M. G. (1966). Lysosome function in the regulation of the secretory process in the cells of the anterior pituitary gland. *J. Cell Biol.* **31**, 319.

Smith, R. E., and Fishman, W. H. (1968). p-(Acetoxymercuric)aniline diazotate, a reagent for visualizing the Naphthol AS BI product of acid hydrolase action at the level of the light and electron microscope. *J. Histochem. Cytochem.* **17**, 1.

Smith, U., Ryan, J. W., Michie, D. D., and Smith, D. S. (1971). Endothelial projections as revealed by scanning electron microscopy. *Science* **173**, 925.

Smithson Tennant. (1805). On the discovery of two new metals in crude platina. *Nicholson's J. Arts & Nat. Philos.* **10**, 24.

Snodgrass, M. J., and Peterson, R. G. (1969). An inexpensive microtome attachment for cutting 50-micron nonfrozen sections for electron microscopic histochemistry. *Stain Technol.* **44**, 151.

Soma, L. R., ed. (1971). "Textbook of Veterinary Anesthesia." Williams & Wilkins, Baltimore, Maryland.

Somlyo, A. P., Devine, C. E., Somlyo, A. V., and Rice, R. V. (1973). Filament organization in vertebrate smooth muscle. *Philos. Trans. R. Soc. London, Ser. B* **265**, 223.

Sommer, J. R., and Hasselbach, W. (1967). The effect of glutaraldehyde and formaldehyde on the calcium pump of the sarcoplasmic reticulum. *J. Cell Biol.* **34**, 902.

Sottocasa, G. L. (1967). An electron transport system associated with the outer membrane of liver mitochondria. A biochemical and morphological study. *J. Cell Biol.* **32**, 415.

Sperelakis, N. (1971). Ultrastructure of the neurogenic heart of *Limulus polyphemus. Z. Zellforsch. Mikrosk. Anat.* **116**, 443.

Sperry, W. M. (1963). Quantitative isolation of sterols. *J. Lipid Res.* **4**, 221.

Spielholz, N. I., and Van der Kloot, W. G. (1973). Localization and properties of the cholinesterase in crustacean muscle. *J. Cell Biol.* **59**, 407.

Staehelin, M. (1958). Reaction of tobacco mosaic virus nucelic acid with formaldehyde. *Biochim. Biophys. Acta* **29**, 410.

Stanley, W. L., Watters, G. G., Kelly, S. H., and Olson, A. C. (1978). Glucoamylase immobilized on chitin with glutaraldehyde. *Biotechnol. Bioeng.* **20**, 135.

Stefanini, M., De Martino, C., and Zamboni, L. (1967). Fixation of ejaculated spermatozoa for electron microscopy. *Nature (London)* **216**, 173.

Stein, O., and Stein, Y. (1967). Lipid synthesis, intracellular transport, storage, and secretion. I. Electron microscopic radioautographic study of liver after injection of tritiated palmitate or glycerol in fasted and ethanol-treated rats. *J. Cell Biol.* **33**, 319.

Stein, O., and Stein, Y. (1971). Light and electron microscopic radioautography of lipids: techniques and biological applications. *Adv. Lipid Res.* **9**, 1.

Stein, O., Stein, Y., Goodman, D., and Fidge, N. (1969). The metabolism of chylomicron cholesteryl ester in rat liver. A combined radioautographic-electron microscopic and biochemical study. *J. Cell Biol.* **43**, 410.

Stelzer, R., Läuchli, A., and Kramer, D. (1978). An improved precipitation technique for intracellular Cl⁻ localization in plant tissues by use of picric acid. *J. Exp. Bot.* **29**, 729.

Stensaas, S. S., Edwards, C. Q., and Stensaas, L. J. (1972). An experimental study of hyperchromic nerve cells in the cerebral cortex. *Exp. Neurol.* **36**, 472.

Steven, F. S., Grant, M. E., Ayad, S., Jackson, D. S., Weiss, J. B., and Leibovitch, S. J. (1971). The action of crude bacterial α-amylase on tropocollagen. A case of strictly limited proteolysis. *Biochim. Biophys. Acta* **236**, 309.

Stevens, C. L., Chay, T. R., and Loga, S. (1977). Rupture of base pairing in double-stranded poly(riboadenylic acid). Poly(ribouridylic acid) by formaldehyde: Medium chain lengths. *Biochemistry* **16**, 3727.

Stirling, C. A. (1978). A simple method for maintaining relative positions of separate tissue elements during processing for electron microscopy. *J. Microsc. (Oxford)* **114**, 107.

Stockert, J. C., and Colman, O. D. (1974). Observations on nucleolar staining with osmium tetroxide. *Experientia* **30**, 751.

Stoeckenius, W., and Mahr, S. C. (1965). Studies on the reaction of osmium tetroxide with lipids and related compounds. *Lab. Invest.* **12**, 458.

Stolinski, C., Breathnach, A. S., and Bellairs, R. (1978). Effect of fixation on cell membrane of early embryonic materials as observed on freeze-fracture replicas. *J. Microsc. (Oxford)* **112**, 293.

Stollar, D., and Grossman, L. (1962). The reaction of formaldehyde with denatured DNA: spectrophotometric immunologic and enzymic studies. *J. Mol. Biol.* **4**, 31.

Stonehill, A. A. (1966). Sporicidal compositions comprising a saturated dialdehyde and a cationic surfactant. U. S. Patent 3,282,775.

Stoner, C. D., and Sirak, H. D. (1969). Osmotically-induced alterations in volume and ultrastructure of mitochondria isolated from rat liver and bovine heart. *J. Cell Biol.* **43**, 521.

Strauss, E. W., and Arabian, A. A. (1969). Fixation of long-chain fatty acids in segments of jejunum from golden hamster. *J. Cell Biol.* **43**, 140a.

Sturgess, J. M., Mitranic, M. M., and Moscarello, M. A. (1978). Extraction of glycoproteins during tissue preparation for electron microscopy. *J. Microsc. (Oxford)* **114**, 101.

Subbaraman, L. R., Subbaraman, J., and Behrman, E. J. (1971). The reaction of osmium tetroxide-pyridine complexes with nucleic acid components. *Bioinorg. Chem.* **1**, 35.

Subbaraman, L. R., Subbaraman, J., and Behrman, E. J. (1972). Studies on the formation and hydrolysis of osmate(VI) esters. *Inorg. Chem.* **11**, 2621.

Subramanian, A. R. (1972). Glutaraldehyde fixation of ribosomes. Its use in the analysis of ribosome dissociation. *Biochemistry* **11**, 2710.

Sugai, N., and Ito, S. (1980). Carbonic anhydrase, ultrastructural localization in the mouse gastric mucosa and improvements in the technique. *J. Histochem. Cytochem.* **28**, 511.

Sumi, S. M. (1969). The extracellular space in the developing rat brain: Its variation with changes in osmolarity of the fixative, method of fixation and maturation. *J. Ultrastruct. Res.* **29**, 398.

Suskind, R. G. (1967). Effect of temperature on aldehyde fixation in the radioautographic localization of ribonucleoprotein in nucleoli of Hela cells. *J. Cell Biol.* **34**, 721.

Suzuki, T., Furusato, M. Takasaki, S., Shimizu, S., and Hataba, Y. (1977). Stereoscopic scanning electron microscopy of the red pulp of dog spleen with special reference to the terminal structure of the cordal capillaries. *Cell Tissue Res.* **182**, 441.

Svendsen, E., and Jorgensen, L. (1978). Focal "spontaneous" alterations and loss of endothelial cells in rabbit aorta. *Acta Pathol. Microbiol. Scand., Sect. A* **86**, 1.

Swift, E. (1967). Cleaning diatom frustules with ultraviolet radiation and peroxide. *Phycologia* **6**, 161.

Swift, J. A. (1966). The electron histochemical demonstration of sulfhydryl and disulfide in electron microscope sections, with particular reference to the presence of these chemical groups in the cell wall of the yeast *Pityrosporum* ovale. *Proc. Int. Congr. Electron Microsc., 6th, 1966* Vol. 2, p. 63.

Swinehart, P. A., Bentley, D. L., and Kardong, K. V. (1976). Scanning electron microscopic study of the effects of pressure on the luminal surfaces of the rabbit aorta. *Am. J. Anat.* **145**, 137.

Symmes, R. M., ed. (1936). (Section 18: Acids, alkalies, and other heavy chemicals). Recovering osmium and ruthenium. Standard Brands Inc. British Patent 449,251; *Chem. Abstr.* **30**, 8542.

Szamier, P. M., Pollard, T. D., and Fujiwara, K. (1975). Tropomyosin prevents the destruction of actin filaments by osmium tetroxide. *J. Cell Biol.* **67**, 424a.

Szmant, H. H. (1975). Physical properties of dimethyl sulfoxide and its function in biological systems. *Ann. N. Y. Acad. Sci.* **243**, 20.

Tahmisian, T. N. (1964). Using of the freezing point to adjust the tonicity of fixing solutions. *J. Ultrastruct. Res.* **10**, 182.

Tandler, C. J., and Solari, A. J. (1969). Nuclear orthophosphate ions. *J. Cell Biol.* **41**, 91.

Tarrant, P. J. V., Jenkins, N., Pearson, A. M., and Dutson, T. R. (1973). Proteolytic enzyme preparation for *Pseudomonas fragi*: Its action on pig muscle. *Appl. Microbiol.* **25**, 996.

Taylor, A. (1961). Attachment and spreading of cells in culture. *Exp. Cell Res.* **8**, 154.

Teichman, R. J., Fujimoto, M., and Yanagimachi, R. (1972). A previously unrecognized material in mammalian spermatozoa as revealed by malachite green and pyronine. *Biol. Reprod.* **7**, 73.

Teichman, R. J., Cummins, J. M., and Takei, G. H. (1974a). The characterization of a malachite green stainable glutaraldehyde extractable phospholipid in rabbit spermatozoa. *Biol. Reprod.* **10**, 565.

Teichman, R. J., Takei, G. H., and Cummins, J. M. (1974b). Detection of fatty acids, fatty aldehydes, phospholipids, glycolipids, and cholesterol on thin-layer chromatogram stained with malachite green. *J. Chromatogr.* **88**, 425.

Temmink, J. H. M., and Spiele, H. (1978). Preservation of cytoskeletal elements for electron microscopy. *Cell Biol. Int. Rep.* **2**, 51.

Ternynck, T., and Avrameas, S. (1972). Polyacrylamide-protein immunoadsorbents prepared with glutaraldehyde. *FEBS Lett.* **23**, 24.

Thiéry, G. (1971). Acid fixation in electron microscopy. *Proc. Int. Congr. Electron Microsc., 7th, 1970* Vol. 1, p. 393.

Thompson, S. W., and Luna, L. G. (1978). "An Atlas of Artifacts." Charles C. Thomas, Springfield, Illinois.

Thornthwaite, J. T., Thomas, R. A., Leif, S. B., Yopp, T. A., Cameron, B. F., and Leif, R. C. (1978). The use of electronic cell volume analysis with the AMAC II to determine the optimum glutaraldehyde fixative concentration for nucleated mammalian cells. *Proc. SEM* **2**, 1123.

Thornton, V. F., and Howe, C. (1974). The effect of change of background color on the ultrastructure of the pars intermedia of the pituitary of the eel (*Anguilla anguilla*). *Cell Tissue Res.* **151**, 103.

Thureson-Klein, A., Klein, R. L., and Chen-Yen, S.-H. (1975). Morphological effects of osmolarity on purified noradrenergic vesicles. *J. Neurocytol.* **4**, 609.

Thyberg, J., and Friberg, U. (1970). Ultrastructure and acid phosphatase activity of matrix vesicles and cytoplasmic dense bodies in the epiphyseal plate. *J. Ultrastruct. Res.* **33**, 554.

Tice, L. W., and Engel, A. G. (1966). Cytochemistry of phosphatases in the endoplasmic reticulum. II. *In situ* localization of the Mg-dependent enzyme. *J. Cell Biol.* **31**, 489.

Tiedmann, K., and Wettstein, R. (1980). The mature mesonephric nephron of the rabbit embryo. I. SEM studies. *Cell Tissue Res.* **209**, 95.

Tilney, L. G. (1975). Actin filaments in the acrosomal reaction of *Limulus* sperm. Motion generated by alterations in the packing of the filaments. *J. Cell Biol.* **64,** 289.

Tilney, L. G., Hatano, S., Ishikawa, H., and Mooseker, M. S. (1973a). The polymerization of actin: Its role in the generation of the acrosomal process of certain Echinoderm sperm. *J. Cell Biol.* **59,** 109.

Tilney, L. G., Bryan, J., Bush, D., Fujiwara, K., Mooseker, M., Murphy, D., and Snyder, D. (1973b). Microtubules: Evidence for 13 protofilaments. *J. Cell Biol.* **59,** 267.

Tippit, D. H., and Pickett-Heaps, J. D. (1977). Cell division in the pennate diatom *Surirella ovalis*. *J. Cell Biol.* **73,** 705.

Tisdale, A. D., and Nakajima, Y. (1976). Fine structure of synaptic vesicles in two types of nerve terminals in crayfish stretch receptor organs: Influence of fixation methods. *J. Comp. Neurol.* **165,** 369.

Tokin, I. B., and Röhlich, P. (1965). Submicroscopic analysis of the genesis of the yolk plates in oocytes of *Rana temporaria. Arch. Anat. Histol. Embryol.* **48,** 106.

Tooze, J. (1964). Measurement of some cellular changes during fixation of amphibian erythrocytes with osmium tetroxide solutions. *J. Cell Biol.* **22,** 551.

Torack, R. M. (1965). The extracellular space of rat membrane following perfusion fixation with glutaraldehyde and hydroxyadipaldehyde. *Z. Zellforsch. Mikrosk. Anat.* **66,** 352.

Torack, R. M. (1966). The penetration of thorotrast into brain spaces following osmium, glutaraldehyde, and hydroxyadipaldehyde fixation. *J. Ultrastruct. Res.* **14,** 590.

Tormey, H. McD. (1963). Fine structure of the ciliary epithelium of the rabbit, with particular reference to "infolded membranes," "vesicles," and the effects of Diamox. *J. Cell Biol.* **17,** 641.

Tormey, J. M. (1964). Differences in membrane configuration between osmium tetroxide-fixed and glutaraldehyde-fixed ciliary epithelium. *J. Cell Biol.* **23,** 658.

Tormey, J. M. (1965). Artifactual localization of ferritin in the ciliary epithelium *in vitro. J. Cell Biol.* **25,** 1.

Traganos, F., Darzynkiewicz, Z., Sharpless, T., and Melamed, M. R. (1975). Denaturation of deoxyribonucleic acid in *in situ* effect of formaldehyde. *J. Histochem. Cytochem.* **23,** 431.

Tranzer, J.-P., and Richards, J. G. (1976). Ultrastructural cytochemistry of biogenic amines in nervous tissue: methodologic improvements. *J. Histochem. Cytochem.* **24,** 1178.

Tranzer, J.-P., and Snipes, R. L. (1968). Fine structural localization of noradrenalin in sympathetic nerve terminals: A critical study on the influence of fixation. *Proc. Eur. Reg. Conf. Electron Microsc., 4th, 1968,* p. 519.

Tranzer, J.-P., da Prada, M., and Pletscher, A. (1972). Storage of 5-hydroxytryptamine in megakaryocytes. *J. Cell Biol.* **52,** 191.

Treffry, T. (1969). Glutaraldehyde fixation and protochlorophyll transformations in etiolated peas. *Planta* **85,** 376.

Trelease, R. N. (1975). Malate synthase. *In* "Electron Microscopy of Enzymes: Principles and Methods" (M. A. Hayat, ed.), Van Nostrand-Reinhold, New York.

Trelstad, R. L. (1969). The effect of pH on the stability of purified glutaraldehyde. *J. Histochem. Cytochem.* **17,** 756.

Trnawska, Z., Sit'aj, S., Grmela, M., and Malinsky, J. (1966). Certain intermediary metabolites and the formation of fibrils from collagen solutions. *Biochim. Biophys. Acta* **126,** 373.

Trump, B. F., and Ericsson, J. L. E. (1965). The effect of the fixative solution on the ultrastructure of cells and tissues. A comparative analysis with particular attention to the proximal convoluted tubule of the rat kidney. *Lab. Invest.* **14,** 1245.

Trump, B. F., and Jones, R. T., eds. (1978). "Diagnostic Electron Microscopy," Vol. 1. Wiley, New York.

Trump, B. F., Goldblatt, P. J., and Stowell, R. E. (1962). An electron microscope study of early cytoplasmic alterations in hepatic parenchymal cells of mouse liver during necrosis *in vitro*. *Lab. Invest.* **11**, 986.

Trusal, L. R., Baker, C. J., and Guzman, A. W. (1979). Transmission and scanning electron microscopy of cell monolayers grown on polymethylpentene coverslips. *Stain Technol.* **54**, 77.

Tsou, K. C. (1975). 5-Nucleotide phosphodiesterase. *In* "Electron Microscopy of Enzymes: Principles and Methods" (M. A. Hayat, ed.), Vol. 4, Van Nostrand-Reinhold, New York.

Tsou, K. C., Hendricks, J., and Gupta, P. D. (1974). A new indigogenic method for the light and electron microscopic demonstration of 5-nucleotide phosphodiesterase. *Histochem. J.* **6**, 327.

Turler, S., and Frei, U. (1969). An apparatus for fixation and continuous dehydration of specimens for paraffin sections. *J. Microsc. (Oxford)* **90**, 79.

Ueki, T., Zalkin, A., and Templeton, D. H. (1965). The crystal structure of osmium tetroxide. *Acta Crystallogr.* **19**, 157.

Ulmer, K. M., and Honjo, S. (1973). Quantitative evaluation of fixation and dehydration methods for scanning electron microscopic preparation of soft sea water organisms. *Proc. SEM.*, p. 365.

Umeda, A., and Amako, K. (1975). Fixation procedure for bacterial cells in SEM. *J. Electron Microsc.* **24**, 210.

Umrath, W. (1974). Cooling bath for rapid freezing in electron microscopy. *J. Microsc. (Oxford)* **101**, 103.

Unger, E., and Buchwalow, I. B. (1975). NAD-pyrophosphorylase. "Electron Microscopy of Enzymes: Principles and Methods" (M. A. Hayat, ed.), Vol. 4, Van Nostrand-Reinhold, New York.

Utsumi, K., and Packer, L. (1967). Glutaraldehyde-fixed mitochondria. I. Enzyme activity, ion translocation, and conformational changes. *Arch. Biochem. Biophys.* **121**, 633.

Valdivia, O. (1971). Methods of fixation and the morphology of synaptic vesicles. *J. Comp. Neurol.* **142**, 257.

van Deenen, L. L. M. (1966). Some structural and dynamic aspects of lipids in biological membranes. *Ann. N. Y. Acad. Sci.* **137**, 717.

van Deurs, B. (1975). The use of a tannic acid-glutaraldehyde fixative to visualize gap and tight junctions. *J. Ultrastruct. Res.* **50**, 185.

van Deurs, B., and Luft, J. H. (1979). Effects of glutaraldehyde fixation on the structure of tight junctions. *J. Ultrastruct. Res.* **68**, 160.

van Duijn, P. (1961). Acrolein-Schiff, a new staining method for proteins. *J. Histochem. Cytochem.* **9**, 234.

Van Harreveld, A. (1972). The extracellular space in the vertebrate central nervous system. *In* "The Structure and Function of Nervous Tissue" (G. H. Bourne, ed.), Vol. 4, p. 449. Academic Press, New York.

Van Harreveld, A., and Fifkova, E. (1972). Release of glutamate from the retina during glutaraldehyde fixation. *J. Neurochem.* **19**, 237.

Van Harreveld, A., and Khattab, F. I. (1968). Perfusion fixation with glutaraldehyde and postfixation with osmium tetroxide for electron microscopy. *J. Cell Sci.* **3**, 579.

Van Harreveld, A., Crowell, J., and Malhotra, S. K. (1965). A study of extracellular space in central nervous tissue by freeze-substitution. *J. Cell Biol.* **25**, 117.

Van Leeuwen, F. W. (1977). Immunoelectron microscopic visualization of neurohypophyseal hormones. Evaluation of some tissue preparations and staining procedures. *J. Histochem. Cytochem.* **25**, 1213.

van Winkle, J. L. (1962). Cited in "Acrolein" (C. W. Smith, ed.), Wiley, New York.

Vassar, P. S., Hards, J. M., Brooks, D. E., Hagenberger, B., and Seaman, G. V. F. (1972). Physicochemical effects of aldehydes on the human erythrocyte. *J. Cell Biol.* **53**, 809.

Veerman, A. J. P., Hoefsmit, E. Ch. M., and Boeré, H. (1974). Perfusion fixation using a cushioning chamber coupled to a peristaltic pump. *Stain Technol.* **49**, 111.

Ventilla, M., and Brown, W. J. (1976). The effect of barbiturates on microtubular assembly. *J. Ultrastruct. Res.* **54**, 325.

Vermeer, B. J., Van Gent, C. M., De Bruijn, W. C., and Boonders, T. (1978a). The effect of digitonin-containing fixatives on the retention of free cholesterol and cholesterol esters. *Histochem. J.* **10**, 287.

Vermeer, B. J., de Bruijn, W. C., Van Gent, C. M., and De Winter, C.P.M. (1978b). Ultrastructural findings on lipoproteins in vitro and in xanthomatous tissue. *Histochem. J.* **10**, 299.

Vial, J. D., and Porter, K. R. (1974). The surface topography of cells isolated from tissues by maceration. *Anat. Rec.* **178**, 502.

Vitale, R. D., Fawcett, D. W., and Dym, M. (1973). The normal development of the blood-testis barrier and the effects of clomiphene and estrogen treatment. *Anat. Rec.* **176**, 333.

Vocel, S. V., Slepneva, I. A., and Backer, J. M. (1975). Influence of Mn^{2+} ion coordination on tRNA macrostructure and determination of some coordination sites of Mn^{2+} in tRNA. *Biopolymers* **14**, 2445.

Vodovar, N., and Desnoyers, F. (1975). Influence du mode de fixation sur la morphologie ultrastructurale des cellules normales et altérées du myocarde. *J. Microsc. (Paris)* **24**, 239.

Voelz, H., and Ortigoza, R. O. (168). Cytochemistry of phosphatases in *Myxococcus xanthus*. *J. Bacteriol.* **96**, 1337.

Vologodskii, A. V., and Frank-Kamenetskii, M. D. (1975). Theoretical study of DNA unwinding under the action of formaldehyde. *J. Theor. Biol.* **55**, 163.

von Barthlott, W., Ehler, N., and Schill, R. (1976). Abtragung Biologischer Oberflächen durch hochfrequenzaktivierten Sauerstoff für die Raster-Elektronenmikroskopie. *Mikroskopie* **32**, 35.

von Feuerstein, H., and Geyer, G. (1971). Zur Entstehung elektronendichter Produkte bei der Chromaffinen Reaktion. *Acta Histochem.* **40**, 73.

von Feustel, E.-M., and Geyer, G. (1966). Zur eignung der acrolein fixierung für histochemische Untersuchungen. II. Lipide, enzyme. *Acta Histochem.* **25**, 219.

von Hesse, G. (1973). Die chemisch-analytische Charakterisierung von Glutaraldehyd in Fixierungslösungen. *Acta Histochem.* **46**, 253.

von Hippel, P. H., and Wong, K. Y. (1964). Neutral salts: The generality of their effects on the stability of macromolecular conformations. *Science* **145**, 577.

von Hippel, P. H., and Wong, K. Y. (1971). Dynamic aspects of native DNA structure: Kinetics of the formaldehyde reaction with calf thymus DNA. *J. Mol. Biol.* **61**, 587.

von Matt, C. A., Fuenfschilling, H., Moppert, J. M., and Gander, E. S. (1971). Comparative determination of liver acid phosphatase activity in decapitated and perfused rats. *Histochemie* **25**, 72.

Vonnahue, F. J. (1980). An improved method for transparenchymal fixation of human liver biopsies for scanning electron microscopy., *Proc. SEM.*, p. 177.

Vorbrodt, A., and Bernhard, W. (1968). Essais de localisation au microscope électronique de l'activité phosphatasique nucléaire dans des coupes à congélation ultrafines. *J. Microsc. (Paris)* **7**, 195.

Voronina, A. S., Bogatyreva, S. A., Rodionova, A. I., and Glinka, A. V. (1977). Fixation of ribonucleoproteins with glutaraldehyde and glyoxal for their analysis by equilibrium centrifugation in a CsCl density gradient. *Biochemistry* **42**, 1243.

Wacker, P., and Forssmann, W. G. (1972). Immersion and perfusion fixed rat adrenal medulla: The problem of mixed cells, clear cells, and the mode of secretion. *Z. Zellforsch. Mikrosk. Anat.* **126**, 261.

Waddell, W. J., and Bates, R. G. (1969). Intracellular pH. *Physiol. Rev.* **49**, 285.

Wagner, R. C. (1976). Tannic acid as a mordant for heavy metal stains. *Proc. 34th Annu. Meet., Electron Microsc. Soc. Am.,* p. 316.

Wakabayashi, T. (1972). Ultrastructure and functional states of mitochondria- effect of fixatives on the stabilization of unstable configuration. *Nagoya J. Med. Sci.* **35,** 1.

Wakabayashi, T., Hatase, O., Allmann, D. W., Smoly, J. M., and Green, D. E. (1970). On the stabilization by fixation of configurational state in beef heart mitochondria. *Bioenergetics* **1,** 527.

Walker, D. G., and Seligman, A. M. (1963). The use of formalin fixation in the cytochemical demonstration of succinic and DPN-and TPN-dependent dehydrogenases in mitochondria. *J. Cell Biol.* **16,** 455.

Walker, J. F. (1964). "Formaldehyde," 3rd ed. Van Nostrand-Reinhold, New York.

Wallach, D. F. H. (1975). Phospholipids. *In* "Membrane Molecular Biology of Neoplastic Cells" (D. F. H. Wallach, ed.), p. 141. Elsevier, Amsterdam.

Wallington, E. A. (1979). Artifacts in tissue sections. *Med. Lab. Sci.* **36,** 3.

Wang, J. H., and Tu, J. (1969). Modification of glycogen phosphorylase b by glutaraldehyde. Preparation and isolation of enzyme derivatives with enhanced stability. *Biochemistry* **8,** 4403.

Warheit, D. B., Salley, S. O., and Barnhart, M. I. (1981). Comparisons of routes of lung fixation on in situ pulmonary alveolar macrophage populations: A morphometric study. *Micron* **12,** 81.

Warmke, H. E., and Edwardson, J. R. (1966). Use of potassium permanganate as a fixative for virus particles in plant tissues. *Virology* **28,** 693.

Waterman, R. E. (1972). Use of the scanning electron microscope for observation of vertebrate embryos. *Dev. Biol.* **27,** 276.

Waterman, R. E. (1974). Embryonic and fetal tissues of vertebrates. *In* "Principles and Techniques of Scanning Electron Microscopy: Biological Applications" (M. A. Hayat, ed.), Vol. 2, Van Nostrand-Reinhold, New York.

Weakley, B. S. (1977). How dangerous is sodium cacodylate? *J. Microsc. (Oxford)* **109,** 249.

Weibel, E. R., and Knight, B. W. (1964). A morphometric study on the thickness of the pulmonary air-blood barrier. *J. Cell Biol.* **21,** 367.

Weiner, L. M., Backer, J. M., and Rezvukhin, A. I. (1975). Participation of manganese ions complexed with RNA in the interaction with amino acids and peptides. *Biochim. Biophys. Acta* **383,** 316.

Weisenberg, R. C. (1972). Microtubule formation *in vitro* in solutions containing low calcium concentrations. *Science* **177,** 1104.

Weissenfels, N. (1960). Präparatveränderungen während der Fixierung. *Proc. Int. Congr. Electron Microsc., 4th, 1958* Vol. 2, p. 60.

Wersäll, J., Kimura, R., and Lundqvist, P.-G. (1965). Early postmortem changes in the organ of Corti (guinea pig). *Z. Zellforsch. Mikrosk. Anat.* **65,** 220.

West, E. S., Todd, W. R., Mason, H. S., and Van Bruggan, J. T. (1966). "Textbook of Biochemistry." Macmillan, New York.

Westbrook, E., Wetzel, B., Cannon, G. B., and Berard, D. (1975). The impact of culture conditions on the surface morphology of cells *in vitro. Proc. SEM.,* p. 351.

Westfall, J. A., and Enos, P. D. (1972). Scanning and transmission electron microscopy of isolated cells of *Hydra littoralis. Proc. 30th Annu. Meet., Electron Microsc. Soc. Am.,* p. 160.

Westrum, L. E., and Broderson, S. H. (1976). Acetylcholinesterase activity of synaptic structures in the spinal trigeminal nucleus. *J. Neurocytol.* **5,** 551.

Westrum, L. E., and Gray, E. G. (1977). Microtubules associated with postsynaptic thickenings. *J. Neurocytol.* **6,** 505.

Wetzel, B. K. (1961). Sodium permanganate fixation for electron microscopy. *J. Biophys. Biochem. Cytol.* **9,** 711.

Wetzel, B. K., Spicer, S. S., Dvorak, H. F., and Heppel, L. A. (1970). Cytochemical localization of certain phosphatases in *Echerichia coli. J. Bacteriol.* **104**, 529.

Wetzel, B. K., Cannon, G. B., Alexander, E. L., Erickson, B. W., and Westbrook, E. W. (1974). A critical approach to the scanning electron microscopy of cells in suspension. *Proc. SEM.,* p. 581.

Wheeler, E. E., Gavin, J. B., and Seelye, R. N. (1975). Freeze-drying from tertiary butanol in the preparation of endocardium for scanning electron microscopy. *Stain Technol.* **50**, 331.

Whipple, E. B., and Ruta, M. (1974). Structure of aqueous glutaraldehyde. *J. Org. Chem.* **39**, 1666.

White, D. L., Andrews, S. B., Faller, J. W., and Barrnett, R. J. (1976). The chemical nature of osmium tetroxide fixation and staining of membranes by x-ray photoelectron spectroscopy. *Biochim. Biophys. Acta* **436**, 577.

White, D. L., Mazurkiewicz, J. E., and Barrnett, R. J. (1980). Authors' reply to Dr. Behrman. *J. Histochem. Cytochem.* **28**, 285.

White, J. G., and Krivit, W. (1967). An ultrastructural basis for the shape changes induced in platelets by chilling. *Blood* **30**, 625.

Whiting, R. F., and Ottensmeyer, F. P. (1972). Heavy atoms in model compounds and nucleic acids imaged by dark field transmission electron microscopy. *J. Mol. Biol.* **67**, 173.

Widnell, C. C. (1972). Cytochemical localization of 5-nucleotidase in subcellular fractions from liver. I. The origin of 5-nucleotidase in microsomes. *J. Cell Biol.* **52**, 542.

Wiesehahn, G., and Hearst, J. E. (1978). DNA unwinding induced by photoaddition of psoralen derivatives and determination of dark-binding equilibrium constants by gel electrophoresis. *Proc. Natl. Acad. Sci. U.S.A.* **75**, 2703.

Wiesehahn, G. P., Hyde, J. E., and Hearst, J. E. (1977). Photoaddition of trimethylpsoralen to *Drosophila melanogaster* nuclei: A probe for chromatin structure. *Biochemistry* **16**, 925.

Wigglesworth, V. B. (1957). Use of osmium in the fixation and staining of tissues. *Proc. R. Soc. Londen, Sec. B* **147**, 185.

Wigglesworth, V. B. (1964). The union of protein and nucleic acid in the living cell and its demonstration by osmium staining. *Q. J. Microsc. Sci.* [N.S.] **105**, 113.

Wilkoff, L. J., Dixon, G. J., Westbrook, L., and Happich, W. F. (1971). Potentially infectious agents associated with shearling bedpads. Part II. *J. Appl. Microbiol.* **21**, 647.

Williams, A. E., Jordan, J. A., Murphy, J. F., and Allen, J. M. (1973). The surface ultrastructure of normal and abnormal cervical epithelia. *Proc. SEM.,* p. 597.

Williams, C. H., Vail, W. J., Harris, R. A., Caldwell, M., Green, D. E., and Valdivia, E. (1970). Conformational basis of energy transduction in membrane systems. VII. Configurational changes of mitochondria *in situ* and *in vitro. Bioenergetics* **1**, 47.

Williams, R.J.P. (1975). The binding of metal ions to membranes and its consequences. *In* "Biological Membranes: Twelve Essays on Their Organization, Properties, and Functions" (D. S. Parsons, ed.), pp. 106–121. Oxford Univ. Press (Clarendon), London and New York.

Williamson, J. R. (1969). Ultrastructural localization and distribution of free cholesterol (3B-hydroxysterols) in tissues. *J. Ultrastruct. Res.* **27**, 118.

Willingham, M. C., and Yamada, S. S. (1979). Development of a new primary fixative for electron microscopic immunocytochemical localization of intracellular antigens in cultured cells. *J. Histochem. Cytochem.* **27**, 947.

Willison, J. H. M., and Rajaraman, R. (1977). 'Large' and 'small' nuclear pore complexes: the influence of glutaraldehyde. *J. Microsc. (Oxford)* **109**, 183.

Winborn, W. B., and Seelig, L. L. (1970). Paraformaldehyde and *s*-collidine—a fixative for preserving large blocks of tissue for electron microscopy. *Tex. Rep. Biol. Med.* **28**, 347.

Wincek, T. J., and Sweat, F. W. (1976). Effects of prostaglandins and catecholamines on rat spleen adenylate cyclase *in vitro. Biochim. Biophys. Acta* **437**, 571.

Windaus, A. (1910). Uber die Quantitative Bestimmung des Cholesterins und der Cholesterinester in

Einigen Normalen und Pathologischen Niern (Quantitative determination of free esterified cholesterol in some normal and pathological kidneys). *Hoppe-Seyler's Z. Physiol. Chem.* **65,** 110.

Wise, G. E., and Flickinger, C. J. (1971). Patterns of cytochemical staining in Golgi apparatus of amoebae following enucleation. *Exp. Cell Res.* **67,** 323.

Witkop, B. (1961). Nonenzymatic methods for the preferential and selective cleavage and modification of proteins. *Adv. Protein Chem.* **16,** 221.

Wohlfarth-Bottermann, K. E. (1957). Die Kontrastierung tierischer Zellen und Gewebe in Rahmen ihrer elektronenmikroskopischen Untersuchung an ultradünnen Schnitten. *Naturwissenschaften* **44,** 287.

Wold, F. (1972). Bifunctional reagents. *In* "Methods in Enzymology" (C. H. W. Hirs and N. Timasheff, eds.), Vol. 25, Part B, p. 623. Academic Press, New York.

Wolfe, S. L., and Martin, P. G. (1968). The ultrastructure and strandedness of chromosomes from two species of *Vicia*. *Exp. Cell Res.* **50,** 140.

Wolff, K. L., Trent, D. W., Karabatsos, N., and Hudson, B. W. (1977). Use of glutaraldehyde-fixed goose erythrocytes in arbovirus serology. *J. Clin. Microbiol.* **6,** 55.

Wolinsky, H. (1972). Endothelial projections. *Science* **176,** 1151.

Wollenzien, P. L., Youvan, D. C., and Hearst, J. E. (1978). Structure of psoralen-crosslinked ribosomal RNA from *Drosophila melanogaster*. *Proc. Natl. Acad. Sci. U.S.A.* **75,** 1642.

Wollman, S. R. (1972). The aortic smooth muscle cell as a possible steroidsecreting cell. *J. Cell Biol.* **55,** 272a.

Wolman, M. (1955). Problems of fixation in cytology, histology, and histochemistry. *Int. Rev. Cytol.* **4,** 79.

Wolman, M. (1957). Histochemical study of changes occuring during the degeneration of myelin. *J. Neurochem.* **1,** 370.

Wolman, M., and Greco, J. (1952). The effect of formaldehyde on tissue lipids and on histochemical reactions for carbonyl groups. *Stain Technol.* **29,** 317.

Wood, D. A. W., and Tristram, H. (1970). Localization of the cell and extraction of alkaline phosphatase from *Bacillus subtilis*. *J. Bacteriol.* **104,** 1045.

Wood, J. G. (1966). Electron microscopic localization of amines in central nervous tissue. *Nature (London)* **209,** 1131.

Wood, J. G. (1973). The effects of glutaraldehyde and osmium on the proteins and lipids of myelin and mitochondria. *Biochim. Biophys. Acta* **329,** 118.

Wood, J. G. (1975). Use of the analytical electron microscope (AEM) in cytochemical studies of the central nervous system. *Histochemistry* **41,** 233.

Wood, J. G. (1976). X-ray analysis of cytochemical reaction products in synaptic areas. *Proc. 34th Annu. Meet., Electron Microsc. Soc. Am.,* p. 6.

Wood, J. G. (1977). Cytochemical studies of biogenic amines. *Proc. 35th Annu. Meet., Electron Microsc. Soc. Am.,* p. 374.

Wood, J. G., and Barrnett, R. J. (1964). Histochemical demonstration of norepinephrine at a fine structural level. *J. Histochem. Cytochem.* **12,** 197.

Wood, L. R., and Dubois, A. (1981). Studies of gastric cell surface epithelium: an evaluation of various methods of preparation. *Virchows Arch. Cell Path.* **35,** 207.

Wood, R. L., and Luft, J. H. (1963). The influence of the buffer system on fixation with osmium tetroxide. *J. Cell Biol.* **19,** 83A.

Wood, R. L., and Luft, J. H. (1965). The influence of the buffer system on fixation with osmium tetroxide. *J. Ultrastruct. Res.* **12,** 22.

Woodward, L. A., and Roberts, H. L. (1956). The Raman and infra-red absorption spectra of osmium tetroxide. *Trans. Faraday Soc.* **52,** 615.

World Health Organization (1970). Passive hemaagglutination test. *W.H.O., Tech. Rep. Ser.* **447**, 23.

Wrench, C. P. (1970). Polymerized glutaraldehyde in the fixation of rat lung tissue. *Proc. 28th Annu. Meet., Electron Microsc. Soc. Am.*, p. 300.

Wrigglesworth, J. M., and Packer, L. (1969). pH-dependent conformational changes in submitochondrial particles. *Arch. Biochem. Biophys.* **133**, 194.

Yamada, A., Akasaka, K., and Hatano, H. (1976). Proton and phosphorus-31 magnetic relaxation studies on the interaction of polyriboadenylic acid with Mn^{2+}. *Biopolymers* **15**, 1315.

Yamamoto, N., and Yasuda, K. (1977). Use of a water-soluble carbodiimide as a fixing agent. *Acta Histochem. Cytochem.* **10**, 14.

Yanoff, M. (1973). Formaldehyde-glutaraldehyde fixation. *Am. J. Ophthalmol.* **76**, 303.

Yazawa, M., and Morita, F. (1974). Electron spin resonance study of the interaction between heavy meromyosin and Mn^{2+}. *J. Biochem. (Tokyo)* **76**, 217.

Yip, D. K., and Auersperg, N. (1972). The dye-exclusion test for cell viability: Persistence of differential staining following fixation. *In Vitro* **7**, 323.

Yokota, K., Suzuki, Y., and Ishii, Y. (1965). Temperature dependence of the polymerization modes of glutaraldehyde. *Chem. Abstr.* **65**, 13835.

Yokota, S., and Fahimi, H. D. (1978). The peroxisime (microbody) membrane: Effects of detergents and lipid solvents on its ultrastructure and permeability to catalase. *Histochem. J.* **10**, 469.

Yoo, B. Y., Oreland, L., and Persson, A. (1974). Effects of formaldehyde and glutaraldehyde fixation on the monoamine oxidase activity in isolated rat liver mitochondria. *J. Histochem. Cytochem.* **22**, 445.

Young, E. G. (1963). "Comprehensive Biochemistry," Vol. 7, p. 25. Elsevier, New York.

Yun, J., and Kenney, R. A. (1976). Preparation of cat kidney tissue for ultrastructural studies. *J. Electron Microsc.* **25**, 11.

Zamboni, L., and De Martino, C. (1967). Buffered picric acid-formaldehyde: A new, rapid fixative for electron microscopy. *J. Cell Biol.* **35**, 148a.

Zeikus, J. A., and Aldrich, H. C. (1975). Use of hot formaldehyde fixative in processing plant-parasite nematodes for electron microscopy. *Stain Technol.* **50**, 219.

Zetterqvist, H. (1956). The ultrastructural organization of the columnar absorbing cells of the mouse jejunum. Thesis, Stockholm.

Ziegler, K., and Liesenfeld, I. (1976). Studies on glutaraldehyde treated wool. *Proc. Int. Wool Text. Res. Conf., 5th, 1975* Vol. III, p. 88.

Zimmerman, L. E., Font, R. L., and Ts'o, M. O. M. (1972). Application of electron microscopy to histopathologic diagnosis. *Trans.—Am. Acad. Ophthalmol. Otolaryngol.* **76**, 101.

Zotikov, L., and Bernhard, W. (1970). Localisation au microscope électronique de l'activité de certaines nuclease dans des coupes à congélation ultrafines. *J. Ulstrastruct. Res.* **30**, 642.

Zoto, G. A. (1973). A rapid method for cleaning diatoms for taxonomic and ecological studies. *Phycologia* **12**, 69.

Index

A

Abdominal aorta, 253, 255
rat, shrinkage, 337
Acetaldehyde, osmolality, 37
Acetone, 4
N-Acetyl cysteine, as mucolytic agent, 331
N-Acetyl-β-glucosaminidase, inhibition by
glutaraldehyde, 345, 347
Acid mucopolysaccharides, 120
Acid phosphatase, 102
cytochemical study of, 344
demonstration, 102
effect of lysosome labilizer on, 359
inhibition by glutaraldehyde, 345, 347
localization, aided by DMSO, 375
preservation, 359
quantitation, 354
in subcellular fractions, 360-361, 362-363
Acid phosphomonoesterase, localization, aided
by DMSO, 375
Acid ribonuclease, 307
Acrolein, 64, 137-144
chemical characteristics, 137-138
formaldehyde mixture, 133

free aldehyde from, 289
glutaraldehyde mixture, 142-143
with DMSO, 144
with formaldehyde, 143
with paraformaldehyde, 143, 144
with potassium permanganate, 144
in glutaraldehyde synthesis, 66
handling precautions, 141-142
hazards, 142
as noncoagulant, 4
osmolarity, 140
polymerization, 141
preparation, 140
protein functional groups available to, 140
reaction
with lipids, 140-141
with proteins, 140
used for scanning electron microscopy, 339
vapor fixation, 139
waste, disposal, 142
Actin
effect of fixation on, 262-264
filaments, in living cells, 277
proteolysis, 309
Actinophrys, 202

473

Adenine, reaction, with osmium tetroxide, 160
Adenosine triphosphatase
  carbodiimide-fixed, 198
  inhibition, 357
  localization, 348
  magnesium-activated
    inhibition by glutaraldehyde, 345
    in intact sections vs. homogenates, 359
    in subcellular fractions, 360-361
  membrane binding, with glutaraldehyde, 351,
    356
  potassium-activated
    inhibition by glutaraldehyde, 345
    in subcellular fractions, 360-361
  preservation, in microsomal fractions, 358
  sodium-activated
    in subcellular fractions, 360-361
    inhibition by glutaraldehyde, 345
  suppression, 92
Adenosine triphosphate, 146
Adipocellulose, solubility, 6
Adipocytes, isotonicity, 25
ADPase, 357
Adrenal cortex, fixation artifact in, 384
Adrenal medulla
  fixation method for, 192
  rat, monoamines in, 185-186
Adrenalin
  identification, 115
  reaction, with potassium permanganate, 185
Adrenalin-storing cells, differentiated from
    noradrenaline-storing cells, 71
Aedes aegypti, 22
Agkistrodon piscivorus, 277
Air drying, artifacts, 385
Alanine
  binding, 101
  increase, with aldehyde fixation, 72, 146
Alanine aminotransferase, inhibition, by
    glutaraldehyde, 345
Albumin
  bovine serum, 68
    reaction with osmium tetroxide, 157
  egg, 3
    solubility, 5
  human, reaction with osmium tetroxide, 157
  liver, artifactual passive movement, 384
Alcian blue
  mixture with glutaraldehyde, 106-107
  staining, differential, after glutaraldehyde
    fixation, 71

Alcohol dehydrogenase, 198
Aldehydes, 64-147, *see also* specific aldehyde
  in calcium binding, 31
  compared, for use in enzyme cytochemistry,
    346-348
  concentration, in enzyme cytochemistry, 351
  effect
    on cell volume, 301
    on enzymes, 348-350
    on membrane fracture, 272-273
    on synaptic vesicles, 292-293
    on tissue physiology, 90-92
  electrolytes added to, 33
  in enzyme cytochemistry, 343, 346-350
  fixation, and simultaneous chromation, 115
  osmolalities, in cacodylate buffer, 37
  prefixation, 189
    effect on cell volume, 304
    effect on enzymes, 355-357
  in quantitative electron microscopy, 145-146
  reaction
    with osmium tetroxide, 205
    with proteins, 65-66
  in scanning electron microscopy, 338-341
  shrinkage caused by, 98-99
Algae, 139, 144, 205
  green, fixation method for, 204, 389-390
  red
    fixation method for, 390, 391
    membrane stabilization, 31
Alkaline phosphatase, 343
  inhibition, 357
  localization, 356
N-Alkyl-2,6-dihydroxypiperidine, 84
Allomyces neomoniliformis, spindle mi-
    crotubules, 276
Ambystoma, eggs, fixation, 139
Amines
  localization, 111, 115, 118
  preservation, 146
Amino acids, *see also* specific amino acid
  binding, by glutaraldehyde, 101
  quantitation, 146
  reaction
    with glutaraldehyde, 80
    with osmium tetroxide, 158
  solubility, effect of electrolytes on, 29
6-Amino hexanoic acid, 84
Amino transferases, inhibition, by glutaral-
    dehyde, 347
Aminophenols, identification, 115

Ammonium sulfate, stabilizer of protein, 26
Amoeba
    actin filaments, preservation, 264
    alcian blue exposure, 106
    enzyme localization in, 348
    fixation, 189
        with glutaraldehyde, 87
        method for, 390
    lipid loss, 167
    phospholipid loss, 4
    preservation of fine structure, 124
Amphibian, body fluid composition, 22
Amphioxus, total osmolality, 35
Amyl acetate, shrinkage caused by, 338
α-Amylase, specimen cleaning with, 329
Anabilysine, 85
Anesthesia, 210, 260-261
    for motile organisms, 332
Anesthetics, effects, 210-211
    on synaptic morphology, 293
Anhydrous fixation, 207-208
Animal tissue
    dense, preservation, 143
    fixation, 205
    fixative penetration, 97-98
    glutaraldehyde fixation
        optimal concentration, 95, 341
        optimal temperature, 94
        pH, 96
    pH of cells, 56
    phenol-containing regions, fixation, 164-166
    swelling, during fixation, 170-171
Animals, freshwater
    anesthetic for, 332
    body fluid composition, 21-22
Animals, marine
    anesthetic for, 332
    body fluid
        composition, 21-22
        pH, 57
    fixative for, 341
Annelid
    freshwater, body fluid composition, 22
    marine, body fluid composition, 21
    pH of body fluid, 57
Anodonta cygnea, 22
Anticoagulants, 211, 212
Antigenicity, effect of fixation on, 364-370
Antigens
    isolation, 69
    localization, procedure for, 367-370

Aorta, *see also* Abdominal aorta
    protein loss, 169
    rabbit, perfusion pressure for, 214
    vascular perfusion, 225-226
Aphelidium, membrane fusion, 383
Arenicola marina, 21
Argentaffin cells, demonstration, 117
Arginine, 65, 158
Arsenate buffer
    effect on staining, 42
    morphological preservation with, 42
    sectioning properties with, 42
Artery
    rabbit, perfusion pressure for, 214
    vascular perfusion, 226
Arthropods
    marine, body fluid composition, 21
    pH of body fluid, 57
Artifacts, 4, 381-388
    from acidification, 39
    aldehyde-induced, 202
    avoidance of, in vascular perfusion, 209-210
    in brain tissue, 268
    from cutting, 257
    in CNS tissue, 257
    dehydration-induced, 388
    diffusion, in unfixed tissue, 346
    dimensional changes, 381, 385-386
    DMSO-induced, 376
    from double fixation, 201-202
    electron-dense granules, 41
    environment-induced, 385
    formaldehyde-induced, 382, 384
    from glutaraldehyde, 384
    from glutaraldehyde decrease, 94, 95
    from glutaraldehyde-digitonin, 108-109
    incubation, reduction by DMSO, 375
    from large specimens, 11
    light and dark cells, 384
    from mechanical treatments, 386
    nonspecific precipitates on alveolar walls, 118
    osmiophilic droplets, 383
    from perfusion pressure, 213-214
    from phosphate buffers, 45
    plasma membrane vesiculation, 285
    precipitation, 33
    procaine-induced, 211
    redistribution, 296
    systematic, analysis for, 8

Artifacts (*cont.*)
in vascular endothelium, 297–298
vehicle-induced, 386
Aryl sulfatase, inhibition, by glutaraldehyde, 345
Ascorbic acid, in specimen cleaning, 331
Asparagine, 65, 158
Aspartate, 72, 146
Aspartate aminotransferase
inactivation, 355
inhibition, by glutaraldehyde, 345
Aspartic acid, 65
Astacus fluviatilis, 22
ATPase, *see* Adenosine triphosphatase
Autolysis, 307–313
effect of temperature on, 313
Autophagocytosis, 308
Autoradiography
artifact in, caused by glutaraldehyde, 101
spurious labeling patterns, 100
Avena coleoptile, preservation, 95
Axon(s), fixation, during electrical stimulation, 10

B

Bacillus licheniformis, 357
Bacillus subtilis, 10
Bacteria
aldehyde-fixed, 70
fixation methods for, 390–394
fixative for, 129
glutaraldehyde-fixed, 67
in intestinal mucosa, 331–332
mesosomes, 274–275
morphology
affected by temperature, 10
influenced by electrolytes during fixation, 32
preservation of electron opaque granules in, 114
Barbital, 260
Barium permanganate, 191
Beeswax, solubility, 5
BES buffer, 49
Bicarbonate buffer, 47–48
effect on staining, 43
microtubule loss with, 41
osmolality, 39
sectioning properties with, 43

Bicarbonate-formaldehyde, in protein fixation, 133
Bicine, 49
Bismuth, staining, effect of fixation on, 288
Blood, removal, 296–297
in vascular perfusion, 211–212
Blood-brain barrier, circumvention, 265–268
Blood cell, fish, effect of fixative osmolarity, 19
Blood plasma, stabilization, 67
Bone
acrolein-fixed, 139
immersion fixation, 258
undecalcified, fixation method for, 394
Brain
artifactual dark neurons in, 268
dissection artifacts, 382
effect of fixation on, 265–269
exclusive fixation of, 229–230
extracellular space
preservation, 230
volume, 265
formaldehyde-fixed, 144
glycogen in, 146
human, vascular perfusion, 210
intraventricular perfusion of, 228, 268
lipids in, 87
monkey, perfusion pressure for, 214
osmolarities, 268
postmortem effects, 308
on subcellular conditions, 312
preservation, 212
rabbit
perfusate osmolality for, 213
perfusion pressure for, 214
rat
amino acids in, 101
lipid extraction in, 140
perfusate osmolality for, 213
perfusion pressure for, 214
shrinkage, 98
swelling, 302
sensitivity to osmolarity of fixative, 17–20
shrinkage, 337
swelling, 27
vascular perfusion of, 211, 227–232
volume, and perfusion pressure, 215
Brome mosaic virus, 90, 283
Buffers, 23, 39–54, *see also* specific buffer
comparative effects, 42–43
concentration of solute, 40
effect on staining, 291

efficiency, factors affecting, 40–41
in enzyme cytochemistry, 353
evaluation, 44
extraction caused by, in glutaraldehyde-fixed
  tissue, 101
ionic composition, effects, 40
osmolalities, 36
osmolarity, effect, on specimen appearance,
  17–20
pH, 39–40
  adjustment after addition of DMSO, 376
  and electrolytes, 33–34
  effects, 54–57
  physiological, 44
  preparation, 50–54
  rinsing, osmolarity, 25
in SEM, 334
tissue reaction to, 41
types, 44–50

C

Cacodylate buffer, 47, 297
  artifacts from, 41
  effective pH, 47
  inhibitor of β-glucuronidase, 353
  morphological preservation with, 41
  osmolality, 36
  in plant fixation, 316
  preparation, 50
  as washing solution, 297
    preparation, 412
Caffeine, reaction, with phenolics, 164, 166
Calcium
  in aquatic animals, 21–22
  denaturant of protein, 26
  effect
    on chromosomes, 33
    on microtubules, 33
  in fixative solution, 29–32
  intracellular, postmortem increase in, 313
  loss, during fixation, 295
  with phosphate buffer, 45
  precipitation, 296
Calcium-binding sites, see Electron-dense spots
Calcium chloride
  adjustment of osmolarity with, 39
  chemical formula, 413
  denaturant of protein, 26
  effects, in fixative, 28–30

inhibitor of myelin figure formation, 88
in minimization of phospholipid loss, 341
molecular weight, 413
physicochemical properties, 413
in prevention of swelling, 171
in retention of free fatty acids, 168
Calcium permanganate, 190
CAPS buffer, 49
Carbodiimides, 198–199
  in enzyme preservation, 348
  preservation of antigenicity with, 366
Carbohydrate
  cross-linking, antigenicity preservation by,
    367
  reaction
    with formaldehyde, 136
    with glutaraldehyde, 90
    with osmium tetroxide, 163–164
  solubility, 5
Carboxypeptidase, inhibition, by glutaral-
    dehyde, 345
Carcinus maenas, 21
Cartilage, calcifying, fixation method for, 394–
    395
Casein, milk, solubility, 6
Catalase, 196
  inhibition by glutaraldehyde, 345, 347
  peroxidatic activity, enhanced by glutaral-
    dehyde, 69, 88
  preservation, 346
D-Catechin, 166
Caulfield buffer, osmolality, 36
Cell
  agglutinability, with glutaraldehyde, 70
  coat, alcian blue in demonstration of, 106–
    107
  components
    appearance, in satisfactory fixation, 379
    solubility, 4–6
  in culture, see Cultured cells
  enzyme inhibition in different components,
    344
  fixation front in, 60
  fixative penetration, 59–60
  fractions, fixation method for, 395
  free, fixation method for, 399
  isolated, shrinkage, 98–99
  mammalian, fixation, 93, 95
  membrane, see also Membrane; Plasma
    membrane
    depolarization, 91

Cell (*cont.*)
impermeability, with osmium tetroxide, 27–28
increased contrast, with tannic acid, 125
ion permeability, 91–92
preservation, 31
resistance, in aldehyde fixation, 67
stabilization, 30
swelling, 29
monolayers, fixation method for, 395
nucleus
autolysis, 310
shrinkage, 98
neurosecretory, staining, with osmium tetroxide, 158
organelles
autolysis, resistance to, 312, 313
circadian variation in, 9
effects of hypertonic fixative on, 304
formaldehyde fixation, 130
glutaraldehyde fixation, 68
isolation for SEM, 321
morphology, 61, 166
postmortem changes, 308
resistance to osmotic changes, 337
response to fixative concentration, 63
sensitivity to buffers, 41–44
ultracentrifugal analysis, 68
whole-mounted, 2
pathological, 66, 202
pH variations, effects, 54–55
ruffling activity, 385
structure, preservation, by vascular perfusion, 4
substrate interaction with, 385–386
surface morphology, 388
artifactual alterations in, 385
wall, appearance in satisfactory fixation, 379
Cellulose, solubility, 5
Central nervous system tissue
appearance, in satisfactory fixation, 380
autolysis, 309
human, postmortem changes in, 310
perfusate preparations for, 227
shrinkage, 337
trauma to, 209
vascular perfusion, 144, 227–232
Cephalin, solubility, 5
Cercosporella herpotrichoides, 374
Cerebral cortex, fixation, 99

Cetylpyridinium chloride, as mucus remover, 325
Chloral hydrate, 260
Chloretone, as relaxant, 332
Chloride
in aquatic animals, 21–22
diffusion, into cells, 300, 302
intracellular, postmortem increase in, 313
precipitation, 296
Chloroform, effect, on enzyme activity, 211
Chlorophyll, extraction
reduction by sucrose, 27
with Zwitterionic buffers, 49
Chloroplasts
external nonelectrolyte concentration for, 28
Hill activity, 61
in aldehyde fixation, 67
lipid fixation in, 167
protein content, 314
shrinkage, 99
spinach, lipid fixation in, 168
Cholesterol, 71
localization, with glutaraldehyde-digitonin, 107
preservation, 107–108, 168
reaction with potassium permanganate, 187–188
solubility, 5, 109
Cholesterol-digitonide complex, 107–109
Cholinesterase
inhibition by glutaraldehyde, 345
localization, 111
preservation, 351–352
Chondroitin, solubility, 6
Chondroitin sulfate, 126
Chromaffin reaction, 115
Chromatin
artifacts in, 382–383
clumping, 100
electron opacity, 183
lipids in, 166
Chromatin-dichromate buffer
effect on staining, 43
morphological preservation with, 43
sectioning properties with, 43
Chromic acid, 130
mixture with formaldehyde, 137
Chromoproteins, solubility, 6
Chromosomes, artifacts in, 382–383

Chymotrypsin
  cleaning of intestinal epithelium by, 329
  inhibition by glutaraldehyde, 345
Cilia, preservation, 339
Ciliates
  fixation method for, 396–397
  fixative for, 339
  nonphysiological membrane fusion in, 383
  preservation, 202
Cinchonine, reaction, with phenolics, 164
Circadian rhythm, effect, on cell morphology, 9
iso-Citric dehydrogenase, inhibition by glutaral-
    dehyde, 345
Cnidaria, pH of body fluid, 57
Cocaine hydrochloride, 332
Cochlea, membranous, shrinkage, 337
Coelenterates, body fluid composition, 22
Collagen
  cross-linking, with aldehydes, 79–80
  effect of salts on, 26
  preservation, 133
  removal by enzymes, 327
  stabilization, with glutaraldehyde, 71
Collagenase, contaminant removal by, 327–329
s-Collidine
  chemical formula, 413
  molecular weight, 413
Collidine buffer
  advantage, 40
  artifacts from, 41
  effect on staining, 42
  effective pH, 46
  morphological preservation with, 41–42
  osmolality, 38
  preparation, 51
  sectioning properties with, 42
Common carotid artery, rabbit, endothelial cell
    surface, 328–329
Concanavalin A, cross-linked with glutaral-
    dehyde, 69
Copper ferrocyanide, 374, 375
Corneal tissue, artifacts in, 387
Cotegonus clupeiodes, 22
Cowpea mosaic virus, 283
Creatinine phosphokinase, inhibition by
    glutaraldehyde, 345
Crista ampullaris
  hair cell, postmortem changes in, 308, 310
  postmortem alterations, 308, 309
Cristae, configurations, 277–282

Critical point drying, 2
  artifacts, 385
  shrinkage caused by, 337, 338
Crotonaldehyde, 346
  osmolality, 37
Crustacea
  body fluid composition, 22
  ostracod, relaxant for, 332
Cryoultramicrotomy, 359–363
Cultured cells
  artifacts, 385
  effect of vehicle osmolality on, 334
  fixation, 144
    method for, 396
  fixative for, in SEM, 339, 340
  localization of intracellular antigens in, 368–
    369
  shrinkage, 337, 338
  tissue, isotonic fixation conditions for, 304
Cutting
  artifacts caused by, 257
  instruments, 12–15
  method, 11, 12–15
Cyanuric chloride, in enzyme preservation, 348
3-(Cyclohexylamino)propanesulfonic acid, see
    CAPS buffer
Cyclostome, body fluid composition, 21
Cysteine, 65, 80, 158
  effect of DMSO on, 374
  solubility, 29
Cysts, fixation method for, 397
Cytochrome oxidase, 344
  preservation, 358
  sensitivity, to glutaraldehyde, 67
Cytoplasm, shrinkage, 98
Cytoplasmic ground substance, appearance, in
    satisfactory fixation, 379

D

DAB, see Diaminobenzidine
Dalton dichromate buffer, osmolality, 36
Dehydration, 3
  artifacts, 387–388
  automatic devices for, 15–16
  continuous, 337
    advantages of, 304–306
  effect of temperature during, 61

Dehydration (*cont.*)
  effect on intercellular spaces, 4
  lipid loss during, 167, 168
  shrinkage caused by, 337, 338
  solution, osmolarity, 93
  step procedure for, effect on cell volume,
    304–306
Dehydrogenases
  coenzyme-linked, effect of DMSO on, 374–
    375
  incubation of, 344
  inhibition, by glutaraldehyde, 345, 347
DEM, *see* Diethylmalonimidate
Denaturation
  defined, 3
  reversibility, 3
Deoxyribonucleic acid
  in bacteria, 283
  clumping, 162, 187
  denaturation, 135
    with sodium salts, 26
  effect of salts on, 26
  effect of trioxsalen on, 199
  in eukaryotic nuclei, reaction with glutaral-
    dehyde, 88–89
  gelation, 89
  in noneukaryotic nuclei, reaction with
    glutaraldehyde, 89
  postmortem retention, 310
  preservation, 129
  reaction
    with formaldehyde, 134–136
    with osmium tetroxide, 161–162
    with potassium permanganate, 187
Deoxyribose, solubility, 5
Dextran, 27, 28
Diaminobenzidine, 69
  in osmium black production, 177
Diastase, cleaning of isolated organelles by, 329
Diatoms
  cleaning, 329
  fixation, 205
Dibucaine hydrochloride, deciliation with, 331
Diethylmalonimidate, preservation of antigenicity with, 366
Digitonin, 366
  artifacts induced by, 108–109
  in cholesterol retention, 168
  effect on cellular membranes, 109
  mixture with glutaraldehyde, 107–110

Digitonin-cholesterol complexes, 108–109
2,6-Dihydroxytetrahydropyran, 84
4,5-Dihydroxythymine, 161
Dimerization, glutaraldehyde-induced, 384
Dimethylsuberimidate, 195–198
  in enzyme preservation, 348
  preservation of antigenicity with, 366
Dimethylsulfoxide
  added to acrolein-glutaraldehyde mixture, 144
  added to glutaraldehyde, 98
  as analgesic, 374
  as cryoprotectant, 373
  effects
    on biological systems, 373
    on enzymes, 374–375
    on protein, 373
    on ultrastructure, 375–377
    on shrinkage, 373
  in fixation, 372–377
  physicochemical characteristics, 372–373
  as radioprotectant, 374
Disaccharides, solubility, 5
Disodium hydrogen phosphate, physicochemical
  properties, 412
Dissection, 386
  deformation caused by, 321
Distal tubule, postmortem effects on, 312
DMS, *see* Dimethylsuberimidate
DMSO, *see* Dimethylsulfoxide
DNA, *see* Deoxyribonucleic acid
Dopamine, 115
Dripping method, fixation by, 258–259
Dye exclusion method, 71

E

Echinoderms
  body fluid
    composition, 21
    pH, 57
  fixation method for, 397
Echinus esculentus, 21
EDTA, *see* Ethylenediaminetetraacetic acid
Eggs
  amphibian, fixation, 139
  fixation method for, 397
Egg white, *see* Albumin, egg
Ehrlich ascites tumor cells, loss of cell components from, 295

Elasmobranchs, body fluid composition, 21
Elastase, 150
  neuronal sheath removal with, 329
Elastic fibers, staining, 291
Electrolytes, *see also* specific electrolyte
  effects, in fixative solution, 28–32
  in osmolarity adjustment, 33–34
Electron-dense particles, 45
Electron-dense spots, 32, 41, 167
Electron microscopy, *see also* Scanning electron
      microscopy; Transmission electron micros-
      copy
  diagnostic application, 206
  quantitative, 145–146
Eledone cirrosa, 21
Embryo, 66, 137
  acrolein fixation, 139
  amphibian, preservation, 144
  chick, 202, 203
    effect of fixative osmolarity on, 325
  fixation method for, 398
  isotonic fixation conditions for, 304
  preservation, 175
  tissue, fixative for, 98
  vascular perfusion of, 232–233
  vertebrate, shrinkage, 337
Endocardium, canine, shrinkage, 337
Endoplasmic reticulum, 67, 184, 189
  diurnal variation, 9–10
  rough
    appearance in satisfactory fixation, 379
    postmortem changes in, 311
  smooth
    appearance in satisfactory fixation, 379
    postmortem changes in, 311
  swelling, 198
Enzyme activity
  effect of fixation on, 342–343
  in homogenates, retention of, 358–359
  localization, 177
    in single cells, 357
  postmortem changes in, 307
  preservation, 111–112, 210
    with cacodylate, 47
    factors affecting, 350–355
    with glutaraldehyde, 78
  retention, 196, 198
Enzyme cytochemistry
  acrolein use in, 139
  effect of temperature on, 352

fixation, 342–363
  mode, 353–355
  fixative, pH, 352–353
  reaction medium, penetration into unfixed tis-
      sue, 346
  specimens
    single-cell, 355–357
    size, 350–351
    subcellular fractions, 357–359
Enzymes
  diffusion, from unfixed tissue, 346
  digestion of mucus by, 327
  effect of aldehydes on, 348–350
  effect of DMSO on, 374–375
  hydrolytic
    fixation, 344
    localization by cryoultramicrotomy,
        362
    released from tonoplast, 318
  immobilization, by glutaraldehyde, 69
  inactivation, by osmium tetroxide, 150
  inhibitors, 26–27
  localization, glutaraldehyde concentration for,
      95–96
  removal of contaminants by, 327–329
  studies, glutaraldehyde in, 101, 357–358
Epidermis, rat, effect of fixative osmolarity on,
      20
Epithelium
  intestinal, preservation of microbial associa-
      tion, 331–332
  mucus removal from, 327
EPPS buffer, 49
Eriocheir sinensis, 22
Erythrocytes, *see also* Red blood cells
  acrolein-fixed, 140–141
  fixation method for, 398
  fixative pH for, 55
  ghosts, postfixative for, 287
  glutaraldehyde-fixed, 67
    agglutinability, 70
  human
    preservation, 29
    shrinkage, 338
  murine, fixative for, 339
  shrinkage, 98
Escherichia coli, morphology, affected by tem-
      perature, 10
Ethanol, 3, 4
  shrinkage caused by, 338

Ether, 260
  effect, on enzyme activity, 211
Ethoxydihydropyran, 66
1-Ethyl-3(3-dimethylaminopropyl)
    carbodiimide-HCl, 198
Ethylenediaminetetraacetic acid, use in speci-
    men cleaning, 331
Ethylene glycol
  for helical protein fixation, 2
  lipid extraction reduced by, 286
Ethylene glycol-cellosolve, 338
Eucaine hydrochloride, 332
Exoskeleton, fixative penetration, 59
Extracellular spaces, in satisfactory fixation, 380
Extraction
  during fixation, 4–6
  effect of fixation duration on, 62
  effect of fixative solution osmolarity on,
      22–23
  in glutaraldehyde-fixed specimens, 101
  increase, with nonelectrolytes, 32–33
  reduction
    by double fixation, 6
    by electrolytes, 29–30
    by nonelectrolytes added to fixative, 27
Eyes, fixation method for, 398

F

Farnesol pyrophosphate, reaction with osmium
    tetroxide, 151
Fatty acids
  reaction with potassium permanganate, 187
  unsaturated, reaction with acrolein, 141
Ferritin, for ultrastructural mapping, 145
Fetal tissue, fixation method for, 398–399
Feulgen hydrolysis, effect of fixatives on, 289
Fibrinogen, reaction with osmium tetroxide, 157
Fish, see also Animals, freshwater; Animals,
    marine; and specific fish
  anesthetic for, 332
  vascular perfusion of, 233–235
Fixation
  anhydrous, 207–208
  artifacts, 381–388
  chemical, advantages, 1
  double, 64, 102, 200–201
    sequential, limitations of, 200–202
    to reduce extraction, 6

  duration, 62
    advantages of varying, 8
    effect on microtubules, 276
    in enzyme cytochemistry, 351–352
    for scanning electron microscopy, 336–337
    with Zwitterionic buffers, 49
  effect
    on actin, 262–264
    dimensional changes, 4
    on kidney, 269–271
    on membrane fracture, 271–274
    on mesosomes, 274–275
    on microfilaments, 275–277
    on microtubules, 275–277
    on mitochondria, 277–282
    on myelin, 282–283
    on plasma membrane, 284–287
    protein denaturation, 3
    on protoplasm, 2
    on staining, 287–291
    on synaptic vesicles, 291–294
    on tight junctions, 294
    on tissue structure and function, 90–92
    on vascular endothelium, 296–298
  for enzyme cytochemistry, 342–363
  for immunoelectron microscopy, 364–371
  methods, 200–208
    advantages of varying, 8
  modes, 209–261, see also specific mode
    artifacts, 386–387
  objectives, 1–2
  pH during, 54–57
  and photoperiod, 9
  physical, vs. chemical, 1
  quality
    and automatic specimen processing, 15–16
    effect of buffers on, 39–54
    effect of osmolarity on, 16–39
    effect of specimen size on, 11–15
    effect of vehicle osmolality on, 23–26
    factors affecting, 9–63
    satisfactory, criteria for, 378–380
    for scanning electron microscopy, 320–341
    simultaneous, for light and electron micros-
      copy, 206–207
    temperature, 60–62
    uniformity, affected by specimen size, 11
Fixative, see also specific fixative
  additive, 4
  choice, and objective of study, 2

coagulant, 3
coefficient of diffusibility, 58
concentration, 63
cross-link formation by, 3
formulations, for light and electron microscopic examination, 206-207
glutaraldehyde-containing, 106-129
hypertonicity
  effect, 301-304
  excessive, 304
isotonic, effects, 301-302
noncoagulant, 3-4
osmolality, 92
penetration, 58-60
penetration rate, 59-61
  effect of temperature on, 60-61
  with addition of electrolytes, 32
  with addition of nonelectrolytes, 32-33
tonicity, 17
vehicle, effect on staining, 291
Fixative solution, *see also* Osmolarity
effects of added substances, 26-33
ionic composition, 20-23
isosmotic, definition, 17
isotonic, definition, 17
osmolarity, effects, on specimen appearance, 17-20
Flagellates
fixation method for, 398
fixative for, 339
Flavoproteins, solubility, 6
Fluorescence
acrolein-induced, 140
formaldehyde-induced, 132
induced by glutaraldehyde-formaldehyde, 111, 228
Formaldehyde, 129-137
chemical characteristics, 129
chromic acid mixture, 137
cross-linking with histones, 72
in differentiation of amines, 115
effect, on antigens, 365
in enzyme cytochemistry, 343
fixation of tissue fractions, 358
fixatives, preparation, 136-137
inhibition of enzyme activity, compared to glutaraldehyde, 346-347
mixture with glutaraldehyde, 110-111
as noncoagulant, 4
oligomers, formation, 129-130

osmolality, 37
penetration, 58-60
picric acid mixture, 136-137
  preparation, 370
  use in immunoelectron microscopy, 365-366
protein functional groups available to, 130
reaction
  with carbohydrates, 136
  with lipids, 133
  with nucleic acids, 134-136
  with proteins, 77, 130-133
  in scanning electron microscopy, 338-339
Formalin, 129-130, 136, 346
with Millonig buffer, 45-46
osmolality, 37
Fowlpox virus, preservation, 35, 57
Freeze-drying, vapor fixation with, 176
Freeze-etching, 378-379
Freeze-fracture, 271
Freezing, 296
ultrarapid, 273-274
Freezing point depression, 23, 35-37
for marine animal body fluid, 21
Freshwater animals, *see* Animals, freshwater
Fructose, solubility, 5
Fructose-1,6-diphosphatase, 198
Fungi, 66, 139, 144, 205
fixation method for, 399
preservation, 143

G

β-Galactosidase, inhibition by glutaraldehyde, 345, 347
Gallbladder, frog
epithelium, 330-331
shrinkage, 303, 304, 306
Gastric mucosa
mouse, carbonic anhydrase fixation in, 349
pH for fixation, 316
Gelatin, use as tissue glue, 14-15
Gliadin, solubility, 5
Globin, solubility, 5
Globulin
reaction with osmium tetroxide, 157
solubility, 5
Glomerulus, postmortem effects on, 312
Glucoamylase, immobilization, 69

Glucose, 146
 chemical formula, 413
 molecular weight, 413
 solubility, 5
Glucose-6-phosphatase, 196, 198, 344
 glutaraldehyde concentration for, 351
 glutaraldehyde-fixed, 348
 hepatic
  effect of fixation duration on, 358
  natural inhibitor, 354
  preservation, 354
 preservation, 355, 358–359
  effect of fixation duration on, 352
 in subcellular fractions, 360–363
Glucose-6-phosphate, 146
 inhibition, by phosphate buffers, 45
Glucose-6-phosphate dehydrogenase, inhibition,
  353
 by phosphate buffer, 45
β-Glucuronidase inhibition, 353
 by glutaraldehyde, 345, 347
Glusulase, removal of mucilaginous material by,
  329
Glutamate, 72, 146
Glutamate dehydrogenase, 196
Glutamic acid, 65
 binding, 101
Glutamic oxalacetic transaminase, localization,
  348, 354
Glutamine, 65, 158
Glutaraldehyde, 2
 acrolein mixture, 142–143
 advantages, 66–67
 alcian blue mixture, 106–107
 as antimitotic agent, 70
 artifacts, 384
 buffered, osmolality, at various concentrations, 36
 cacodylate-buffered, 47
 carbodiimide mixture
  localization of antigens with, 366–367
  preparation, 370–371
 chemical characteristics, 66
 commercial, 72–77
 concentration, 63, 94–96
  determination, 104–105
  in enzyme cytochemistry, 351
  for SEM, 333–334
 contribution to freezing-point depression, 38

 cross-linking of protein by, 314
 digitonin mixture, 107–110
  preparation method, 110
 effect of pH on, 76–77
 effect of temperature on, 76–77
 effects
  on actin, 262–263
  on antigens, 364–365
  on enzyme displacement, 356
  on membrane fracture, 272–273
  on membrane proteins, 31
  on plasma membrane, 284–287
 as enzyme cross-linking agent, 347
 enzyme localization, 347
 extraction with, 22–23
 fixation
  at higher pH, 96–97
  of tissue fractions, 357–358
  time, 62
 fixatives, 106–129
 formaldehyde mixture, 110–111
  preparation methods, 110–111
   with parabenzoquinone preparation, 370
  used in SEM, 338–339
 free aldehyde introduced by, 289
 heated, 73, 99
  method for using, 99
 hydrates
  role in protein cross-linking, 73–77, 79
  types, 73
 hydrogen peroxide mixture, 111–113
  preparation, 113
  preservation of enzymes by, 354
 impurities, 73, 92–93, 102, 105
 inhibitor, of enzymes, 101, 344, 345, 346–
  347
 lead acetate mixture, 113
 limitations, 99–102, 201
 malachite green mixture, 113–114
 membrane blebs induced by, 387
 molecule size, effect on protein cross-linking,
  81
 monomeric, in protein cross-linking, 73–75,
  78–80
 osmolality, 37, 92
  in various vehicles, 36–39
 osmolarity, 92–94
 osmotic contribution, 23–24
 osmotic effect of vehicle, 23–25

osmium tetroxide mixture, 201–206
  preparation, 206
  used in SEM, 339
paraformaldehyde mixture, preparation, 110
penetration, 58–60
  rate, 97–98
  of single cells, 59
  and specimen size, 11
pH, 72
phosphate-buffered, osmolality, 93
phosphotungstic acid mixture, 114
polymeric, in protein cross-linking, 73–76,
  78–80
polymerization, 72–77
  at alkaline pH, 102–103, 105–106
  effect of pH on, 96
  effect on osmolarity, 93
  role of pH, 74
  temperature range, 76–77
potassium dichromate mixture, 115–118
  with formaldehyde, 117–118
  method for use, 117–118
protein cross-linking with, 112–113
protein functional groups available to, 80–
  81
purification, 102–104
  charcoal method, 103
  distillation method, 103–104
purified
  in cross-linking, 78–79
  ultraviolet absorption maximum, 102
reaction
  with carbohydrates, 90
  with collagen, 79–80
  with lipids, 87–88
  with lysine, 80–81
  with nucleic acids, 88–90
  with proteins, 69, 77–87
  with simple alkyl amines, 84–85
ruthenium red mixture, 120–124
spermidine phosphate mixture, 124
storage, 105–106
tannic acid mixture, 124–128
  preparation, 128
  with paraformaldehyde, 128
trimers, 75
trinitro compound mixtures, 114–115
ultraviolet absorption spectra, 76
α,β-unsaturated dimer, 75–76

in enzyme cross-linking, 347
in epoxy aldehyde synthesis, 112
uranyl acetate mixture, 129
uses, 68–72
  in collagen stabilization, 71
  in inducing tumor immunity, 70
  in labeling polypeptide binding sites, 70
  in quantitating specimen constituents, 72
  in scanning electron microscopy, 338–341
with potassium ferricyanide-osmium tetroxide
  mixture, 118–119
  methods for use, 119
with potassium ferrocyanide-osmium tetroxide mixture, 120
Glutaric acid dialdehyde, *see* Glutaraldehyde
Glutelin, solubility, 5
Glycerol, 71
Glycine, solubility, 29
Glycogen
  appearance, in satisfactory fixation, 379
  in brain, preservation, 146
  effect of uranyl acetate on, 289
  lead stained, 287–288
  postmortem effects on, 307, 309
  preservation, 120, 133, 136, 188
  reaction
    with osmium tetroxide, 163
    with ruthenium tetroxide, 195
  retention, 196
  solubility, 5
  staining, 90
Glycogen phosphorylase *b*, 80
  inhibition, by glutaraldehyde, 345
Glycolipids, fixation, 168, 187
Glycolysis, postmortem, 308
Glycoproteins, solubility, 6
Glycosaminoglycans, preservation, 126
Glycosyltransferase, 47
Glyoxal, 346
Goldfish, spinal cord, vascular perfusion, 251–
  252
Golgi, appearance, in satisfactory fixation,
  379
Golgi apparatus, 184
  swelling, 198
  morphology, circadian variation in, 9
  postmortem changes in, 311
Gram-negative organisms, enzyme localization
  in, 357

Gram-positive organisms, enzyme localization in, 357
Granularity, increase, with electrolytes, 33
Granulocytes, human, 95
Gray matter, perfusion, 269

H

Hair, solubility, 5
Hair cell
    fixative penetration, 59
    plant, glutaraldehyde fixation, 100
Halothane, 260
Hank's balanced salt solution, in washing, 297
Heart
    artifactual contraction bands, 384–385
    frog, fixation, 259
    perfusate preparations for, 235
    rat, 174
        DMSO-induced artifacts in, 376
        perfusate osmolality for, 213
        perfusion pressure for, 214
    vascular perfusion of, 235–237
HeLa cells, 10
    fixation method for, 399–401
    PM buffer for, 48
Hemoglobin
    cross-linking, 81
    fixative pH for, 55, 57
    solubility, 6
Heparin, 211, 212
Hepatocytes
    albumin loss, 384
    blebs, DMSO-induced, 376
    isotonicity, 25
HEPES buffer, 48–50
    artifacts from, 41
Heteropolysaccharides, nitrogenous neutral, solubility, 6
Histidine, 65, 158
    reaction, with glutaraldehyde, 80
Histones
    cross-linking, 72
    solubility, 5
    thymus, solubility, 5
Holothuria tubulosa, 21
Homarus vulgaris, 21
Horn, solubility, 5

Hyaluronic acid, 126
    solubility, 6
Hyaluronidase, as mucus remover, 325
Hydra viridis, 22
Hydrazine, termination of protein cross-linking by, 352
Hydrochloric acid
    chemical formula, 413
    molecular weight, 413
    in specimen cleaning, 327, 329–331
Hydrogen peroxide
    mixed with glutaraldehyde, 111–113, 354
    osmium oxide removal by, 289
Hydroxyadipaldehyde, 346
α-Hydroxybutyrate dehydrogenase, inhibition, by glutaraldehyde, 345
N,N-bis(2-Hydroxyethyl)-2-aminoethanesulfonic acid, *see* BES buffer
N,N-bis(2-Hydroxyethyl)glycine, *see* Bicine
N-2-Hydroxyethylpiperazine-N'-2-ethanesulfonic acid, *see* HEPES buffer
N-(2-Hydroxyethyl)piperazine sulfonic acid, *see* EPPS buffer
δ-N-Hydroxymethyl glutamine, 131

I

Immersion fixation, 256–258
    artifacts, 384, 387
        in vascular endothelium, 335
    disadvantages, 256–257
        in scanning electron microscopy, 334–335
    effects, dimensional changes, 4
    limitation, in enzyme studies, 354
    nonuniformity, 11–12
    for plants, 316
    pretreatment for, with osmium tetroxide vapor, 175–176
    temperature for, 336
Immunoadsorbents, preparation, with glutaraldehyde, 69
Immunoelectron microscopy
    fixatives, preparations, 370–371
    fixation for, 364–371
Immunoreactivity, effect of fixation on, 364–365
Inactin, 260, 261

Incubation, 354-355
  procedure, for preservation of insoluble de-
    hydrogenase, 344
  without prior fixation, 346
Indigo dyes, osmication, 177
Infiltration, continuous, advantages of, 306
Injection method, fixation by, 259-260
Inner ear tissue, 205
  cleaning, 331
  postmortem changes in, 308
Insect
  body fluid composition, 22
  epithelial cells, artifact, 11
Intercellular spaces, in satisfactory fixation, 380
Intestine, lipid loss from, 167, 168
Invertebrates, *see* Animals
Iron hematoxylin, staining, 290-291
Ischemia, postmortem, 307, 309, 310
Isocitrate dehydrogenase, effect of DMSO on,
  374
Isoelectric pH, of proteins, 54
Isoelectric point, definition, 54
Isoleucine, binding, 101
Isosmotic solutions, definition, 17
Isotonic solution, definition, 17

J

Jet fixation, 335

K

Keratin, 174
Ketoglutarate, effect, on enzymes, 355
Kidney, 137, 174
  artifacts, from transition in perfusion pres-
    sure, 214
  autolysis, 309
  biopsy specimens, fixation, 271
  cat, vascular perfusion of, 237-238
  criteria of satisfactory fixation, 380
  effect of fixation on, 269-271
  immersion fixation, 257
  osmolalities, 34
  perfusate preparation for, 237, 238
  pig, vascular perfusion, 238-240
  rabbit, ultrastructural effects of DMSO in,
    376

rat
  perfusate osmolality for, 213
  perfusion pressure for, 214
  shrinkage, 337
  ultrastructural effects of DMSO in, 376
  vascular perfusion of, 237-240
Krebs buffer, osmolality, 38

L

Labyrinth hair cells, 210
Lactate, 146
Lactate dehydrogenase
  effect of DMSO on, 374
  formaldehyde preservation of, 346
  inhibition by glutaraldehyde, 345
  inhibition by Tris-HCl, 353
β-Lactoglobulin, isoelectric point, 55
Lactose, solubility, 5
Lanthanum permanganate, 190-191
  preparation, 192
LDH, *see* Lactate dehydrogenase
Lead
  binding, to glycogen complex, 289
  staining
    effect of fixation on, 287-288
    with various buffers, compared, 42-43
    with Zwitterionic buffers, 49
Lead acetate, mixture with glutaraldehyde, 113
Lead hydroxyapatite, 113
Leaves, fixation method for, 401
Lecithin
  dipalmitoyl
    preservation, 118-119
    staining, 156-157
  reaction, with potassium ·permanganate,
    188
  solubility, 5
Lectin, cross-linked with glutaraldehyde, 69
Leucine
  binding, 101
  solubility, 29
Leukemia cells, human, fixative for, 339
Leydig cells, preservation, 120, 122
Lichens, 205, 207
Lidocaine chloride, 212, 332
Light microscopy, fixation for, simultaneously
    with electron microscopy fixation, 206-207

Limnaea eggs, 91
  effect of fixative osmolarity on, 20
  swelling, 170-171
Limnea peregra, 22
Limulus, 304
Linoleic acid, 141
Linseed oil, in lung fixation, 119
Lipase, preservation, 351
Lipid droplets
  identification, 190
  postmortem accumulation, 307
  preservation, 188
Lipids
  aldehyde-fixed, 202
  appearance, in satisfactory fixation, 379
  extraction
    during fixation, 4-6
    effect of electrolytes on, 30
    in osmium tetroxide fixation, 166-168
  neutral, retention, 167
  osmium tetroxide staining of, 288
  postmortem effects on, 307
  preservation, 118
  reaction
    with acrolein, 140-141
    with formaldehyde, 133
    with glutaraldehyde, 87-88
    with ruthenium tetroxide, 195
  saturated, reaction with osmium tetroxide,
    156-157
  solubility, 5
    in dehydration agents, 337
  unsaturated, reaction with osmium tetroxide,
    150-156
Lipofuschin, 102
Lipoproteins
  reaction with osmium tetroxide, 159-160
  response to autolysis, 309
  solubility, 6
Lithium, denaturant of protein, 26
Lithium permanganate, 8, 191
  fixation procedure, 193
Liver
  artifactual albumin movement in, 384
  biopsy, vascular perfusion for, 244-245
  cat
    perfusate osmolality for, 213
    perfusion pressure for, 214
  chicken

    perfusate osmolality for, 213
    perfusion pressure for, 214
  effect of malachite green on, 114
  embryonic, vascular perfusion, 244
  fetal pig, fixative penetration, 11
  fish, effect of fixative osmolarity, 18
  fixative penetration, 97
  frog, perfusate osmolality for, 213
  glutaraldehyde penetration, 97
  hypertonic fixative for, 92
  lipid loss from, 167-168
  mouse, 128
  osmium uptake, 174
  perfusate preparation for, 240, 242, 244
  postmortem changes in, 310
    in subcellular conditions, 312
  rat
    dimethylsuberimidate-fixed, 196
    enzyme studies in, 360-363
    lipid loss, 168
    perfusate osmolality for, 213
    perfusion pressure for, 214
    preservation, 190
    protein loss from, 169
    shrinkage, 98
    vascular perfusion of, 240-243
  shrinkage, 337
  swelling, during fixation, 170-171
  vascular perfusion of, 240-245
Lumbricus terrestris, 22
Lung
  alveolar surface, artifactual precipitates on,
    118
  baboon, perfusion pressure for, 214
  dipalmitoyl lecithin in, 157
  dissection artifacts, 382
  fixation, with polymerized glutaraldehyde, 78
  glutaraldehyde concentration for, 95
  human
    injection of fixative into, 260
    vascular perfusion of, 248-249
  isotonic fixation conditions, 304
  lipid loss from, 167
  perfusate preparation for, 245, 246
  postmortem changes in, 308, 309
  preservation, with collidine, 46
  rat
    perfusate osmolality for, 213
    vascular perfusion of, 246-247

shrinkage, 23, 337
surfactants, preservation, 118–119
vascular perfusion of, 244–249
Lymphocytes
fixative for, in SEM, 339
human, 95
shrinkage, 338
interaction with substrate, 386
microvilli, effect of temperature on, 386
Lysine, 65
cross-linking reaction with formaldehyde, 132
mixed with periodate and paraformaldehyde
for preservation of antigenicity, 367
preparation, 370
reaction
with glutaraldehyde, 80–81
with osmium tetroxide, 158
Lysosomes
appearance, in satisfactory fixation, 379
effect of digitonin on, 109
enzyme loss from, 88
postmortem effects on, 307–308
Lytic peptidase, inhibition, by glutaraldehyde,
70–71

M

Macrophage
flattening, 386
lipid loss from, 167
Magnesium
in aquatic animals, 21–22
intracellular, postmortem decrease in, 313
loss, during fixation, 295
precipitation, 296
Magnesium chloride
in osmium tetroxide fixative, 30
as relaxant for motile organisms, 332
Magnesium sulfoxide, effect, in fixative, 30
Malachite green, mixture with glutaraldehyde,
113–114
Malate synthase, 358
Malt diastase, 150
Maltose, solubility, 5
Manganese dioxide, 184
MAO, *see* Monoamine oxidase
Marek's disease, 70
Marine animals, *see* Animals, marine

Mastocytoma tumor cells, fixation, 95
Megakaryocytes, fixation of, 128
Membranes
artifactual fusion, 383
aldehyde-fixed, osmotic activity, 25
blebs, in aldehyde-fixed tissue, 202
blisters, 272
ciliary epithelium, vesicles and tubular system
formation in, 200
continuity, as criterion for satisfactory fixation, 379
fracture, effect of fixation on, 271–274
fracture faces, effects of osmium tetroxide on,
151
permeability, 175
effect of electrolytes on, 29–30
effect of nonelectrolytes on, 28
preservation, aided by DMSO, 376
reaction, with potassium permanganate,
184–185
trilaminar appearance, in osmium tetroxide fixation, 154–156
triple-layered structures, 194–195
vacuolar, as indicator of fixation quality, 318
Membrane-virus interactions, 191, 193
Meromyosin, in preservation of actin, 264
MES buffer, 50
Mesosome
deterioration, 10
effect of fixation on, 274–275
Methionine, 158
ϵ-N,ϵ-N'-Methylenedilysine, 131
Micoplasma laidlawil, 151
Microbes, in intestinal epithelium, 331–332
Microbodies, appearance, in satisfactory fixation, 379
Micrococcus lysodeikticus, 356
Microfilaments
effect of fixation on, 275–277
fixation method for, 401–402
postmortem changes in, 311
preservation, 120, 126, 128
stabilization, 125
Microtome
hand, 13
rotary, 13
Microtubules
cross-bridges, and temperature, 10
destabilization, 48

Microtubules (*cont.*)
  effect of DMSO on, 377
  effect of fixation on, 275–277
  effect of fixation temperature on, 172
  fixation method for, 401–402
  loss, 100
  neuronal, loss, 41
  postmortem changes in, 311
  preservation, 139, 175
  stabilization, 30, 66
  visualization, with tannic acid, 126
  volume changes, with different fixatives, 8
Microvilli, postmortem changes in, 311
Millonig buffer
  artifacts from, 41
  osmolality, 38
Mitochondria
  appearance, in satisfactory fixation, 379
  autolysis, resistance to, 312, 313
  configuration, preservation of, 358
  cristal configurations, 277–282
  effect of fixation on, 277–282
  effect of fixative osmolarity on, 20
  enzymes in, fixation of, 346
  fixation method for, 281–282
  hypertonic fixative for, 302
  lipid extraction from, 166
  postmortem alterations in, 308, 310–311
  potassium-permanganate fixed, 183, 184, 189
  protein loss from, 169
  swelling, 301
  ultrastructural changes, affected by physiological state, 10
Mitosis, arrest, with glutaraldehyde, 70
Mitotic spindles
  fixation, 205
  stabilization, 67
Molality, definition, 16
Molarity, definition, 16
Mollusks
  freshwater, body fluid composition, 22
  marine, body fluid composition, 21
  pH of body fluid, 57
Monoamine oxidase
  effect of DMSO on, 374–375
  effect of fixation duration on, 358
  preservation, 358
Monoamines, reaction, with potassium permanganate, 185–186

Monocytes, human, 95
Monosaccharides
  reaction, with ruthenium tetroxide, 195
  solubility, 5
MOPS buffer, 50
2-(*N*-Morpholino)ethane sulfonic acid, *see* MES buffer
2-(*N*-Morpholino)propanesulfonic acid, *see* MOPS buffer
Mucopolysaccharides, solubility, 6
Mucus
  removal, 323–327
  and preservation of intestinal microbes, 331–332
Muraena helena, 21
Muscle
  acrolein-fixed, 139
  cardiac, 114
    mitochondrial morphology postmortem, 310
    postmortem changes in subcellular conditions, 312
  frog, skeletal, preservation, 139
  hypertonic fixative for, 92
  immersion fixation, 257
  osmium uptake, 174
  postmortem changes in, 308
  protein loss from, 169
  rabbit
    enzyme study in, 360–361
    skeletal, 127, 128
  rat
    effect of fixative osmolarity on, 20
    skeletal, vascular perfusion, 213, 214, 249–251
  skeletal, postmortem changes in subcellular conditions, 312
  smooth, fixation method for, 406, 407
  striated
    A band, postmortem, 309
    I band degradation postmortem, 309
    Z band degradation postmortem, 309
Mustelis canis, 21
Myelin
  effect of DMSO on, 376
  effect of fixation on, 282–283
  preservation, 27
  rat, fixation, 87–88
Myelin figures, 96, 100, 109, 110, 124, 166–167

formation, postmortem, 310, 311
prevention, 88
Myelin sheath
potassium permanganate as fixative for, 184
preservation, 120, 123, 269
Myocardium
autolysis in, 310
rat, immersion fixation, 257
Myofibrils, 282
postmortem degradation, 309
stimulation, by calcium, 33
Myosin
effect of salts on, 26
proteolysis, 309
Mytilus edulis, 21
Myxine glutinosa, 21

N

NAD-pyrophosphorylase, 358
NADH dehydrogenase, 356
preservation, 351
NADH-terricyanide reductase, in subcellular fractions, 362-363
Nails, solubility, 5
Nematodes, fixation method for, 402
Nembutal, 260, *see also* Sodium pentobarbital
Nerve, *see also* Central nervous system
cat, preservation, 259-260
extracellular space, contraction, during glutaraldehyde fixation, 100
frog median emminence, 8
frog sciatic, lipid loss from, 167
glutaraldehyde concentration for, 95
preservation, 189, 192, 193
snake, 10
Neuropil, postmortem changes in subcellular conditions, 312
Neurospora, fixation method for, 399, 400
Nicotine, reaction, with phenolics, 164
Node of Ranvier, preservation, 27
Nonelectrolytes
effects, in fixative solution, 27-28
in osmolarity adjustment, 33-34
Noradrenalin
preservation and staining, 71
reaction

with glutaraldehyde, 115-117
with potassium permanganate, 185
Norepinephrine, demonstration, 117
Novikoff tumor cells, 367
Nuclear contents, appearance, in satisfactory fixation, 379
Nuclear envelope, appearance, in satisfactory fixation, 379
Nuclear sap
artifacts in, 41
artifactual network in, 382-383
in glutaraldehyde fixation, 100
Nucleases, localization, by cryoultramicrotomy, 362
Nucleic acids
heated, 89
reaction
with formaldehyde, 134-136
with glutaraldehyde, 88-90
with osmium tetroxide, 160-163
retention, 88-89
Nucleohistone, solubility, 6
Nucleolus, electron opacity, 183
Nucleoplasm, 188
Nucleoproteins, solubility, 6
Nucleoside diphosphatase, 354
5'-Nucleotidase
inactivation, 357
localization, by cryoultramicrotomy, 362
in subcellular fractions, 360-363
5'-Nucleotide phosphodiesterase
demonstration, 177
inhibition, by glutaraldehyde, 344
Nucleotides, reaction with potassium permanganate, 186-187

O

Olefin, oxidation, by osmium tetroxide, 151-153
Oleic acid, 141
reduction of osmium tetroxide by, 151
Om U Reichert microtome, 350
Oocytes, *see also* Eggs
amphibian, effects of osmium tetroxide fixation, 200
fixation method for, 402-403
Ornithine transcarbamylase, 307
Orthophosphate, localization, 113

Osmeth, 177–178
Osmic acid, 148
Osmium
  bound, removal of, 176
  oxidation states, 150, 153
Osmium blacks, 120, 157, 176–177, 190,
  205
  formation, 72
    in glutaraldehyde-hydrogen peroxide fixed
      specimens, 112
Osmium tetroxide, 148–182
  buffer osmolarity, 25
  with calcium added, 31
  chemical characteristics, 148–150
  with collidine, 46
  color, as indicator of reaction, 159
  compared to potassium permanganate, 189–
    190
  concentration, 63, 171, 180
  contribution to freezing-point depression, 38
  diesters, 151–153
  with digitonin, 2
  disadvantages, 149
  effect
    on actin, 262–264
    on cell membrane, 27–28
    on cell volume, 302, 304
    on membrane fracture, 272
    on microtubules, 275–276
    on plasma membrane, 284–287
  in enzyme cytochemistry, 343
  enzyme inactivator, 150
  extraction with, 22–23
  fixation
    duration, 62, 174–175
    temperature, 171–172
  handling, precautions, 178–180
  hazards, 178–179
  lipid extraction with, 166–168
  mercuric chloride mixture, used in SEM,
    339
  mixture with glutaraldehyde, 201–206
    preparation, 206
  monoesters, 151–153
  α-naphthylamine reaction, 158
  as noncoagulant, 4
  osmolality, in various vehicles, 38
  osmotic contribution of buffer, 25
  penetration, 58–60
    and specimen size, 11

penetration rate, 172–173
  and nonelectrolytes, 32–33
postfixation, 149
  in glutaraldehyde fixation, 102
  in SEM, 339
  to neutralize shrinkage caused by al-
    dehydes, 99
  as primary fixative, 175, 200
  protein extraction with, 168–170
  protein functional groups available to, 158–
    159
  reaction
    with aldehydes, 205–206
    with carbohydrates, 163–164
    with lipoproteins, 159–160
    with nucleic acids, 160–163
    with phenolic compounds, 164–166
    with proteins, 157–159
    with saturated lipids, 156–157
    with sulfhydryl groups, 10, 158
    with unsaturated lipids, 150–156
  secondary fixation with, 64
  solubility, 148, 150
  2% solution, preparation, 181–182
  staining, 288–289
  swelling caused by, 170–171
  used, regeneration of, 180–182
  as vapor fixative, 175–176
Osmolality
  definition, 17
  measurement, 35–39
  vehicle, effect on synaptic vesicles, 292–293
Osmolarity, see also Fixative solution
  adjustment, methods for, 33–34
  definition, 17
  effect, on fixation quality, 16–39
  recommended, 34–35
Osmotic pressure
  effective
    definition, 23
    glutaraldehyde, 24–25
  total, 23
OTAN reaction, 158
Ovary, vascular perfusion of, 251
Overfixation, 62
Oxidases, preservation, 346
Oxidoreductase
  demonstration, 177
  preservation, 346
Oxygen, role, during fixation, 112–113

P

Pachygrapsus marmoratus, 21
Palaemon serratus, 21
Palmitic acid, 151
Palynomorphs, cleaning, 329
Pancreas, 174
    postmortem changes in, 308, 309, 310
    swelling, 27
Pancreatic amylase, 150
Paneth cell granules, mouse, double fixation of,
    201
Papain
    contaminant removal by, 327
    insolubilization, 67
Papaverine, 212
Parabenzoquinone, in enzyme preservation, 348
Parachlorophenylalanine, 115
Paraformaldehyde
    with acrolein-glutaraldehyde mixture, 143,
        144
    with collidine, membrane damage from, 46
    extraction with, 22–23
    mixed with periodate and lysine
        for preservation of antigens, 367
        preparation, 370
    preparation, 136
Paramecium
    fixation artifact in, 383
    fixation method for, 403
PBS, *see* Saline, phosphate-buffered
Pelmatohydra oligactis, 22
Pentazocine, 260, 261
Pentobarbital, 260, 261
Pentobarbitone, effects, on CNS, 210
Peptide, 80
    reaction with osmium tetroxide, 158
Perfusate
    osmolality, for selected tissues, 213
    osmotic pressure, 212
    pH, 212
    temperature, 212
Perfusion fixation, DMSO-induced artifacts in,
    377
Perfusion pressure, *see* Vascular perfusion, per-
    fusion pressure
Periodate-lysine mixture, with paraformal-
    dehyde
    preparation, 370
    use in preservation of antigens, 367

Periodic acid oxidation, effect of fixatives on,
    289
Permanganates, 183–193
Peroxidase, localization, 348
Peroxisomes
    effect of digitonin on, 109
    enzymes in fixation of, 346
    enzyme loss from, 88
    rat liver, fixation, 69
pH
    acidic, desirability, 56–57
    alkaline, recommended, 57
    buffer, effect on fixation quality, 54–57
    effect
        in enzyme cytochemistry, 352–353
        on glutaraldehyde, 76–77
        on glutaraldehyde reaction with protein,
            96
    for hydrated tissue, 56
    importance, in fixation, 39–40
    of perfusate, 212
Phage, stabilization, 130
Phenolic compounds, reaction with osmium tet-
    roxide, 164–166
Phenols, identification, 115
Phenylalanine, 65
    reaction with glutaraldehyde, 80
    solubility, 29
Phosphatases
    preservation, method for, 355
    in tissue fractions, 357
Phosphate, precipitation, 296
Phosphate buffers, 45–46, *see also* Millonig
        buffer
    artifacts, 45
    effect on lactate dehydrogenase, 353
    effect on staining, 42
    inhibitor of glucose-6-phosphate dehyd-
        rogenase, 353
    isotonic, osmolality, 36
    morphological preservation with, 41–42
    osmolality, 36, 39
    pH, 45
    in plant fixation, 316
    preparation, 51–52
        chemicals recommended for, 46
    sectioning properties with, 42
Phosphatidyl choline, 126
    fixation, 168, 187
    reaction, with glutaraldehyde, 88

Phosphatidyl ethanolamine
  reaction
    with glutaraldehyde, 87
    with osmium tetroxide, 156
Phosphatidyl glycerol, fixation, 168, 187
Phosphatidylserine, reaction, with glutaraldehyde, 87
Phosphocreatine, 146
Phospholipid bilayer, effect of glutaraldehyde on, 87
  extraction
    in acrolein fixation, 141
    during dehydration, 341
    effect of calcium on, 30–31
    with Zwitterionic buffers, 49
  preservation, 118–119
  reaction
    with glutaraldehyde, 87–88
    with potassium permanganate, 187
  redistribution, in glutaraldehyde-fixed cells, 100
  retention, 167
  saturated, reaction with tannic acid, 125–126
  solubility, 5
Phosphoproteins, solubility, 6
Phosphorus
  extraction, 295
  intracellular, postmortem levels, 313
Phosphotungstic acid, mixture with glutaraldehyde, 114
Photoperiod, effect, on fixation, 9
Physiological buffer systems, definition, 44
Phytoplanktons, cleaning, 329
Picric acid formaldehyde mixture, 136–137
  preparation, 370
  in immunoelectron microscopy, 365–366
Pinocytotic vesicles, 67
  in CNS, as artifacts, 383
3-(2-Piperidyl) pyridinium, *see* Anabilysine
PIPES buffer, 48–50
  osmolality, 39
  preparation, 52
Pituitary gland, rat, anterior, 137
Plant cells
  fixation method for, 403
  internal osmotic pressure, 56
Plant specimens, 314–319
Plant tissue
  acrolein-glutaraldehyde fixation, 139

acrolein-induced fluorescence in, 140
air in, 316
artifacts, from double fixation, 201–202
bicarbonate buffers for, 47
dense, preservation, 143
fixation, 205
fixative for, in SEM, 339
formaldehyde-fixed, 135
glutaraldehyde concentration for, 95, 341
glutaraldehyde fixation
  optimal concentration, 95
  optimal temperature, 94
  penetration rate, 97–98
  pH, 96
glutaraldehyde-fixed, weight gain, 66
hydrated, preservation, 143
osmiophilia, after aldehyde fixation, 88
osmium tetroxide penetration, 32
osmolalities, 34–35
pH, 315–316
  effects of, during fixation, 56
phenol-containing regions, fixation, 164–166
postmortem desiccation, 316
proteins, 314
senescence, 318–319
Plasma membrane
  antibody penetration through, 366
  appearance, in satisfactory fixation, 379
  artifacts, osmium tetroxide-induced, 31
  blebs, 387
  effect of fixation on, 284–287
  effect of glutaraldehyde on, 101
  ischemic changes in, 310
  permeability, 300–302
  postmortem changes in, 311
  sodium permanganate fixation, 190
  structural reorganization during fixation, 285–286
  trilaminar appearance, 286
  vesiculation, 285
Plasmodesmata, 316
Plastids
  appearance, in satisfactory fixation, 379
  from corn plants, swelling, 9
  potassium-permanganate fixed, 183, 184, 188, 189
Platelets, 67
  fixation, 95, 128, 403–405
PM buffer, 48

Polarization microscope, 378
Pollen wall preservation, 126
  with glutaraldehyde-alcian blue, 107
Polyaldehyde, 145
  preparation, 145
  supravital, 145
Polyamines, identification, 115
Polyethyleneimine, as tracer for anionic sites, 114
Polynucleotide phosphorylase, 356
Polyribosomes, preservation, 205
Polysaccharides, solubility, 5
Polyvinylpyrrolidone, 28, 33
  enzyme inhibitor, 26
Poriferia, body fluid composition, 22
Postmortem changes
  in plant tissue, 316
  in tissues and cells, 307–313
Potassium
  effects, in fixative, 31
  intracellular, postmortem decrease in, 313
  loss, 300, 303
    during fixation, 295
Potassium dichromate
  mixture with glutaraldehyde, 115–118
  with osmium tetroxide, 174
Potassium dihydrogen phosphate, stabilizer of protein, 26
Potassium ferricyanide, mixture with osmium tetroxide, used with glutaraldehyde, 118–119
Potassium ferrocyanide, mixture with osmium tetroxide, used with glutaraldehyde, 120
Potassium permanganate, 8
  acid, fixation procedure, 192
  with acrolein-glutaraldehyde mixture, 144
  compared to osmium tetroxide, 189–190
  contrast, 188
  fixation, cellular materials lost during, 189
  neutral, fixation procedure, 192
  penetration, 183
  reaction
    with membranes, 184–185
    with monoamines, 185–186
    with proteins, 184–185, 187
  reduction, 188
Potassium persulfate, in cleaning diatoms, 329
Potassium thiocyanate, denaturant of protein, 26
Precipitation, artifactual, 33

Procaine, 211, 212
Procaine hydrochloride, 332
Prokaryotes, 64, 66
  enzymes in, 357
Prolamine, solubility, 5
Proline, 158
Pronase, neuronal sheath removal with, 329
Protamin, solubility, 5
Protease, immobilization, 69
Protein
  cleavage, 171
    by osmium tetroxide, 169–170
  conformational change, 55
    in aldehyde fixation, 67
  cross-linking
    by dimethylsuberimidate, 195
    by glutaraldehyde, molecular factors affecting, 80–81
    by WSC, 198
  effect of glutaraldehyde concentration on, 94–96
  effect of pH on, 96
  with glutaraldehyde, 69, 72–87
    mechanism, 83–87
    Schiff bases in, 82
    termination by hydrazine, 352
  denaturation, 3
    by salts added to fixative, 26
  effect of DMSO on, 373
  extraction
    with collidine buffer, 46
    effect of electrolytes on, 30–31
    in osmium tetroxide fixation, 149, 168–170
    with Zwitterionic buffers, 49
  helical, fixative for, 2
  inhibitors, of actin destruction, 263–264
  isoelectric pH, 54
  isoelectric point, lowered by osmium tetroxide, 54
  oxidative deamination, by osmium tetroxide, 158
  in plant tissues, 314
  polymers, as molecular weight markers, 70
  reaction
    with acrolein, 140
    with aldehydes, 65–66
    with formaldehyde, 77, 130–133
    with glutaraldehyde, 77–87
    with osmium tetroxide, 157–159

Protein (*cont.*)
    reaction (*cont.*)
        with potassium permanganate, 184–185, 187
        with ruthenium tetroxide, 195
    response to autolysis, 309
    solubility, 5
Proteolysis, postmortem, 308
Protoplasm, changes, during fixation, 2
Protozoa
    body fluid composition, 22
    fixative for, 339
    pH for fixation, 316
Proximal tubule, postmortem effects on, 312
Pseudomicrothorax, 383
Pseudomonas aeruginosa, 357
Pseudomonas fragi, proteolytic enzyme, 309
Puncture perfusion, 322–323
PVP, *see* Polyvinylpyrrolidone
Pyrogallol, 151
Pyruvate, 146

R

Rana esculenta, 22
Red blood cells, *see also* Erythrocytes
    ghosts
        enzyme study in, 360–361
        fixative penetration, 71
    human
        effect of glutaraldehyde concentration on, 322–323
        fixative pH for, 55
        protein cross-linking in, 81
    membrane protein, reaction with osmium tetroxide, 158, 159
    mouse, effect of fixative concentration on, 333–334
    pH, 56
    potassium loss from, 284
    preservation, 28–29
    rat, phosphate buffers for, 46
Renal cortex
    postmortem changes in, 310
    rat
        fixation, 270
        perfusate osmolality for, 213
Renal medulla
    fixation, 270–271
    rat, perfusate osmolality for, 213

Rhodepsin, solubility, 6
Rhodnius prolixus, 24, 25
Ribonuclease
    bovine, antigenicity, 365
    effect of salts on, 26
    inhibition by glutaraldehyde, 345
Ribonucleic acid
    cytoplasmic, postmortem loss, 310
    effect of trioxsalen on, 199
    nuclear, postmortem retention, 310
    reaction
        with formaldehyde, 134–135
        with glutaraldehyde, 89–90
        with osmium tetroxide, 162
        with potassium permanganate, 187
Ribose, solubility, 5
Ribosomes, 190
    dispersion, in glutaraldehyde fixation, 100
    fixation, with glutaraldehyde, 68–69
    particle aggregation, 384
Ringer solution, 297
    amphibian, preparation, 411–412
    mammalian, preparation, 412
    with heparin, as washing solution, 297
RNA, *see* Ribonucleic acid
Roots, fixation method for, 405
Rubus fruticosus, 201
Ruthenium red, 106
    mixture with glutaraldehyde, 120–124
Ruthenium tetroxide, 4, 120, 194–195

S

Saline
    artifacts induced by, 268
    with heparin, 297
    phosphate-buffered, 297
        as washing solution, 297
        preparation, 412
    in vascular perfusion, 211–212
Salt solutions, balanced, 411–412
Salts, *see also* specific salt
    denaturants of proteins, 26
    effect on osmium tetroxide penetration rate, 172–173
    in fixative, 17, 26
Saponin, effect, on cell membrane, 366
Scanning electron microscopy
    artifacts, 385–388

fixation for, 320–341
  duration, 336–337
  mode, 334–335
  process, 333–334
  temperature, 336
fixatives, 338–341
  concentration, 333–334
  isotonicity, 333
specimens, 320–321
  cleaning, 321–332
  relaxation procedures for, 332
  shrinkage, 337–338
  size, 335–336
vehicle osmolality, 35, 334
Scenedesmus, membrane fusion, 383
Schiff base
  color, as indicator of fixative penetration, 97
  formation, 82
  in protein cross-linking, 82, 86
Scleroprotein, solubility, 5
Sea urchin eggs, swelling, during fixation,
    170–171
Secondary blackening, 157
Sectioning, 386
  deformation caused by, 321
Seeds, 207
  fixation method for, 405
  fixative for, 130, 144
  preservation, 139, 189
SEM, *see* Scanning electron microscope
Septate junctions, artifactual, 384
Serine, 65
Serotonin, 115
Shrinkage, 385, 386
  caused by aldehydes, 98–99
  caused by fixative hypertonicity, 301–304
  control, by nonelectrolytes, 27–28
  effect of DMSO on, 373
  increase, with electrolytes, 33
  reduction, 33
  in specimens for scanning electron micros-
      copy, 337–338
Sialyltransferase, 47
Silver methenamine reaction, 288, 289
Sipunculida, pH of body fluid, 57
Skin, 174
  immersion fixation, 258
  preservation, 126
  protein loss, 169
  shrinkage, 337

Slime, fixation, alternatives for, 8
Slime mold, fixation method for, 405–406
Sodium
  in aquatic animals, 21–22
  diffusion, into cells, 300, 302
  effects, in fixative, 31
  intracellular, postmortem increase in, 313
  precipitation, 296
Sodium acetate
  chemical formula, 413
  molecular weight, 413
  physicochemical properties, 413
Sodium bicarbonate buffer, *see* Bicarbonate buf-
    fer
Sodium borohydride, free aldehydes blocked by,
    289
Sodium cacodylate
  chemical formula, 413
  hazards, 47
  molecular weight, 413
  physicochemical properties, 412
Sodium chloride
  adjustment of osmolarity with, 33
  effect on osmium tetroxide penetration rate,
      173
  in prevention of swelling, 171
Sodium dihydrogen phosphate, physicochemical
    properties, 413
Sodium hydroxide
  chemical formula, 413
  molecular weight, 413
Sodium hydroxide-PIPES buffer, in plant fixa-
    tion, 316
Sodium nitrite, 212
Sodium pentobarbital, 296
Sodium permanganate, 190
  preparation, 192
Sodium phosphate
  dibasic
    chemical formula, 413
    molecular weight, 413
  monobasic
    chemical formula, 413
    molecular weight, 413
Sonication, for specimen cleaning, 331
Sorbitol, 27
Sörensen buffer, osmolality, 36
Soybean, agglutinin, 69
Spheroplasts, glutaraldehyde pretreatment,
    71

Specimen
  aldehyde-fixed, storage of, 146–147
  biopsy, artifacts in, 384
  damage, prior to fixation, 382
  floatation, 59
  preservation, satisfactory, criteria for, 378–380
  single-cell, for enzyme studies, 355–357
  size
    effect on glutaraldehyde penetration rate, 97
    effect on osmium tetroxide penetration rate, 173
    in enzyme cytochemistry, 350–351
    and fixation quality, 11–15
    and storage fixative, 147
    for scanning electron microscopy, 335–336
  soft, artifacts in, 385
  storage, 146–147
  volume, 299–306, *see also* Shrinkage; Swelling
Specimen processing
  automatic, affecting fixation quality, 15–16
  manual, 15
Specimen processor, automatic, 15–16
Spermatids
  fixation method for, 406
  grasshopper, effect of fixative osmolarity on, 20
Spermatocytes, grasshopper, effect of fixative osmolarity on, 20
Spermatozoa
  fixation method for, 406–408
  formaldehyde-fixed, 137
  mammalian, mucus removal from, 325–327
  marine invertebrate, mucus removal from, 325
  preservation of lipid-containing granules in, 113
Spermidine phosphate, mixture with glutaraldehyde, 124
Spinal cord
  dissection artifacts, 382
  vascular perfusion of, 251–252
Spirostomum ambiguum, 22
Spleen
  dog, vascular perfusion, 326–327
  rabbit, perfusion pressure for, 214
  shrinkage, 337
  vascular perfusion, 252
Spongilla, 22

Spores, 207
  fixation method for, 408
Squalene, reaction with osmium tetroxide, 151
Staining
  artifactual precipitates, 383
  of cell nuclei, by osmium tetroxide, 162–163
  effect of buffers on, 42–43
  effect of fixation on, 287–291
  enzymes, effect of DMSO on, 375
Staphylococcus aureus, mesosomes, 274
Starch, solubility, 5
Stearic acid, 151
Steroids, solubility in CO_2, 337
Sterols, solubility, 5
Stirring, effect on fixative penetration, 59
Stria vascularis, 205
Stripping, 386
Subtilisin
  insolubilization, 67
  inhibition by glutaraldehyde, 345
Succinate reductase, localization, 356
Succinic dehydrogenase, 343
  effect of DMSO on, 374
  effects of various aldehydes on, 346
  preservation, 344
Sucrose
  addition to osmium tetroxide, 34
  concentration in glutaraldehyde, 92
  effect on chlorophyll extraction, 27
  enzyme inhibitor, 26–27
  reaction
    with osmium tetroxide, 163
    with potassium permanganate, 188
  replacement of extracellular fluid in brain, 265–268
  solubility, 5
  in specimen cleaning, 331
  in unbuffered osmium tetroxide, 28
Sudanophilia, demonstration, 150–151
Sulfur, intracellular, postmortem levels, 313
Sulfuric acid, solubility, 6
Sulfur tetroxide, in aquatic animals, 21–22
Swelling, 20, 299–301, 386
  caused by osmium tetroxide, 170–171
  caused by perfusion pressure, 214
  control
    by electrolytes, 28–29
    by nonelectrolytes, 27–28
  effect of fixation temperature on, 60, 336
  mesengial, 269
  in potassium-permanganate fixation, 183, 189

Swiss 3T3 cells, antigen localization in, 368-369
Synaptic vesicles
  effect of fixation on, 291-294
  flattening, 67

T

Tannic acid, 166
  mixed with glutaraldehyde, 124-128
  reaction
    with osmium tetroxide, 128
    with saturated phospholipids, 125-126
  use
    before or after osmication, 126
    use in minimization of shrinkage, 338
TC-2 tissue sectioner, 13-14, 350-351
Teleosts
  freshwater, body fluid composition, 22
  marine, body fluid composition, 21
Telphusa fluviatilis, 22
TEM, *see* Transmission electron microscope
Temperature
  effect
    on bacterial morphology, 10
    on buffer efficiency, 40
    on cell morphology, 61
    during dehydration, 61
    on DMSO, 373
    in enzyme cytochemistry, 352
    on fixation, 60-62
    on glutaraldehyde, 76-77, 89, 94, 97
    in osmium tetroxide fixation, 171-172
    in scanning electron microscopy, 336
    on surface morphology, 386
  of perfusate, 212
Tendon, protein loss from, 169
Terpenes, solubility in CO_2, 337
TES buffer, 48-50
Testes
  guinea pig, vascular perfusion of, 253-254
  perfusion preparation for, 253
  rat, 128
  vascular perfusion of, 253-256
Tetrahymena
  fixation method for, 408
  myoblasts, 202
Thiamine pyrophosphatase, 196, 354
Thiocarbohydrazide, for scanning electron microscopy, 339
Thoracic aorta, rat, shrinkage, 337-338

Threonine, 65
Thymidine, reaction with osmium tetroxide, 160-161
Thymine, reaction with osmium tetroxide, 160-161
Tight junctions, effect of fixation on, 294
Tissues
  autopsy, 311
    ion shifts in, 312-313
  blocks
    aldehyde-fixed, 64
    core, 12
    preparation of, 12-15
    shrinkage, 337, 338
  density, effect of osmolarity on, 17
  fine structure, preservation, effect of vehicle osmolality on, 25
  fractions, in enzyme studies, 357-359
  pathological, 11-12
  sectioners, 13-15, 350-351
  storage, 146-147
  structure-function correlation, 90-92
  volume changes, and perfusion pressure, 215
Tobacco yellow mosaic virus, 283, 284
Tonoplast, indicator of fixation quality, 318
Transmission electron microscopy
  artifacts, 382-385
  fixative, osmolarity, 333
  specimen size, 11-15
  tissue blocks for, preparation, 12-15
  total osmolality recommended
    for mammalian tissue, 35
    for plants, 35
  vehicle osmolality, 334
Transverse tubules, postmortem effects on, 309
Triakis scyllia, 18, 19
Tricaine methane sulfonate, anesthetic for fish, 332
Tricine, 49-50
Triglycerides, solubility, 5
4,5',8-Trimethylpsoralen, *see* Trioxsalen
Trinitro compounds, mixture with glutaral-dehyde, 114-115
2,4,6-Trinitrocresol, in fixation, 114
Trioxsalen, 199
Tripalmitin, solubility, 5
Tris acid maleate
  chemical formula, 413
  molecular weight, 413
Tris buffer, 47
  with dimethylsuberimidate, 196

Tris buffer (*cont.*)
  effective pH, 47
  pH, 40
    effect of temperature on, 40
  preparation, 52–53
  reaction with glutaraldehyde, 47
Tris-HCl, 47
Tris (hydroxymethyl)aminomethane, *see* Tris
  buffer
Tris-maleate, 47
Tristearin, solubility, 5
Triton X-100, 359, 366
Tropomyosin, in preservation of actin, 263–264
Trypsin
  cleaning of chromosomes by, 329
  cleaning of mucosa by, 329
Trypsinogen, bovine, antigenicity, 365
Tryptophan, 65, 158, 162
  reaction
    with glutaraldehyde, 80
    with osmium tetroxide, 157
  solubility, 29
Tryptophan hydroxylase, inhibition, 115
Tyrode solution, 297
  phosphate-buffered, as washing solution, 297
  preparation, 411
Tyrosine, 65
  binding, 101
  cross-linking reaction with formaldehyde, 132
  reaction, with glutaraldehyde, 80

U

Ultrastructure
  effect of buffer on, 44
  effect of DMSO on, 375–377
  postmortem alterations in, 307–313
  preservation
    by double fixation, 200–201
    with formalin, 45–46
Uracil, reaction with osmium tetroxide, 160
Uranyl acetate
  compatibility, with bicarbonate buffers, 48
  effect on glycogen, 289
  incompatibility with cacodylate, 47
  incompatibility with phosphate buffer, 95
  minimization of shrinkage with, 338
  mixture with glutaraldehyde, 129
  negative staining of actin, 264
  shrinkage caused by, 304–306

staining
  with various buffers, compared, 42–43
  with Zwitterionic buffers, 49
  with veronal acetate, 47
Urethane, 260

V

Vacuoles
  appearance, in satisfactory fixation, 379
  artifactual, membranous structures in, 383
  central, effect of fixation on, 314–316
  membrane, indicator of fixation quality, 318
Valine, binding, 101
Vascular endothelium
  artifacts, 297–298
    from immersion fixation, 335
  effect of fixation on, 296–298
Vascular perfusion, 209–256
  advantages, 209–210
  in enzyme studies, 354
  apparatus, 215–223
  artifacts, 387
  with glutaraldehyde, 94
  limitations, 210–211
  methods, 223–256
    for aorta, 225–226
    for arteries, 226
    for central nervous system, 227–232
    for embryo, 232–233, 234
    for fish, 233–235
    for heart, 235–237
    for kidney, 237–240
    for liver, 240–245
    for lung, 245–249
    for muscle, 249–251
    for ovary, 251
    for spinal cord, 251–252
    for spleen, 252
    for testes, 253–256
  parameters for success, 211–215
  perfusion pressure, 213–215
    recommended, for selected tissues, 214
  in scanning electron microscopy, 334–335
  temperature for, 336
Vasodilators, 211, 212
Vasopressin, antigenicity, 365
Vehicle, *see also* Buffer
  approximate osmolalities, 36
  effects of added substances, 26–33

ionic composition, effect on extraction, 22–23
osmolality
    effect on fixation quality, 23–26
    effect on glutaraldehyde fixation, 23–25
osmolarity, effect on cell volume, 301
Veronal acetate, 40, 46–47
    effective pH, 46
    effect on staining, 43
    preparation, 53–54
    sectioning properties with, 43
Vesicles
    artifactual formation, 383
    coated, 383
Vestopal, lipid extraction reduced by, 286
Vibratome, 13–15, 350–351
Villi, intestinal, postmortem changes in, 309
Virus, *see also* specific virus
    lithium permanganate fixation, 191
    mosaic, 137, 138
    plant
        effect of fixation on, 283–284
        inability of glutaraldehyde to preserve, 100
    preservation, 139
    stabilization, 130
        with formaldehyde, 134–135
Volvox, 204

W

Washing solutions, 297
    for cleaning of specimens for SEM, 321–322
    osmolarity, 93–94
Water, effect of DMSO on, 373
Wax
    on plant surface, 316
    solubility, 5

White blood cells, 202
White matter, perfusion, 228, 268, 269
Wood, fixation method for, 408
WSC, *see* 1-Ethyl-3(3-dimethylaminopropyl)
    carbodiimide-HCl

X

X-ray diffraction, 378–379
X-ray microanalysis
    of autopsy specimens, ion shifts observed in,
        313
    specimens for, effect of fixation on, 294–296
    vapor fixation for, 176
Xylans, wood, solubility, 5
Xylene, osmium solubility in, 167
Xylocaine, 296

Y

Yeast
    acrolein-fixed, 139
    fixation method for, 408–409

Z

Zinc permanganate, 191
Zwitterionic buffers, 48–50
    advantages, 48–49
Zygotes, fixation method for, 397
Zymogen granules, 188
    autolysis in, 309
    fraction, enzyme study in, 360–361
    protein loss, 170